physica status solidi (a)

AKADEMIE-VERLAG · BERLIN

VOLUME 10 · NUMBER 2 · 1972

physica status solidi (a)

applied research

Volume 10 · Number 2 · Pages 307 to 672, K101 to K154, and A9 to A16

April 16, 1972

AKADEMIE-VERLAG · BERLIN

Subscriptions and orders for single copies should be addressed to

AKADEMIE-VERLAG GmbH, 108 Berlin, Leipziger Straße 3 – 4 or to Buchhandlung KUNST UND WISSEN, Erich Bieber, 7 Stuttgart 1, Wilhelmstr. 4 – 6 or to Deutsche Buch-Export und -Import GmbH, 701 Leipzig, Postschließfach 160

Editorial Note:

"physica status solidi (a)" undertakes that an original paper *accepted for publication* before the 23rd of any month will be published within 50 days of this date unless the author requests a postponement. In special cases there may be some delay between *receipt* and *acceptance* of a paper due to the review and, if necessary, revision of the paper.

Schriftleiter und verantwortlich für den Inhalt: Professor Dr. Dr. h. c. P. Görlich, 102 Berlin, Neue Schönhauser Str. 20 bzw. 69 Jena, Humboldtstr. 26. Chefredakteur: Dr. H.-J. Hänsch. Redaktionskollegium: Prof. Dr. E. Gutsche, Dr. H.-J. Hänsch, Dr. H. Lange, Dr. S. Oberländer. Anschrift der Schriftleitung: 102 Berlin, Neue Schönhauser Str. 20, Fernruf: 423380. Verlag: Akademie-Verlag GmbH, 108 Berlin, Leipziger Str. 3–4, Fernruf: 220441, Telex-Nr. 112020, Postscheckkonto: Berlin 35021. — Die Zeitschrift „physica status solidi (a)" erscheint jeweils am 16. des Monats. Bezugspreis eines Bandes 120,— M. Bestellnummer dieses Bandes: 1085/10. Jeder Band enthält zwei Hefte. Gesamtherstellung: VEB Druckerei „Thomas Müntzer" Bad Langensalza. — Veröffentlicht unter der Lizenznummer 1310a des Presseamtes beim Vorsitzenden des Ministerrates der Deutschen Demokratischen Republik.

Contents

Short Notes

Pre-printed Titles

Systematic List

The Author Index of Volume 10 Begins on Page 673

(It will be delivered together with Volume 11, Number 1.)

Review Article

phys. stat. sol. (a) **10**, 315 (1972)

Subject classification: 20.1 and 20.3; 6; 13.4; 19; 20.2; 22.6

Physics Department, Oklahoma State University, Stillwater (a),
and Physics Department, Boston College, Chestnut Hill, Massachusetts (b)

Optical Properties of Heavily Doped Ruby

By

R. C. Powell (a) and B. DiBartolo (b)

Contents

1. Introduction

During the last ten years a great amount of time and effort has been expended in attempting to understand the optical properties of heavily doped ruby. This paper is an attempt to review the voluminous amount of literature which has appeared on the subject. A review of this nature is important, since the interesting physical phenomena which have been studied in heavily doped ruby are now being observed in other materials [1] and the extensive knowledge gained from the investigations reviewed here will undoubtedly serve as a basis for interpretation of the new results. However, the literature on heavily doped ruby can be quite confusing to someone trying to read it for the first time. The reason for this is twofold: First, a number of authors have published their work in a series of papers, in which it is common to find the more recent papers contradicting the conclusions drawn in earlier ones. Secondly, several different types of experimental investigations have been performed, and the results obtained by one technique sometimes lend themselves to different interpretation than those obtained by some other technique. In at least one case, we feel that the conclusions drawn in the more recent work are incorrect. Thus, there appears to be a need to summarize the present state of our knowledge about heavily doped ruby and, since our understanding of this system has reached some degree of completeness, this seems like an ideal time for such a review.

To begin with, it is necessary to define what is meant by heavily doped ruby. The familiar laser material is "lightly doped ruby", which consists of Al_2O_3 with $<0.1\%$ (atomic) Cr^{3+} ions substituted for aluminum ions. Aluminum oxide with a higher percentage of chromium impurity than this is generally referred to as "heavily doped ruby". The optical absorption spectrum of lightly doped ruby consists of two broad bands and three sets of sharp lines, while the dominant feature of the emission spectrum is the R-line doublet. These same features appear in the spectra of heavily doped ruby along with numerous other sharp lines, a third absorption band, and a broad fluorescence band. These extra features can all be attributed to the interaction of chromium ions.

There are several important motivations for studying heavily doped ruby. One is the need to understand concentration-dependent loss mechanisms in fluorescent materials, especially those used as lasers. Also, there has been some interest in obtaining laser action at frequencies in heavily doped ruby where no emission occurs in lightly doped ruby. Finally, heavily doped ruby is a good material for studying the basic physics of the interaction of impurity ions in a crystalline host; the importance of this lies in the hope that knowledge obtained on the magnetic interactions between two chromium ions in the Al_2O_3 lattice can be extrapolated to the isomorphic crystal Cr_2O_3 and thus increase our understanding of the antiferromagnetic nature of this latter material.

The investigations of the optical properties of heavily doped ruby can generally be divided into two categories: 1. attempts to identify the different types of optically active centers in the system and to associate the transitions observed in the absorption and fluorescence spectra with specific types of centers, and 2. attempts to elucidate the dynamics of the pumping of these different types of optically active centers. Investigations of the first category are discussed in the following three sections and include those which establish nonequivalent pair types, yield information on pair coupling interactions, establish energy levels for the different centers, and assign observed spectral lines to transitions between specific levels of these centers. In the second category, we place observations of stimulated emission, studies of the interaction between the different fluorescent systems, and investigations of relaxation phenomena within a specific system. Work in this category is discussed in Sections 5 through 7.

2. N-Lines

2.1 Line spectra of heavily doped ruby — concentration dependence

Even in the very early observations of the optical spectra of ruby, a group of lines was observed to the low energy side of the R-lines [2, 3]. These lines were labelled "Nebenlinien" (neighbor lines) or N-lines by the early workers, but no attempt was made to determine their origin. A careful enumeration of all the lines in the spectra of heavily doped ruby was made by Jacobs in 1956; he listed 110 lines [4].

The fluorescence spectrum of ruby with 2.1% chromium in the region of the R- and N-lines is shown in Fig. 1 at several temperatures [5]. At low temperatures the R_1- and R_2-lines are located at 6935 and 6920 Å, respectively. The two most intense N-lines are generally referred to as N_1 and N_2 and at low temperatures are located at 7041 and 7009 Å, respectively. Lines which are vibrational–electronic (vibronic) in nature appear between 7000 and 7400 Å in lightly doped ruby, and are sometimes difficult to distinguish from the N-lines in heavily doped crystals.

It was first suggested that the N-lines arose from vibronic transitions [3, 6]. However, the satellite lines in emission have different spacings than those in absorption contrary to what one would expect for vibronic transitions. Schawlow et al. [7] showed that the ratio of the fluorescence intensity of an N-line to an R-line varies linearly with chromium concentration at low levels of concen-

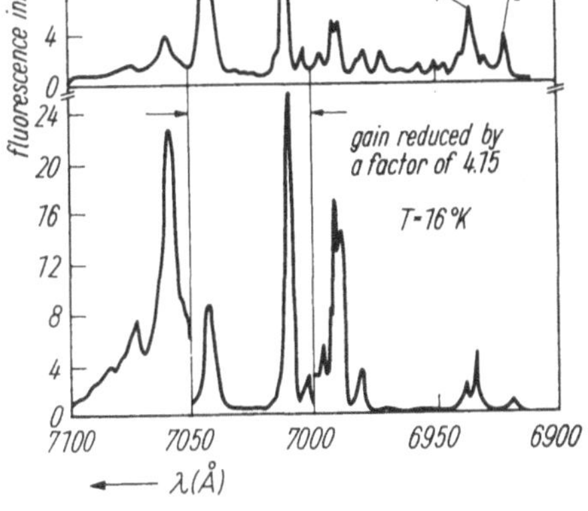

Fig. 1. Fluorescence spectrum of ruby with 2.1% chromium in the region of the R- and N-line at serveal temperatures (after [5])

Fig. 2. Intensity ratio of the N_2- to the R_1-line in ruby as a function of chromium concentration (after [7])

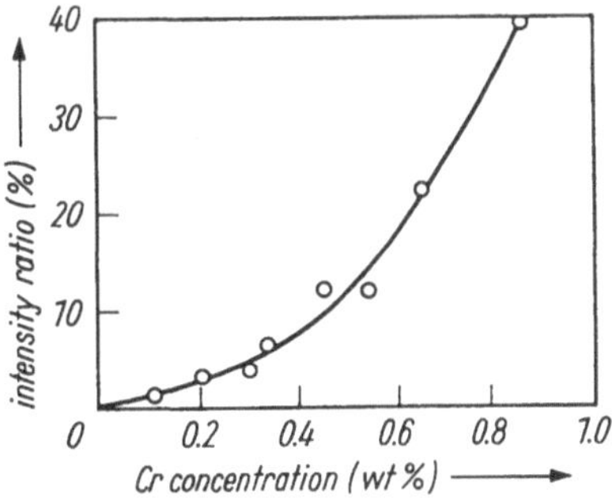

tration. Their data for the I_{N_1}/I_{R_1} ratio are shown in Fig. 2. The greater than linear increase at higher concentrations will be discussed in a later section. Since the R-line intensity was known to vary linearly with chromium concentration, these results indicated that the N-line intensity varies quadratically with concentration. Schawlow et al. [7] deduced from this that the origin of the N-lines is the presence of coupled pairs of Cr^{3+} ions. Tolstoi and Abramov [8] investigated the concentration dependence of the intensities of several different satellite lines and obtained results consistent with those of reference [7].

The position of the N-lines as a function of concentration of chromium has been monitored and found to be linear [9]. This effect is attributed to the lattice expansion which occurs with increasing impurity concentration.

2.2 Pair coupling theory

Because of the antiferromagnetic nature of Cr_2O_3, Clogston [10] suggested exchange interaction as the coupling mechanism for the ion pairs in heavily doped ruby. He found that a very simplified theory of isotropic exchange interaction with an adjustable coupling coefficient J could accurately describe the ground state splittings of the pair systems, but not the more complicated excited state splittings. In this theory the interaction Hamiltonian is

$$H_{\text{int}} = \frac{J}{\hbar^2} \boldsymbol{S}_1 \cdot \boldsymbol{S}_2 , \tag{1}$$

where S_1 and S_2 represent the spins of the single ions and $J < 0$ corresponds to ferromagnetic coupling, whereas $J > 0$ corresponds to antiferromagnetic coupling. The zero-field ground state splitting of ruby (≈ 0.38 cm^{-1}) is small compared with the exchange splitting and is therefore neglected. However, matters are complicated in the excited metastable state where one of the ions in a pair is in the ^2E-level, since this level is split by 29 cm^{-1} by spin–orbit interaction and the trigonal crystal field. This is comparable to the magnitude of the exchange splitting.

Substituting the expression $\boldsymbol{S}_1 \cdot \boldsymbol{S}_2 = (\boldsymbol{S}^2 - \boldsymbol{S}_1^2 - \boldsymbol{S}_2^2)/2$ in equation (1) and using linear combinations of the products of single-ion wave functions for the pair wave functions, the energies of the components of the ground state manifold of the pair systems are

$$\begin{aligned} E_0^{\text{pair}} &= \langle \psi_0^{\text{pair}} | \, H_{\text{int}} \, | \psi_0^{\text{pair}} \rangle = \\ &= \frac{J}{2} \left[S\,(S+1) - S_1\,(S_1+1) - S_2\,(S_2+1) \right] , \end{aligned} \tag{2}$$

where the single-ion energies which simply shift the levels uniformly have been neglected. For the ruby ground state $S_1 = S_2 = 3/2$ and the coupling of these two angular momenta states gives $S = 3, 2, 1, 0$. Substituting these values into equation (2) gives

$$E_0^{\text{pair}} = -\frac{15}{4} J + \frac{J}{2} S\,(S+1) . \tag{3}$$

The first term shifts the energy levels without splitting them. The second term gives the following levels:

$$E_0^{\text{pair}} = 0, J, 3\,J, 6\,J . \tag{4}$$

An important aspect of this theory with regard to comparison with experimental results is the predicted temperature dependence of the population of a ground state level E_i, which is given by

$$n_i = \frac{N}{Z} \exp\left(\frac{E_i}{kT}\right) \quad (5)$$

with the partition function given by

$$Z = \sum_i \exp\left(\frac{E_i}{kT}\right). \quad (6)$$

A summary of the temperature dependences of the populations of ground state levels for antiferromagnetically and ferromagnetically coupled pairs is given in Table 1 [11].

Several attempts have been made to add sophistication to the simplified theory outlined above. One of these has been to add a higher-order term to the interaction Hamiltonian given in equation (1). This is a biquadratic term given by

$$H'_{\text{int}} = j(S_1 \cdot S_2)^2 \quad (7)$$

and is generally thought to arise from superexchange interaction [12, 13]. This can be thought of as exchange interaction between a pair of ions taking place through the polarization effects on a nonmagnetic ion located between the pair. Since oxygen ions are located between chromium ions in ruby, it is reasonable to assume that this effect is present.

Theoretical considerations have been given to the effects of anisotropic exchange, dipolar interactions, and covalency on the energy levels of ion pairs

Table 1

Energy levels of Cr^{3+} ions coupled by isotropic exchange (after [11])

antiferromagnetic coupling			ferromagnetic coupling		
$H_{\text{exch}} = J\,(\mathbf{S}_1 \cdot \mathbf{S}_2)$ $Z_a = 1 + 3\,e^{-J/kT} + 5\,e^{-3J/kT} + 7\,e^{-6J/kT}$			$H_{\text{exch}} = -J\,(\mathbf{S}_1 \cdot \mathbf{S}_2)$ $Z_f = 7 + 5\,e^{-3J/kT} + 3\,e^{-5J/kT} + e^{-6J/kT}$		
energy	$S = \|\mathbf{S}_1 + \mathbf{S}_2\|$	temperature dependence	energy	$S = \|\mathbf{S}_1 + \mathbf{S}_2\|$	temperature dependence
$6J$	3	$\frac{e^{-6J/kT}}{Z_a}$	$6J$	0	$\frac{e^{-6J/kT}}{Z_f}$
			$5J$	1	$\frac{e^{-5J/kT}}{Z_f}$
$3J$	2	$\frac{e^{-3J/kT}}{Z_a}$	$3J$	2	$\frac{e^{-3J/kT}}{Z_f}$
J	1	$\frac{e^{-J/kT}}{Z_a}$			
0	0	$\frac{1}{Z_a}$	0	3	$\frac{1}{Z_f}$

J is the magnitude of the coupling parameter, $J = |J|$.

[14, 15]. It has been shown experimentally through high-resolution spectroscopy that these added effects cause splittings of less than 1 cm^{-1} of the ground state levels derived from the simple theory outlined above [14].

The calassification of the energy levels of chromium ion pairs and the determination of their splittings through the use of group theory has also been attempted [16 to 19]. So far, this work has not proven to be especially useful in contributing to the understanding of pair systems.

Although simple isotropic exchange theory is capable of successfully explaining the ground state energy levels of Cr^{3+} pairs, it does not give the correct predictions for the excited state splittings, even with the various refinements discussed above. Nikiforov and Cherepanov [20] have obtained a fit to the excited state splitting of fourth-nearest neighbor Cr^{3+} pairs using a semi-empirical theory with an "effective Hamiltonian" having eight adjustable parameters. It has been pointed out that simple isotropic exchange theory fails because the Hamiltonian given in equation (1) is not valid if an electron on one of the ions of the pair is in the excited ^{2}E-state [21]. This form of H_{int} must be modified to allow for different exchange integrals between different orbitals on the two coupled ions. To do this it is necessary to use a more general Hamiltonian of the form

$$H''_{\text{int}} = \sum_{i,j} J_{ij}\, \mathbf{s}_{1i} \cdot \mathbf{s}_{2j}\,, \tag{8}$$

where, for example, $\mathbf{s}_{1i}$ represents the spin of the electron in the i-th t_{2g} orbital on ion 1. Calculations based on this Hamiltonian are now being carried out by Pryce [22].

The association of spectral lines with specific types of ion pairs is complicated in the Al_2O_3 lattice by the fact that the fifth-nearest neighbor pair separation is only about 1 Å greater than that of first-nearest neighbors. Fig. 3 shows the relative positions of some of the near neighbor pairs [23]. There is one nearest neighbor pair, three equivalent second-nearest neighbor pairs, three equivalent third-nearest neighbor pairs, and six equivalent fourth-nearest neighbor pairs. The spin orientations in the isomorphic crystal Cr_2O_3 have been determined from neutron scattering to be antiferromagnetic for the first three nearest neighbor pairs and ferromagnetic for the fourth one [24, 25]. These properties are summarized in Table 2.

Before leaving our discussion of the theory of the coupling of ions it should be mentioned that J and j are treated as strictly adjustable parameters whose absolute magnitudes have no specific theoretical connotations. Although the need for a biquadratic term in the interaction Hamiltonian, the relative magnitudes of j to J, the observation by ENDOR experiments of covalent bonding effects between the metal ions in ruby [23], and the comparisons with the

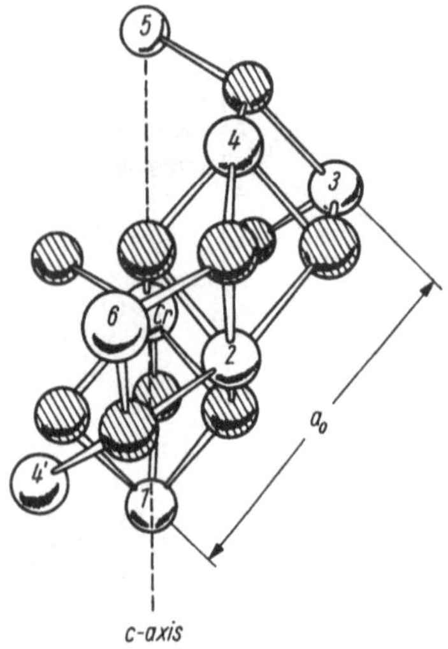

Fig. 3. Nearest-neighbor pair sites for a chromium ion in an Al_2O_3 lattice. The dark balls are oxygen ions (after [23])

Table 2
Types of exchange-coupled pairs in ruby

type of nearest neighbor	pair separation (Å) *)	number of equivalent sites	type of coupling in the ground state
first	2.733	1	antiferromagnetic
second	2.809	3	antiferromagnetic
third	3.185	3	antiferromagnetic
fourth	3.504	6	ferromagnetic

*) Data from reference [23].

magnetic properties of Cr_2O_3 [26, 27] all qualitatively point toward superexchange as the coupling mechanism, the magnitudes of the values that have been obtained for J and j are much greater than the theories of exchange or superexchange interaction predict. Other spin coupling mechanisms involving optical phonons have been suggested, but theoretical analysis would be extremely complicated and thus far has not been carried out [28].

2.3 Establishment of energy levels and assignment of transitions

Several different types of experimental techniques have been employed in an attempt to establish energy levels for pair systems and to assign observed spectral lines to transitions between specific levels. These include electron paramagnetic resonance (EPR), optical excitation spectra, temperature dependence of optical absorption and fluorescence lines, and infrared absorption. These investigations are described in this section. By far the most successful studies are those involving the optical spectra. The resulting data obtained on the pair systems are summarized in Tables 3 and 4 of Section 4, and the reader may find it helpful to refer to this summary while reading the present section.

2.3.1 Electron paramagnetic resonance observations

A number of attempts have been made to understand the electron paramagnetic resonance spectra exhibited by heavily doped ruby [14, 29 to 38]. However, the interpretation of results proved to be an extremely difficult job, since the room temperature spectrum consists of hundreds of lines mostly unresolved. The main purpose of this work was to establish values of the exchange coupling parameter J for the different types of ion pair systems. The first important work in this area was done by Rimai et al. [29 to 31]. They studied the angular dependence of the positions and the temperature dependence of the intensities of the lines in the EPR spectra of heavily doped ruby, and interpreted their data using the effective spin Hamiltonian

$$H_s = J\,\boldsymbol{S}_1 \cdot \boldsymbol{S}_2 + g\,\beta\,\boldsymbol{H} \cdot (\boldsymbol{S}_1 + \boldsymbol{S}_2) + D\,(S_{Z1}^2 + S_{Z2}^2) + \\ + \frac{4\,\beta^2}{r_{12}^3}\left[\boldsymbol{S}_1 \cdot \boldsymbol{S}_2 - 3\,\frac{(\boldsymbol{r}_{12} \cdot \boldsymbol{S}_1)\,(\boldsymbol{r}_{12} \cdot \boldsymbol{S}_2)}{r_{12}^2}\right] + \text{anisotropic terms}\,, \quad (9)$$

where it is assumed that $|J| \gg |D|$ so that the last four terms act to split the states derived previously for the exchange-coupled pair. The second and third terms represent the sum of the spin Hamiltonians for the two individual ions, where D is the zero-field splitting parameter and the fourth term represents dipolar interaction. Generally the last two terms are of much less importance than the first three. Various lines were identified by the angular dependence of their EPR spectra and J-values were obtained from the line splittings and the temperature dependence of their intensities. For first-nearest neighbor pairs, this work indicated that $J = (390 \pm 50)$ cm^{-1}. Both third- and fifth-nearest neighbor pairs were found to have $J \approx 1$ cm^{-1}, while no second-nearest neighbor pair lines were identified, and the fourth-nearest neighbor assignments were uncertain but appeared to imply ferromagnetic coupling. For sixth- through eleventh-nearest neighbor shells, the J-values were found to be on the order of 1 cm^{-1}. However, the difficulty of this work makes these results uncertain [32].

Jelenski et al. [34] later identified from the temperature dependence of its intensity an EPR line with $J \approx -7$ cm^{-1} which they attributed to the ferromagnetically coupled fourth-nearest neighbors.

Nikiforov and Cherepanov [35] discussed the theoretical interpretation of the EPR spectra of heavily doped ruby in more detail, and Shelekhin et al. [36] did further experimental work in determining values for g and D for first-nearest neighbors. Although this work is important in understanding EPR spectra, it does not contribute significantly to the interpretation of the optical spectra and therefore is not discussed further.

Berggren et al. [14] have made recent investigations of the EPR lines assigned to fourth-nearest neighbor pairs and established values for the zero-field splitting parameters. These authors also made use of the technique of optical detection of the EPR resonances of fourth-nearest neighbor pairs. In this technique the intensity of the 7009 Å fluorescence line was monitored while the crystal was also exposed to a sweeping magnetic field and microwave radiation. The microwave pumping altered the population distribution within the ground state manifold which effects the optical pumping of the excited state. This technique allows EPR transitions belonging to the fourth-nearest neighbor pairs to be readily identified from the complex normal EPR spectra. They obtained data on the positions of the resonance lines as a function of field, angle, and microwave frequency and on the intensities of the lines as a function of microwave power. Their results could be explained by a simple spin Hamiltonian and lead to information on the fine structure of the fourth-nearest neighbor pair lines such as values for zero-field splitting parameters, dipole–dipole interaction strength, and importance of anisotropic interactions.

Microwave transitions within the excited state manifolds of the pair systems have also been identified in the optically detected EPR spectra [14, 37]. One important result of this work is the evidence that the fourth-nearest neighbor pairs are antiferromagnetically coupled in the excited state instead of being ferromagnetically coupled as they are in the ground state.

Theoretical predictions have also been made as to the splittings of EPR lines of heavily doped ruby in an external electric field [38]. Although this technique may be useful in helping to unravel the complicated EPR spectra of exchange-coupled pairs, no experimental data have yet been reported.

2.3.2 *Investigations of the optical spectra*

The first attempt to establish energy level diagrams for the pair systems and to assign N-lines to specific transitions between these levels was made by Daly [39] in 1961. He excited the sample with monochromatic light of variable frequency and monitored the fluorescence at specific N-lines. Since these experiments were performed at 4 °K, the fluorescence transitions were assumed to originate in the lowest components of the excited state manifolds, and the ab sorption transitions to originate in the lowest components of the ground state manifolds. Therefore, fluorescence at a particular frequency was observed only when the excitation energy was at a frequency corresponding to a transition from the ground state to some component of the excited state manifolds. From these experiments the lowest lying components of the metastable states and the various components of the ground state manifolds were obtained, along with the assignments of several transitions. However, the low intensity of the fluorescence caused these measurements to be made with monochromator slit widths which were too wide for good resolution.

Kisliuk et al. [11] in 1964 measured the temperature dependence of the absorption strength of the two strongest satellite lines, N_1 and N_2. Using an isotropic exchange interaction, they were able to determine values of J for the ground states and to establish the initial levels of the various absorption transitions. The temperature dependence of the intensity of a line starting from level E_i is the same as the temperature dependence of the population of E_i and is given by equation (5). Using J as an adjustable parameter, all of the predicted temperature dependences listed in Table 1 were compared to the data obtained on the N_1- and N_2-lines. It was found that three of the ferromagnetic curves and one antiferromagnetic curve could give a good fit to the N_2-line data. To determine the correct assignment, the values of J needed to fit these curves to the experimental data were used to construct energy level diagrams for each case. The transitions predicted from these diagrams were compared with spectral lines seen experimentally, and only one set was found to agree. Fig. 4 shows the observed temperature dependence of these lines and the theoretical fittings. This work established that the 7009 Å (N_2) line was part of a ferromagnetically coupled pair system with $J = -7$ cm^{-1}. The lower level of the transition giving rise to this line was found to be the $S = 1$ component of the ground state manifold, which is located 35 cm^{-1} above the ground state. Similar investigations led originally to the assignment of the 7041 Å (N_1) line to an antiferromagnetically coupled pair system with $J = 32$ cm^{-1}. The lower level of the transition was identified as the $S = 1$ component of the ground state manifold, which is 32 cm^{-1} above the ground state. However, they noted that the transition terminating on the $S = 2$ component of the ground state of an antiferromagnetically coupled pair system is also consistent with experimental observations and, in later work, this was shown to be the correct assignment for the N_1-line as discussed below. From this work a total of eight lines was assigned to transitions within the two different pair systems.

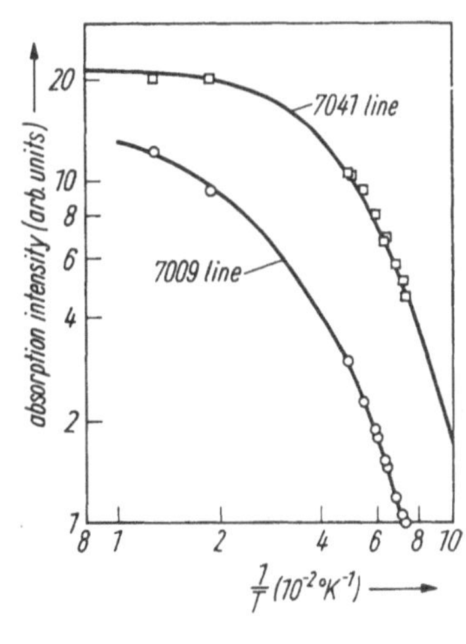

Fig. 4. Temperature dependence of the N-line absorption intensities. The plotted points represent experimental data and the curves represent the calculated temperature dependence of the populations of the lower states of the transitions assuming that the N_1-line originates from an $S = 1$ state of an antiferromagnetically coupled pair and the N_2-line originates from an $S = 1$ level of a ferromagnetically coupled pair (after [11])

Kisliuk et al. [11] suggested that, for the J-values of their results to be consistent with those of Statz et al. [30, 31] for the EPR spectra, the 7009 Å complex should arise from fourth-nearest neighbor pairs and the 7041 Å set from second-nearest neighbor pairs. Later work showed this latter assignment to be incorrect, but several subsequent papers referred to the 7041 Å set of lines as second-nearest neighbor pair lines. The correct assignment is the third-nearest neighbor pair system as discussed in Section 2.4.1.

Kisliuk and Krupke [40, 41] found improved values for the ground state splittings of the pair systems by including a higher-order term in the exchange Hamiltonian. This is the biquadratic term given in equation (7) where $j \approx 0.02\ J$. By extending their measurements to 4.2 °K and using Daly's [39] excitation data and the recurrence of the established ground state energy differences, they were able to assign 28 of the lines listed by Jacobs [4] to the two main nearest neighbor pair systems. This work corrected some assignments made by Kisliuk et al. [11]. One important change was the assignment of the 7041 Å line to a transition in an antiferromagnetically coupled pair system with $J \approx 11\ \mathrm{cm}^{-1}$ which had, as its lower level, the $S = 2$ component of the ground state manifold located 32.7 cm^{-1} above the ground state.

Powell et al. [5, 42] established energy level diagrams for the same two pair systems and assigned transitions within these systems by studying the temperature dependence of the relative intensities of the fluorescence lines. Here they used the fact that the ratio of the fluorescence intensities of two lines is given by $I_1/I_2 = \exp(-\Delta E_{12}/kT)$, where ΔE_{12} is the separation of the upper levels of the transitions. The results of this work are consistent with those of Kisliuk and Krupke [41]. One new line was assigned in the N_1 pair system (6978.4 Å). At least seven lines not assigned previously were observed near to the R-lines. These lines were too closely spaced to obtain accurate intensity measurements but, by virtue of their general change with temperature and their energy splittings, an assignment was suggested for three of them. This lead to the establishment of a new excited level in the fourth-nearest neighbor pair system lying between the R_1 and R_2 metastable levels which is consistent with Daly's [39] observation of fluorescence at the N_2-line while exciting at a frequency between the R_1- and R_2-lines. More recent work tends to show that this assignment is incorrect [21].

Mollenauer and Schawlow [43] were able to identify several lines belonging to first- and second-nearest neighbor pair systems by studying the temperature dependence of their absorption strengths. In order to detect these weak lines, they used a sample 15 cm long and took an average of more than 100 scans of the spectrum. They found a series of five lines whose temperature dependences indicated that they belonged to the same pair system. This type of pair was antiferromagnetically coupled with $J = 225\ \mathrm{cm}^{-1}$ and was identified as the first-nearest neighbor pair system, as discussed in the next section. These absorption measurements were much less successful in identifying second-nearest neighbor pair lines. Only the 7452 Å line was unambiguously associated with this pair type and a large uncertainty still remained in the value of J.

Kisliuk et al. [21] extended their previous work on the temperature dependence of the intensities of absorption lines and, from this extensive investigation, were able to assign nearly all the lines in the spectrum of heavily doped ruby to transitions in first- through fourth-nearest neighbor pairs. They were able to assign 13 lines to transitions in the second-nearest pair system and to determine

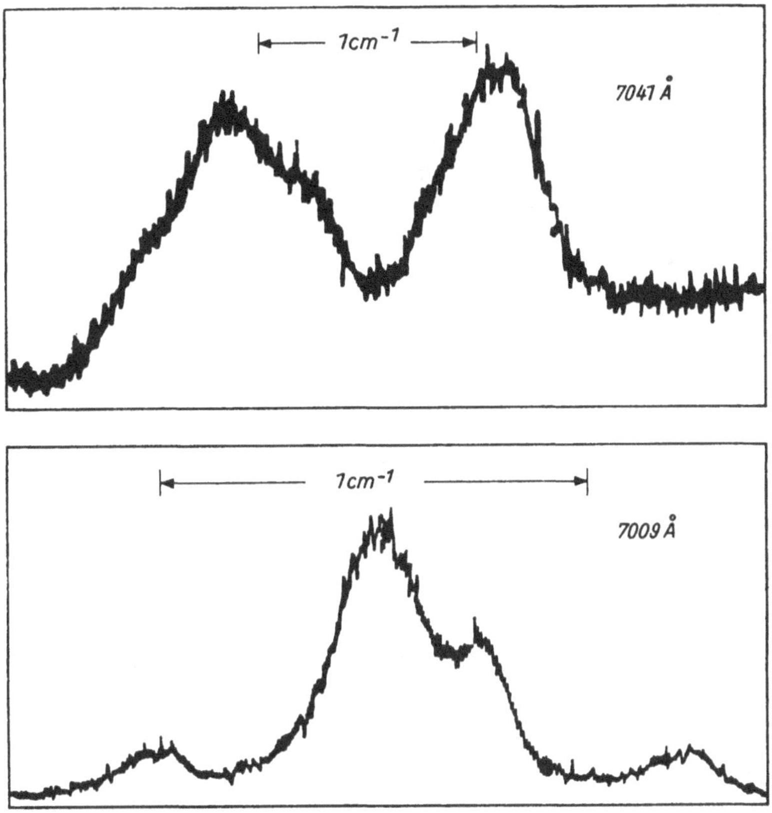

Fig. 5. Fine structure of the N-lines at 4 °K (after [44])

an energy level scheme with $J = 83.6\ \text{cm}^{-1}$ and $j = -9.7\ \text{cm}^{-1}$. Their results for first-nearest neighbor pairs were consistent with those reported earlier from piezospectroscopic work, which is discussed in the next section. Their results on third- and fourth-nearest neighbor pair lines are generally in agreement with those reported earlier, but more extensive. They do find that the three lines in the vicinity of the R-lines, which Powell et al. [5, 42] suggested might belong to fourth-nearest neighbor pairs, are more probably associated with more distant pairs.

Before leaving our discussion of the optical spectra of heavily doped ruby, it should be mentioned that a fine structure has been observed in the most intense pair lines [14, 44]. This is shown in Fig. 5 and is seen to be on the order of 1 cm^{-1} or less and, therefore, is generally not resolved. This structure is consistent with the zero-field splittings determined from EPR investigations as discussed in the previous section.

2.3.3 Infrared absorption spectrum

There have been reports of the far infrared absorption spectrum of heavily doped ruby in which observed lines have been assigned to transitions within the ground state manifolds of the pair systems [45, 46]. However, Kisliuk et al.

[21] point out that such transitions are forbidden, and therefore these assignments are probably erroneous.

2.4 Identification of pair types

The work described in the previous section was successful in establishing four different sets of energy levels and transitions but, in order to identify the specific types of pairs associated with each of these sets, it is necessary to observe the effects of an external perturbation on the optical spectra. Three types of perturbations have been used: uniaxial stress, magnetic field, and electric field. The observed effects of these perturbations are dependent on the orientation of a pair type with respect to the external field.

2.4.1 Piezospectroscopic investigations

Piezospectroscopic studies have proven to be the most valuable work in determining the type of pairs responsible for the various sets of lines observed in the optical spectra. A specific type of cation pair can exist in several different orientations in the Al_2O_3 lattice. The pairs in these different orientations are affected differently by the application of uniaxial stress on the crystal and are no longer equivalent. Thus the pair lines can be observed to split under uniaxial stress as the orientational degeneracy is lifted. By observing the number and magnitude of the splittings at different directions of applied stress, it is possible to deduce the type of pair from which a specific line comes.

The highest point symmetry of a site in the corundum lattice is D_3. The first-nearest neighbor cation pairs are parallel to the C_3-axis, while the second, third, and fourth neighbor pairs are oriented at various angles to C_3 (see Fig. 3). Therefore, spectral lines originating from first-nearest neighbor pairs will not be split by applied stress, while all other lines will exhibit splitting. The third-nearest neighbor pairs have a twofold symmetry axis through the center of the pair axis and at right angles to it, while a C_1-axis is found at similar locations for the second- and fourth-nearest neighbor pairs. These are the important symmetry considerations for identifying lines associated with different types of pairs.

The first important piezospectroscopic investigations of heavily doped ruby were made by Kaplyanskii and Przhevuskii [47 to 49]. In their first paper, they observed stress-induced splittings of several pair lines and suggested possible associations with specific pairs [47]. In their later papers, they established the existence of C_2-symmetry for the N_1 set of lines, and thus unambiguously assigned this set to third-nearest neighbor pairs [48, 49]. The N_2 series of lines exhibited the C_1-symmetry associated with both second- and fourth-nearest neighbor pairs. The former is known to be antiferromagnetically coupled and the latter ferromagnetically coupled in Cr_2O_3. Since it has been previously established that this series of lines is consistent with ferromagnetic coupling, they are assigned to fourth-nearest neighbor pairs. These authors also tentatively suggested the assignment of the series of lines 7214, 7290, 7451, and 7707 Å to second-nearest neighbor pairs with $J = 147\ cm^{-1}$, but this has been shown to be incorrect. They also report splittings of other lines both at longer wavelengths and between the R-lines, but are not able to make unambiguous assignments of these lines to specific pair types. The stress-induced splitting of the lines in the neighborhood of the R-lines was also observed by Spitale [50] but,

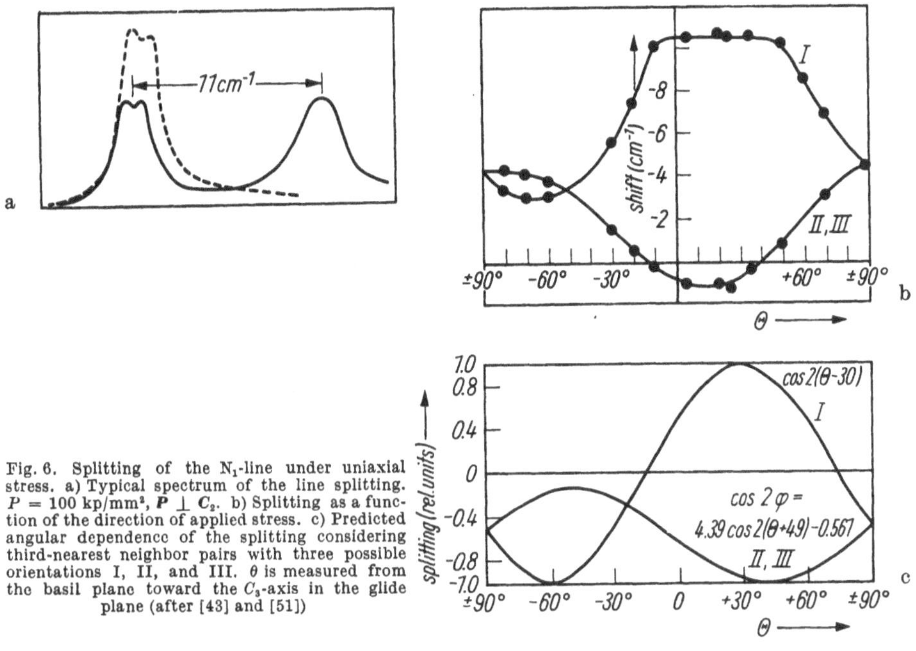

Fig. 6. Splitting of the N_1-line under uniaxial stress. a) Typical spectrum of the line splitting. $P = 100$ kp/mm², $\boldsymbol{P} \perp \boldsymbol{C}_2$. b) Splitting as a function of the direction of applied stress. c) Predicted angular dependence of the splitting considering third-nearest neighbor pairs with three possible orientations I, II, and III. θ is measured from the basil plane toward the C_3-axis in the glide plane (after [43] and [51])

again, no important conclusions could be drawn as to the specific system giving rise to these lines.

The most recent and, perhaps, the definitive work on the effects of pressure on the fluorescence spectra of heavily doped ruby is that of Mollenauer and Schawlow [43]. They investigated at 77 °K a set of oriented samples cut at intervals of 10° to 15° spanning a full 180° in a glide plane, which is a plane containing the C_3-axis and perpendicular to a C_2-axis. This gave detailed plots of the splittings of various lines as a function of direction of applied stress. By knowing the orientations of the different pair types with respect to the various directions of applied stress and assuming that the maximum splittings of lines from a specific pair occur when the direction of applied stress is parallel to the axis of that pair, they were able to theoretically predict the angular dependences of the splittings of lines from different pair systems. The results obtained on the N_1- and N_2-lines are shown in Fig. 6 and 7, respectively [43, 51]. Their results confirm that the N_1 series of lines originate from third-nearest neighbor pairs. Also, this detailed study is able to directly distinguish between second- and fourth-nearest neighbor pair lines, and the results demonstrate conclusively, without any assumptions as to the type of exchange interaction, that the N_2 set of lines belongs to the fourth-nearest neighbor pair system. Note that pairs II and III for third-nearest neighbors are equivalent for stress applied in the glide plane and therefore do not split in Fig. 6. This is not true for fourth-nearest neighbor pairs II and III. For these pairs an oxygen ion partially intervenes between the two chromium ions, giving V-shaped Cr–O–Cr complexes which are not spatially equivalent and therefore exhibit splitting under applied stress [51].

Mollenauer and Schawlow [43] also identified lines belonging to first-nearest neighbor pairs at 7540 and 7302 Å, and a second-nearest neighbor line at

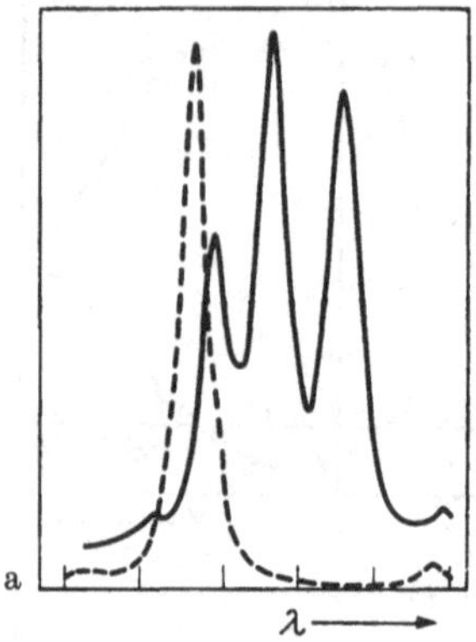

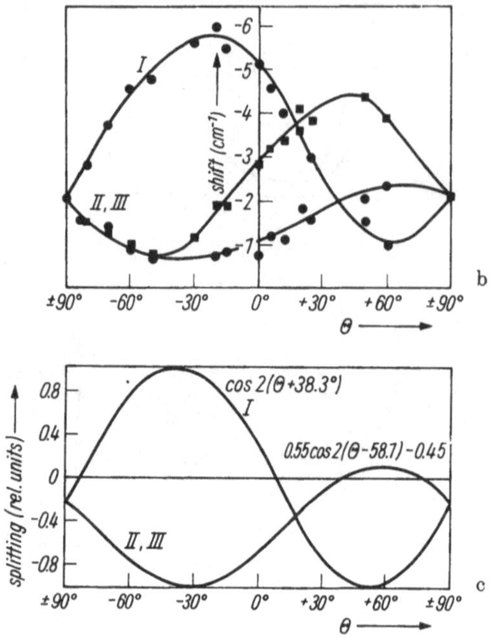

Fig. 7. Splitting of the N_2-line under uniaxial stress. a) Typical spectrum of the line splitting. b) Splitting as a function of the direction of applied stress. c) Predicted angular dependence of the splitting considering fourth-nearest neighbor pairs with three possible orientations I, II, and III. θ is measured from the basil plane toward the C_3-axis in the glide plane (after [43] and [51])

7452 Å. Mollenauer [43] went on to develop a dynamic piezospectroscopic technique in which stress modulation was used to increase the sensitivity of the measurements. This allowed three other second-nearest neighbor pair lines to be identified: 7159, 7214, and 7290 Å.

2.4.2 *Observations of Zeeman and Stark splittings*

The Zeeman effect in the fluorescence spectra of heavily doped ruby has also been studied. The splittings of fluorescence lines of heavily doped ruby in a magnetic field were first observed by the earlier workers [2, 3], and were recently studied at 77 and 4.2 °K by Kanskaya and Przhevuskii [52]. This latest investigation showed that the lowest components of the excited state manifolds of the N_1 and N_2 pair systems are of spin $S = 2$ and $S = 1$, respectively. This explains why the N_1- and N_2-transitions are the most intense lines in their respective pair systems, since they represent transitions with the spin-allowed selection rule $\Delta S = 0$. These assignments indicate that the sign of the exchange integral changes in going from the ground to the excited state, thus making the N_1-pairs ferromagnetically coupled and the N_2-pairs antiferromagnetically coupled in the excited state. No explanation was offered as to why this change in coupling occurs, but the same conclusion was reached by Imbusch and Graifman [37].

Berggren et al. [14] have also reported Zeeman studies of the fourth-nearest neighbor pair system in ruby. They confirm the findings of reference [52], and they make use of these results to help interpret the EPR spectra they observe (see Section 2.3.1).

The "pseudo-Stark" splittings of the fluorescence lines in the spectrum of heavily doped ruby have also been observed [53]. This is the electric-field-induced splitting due to the lifting of the degeneracy of pairs located at sites having equal crystal fields with opposite signs. The sign of the odd component of the trigonal crystal field is opposite for each ion of the first- through third-nearest neighbor pairs, which indicates that electric field effects will cancel and no splitting will be observed. For fourth-nearest neighbor pairs, the sign of the odd component of the trigonal crystal field is the same for each ion of the pair. The field at these pairs may be either parallel or antiparallel to an external electric field applied parallel to the *C*-axis, and thus the lines originating from these pairs will split into doublets when such a field is present. For measurements made at 77 and 4.2 °K with fields up to 200 kV/cm, no splittings were observed when the electric field was applied perpendicular to the *C*-axis of the crystal. For fields parallel to the *C*-axis, the lines of the N_1-series did not split, but those of the N_2-series split into doublets. The splitting was slightly different for each line and approximately linear with field intensity. These observations of Kaplyanskii et al. [53] are consistent with the previous assignments of the N_1 and N_2 series of lines to third- and fourth-nearest neighbor pairs, respectively.

Perhaps the most important aspect of this work is that electric-field-induced splitting can be used to directly identify fourth-nearest neighbor pair lines without resorting to indirect arguments about exchange interactions. The theory predicts that, for the first-, second-, third-, and fourth-nearest neighbor pair lines, only the last one will exhibit splitting in an axial electric field, and thus the observed splittings of the N_2 line series unambiguously associates this set to fourth-nearest neighbor pairs.

3. Absorption and Fluorescence Bands

Along with the N-lines, both the absorption and emission spectrum of heavily doped ruby exhibit bands that are not present in the spectra of lightly doped samples. However, in comparison with the N-lines, relatively little work has been done in attempting to understand the origin of these bands.

3.1 Absorption band of heavily doped ruby

An absorption band appears in the near ultraviolet region of the absorption spectrum of heavily doped ruby which is not present in the spectrum of lightly doped ruby. This band is shown in Fig. 8, and is located on the shoulder of the well-known ruby absorption band which peaks at about 4000 Å [54].

The uv absorption band of heavily doped ruby was first reported by Linz and Newnham [55]. They noted that the a-spectrum of this band (***E*** ⊥ ***c***, ***H*** ⊥ ***c***) is identical to the σ-spectrum (***E*** ⊥ ***c***, ***H*** || ***c***), which indicates that the transitions are electric multipole in origin (probably electric dipole, since electric

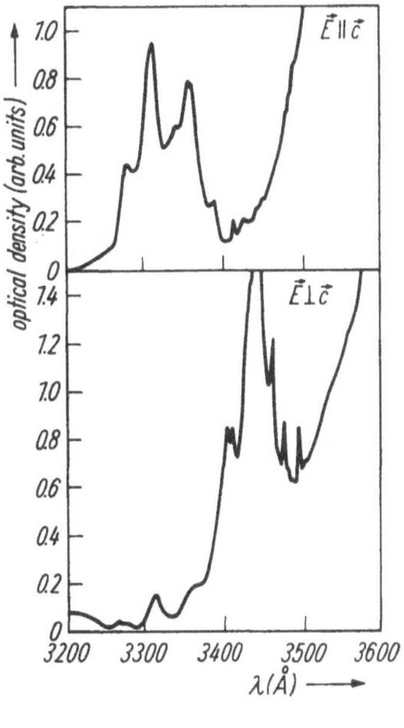

Fig. 8. Absorption spectra of ruby 4.26 mm thick with 2.1 at% Cr in the region of the "pair band" at 77 °K (after [54])

quadrupole transitions are expected to be weaker than electric dipole transitions). They also measured the integrated intensity of this band as a function of chromium concentration and found that it increased quadratically. In analogy with the N-lines, they interpreted these results as showing that this band is associated with pairs of interacting chromium ions.

Naiman and Linz [56] investigated the structure of the pair band and found a peak which varied linearly with chromium concentration. They attributed this to the forbidden transition from the ground state to the $^2A_1(^1E)$ state of isolated Cr^{3+} ions, which is consistent with the results of excited-state absorption measurements [57]. They felt that the remainder of the pair band could not be simply attributed to exchange-coupled pairs, since this would necessitate an unreasonable value for the strength of the exchange interaction. Instead, they suggested two possible explanations for the band. The first is a double charge transfer complex which has as its final state $Cr^{2+}(d^4)$–$Cr^{4+}(d^2)$. The second is a double ion excitation of the type observed by Varsanyi and Dieke [58]. No work has yet been done to verify either of these suggestions.

3.2 Fluorescence band of heavily doped ruby

Fig. 9 shows the fluorescence spectrum of heavily doped ruby between 6000 and 8500 Å at several temperatures [5]. The region of identified N-lines extends to about 7400 Å, as does the main vibronic band of lightly doped ruby. As seen in Fig. 9, there is a prominent, structured band in the emission spectrum of heavily doped ruby extending from approximately 7400 to 8000 Å, with its main peak located at about 7750 Å.

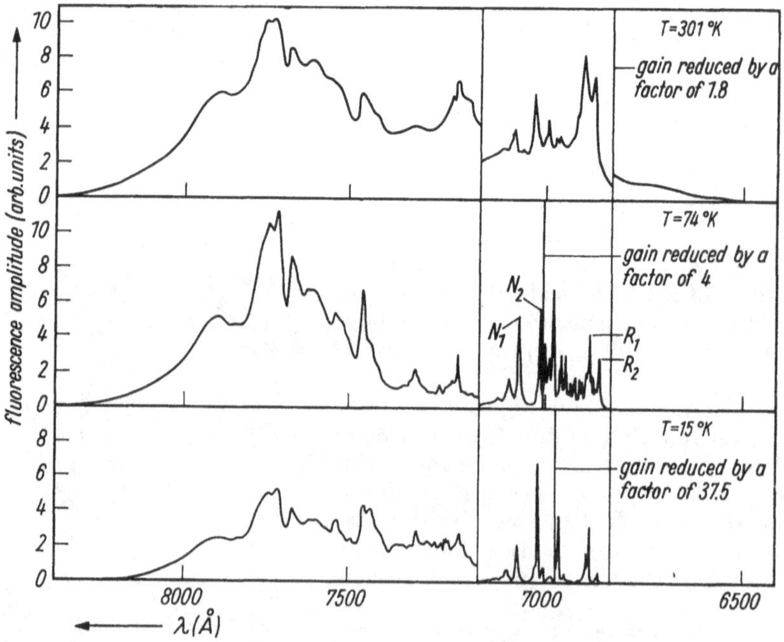

Fig. 9. Fluorescence spectrum of ruby with 2.1 at% Cr at several temperatures (after [5]). In the middle part of the spectrum the λ-scale is expanded

This band appeared in the emission spectrum reported in some of the early investigations of ruby [3]. Tolstoi et al. [59] investigated the intensity of this band as a function of chromium concentration and found that it increased greater than quadratically. They concluded from this that the band is due to more than two interacting Cr^{3+} ions.

Kaplyanskii and Przhevuskii [49] investigated the splittings of some of the structure in this band under uniaxial stress. They found that the peak at 7707 Å exhibited the same splitting as the 7452 Å second-nearest neighbor pair line. This is consistent with the suggestion of Mollenauer and Schawlow [43] that this peak is a vibrational sideband of the 7452 Å line. Since it is energetically possible that some of the structure in this band is associated with N-line vibronic transitions, the band may partially be due to the interaction of more than two Cr^{3+} ions and partially vibronic in nature.

4. Summary of Spectra of Heavily Doped Ruby

Table 3 lists the exchange parameters for the first four-nearest neighbor pair types as determined from optical measurements. Figure 10 shows the energy level schemes and the various observed transitions for these four pair systems. The direction of the arrows indicates whether the lines were observed in absorption or emission. Table 4 lists the remaining lines observed in the spectrum of heavily doped ruby and their suggested assignments to more distant neighbor pair systems, vibronic transitions, multi-ion systems, or absorption pair band.

Table 3

Coupling parameters for chromium ion pairs in ruby

type of pair	exchange interaction J (cm^{-1})	biquadratic exchange interaction j (cm^{-1})
first-nearest neighbor	240	—
second-nearest neighbor	83.6	−9.7
third-nearest neighbor	11.59	0.06
fourth-nearest neighbor	−6.99	0.14

5. Interaction with the Phonon Field

Now that we have reviewed the steady-state features of the spectra of heavily doped ruby, let us examine the dynamic aspects of the pair systems. In this section we consider electron–phonon interaction which manifests itself in two different ways: spin–lattice relaxation in the ground state manifolds of the pair systems, and thermal broadening and shifting of the lines in the optical spectra.

5.1 Spin–lattice relaxation

Since the ground state splittings of the ion pair systems are comparable to phonon energies, transitions within these manifolds are in thermal communication with the lattice vibrations. There are three possible mechanisms for causing spin–lattice relaxation. These are the phonon modulation of the crystal field, of the exchange interaction, or of the magnetic dipole–dipole coupling. There are also three types of transitions by which relaxation can occur: direct, Raman, and Orbach. It is possible to distinguish between these processes through the temperature dependences of their rates which are proportional to T, T^7, and $[\exp(\Delta E/kT) - 1]^{-1}$, respectively.

Table 4

Transitions observed in the spectrum of heavily doped ruby other than those assigned to single ions and the first four nearest-neighbor pair systems

transition (cm^{-1})	assignment	reference
30660	absorption pair band, $\boldsymbol{E} \perp \boldsymbol{c}$	[55]
30600	absorption pair band, $\boldsymbol{E} \parallel \boldsymbol{c}$	
30350	absorption pair band, $\boldsymbol{E} \parallel \boldsymbol{c}$	
30300	absorption pair band, $\boldsymbol{E} \parallel \boldsymbol{c}$	
30280	absorption pair band, $\boldsymbol{E} \perp \boldsymbol{c}$	
30060	absorption pair band, $\boldsymbol{E} \parallel \boldsymbol{c}$	
29910	absorption pair band, $\boldsymbol{E} \parallel \boldsymbol{c}$	
29880	absorption pair band, $\boldsymbol{E} \parallel \boldsymbol{c}$	
29870	absorption pair band, $\boldsymbol{E} \perp \boldsymbol{c}$	
29610	absorption pair band, $\boldsymbol{E} \parallel \boldsymbol{c}$	
29480	absorption pair band, $\boldsymbol{E} \perp \boldsymbol{c}$	
29430	absorption pair band, $\boldsymbol{E} \perp \boldsymbol{c}$	
29390	absorption to $^2A_1(^1E)$ level, $\boldsymbol{E} \parallel \boldsymbol{c}$	[56, 57]
29280	absorption pair band, $\boldsymbol{E} \parallel \boldsymbol{c}$	[55]
29240	absorption pair band, $\boldsymbol{E} \perp \boldsymbol{c}$	
29160	absorption pair band, $\boldsymbol{E} \perp \boldsymbol{c}$, $\boldsymbol{E} \parallel \boldsymbol{c}$	
28990	absorption pair band, $\boldsymbol{E} \parallel \boldsymbol{c}$	
28970	absorption pair band, $\boldsymbol{E} \perp \boldsymbol{c}$	
28840	absorption pair band, $\boldsymbol{E} \perp \boldsymbol{c}$	
28690	absorption pair band, $\boldsymbol{E} \perp \boldsymbol{c}$	
14456.2	distant pairs	[21]
14455.7	distant pairs	
14454.8	distant pairs	
14454.0	distant pairs	
14452.6	distant pairs	
14451.4	distant pairs	
14450.7	distant pairs	
14443.8	distant pairs	
14443.5	distant pairs	
14442.5	distant pairs	
14442.0	distant pairs	
14440.6	distant pairs	
14439.7	distant pairs	
14438.8	distant pairs	
14438.1	distant pairs	
14437.3	distant pairs	
14436.5	distant pairs	
14435.6	distant pairs	
14424.8	distant pairs	
14423.2	distant pairs	
14421.6	distant pairs	
14414.0	distant pairs	
14413.0	distant pairs	
14410.5	distant pairs	
14410.0	distant pairs	
14408.9	distant pairs	
14407.7	distant pairs	

Table 4 (continued)

transition (cm^{-1})	assignment	reference
14406.1	distant pairs	[21]
14404.9	distant pairs	
14189.8	vibronic	
14180.5	vibronic	
14168.1	vibronic	
14152.4	vibronic	
14141.3	vibronic	[54]
14137.3	vibronic	
14133.3	vibronic	
14122.3	vibronic	
14111.3	vibronic	
14099.4	vibronic	
14065	vibronic	
14035	vibronic	
13996	vibronic	
13963	vibronic	
13930	vibronic	
13912	vibronic	
13898	vibronic	
13856	vibronic	
13822	vibronic	
13742	vibronic	
13721	vibronic	
13712	vibronic	
13686	vibronic	
13652	vibronic	
13495	fluorescence band, more than two ions	
13450	fluorescence band, more than two ions	
13417	fluorescence band, more than two ions	
13333	fluorescence band, more than two ions	
13294	fluorescence band, more than two ions	
13266	fluorescence band, more than two ions	
13173	fluorescence band, more than two ions	
13148	fluorescence band, more than two ions	
13116	fluorescence band, more than two ions	
13045	fluorescence band, more than two ions	
12970	vibronic sideband of 7452 Å line	[43, 49]
12925	fluorescence band, more than two ions	[54]
12872	fluorescence band, more than two ions	
12747	fluorescence band, more than two ions	
12647	fluorescence band, more than two ions	
12633	fluorescence band, more than two ions	

5.1.1 Relaxation in the pair ground state manifolds

Statz et al. [30, 31] observed a sharp decrease in the intensity of some of the EPR lines in heavily doped ruby at high temperatures. Their results indicate the presence of strong spin–lattice relaxation and can be attributed to either Raman- or Orbach-type processes due to the modulation of either the exchange interaction or the crystal field.

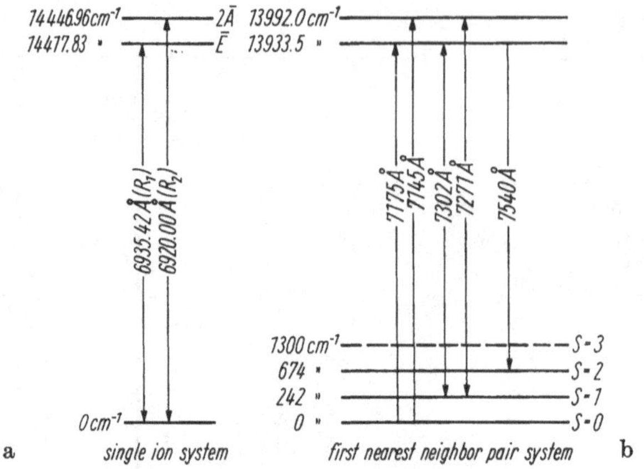

Fig. 10. Energy level diagrams and transitions associated with the sharp line spectra of ruby.
a) Single-ion R-lines: b) first-nearest neighbor pair system; c) second-nearest neighbor pair system; d) third-nearest neighbor pair system; e) fourth-nearest neighbor pair system

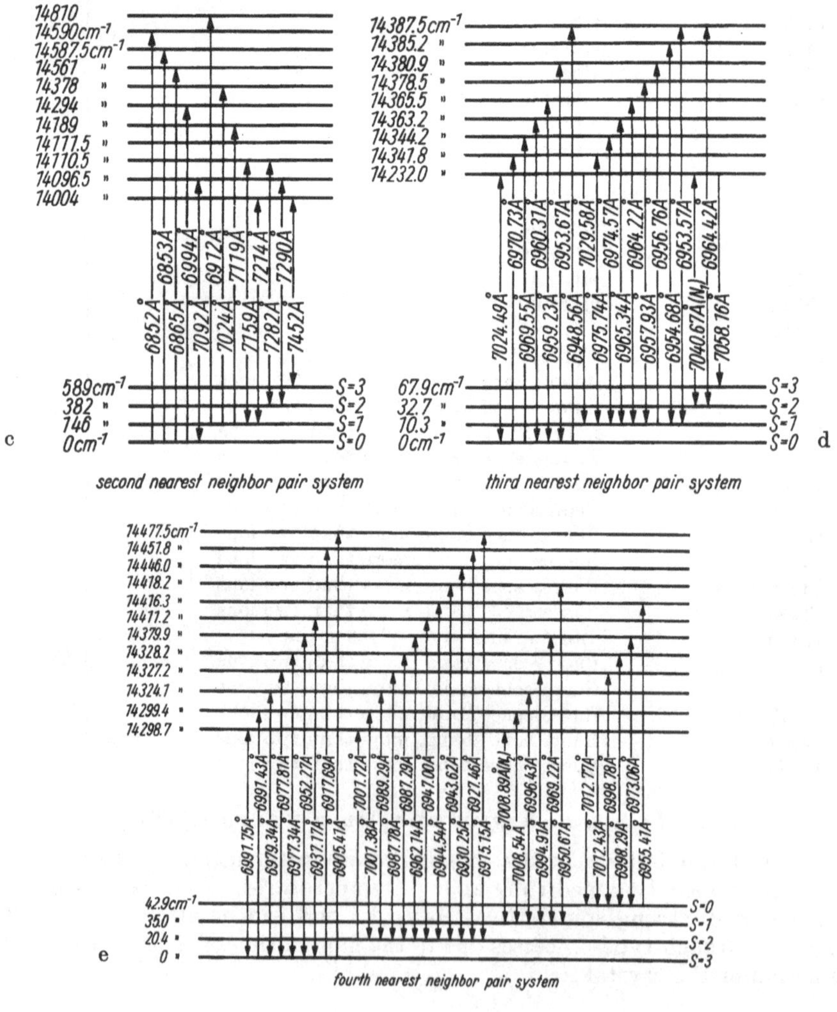

Gill [60] measured the temperature dependence of the relaxation times of several lines of heavily doped ruby after saturation by a pulse of resonant radiation. He used the spin Hamiltonian given in equation (9) to describe the pairs and assumed $J \gg D$. In general the decays observed by Gill were exponential. Small deviations from pure exponential behavior at short times were attributed to spin–spin relaxation. The relaxation times of the various lines investigated by Gill exhibited different variations with temperature and chromium concentration. Some became independent of temperature at low temperatures, while others exhibited one of the variations discussed above for the three different types of relaxation mechanisms. One of the lines originating on an $S = 3$ level was found to have a relaxation time which varied approximately as $1 - \exp(-\Delta E/kT)$ and became independent of temperature at low temperature, which is consistent with an Orbach-type two-step relaxation process involving an intermediate lower level. This type of transition can be caused by the thermal modulation of the exchange parameter as expressed by the operator $H_{\text{int}} = (\mathrm{d}J/\mathrm{d}r)\, \boldsymbol{S}_1 \cdot \boldsymbol{S}_2$ which connects states differing by $\Delta S = \pm 2$. This operator is diagonal in spin states if both dipole–dipole interaction and crystal field anisotropy terms are neglected in equation (9). The relatively large zero-field splitting in ruby ($\approx 0.38\ \mathrm{cm}^{-1}$) indicates that D gives the greatest off-diagonal elements of H_{int}. Gill's estimates showed that this type of interaction can lead to the right order of magnitude for the relaxation time, whereas similar transitions due to thermal modulation of the crystal field will not lead to a relaxation time as fast as that observed experimentally [60].

The relaxation times of other lines were observed to vary as T^{-7} and T^{-3}, and Gill suggested that Raman scattering of phonons is the dominant relaxation transition in these cases [60]. He again ascribed the interaction mechanism to the thermal modulation of the exchange interaction. In this case, the interaction Hamiltonian has two terms: $H_{\text{int}} = H_1^{\mathrm{R}} + H_2^{\mathrm{R}}$, where $H_1^{\mathrm{R}} = (\mathrm{d}^2J/\mathrm{d}r^2) \times \boldsymbol{S}_1 \cdot \boldsymbol{S}_2$ is used with first-order perturbation theory and $H_2^{\mathrm{R}} = (\mathrm{d}J/\mathrm{d}r)\, \boldsymbol{S}_1 \cdot \boldsymbol{S}_2$ is used with second-order perturbation theory. The former leads to a T^{-7}-dependence for the relaxation time, while the latter causes a T^{-3}-variation. However, Gill [60] found that the absolute magnitude of the relaxation times estimated from this theory is much greater than the observed values and concluded that the mechanism leading to Raman relaxation processes is the modulation of the crystal field.

5.1.2 Cross-relaxation between single-ion and pair systems

Before extensive measurements were made on the spin–lattice relaxation times of chromium ion pairs, it had been observed that the single-ion relaxation times in ruby decreased with increasing chromium concentration; no theoretical explanation had been given for this effect [60 to 65]. Gill's measurements showed that the spin–lattice relaxation times for pairs were one to three orders of magnitude faster than that for single ions [60]. It was suggested that the concentration dependence of the relaxation time of single ions could be due to the interaction of single-ion and pair systems [65]. Such an effect could take place through spin–spin interaction along the lines of the cross-relaxation theory discussed by Bloembergen et al. [66].

If cross-relaxation between single ions and pairs is fast enough for thermal equilibrium to be established, then the two kinds of systems will exhibit the same spin–lattice relaxation time given by [67]

$$T_1^{-1} = w_s + w_p\, w_{cr} \left(w_p + \frac{N_s}{N_p}\, w_{cr} \right)^{-1}, \tag{10}$$

where w_s, w_p, and w_{cr} represent the rates of single-ion relaxation, pair relaxation, and cross-relaxation, respectively; N_s is the number of single ions and N_p the number of pairs with resonances close to those of single ions. For heavily doped ruby, the observed decrease in T_1 with increasing chromium concentration can be attributed to cross-relaxation between single-ion and pair systems assuming $w_p > w_s$. The spin–lattice relaxation time for the two extreme cases of large and small cross-relaxation rates is given by

$$\left.\begin{aligned} &T_1 \approx w_s^{-1}; && w_s \gg w_{cr} \\ \text{and}\quad &T_1 \approx w_{cr}^{-1} + \frac{N_s}{N_p}\, w_p^{-1}; && w_s \ll w_{cr}\,. \end{aligned}\right\} \tag{11}$$

The concentration dependence of T_1 can be predicted by noting that w_{cr} is a function of the number of pairs in the lattice and thus proportional to C^2 where C is the chromium ion concentration. There are three cases to be considered: First, for $w_s \gg w_{cr}$, equation (11) indicates that T_1 will be independent of temperature. Secondly, if the relaxation bottleneck is due to the cross-relaxation term in equation (10), then $w_s \ll (N_s/N_p)\, w_{cr} \ll w_p$ and $T_1^{-1} \approx w_{cr} \sim C^2$. Finally, if the relaxation bottleneck is due to the pair relaxation, then $(N_s/N_p)\, w_{cr} \gg w_p \gg w_s$ and $T_1^{-1} \approx (N_p/N_s)\, w_p \sim C$.

This is an extremely simplified picture of single ion–pair cross-relaxation. However, at low temperatures T_1^{-1} has been found to vary approximately as C^2, indicating that cross-relaxation is the limiting process in the spin–lattice relaxation [62]. The magnitude of the cross-relaxation rate necessary to give these results is in reasonable agreement with estimated cross-relaxation rates of between 10^4 and $10^6\ s^{-1}$ [67 to 69].

The magnitudes of the spin–lattice decay times reported in references [60] to [65] do not coincide even when measured at the same temperature. This is probably because the measurements were made using different techniques, different frequencies, and different sample orientations [67]. However, at 77 °K and below for all measurements, T_1 is found to decrease strongly at chromium concentrations between 0.01% and 0.1%. For example, Pace et al. [62] find that, at 1.4 °K, T_1 decreased from about 6×10^{-2} s for $\approx 0.03\%$ chromium to about 10^{-4} s at $\approx 0.4\%$ chromium.

The spin–lattice relaxation time of the pairs has been obtained indirectly from its contribution to the linewidths of optical transitions in the pair systems [70, 71]. This is discussed in more detail in the next section. From these measurements, Powell et al. [70] report typical values of w_p^{-1} of 2.7×10^{-11} s. This is certainly much less than the typical values reported for both w_s^{-1}, which is on the order of 10^{-1} s at 4.2 °K [61], and w_{cr}^{-1}, which is estimated to be 10^{-4} to 10^{-6} s [67 to 69]. Thus these findings are consistent with the inequality $w_s \ll (N_s/N_p)\, w_{cr} \ll w_p$, which gives the second case discussed above and predicts the observed C^2-dependence of T_1^{-1}.

It has been suggested that these transitions in the ground state manifolds of the pair systems might make heavily doped ruby a good material for quantum electronic devices in the submillimeter spectral region [70, 72].

5.2 Temperature dependence of the widths and positions of the R- and N-lines of heavily doped ruby

The interaction between the optically active electrons and the thermal vibrations of the lattice causes a broadening and shifting of the lines in the optical spectra as the temperature is raised. Powell et al. [70, 71] measured the temperature dependence of the widths and positions of the two R-lines and seven of the N-lines in the fluorescence spectrum of heavily doped ruby. The specific lines they investigated are listed in Table 5. The levels between which these transitions take place can be seen in Fig. 10.

5.2.1 Thermal variation of linewidths

The widths of all the lines which were investigated were observed to be approximately constant below about 100 °K and to increase sharply above this temperature [70, 71]. Fig. 11 shows typical results obtained for the N_1- and N_2-lines. These results were interpreted using the following equation:

$$\Delta\tilde{\nu} = \Delta\tilde{\nu}_0 + \bar{\alpha}\left(\frac{T}{T_D}\right)^7 \int_0^{T_D/T} \chi^6\, e^{\chi}\, (e^{\chi} - 1)^{-2}\, d\chi + \sum_{j<i} \beta_{ij}\, e^{\Delta E_{ij}/kT}\, (e^{\Delta E_{ij}/kT} - 1)^{-1} + \sum_{j>i} \beta_{ij}\, (e^{\Delta E_{ij}/kT} - 1)^{-1}\,. \quad (12)$$

Here $\Delta\tilde{\nu}$ is the linewidth, $\Delta\tilde{\nu}_0$ is the temperature-independent contribution due to microscopic strains, $\bar{\alpha}$ and β_{ij} are coupling parameters, and T_D is an effective Debye temperature. The first term describes broadening due to Raman scattering of phonons in the Debye approximation, while the last two terms describe the contributions due to direct phonon transitions between the levels of the pair manifolds.

The solid lines in Fig. 11 are the results of fitting (12) to the experimental data. The values of the adjustable parameters needed to obtain the best fits for all the lines investigated are listed in Table 5. The rapid increase in linewidth above ≈ 100 °K was attributed to the Raman scattering processes and was found to be the same for the four lines of the N_2-system which were investigated and for the three lines observed in the N_1-system. Both microscopic strains and phonon emission processes can contribute to the low-temperature linewidth, and it is possible to distinguish between the two types of contributions through

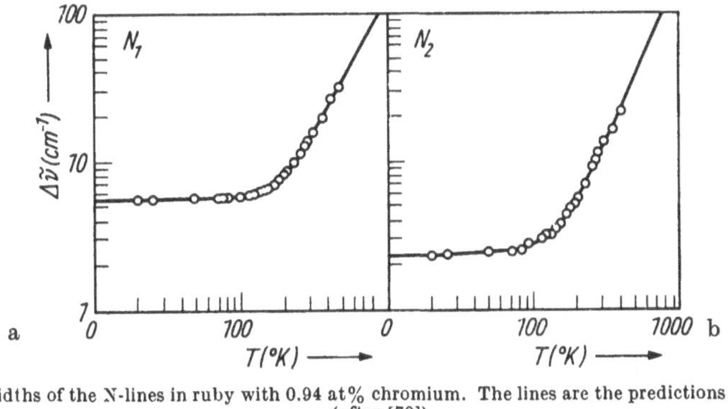

Fig. 11. Linewidths of the N-lines in ruby with 0.94 at% chromium. The lines are the predictions of equation (12) (after [70])

Table 5

Linewidth and lineshift parameters for heavily doped ruby (from [70] and [71])

sample (at% chromium)	system	line	linewidth				lineshift	
			T_D (°K)	$\bar{\alpha}$ (cm^{-1})	β (cm^{-1})	$\Delta\tilde{\nu}_0$ (cm^{-1})	T_D (°K)	α (cm^{-1})
2.1	single ion	R_1	760	1440	0	2.9	450	−177
		R_2	760	640	0	3.1	450	−177
	third neighbor	N_1	935	1681	0.2	6.73	450	−160
	fourth neighbor	N_2	935	1166	0.24	3.06	450	−150
0.94	single ion	R_1	760	1071	0.24 ($\bar{\bar{E}} \to 2\bar{A}$) *)	2.45	450	−177
		R_2	760	475	0.36 ($2\bar{A} \to \bar{E}$) *)	2.30	450	−177
	third neighbor	N_1	935	731	0.20 ($\Delta S = 2$)	5.24	450	−160
		7058.2 Å	935	731	0.50 ($\Delta S = 2$) 0.40 ($\Delta S = 1$)	5.30	450	−125
		6948.6 Å	935	731	0.10 (3 level) *) 0.30 (2 level) *) 0.05 (1 level) *)	0.90	450	−125
	fourth neighbor	N_2	935	777	0.2 ($\Delta S = -2$)	2.14	450	−150
		7001.8 Å	935	777	0.05 ($\Delta S = -1$)	2.14	450	−125
		6989.4 Å	935	777	0.05 ($\Delta S = -1$)	1.95	450	−125
		6987.7 Å	935	777	0.04 ($\Delta S = -1$) 0.015 (1 level) *)	2.20	450	−125

*) Direct phonon transition in the excited state.

lineshape analysis. The former gives a Gaussian lineshape and the latter a Lorentzian shape. Powell et al. [70, 71] were able to deconvolute the observed Voigt profiles to separate the broadening due to the two mechanisms. The results showed that the third-nearest neighbor (N_1) pair lines were broadened more by microscopic strains than the fourth-nearest neighbor (N_2) pair lines, and, in general, the broadening was the same for the lines within the same pair system. The one exception to this was that the one line studied which terminated on the lowest component of the ground state manifold exhibited much less strain broadening than the other lines which terminated on upper components of the ground state manifold. Direct phonon processes in the ground state manifolds between levels with $\Delta S = \pm 2$ and $\Delta S = \pm 1$ were found to contribute to the line broadening. The transitions with the former selection rule were found to have a greater probability of occurrence than those with the latter selection rule. This observation and the magnitudes of β_{ij} are consistent with the spin–lattice relaxation observations of Gill [60] and their interpretation through cross-relaxation theory as discussed in the previous section. Broadening of the R-lines and two of the N-lines due to direct phonon processes in the excited state manifolds was also observed [70, 71].

5.2.2 Thermal variation of line positions

All of the lines investigated were observed to shift toward lower energies as temperature was increased [70, 71]. Typical results obtained on the N_1- and N_2- lines are shown in Fig. 12.

These results were interpreted using the following equation:

$$\delta\tilde{\nu} = \alpha \left(\frac{T}{T_D}\right)^4 \int_0^{T_D/T} \chi^3 (e^\chi - 1)^{-1} d\chi +$$

$$+ \pi \sum_{j<i} \beta_{ij} \left(\frac{kT}{\Delta E_{ij}}\right)^2 P \int_0^{T_D/T} \chi^3 (e^\chi - 1)^{-1} \left[\chi^2 - \left(\frac{\Delta E_{ij}}{kT}\right)^2\right]^{-1} d\chi -$$

$$- \pi \sum_{j>i} \beta_{ij} \left(\frac{kT}{\Delta E_{ij}}\right)^2 P \int_0^{T_D/T} \chi^2 (e^\chi - 1)^{-1} \left[\chi^2 - \left(\frac{\Delta E_{ij}}{kT}\right)^2\right]^{-1} d\chi \,. \qquad (13)$$

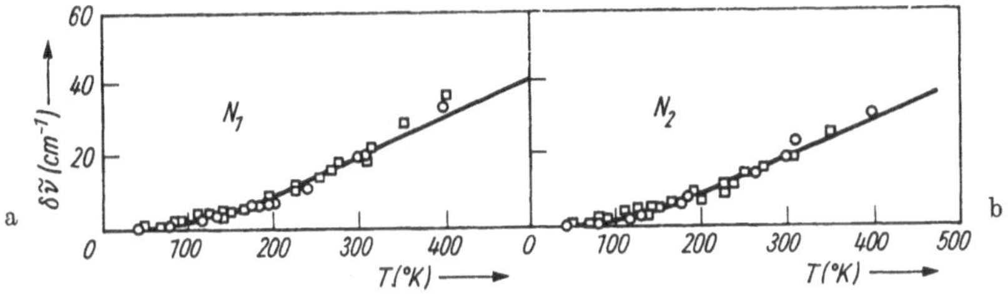

Fig. 12. Energy shift of the N-lines below their values at 0 °K. The lines are the predictions of equation (13) (after [70]). ○ 2.1% Cr; □ 0.94% Cr. ΔE in the exponentials is given in cm^{-1}

Here $\delta\tilde{\nu}$ is the shift in the line from its position at 0 °K, α is a coupling parameter, and P denotes the principal value of the integral. The first term describes the contribution to the lineshift caused by the continual absorption and re-emission of virtual phonons by the impurity ion in the phonon field, while the last two terms describe the contributions due to direct phonon transitions between the levels of the pair manifolds.

The solid lines in Fig. 12 represent the best fit to the experimental data using equation (13). The adjustable parameters needed to obtain this fit are listed in Table 5. For all of the lines investigated, it was found that the best fit to the data was obtained considering only the first term in (13) [70, 71]. Different effective Debye temperatures were needed to fit the linewidth and lineshift data, which indicates that different effective phonon distributions are active in the thermal broadening and shifting processes.

6. Interaction between Single-Ion and Multi-Ion Systems

When two different types of optically active systems are present in the same crystal, interactions can occur between these systems which manifest themselves in their optical spectra. In this section we discuss the interaction between the single-ion system and the various pair and multi-ion systems in heavily doped ruby. Investigations of the interaction mechanism are first considered and then the effects of this interaction on the temperature dependence of the fluorescence intensities and decay times are reviewed.

The presence of energy transfer in heavily doped ruby has been detected in several different ways. As mentioned in Section 2.1 and shown in Fig. 2, the ratio of the fluorescence intensity of an N-line to that of an R-line increases greater than linearly with chromium concentration at high levels of doping [7, 8]; it has been suggested [7] that this might be due to energy transfer from single ions to pairs. Wieder and Sarles [73] explicitly demonstrated energy transfer from single ions to pairs by exciting a heavily doped ruby crystal with R_1-line laser emission from a lightly doped ruby and observing the fluorescence in the N_1- and N_2-lines. The excitation spectra of the N_1- and N_2-lines have also been obtained [54]. These spectra closely resembled that of the R-line with three additional features: 1. Pumping of the N-lines through the absorption pair band near 3600 Å was observed. 2. Strong pumping was exhibited in the region of the S-lines. 3. Excitation occurred at the wavelengths of the R-lines. This last observation is another explicit demonstration of energy transfer from the single ion to the pair systems. The effects of single- ion to pair energy transfer have also been observed in optically detected spin resonance experiments [14].

6.1 Mechanism of interaction between single-ion and pair sytems

The first question to be answered is whether the energy transfer between single ions and pairs is a radiative or a radiationless process. Wieder and Sarles [73] suggested radiative re-absorption as the mechanism of energy transfer which best explained their data. Tolstoi et al. [59] demonstrated the importance of sample size on the concentration dependence of the relative intensity ratios of single-ion and pair lines, an effect which is consistent with re-absorption. However, they also consider the possibility of radiationless energy transfer [8, 59, 74].

Imbusch [75] made an extensive investigation of energy transfer from single ions to pairs in heavily doped ruby. He argued that the mechanism of energy transfer is nonradiative because of several observations. If the process were radiative, the pair transitions which are close to being in resonance with the R-lines would have to have an anomalously high absorption coefficient in order to compete favorably with re-absorption by the more numerous single ions. Radiative re-absorption was also ruled out by the observation that the ratio of intensities of pair fluorescence to single-ion fluorescence is the same for R_1-line laser excitation and blue-light excitation. This ratio should be greater for the former type of excitation if the ion pairs could directly absorb the R_1-radiation. Finally, a concentration quenching of the R-line fluorescence decay time is observed; an effect cannot be caused by radiative energy transfer but is characteristic of nonradiative energy transfer. Figure 13 shows the dependence of the R-line decay time on chromium concentration [74, 75]. The increase in decay time at low levels of doping is due to re-absorption by isolated Cr^{3+} ions, while the sharp decrease at high levels of doping is caused by the presence of efficient energy transfer to pair systems. The data of Tolstoi and Liu Shun-Fu [74] were taken on thin samples where no re-absorption was present. It shows a constant value for the decay time at small concentrations which is close to its intrinsic value of 4.2 ms [76]. The decay time at high concentrations falls below the intrinsic value which is consistent with the assumption of radiationless energy transfer.

The rate of radiationless energy transfer between two ions in a crystal decreases rapidly with ionic separation. Therefore, energy transfer takes place only from Cr^{3+} ions located close to pairs. However, the number of single ions located in such a favorable position is too small to account for the efficiency of the observed energy transfer. Imbusch [75] proposed that energy migrates among the isolated chromium ions until it gets enough to an ion pair to transfer its energy. Such a migration process among single ions must be radiationless, since radiative migration would be too slow. This hypothesis is substantiated by the fact that the N-lines exhibit decays with tails having the same decay time as that measured for the R-lines. If energy transfer occurred only from single ions located close to pairs, the decay would show the presence of an exponential with a time constant shorter than the R-line lifetime; this exponential would be due to those ions which take place in the energy transfer and therefore decay faster than the rest of the chromium ions.

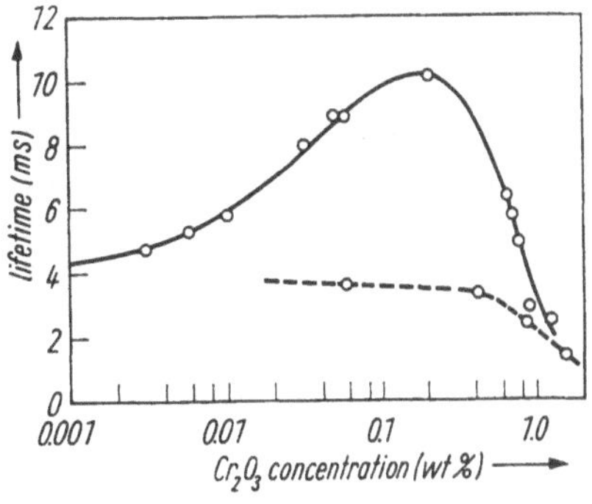

Fig. 13. Fluorescence decay time of the R-lines in ruby at 77 °K as a function of chromium concentration. The solid line is the data of Imbusch [75] on samples about 0.1 × × 0.1 × 0.4 in.³. The dashed curve is the data of Tolstoi and Liu Shun-Fu [74] taken on very small samples with negligible re-absorption

Imbusch also discussed various possibilities for the interaction mechanism causing both the single ion–single ion and single ion–pair energy transfer [75]. According to the theory of long-range resonant energy transfer developed by Förster [77] and Dexter [78], the energy transfer rate is given by

$$P = \frac{2\pi}{\hbar} |V_{fi}|^2 \int g_1(E)\, g_2(E)\, \mathrm{d}E\,, \tag{14}$$

where V_{fi} is the matrix element for the electromagnetic radiation fields of the two ions, and $g_1(E)$ and $g_2(E)$ are normalized line shape functions for ions 1 and 2. Imbusch considered exchange, electric dipole–dipole, magnetic dipole–dipole, electric dipole–quadrupole, and electric quadrupole–quadrupole interactions as the possible energy transfer mechanisms. The estimated rates of energy transfer for the first four of these mechanisms were found to be too slow to account for the observed energy transfer efficiency. The crude estimate of electric quadrupole–quadrupole interaction gave a fast enough rate. The concentration dependence of the ratio of the fluorescence intensity of a pair line to the total fluorescence intensity was also found to be consistent with quadrupole–quadrupole energy transfer [75].

Instead of having resonant energy transfer from the single-ion metastable levels to upper components of the excited pair manifolds, the possibility of phonon-assisted energy transfer to the lowest components of the pair excited states was considered independently by Orbach [79] and by Powell et al. [42]. The latter work is discussed in the following section. Orbach derived an expression for the rate of energy transfer from ion 1 to ion 2 in this model which is

$$P_{\text{phonon-assisted}} = \frac{2\pi}{\hbar^2} |V_{fi}|^2 \left\{\frac{\omega_{\mathrm{p}}}{4\pi^2 \hbar \varrho v^5} [\langle 2^*| V_{01} |2^*\rangle - \langle 1^*| V_{01} |1^*\rangle + \right.$$
$$\left. + \langle 1| V_{01} |1\rangle - \langle 2| V_{01} |2\rangle]^2 \binom{n_{\mathrm{p}} + 1}{n_{\mathrm{p}}}\right\}, \tag{15}$$

where ϱ is the density, v is the velocity of sound, ω_{p} is the phonon frequency, V_{01} is the electronic part of the electron–lattice interaction; the * indicates an excited state, n_{p} is the phonon occupation number, and the upper term in the () is for phonon emission while the lower term is for phonon absorption. A very rough estimate for heavily doped ruby indicates that resonant single-ion to single-ion energy transfer is three orders of magnitude faster than phonon-assisted quadrupole–quadrupole single-ion to pair transfer [79]. This is consistent with the observations of Imbusch [75].

Birgeneau [80] considered the theoretical aspects of the interaction mechanisms for energy transfer in much more detail. His calculations, using the phonon-assisted model of Orbach [79], show that all of the electric and magnetic multipole interactions are too weak by several orders of magnitude to account for the efficient energy transfer observed in heavily doped ruby. This conclusion is in disagreement with the conclusions reached in references [75] and [79]. However, the assumptions leading to his calculated transfer rates are much sounder than the very rough estimates made by the earlier workers. He found that an orbitally-dependent exchange interaction gave the right order of magnitude for single ion–single ion transfer rate to explain Imbusch's results. The concentration dependence of the energy transfer rate from single ions to pairs by exchange interaction should be stronger than that observed experimentally.

Birgeneau attributed this to the presence of single-ion to single-ion migration which acts to enhance short-range interactions such as exchange between single ions and pairs. The fact that the observed concentration dependence is characteristic of quadrupole–quadrupole interaction is simply fortuitous. Moskvin and Dryzhinin [81] recently obtained exact expressions for the energy transfer matrix elements using Racah algebra. Appyling their results to ruby again indicates that exchange interaction is strong enough to give the observed fast rate of energy transfer.

A recent observation of laser-induced fluorescence line narrowing [82] is consistent with the conclusions of Birgeneau. It was found in this experiment that R_1-line fluorescence linewidths as small as 0.002 cm^{-1} can be observed using resonance ruby laser excitation and scanning the spectrum with a Fabry-Perot interferometer. For resonance excitation, inhomogeneous line broadening does not occur and the observed linewidth is due only to homogeneous broadening processes. Such an effect cannot occur if single ion–single ion energy migration takes place before fluorescence emission, since such migration gives rise to inhomogeneous line broadening. The experiment showed this line narrowing effect persisting over times of the order of 10 ms. For ruby with 1% chromium at 77 °K the energy transfer rate is estimated to be $P \approx 10^7$ s^{-1} for single-ion to single-ion transfer [75]. For electric quadrupole–quadrupole interaction, this value extrapolates to $P \approx 10^5$ s^{-1} for 0.05% ruby at 4.2 °K which are the conditions of the line narrowing experiment [82]. This extrapolation accounts for the $C^{10/3}$ concentration dependence predicted by electric quadrupole–quadrupole interaction, and the change in homogeneous linewidth as temperature is lowered from 77 to 4.2 °K. This latter value of energy transfer rate is large enough to lead to a significant amount of single-ion migration and therefore inhomogeneous line broadening in a time of 10 ms. Since this is not observed, it is concluded that the concentration dependence used in obtaining the value for P is incorrect; it is necessary to use a stronger concentration dependence indicative of a shorter-range interaction such as exchange.

Lyo [83] has recently applied Anderson's [84] diffusion model to the energy migration among the single Cr^{3+} ions in heavily doped ruby. In this model the mechanism for single ion–single ion interaction is considered to be superexchange, and the efficiency of single ion–single ion transfer depends on two factors: 1. chromium concentration, which determines the average transfer distance, and 2. the inhomogeneous width of the R_1-line, which detunes the resonance of interacting sites. This latter criteria implies that a Cr^{3+} ion must find another ion "in tune" with it in order to transfer its energy, and this ion may be several nearest neighbor sites away. For superexchange interaction, the energy transfer rate decreases as a power of the overlap integral equal to twice the number of oxygen ions separating the two exchange-coupled ions. In ruby, this interaction decreases fairly slowly with increasing nearest neighbor sites since even ninth-nearest neighbors are separated by only oxygen ions. Such a model predicts that the single-ion excitation energy is localized below a critical concentration of $\approx 0.4\%$ (Cr_2O_3/Al_2O_3) and is mobile above this concentration. Since Imbusch [75] has demonstrated that energy transfer from single ions to pairs depends strongly on the migration of energy in the single-ion system, this result implies that energy transfer from single ions to pairs will be very efficient above $\approx 0.4\%$ (Cr_2O_3/Al_2O_3) and negligible below this concentration. This is consistent with the sharp decrease in the fluorescence lifetime of the R-lines observed above $\approx 0.4\%$ (Cr_2O_3/Al_2O_3) [74, 75] (see Fig. 13) and with the sharp increase in the I_{N_2}/I_{R_1} fluorescence intensity ratio observed above $\approx 0.4\%$ (Cr_2O_3/Al_2O_3) (see Fig. 2). However, as mentioned previously, Imbusch [75] observed a double exponential in the decay pattern of the N-lines at only 0.2% (Cr_2O_3/Al_2O_3), which implies the existance of single-ion to pair energy transfer below the critical concentration. Lyo [83] speculates that this may be due to longer-range interactions such as magnetic dipole.

6.2 Temperature dependence of fluorescence intensities and lifetimes in heavily doped ruby

Energy transfer between single ions and pairs should have pronounced effects on the fluorescence intensities and lifetimes observed for heavily doped ruby. Since the results of the last section suggest that energy transfer is a phonon-assisted process in this case, we would expect these effects to exhibit strong temperature dependences. Investigations substantiating these assumptions are discussed below.

6.2.1 Thermal variations of fluorescence intensities

The first important investigation of the temperature dependence of fluorescence intensities of lines in the spectrum of heavily doped ruby was made by Tolstoi and Liu Shun-Fu [74]. They measured the intensities of the R_1-, R_2-, N_1-, and N_2-lines between about 90 and 420 °K. Both pair lines decreased slightly as temperature was increased, while both R-lines exhibited an initial increase in intensity as temperature was raised from 90 to 200 °K and then a decrease with increasing temperature. Tolstoi and Abramov [8] found that the ratio of the fluorescence intensities of the N_2-line to the R_1-line does not go superlinear with chromium concentration as strongly at 170 °K as it does at 77 °K. Although they refer to this as a curious fact, it is now apparent that this simply indicates that less energy is being transferred to the pair systems at high temperatures than at low temperatures.

Imbusch [75] also measured the fluorescence intensities of the N_1- and N_2-lines as a function of temperature, and his results are shown in Fig. 14. The decrease in intensities as temperature was raised above about 50 °K was attributed to the increased probability of transferring energy back from the pair to single-ion systems with the accompanying absorption of a phonon. The decrease which was observed as temperature was lowered below ≈ 50 °K was attributed to the concurrent decrease in the R_2-line fluorescence relative to the R_1-fluorescence

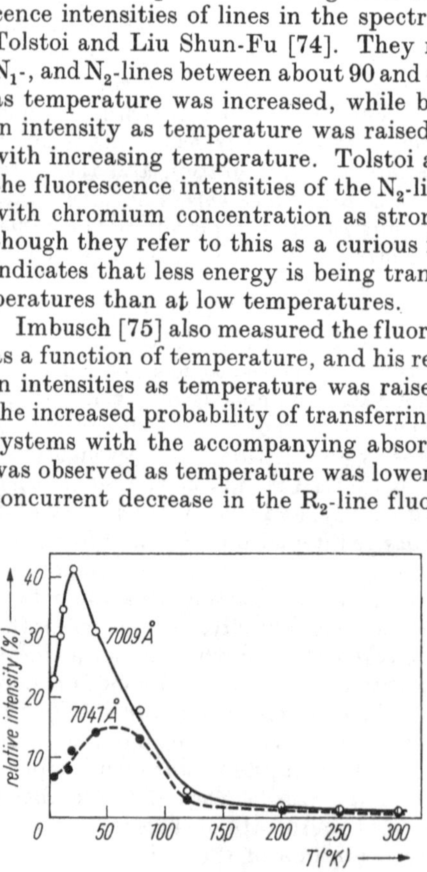

Fig. 14. Temperature dependence of the relative intensities of the 7009 Å (N_2) line and 7041 Å (N_1) line (after [57])

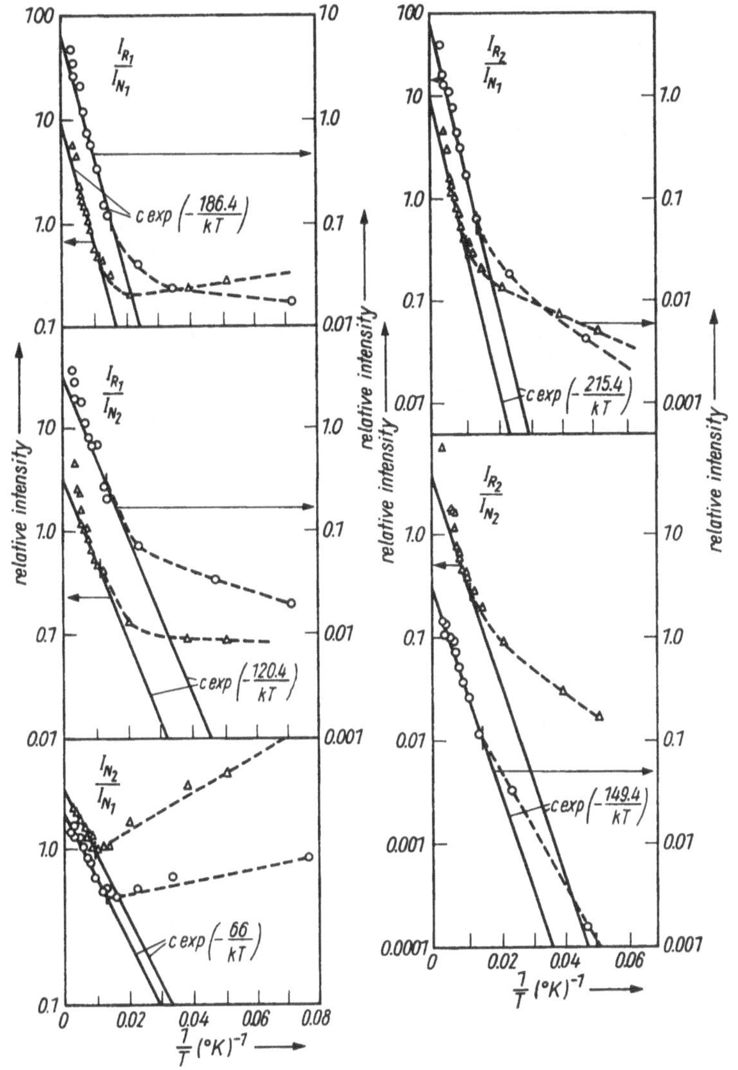

Fig. 15. Temperature dependence of the relative fluorescence intensities of the R- and N-lines (after [5]). ○ 2.1% Cr; △ 0.94% Cr. ΔE in the exponentials is given in cm^{-1}

in this temperature region. Imbusch concluded that energy transfer from the single-ion $2\bar{A}$ level to the pairs may be more efficient than from the $\bar{E}$-level. The R_2-line originates from the former level and the R_1-line from the latter. Since the $2\bar{A}$ level is at higher energy than the $\bar{E}$-level and their populations are in thermal equilibrium, the $2\bar{A}$ level becomes depopulated in this temperature region.

The temperature dependence of the intensity ratios obtained by Powell et al. [5] for the R- and N-lines of samples containing 2.1% and 0.94% chromium is shown in Fig. 15. At high temperatures the experimental data can be fitted to an exponential $e^{-\Delta E/kT}$, where ΔE is the energy difference of the initial levels of the two transitions involved. This reflects a condition of thermal equilibrium of the populations of the excited states of the transitions. The intensity ratios deviate from an exponential temperature dependence at low temperatures indicating the populations of the excited levels of the transitions are no longer thermalized. The intensities of lines in the same system (e.g. R_1 and R_2) are observed to remain thermalized at all temperatures. The observed thermalization of the populations of excited levels of different systems is additional proof that the energy transfer among these systems is nonradiative. The deviation from thermalization occurs at about 68 °K for the 2.1% sample and at approximately 91 °K for the 0.94% sample. The fact that the departure from thermalization occurs at a higher temperature in the more lightly doped sample and that for each line the amount of deviation from thermalization is greater in the more lightly doped sample, shows that the coupling of the single-ion and pair systems is greater for the more heavily doped sample. The same is true for the coupling between the N_1- and N_2-systems. Results obtained on the temperature dependences of the fluorescence lifetime support the conclusions reached from intensity measurements as discussed in the next section.

Considering these facts, the results shown in Fig. 15 can be explained by formulating energy transfer in a cross-relaxation model involving two systems [5]. The transition rates for such a model are shown in Fig. 16. The rate equations for the populations of the various levels can be written as

$$\left.\begin{aligned} \dot{N}_2 &= W N_1 - A_{21} N_2 + \frac{n_2 N_1 \omega'_{cr}}{n} - \frac{N_2 n_1 \omega_{cr}}{n}, \\ \dot{N}_1 &= -\dot{N}_2, \\ \dot{n}_2 &= -n_2 A'_{21} - \frac{n_2 N_1 \omega'_{cr}}{n} + \frac{N_2 n_1 \omega_{cr}}{n}, \\ \dot{n}_1 &= -\dot{n}_2 . \end{aligned}\right\} \quad (16)$$

Here N and n are the total number of ions in systems A and B, respectively; N_1 and n_1 are the number of ions in the ground states of these systems; N_2 and n_2 are the number of ions in the excited metastable states of these systems; W is the pumping rate for system A; A_{21} and A'_{21} are the spontaneous rates for the metastable levels of systems A and B, respectively; ω_{cr}/n is the energy transfer rate from system A to system B; ω'_{cr}/n is the energy transfer rate from system B to system A. In writing equations (16), it has been assumed that the relaxation from level 3 to level 2 in system A (see Fig. 16) is so fast that W essentially pumps level 2.

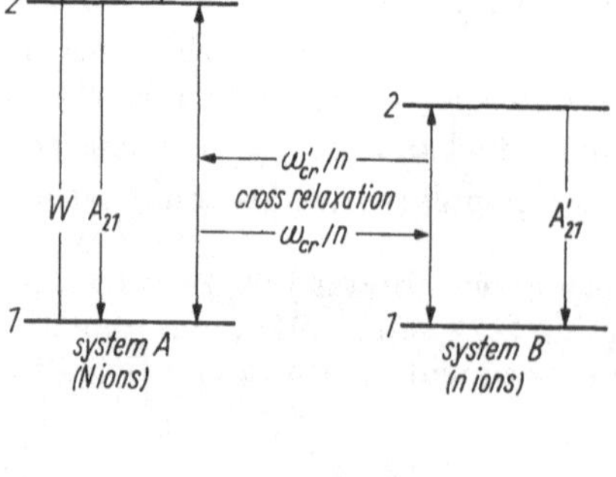

Fig. 16. Energy level diagram for energy transfer between two different fluorescent systems (after [5])

Since the cross-relaxation processes for single ion–pair energy transfer must be accompanied by the emission or absorption of phonons, it is expected that

$$\frac{\omega_{cr}}{\omega'_{cr}} = e^{\Delta E_{22}/kT} . \tag{17}$$

Assuming that $N_1 \approx N$ and $n_1 \approx n$, the steady-state solution of (16) is

$$\frac{N_2}{n_2} = \frac{A'_{21}}{\omega_{cr}} + \frac{N}{n}\frac{\omega'_{cr}}{\omega_{cr}} . \tag{18}$$

Associating the A-system with the R-lines and the B-system with the pair lines, then the population ratio given in (18) should be proportional to the fluorescence intensity ratios of lines from these systems. At high temperatures the second term will dominate, and using equation (17) predicts the observed exponential temperature dependence. As temperature is lowered, the first term will become increasingly important, and any observed temperature dependence should be mainly due to variations of ω_{cr}. In this case the deviations from thermalization shown in Fig. 15 indicate the following ordering of the energy transfer rates:

$$\omega_{cr}(R_1 \to N_1) < \omega_{cr}(R_1 \to N_2) < \omega_{cr}(R_2 \to N_1) < \omega_{cr}(R_2 \to N_2) . \tag{19}$$

Thus the N_2-system is pumped more efficiently than the N_1-system at low temperatures and the pumping of the pair systems through the metastable level of the R_2-line is greater than the pumping through the metastable level of the R_1-line. This latter observation verifies the assumption to this effect made by Imbusch [75].

In a recent paper by Gerlovin and Tolstoi [85], the temperature dependences of absolute fluorescence intensities of the R_1-, N_1-, and N_2-lines are discussed. According to these authors, the R_1-line fluorescence intensity is essentially constant between room temperature and 4.2 °K, while the intensities of the N-lines increase by about two orders of magnitude. They interpret reference [5] as attributing the decrease in fluorescence intensity of the N-lines with increasing temperature to energy transfer from pairs to single ions, and claim that their results preclude any such interpretation since the total fluorescence intensity $(I_{R_1} + I_{R_2} + I_{N_1} + I_{N_2})$ changes. Instead, they feel this quenching is due to energy transfer from the pair systems to the multi-ion system with fluorescence emission centered at about 7750 Å.

The conclusions of Gerlovin and Tolstoi seem doubtful for the following reasons: First, the lack of temperature dependence of the R_1-line intensity is an unusual result. In general, both the R_1- and R_2-lines in lightly doped ruby increase by about a factor of five as temperature is decreased from ≈ 300 to ≈ 77 °K due to the decrease in importance of vibronic emission processes [55]. In heavily doped ruby it decreases by about a factor of four in the same temperature range. Thus, when the temperature dependence of both vibronic processes and energy transfer are considered, it is difficult to predict the variation of the intensities of individual lines and total fluorescence. Second, they appear to have misinterpreted the conclusions of reference [5]. The authors of this reference did not conclude that the quenching of the N-line fluorescence as temperature is raised above 77 °K is simply due to energy transfer from pairs to single ions. Instead, they found that the populations of the excited state levels of the single-ion, pair, and multi-ion systems were all in thermal equilibrium in this temperature range, and this is proven conclusively by the exponential temperature dependence of the intensity ratios of lines from these systems and by lifetime data discussed below. In order to achieve such a population distribution it is necessary to have efficient relaxation processes going both directions between these various systems.

6.2.2 *Thermal variations of fluorescence lifetimes*

The first extensive investigations of the fluorescence lifetimes in heavily doped ruby were made by Tolstoi et al. [59, 74, 86] who measured the fluorescence lifetimes of the R- and N-lines as a function of chromium concentration and temperature. They found that the R- and N-lines have different lifetimes at low temperatures but the same lifetimes at high temperatures. The R-line lifetime for heavily doped ruby was found to increase as temperature was lowered from 400 to about 200 °K, and then either to remain constant or decrease, depending on the chromium concentration as temperature was lowered to 77 °K. The lifetimes of the N-lines showed a slight increase when temperature was lowered from 200 to 77 °K. They concluded that, since at high temperatures all parts of the spectrum exhibit the same decay pattern, the populations of the excited states of the various fluorescent systems are in thermal equilibrium. At low temperatures this is not true, and the situation is very complex.

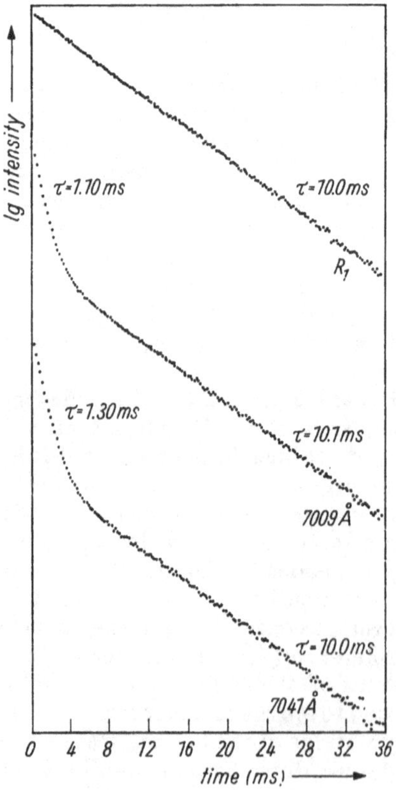

Fig. 17. Decay patterns of the R_1-, N_1-, and N_2-lines in ruby with 0.2% Cr_2O_3 at 77 °K (after [75])

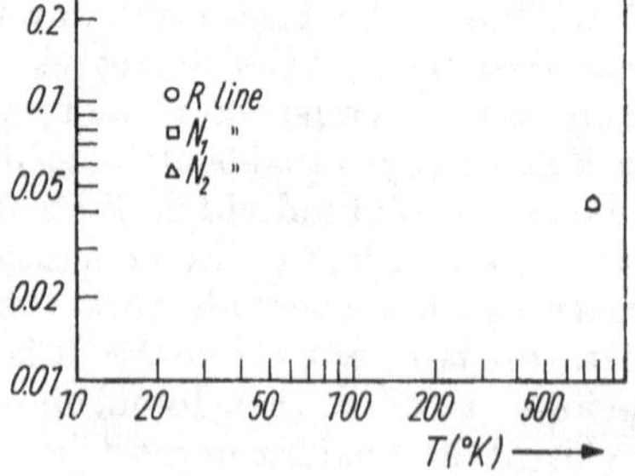

Fig. 18. Temperature dependence of the fluorescence decay times of the R- and N-lines in ruby with 2.1% chromium (after [5])

Imbusch [75] found that the intrinsic decay times of the R-, N_1-, and N_2-lines at 77 °K are 4.2, 1.3, and 1.1 ms, respectively. At intermediate concentrations ($\approx 0.2\%$ Cr_2O_3), the N-lines exhibit double exponential decays with the initial exponential having their intrinsic decay times of the N_1- and N_2-levels, and the second exponential having the decay time of the R-lines. These results are shown in Fig. 17. This is an indication of the pumping of the pair systems by the single-ion system. At concentrations of $\approx 1\%$, the N-line fluorescence exhibits an initial rise, with the subsequent decay having the same decay time as the R-lines. These results indicate that the dominant excitation mechanism of the pair systems is energy transfer from the single-ion system.

Powell et al. [5] measured the temperature dependence of the fluorescence lifetimes of the R- and N-lines in ruby with 0.94% and 2.1% chromium. Their data for the more heavily doped sample are shown in Fig. 18. They found that, for both samples above 100 °K, all the lines decayed exponentially with the same time constant. This is consistent with the conclusion they drew from their relative intensity measurements that the populations of the excited states of the single-ion and pair systems are in thermal equilibrium in this temperature range. The magnitude of the decay time in this range increased as temperature was lowered from 700 to about 300 °K and then decreased between 300 and 100 °K. As temperature was lowered to 10 °K, the decay times of the R_1-, N_1-, and N_2-lines in the heavily doped sample were observed to be different, showing that the different systems are no longer in thermal equilibrium. In both samples the R-lines continued to exhibit an exponential fluorescence decay with a lifetime less than their intrinsic value, while the N-lines presented an initial rise in their fluorescence. The time of the maximum of the fluorescence, t_m, increased as temperature was lowered and was larger for the 0.94% sample. The subsequent decays of the N-line fluorescence in the 0.94% sample had the same decay time as the R-lines, while in the 2.1% sample their decay times tended toward their intrinsic values. The deviation from pure exponential decay again shows that thermalization among the various systems does not exist at low temperatures.

These fluorescence decay time results can be explained with the same cross-relaxation model used to explain the relative intensity results and shown in Fig. 16. Equation (16) can be solved in the dynamic case and, assuming radiationless processes are faster than radiative processes so that thermal equilibrium can be established, the result is

$$N_2(t) = N_2(0)\, e^{-t/\tau}\,, \qquad n_2(t) = n_2(0)\, e^{-t/\tau}\,, \tag{20}$$

where

$$\tau^{-1} = [N_2(0) + n_2(0)]^{-1} \left[\frac{N_2(0)}{\tau_R} + \frac{n_2(0)}{\tau_N}\right]$$

and τ_R and τ_N are the intrinsic decay times of the R- and N-lines, respectively. This shows that, when the populations of the two systems are in thermal equilibrium, their fluorescence will have the same decay time. The increase in decay time at high temperatures is attributed to the decrease in vibronic emission processes. The decrease in decay time between 300 and 100 °K can be understood by noting that in this region the population of pair excited states is increasing at the expense of the single-ion excited states, thus weighting the second term in equation (20) more than the first term. Since the intrinsic decay times of the N-lines are less than that of the R-lines, this causes the observed decay time to decrease.

At lower temperatures, energy transfer is assumed to go in only one direction, and thus ω'_{cr} is set equal to zero (see Fig. 16). Taking $t = 0$ at the end of the exciting pulse, the solution to the rate equations is then

$$\left.\begin{aligned} N_2(t) &= N_2(0)\,\mathrm{e}^{-p_3 t}\,,\\ n_2(t) &= \left[n_2(0) + \frac{\omega_{cr}\,N_2(0)}{p_3 - A'_{21}}\right]\mathrm{e}^{-A'_{21}t} - \frac{\omega_{cr}\,N_2(0)}{p_3 - A'_{21}}\,\mathrm{e}^{-p_3 t}\,,\end{aligned}\right\} \tag{21}$$

where $p_3 = \omega_{cr} + A_{21}$.

We again associate N_2 with the single-ion metastable state and n_2 with the excited state of one of the pair systems. Two cases are of interest:

First, if $A'_{21} < \omega_{cr}\,[N_2(0)/n_2(0)]$, then equation (21) predicts an initial rise in the N-line fluorescence with the maximum occurring at time t_m given by

$$t_m = (A'_{21} - p_3)^{-1} \ln\left[\frac{A'_{21}}{p_3} + A'_{21}\,(p_3 - A'_{21})\,(p_3\,\omega_{cr})^{-1}\,\frac{n_2(0)}{N_2(0)}\right]. \tag{22}$$

Equation (21) predicts that the R-line decay time should be less than its intrinsic value due to energy transfer, as is observed experimentally. Equations (21) and (22) predict the observed initial rise in the fluorescence of the N-lines; and the former equation shows that the decay after the initial rise can have either the decay time of the R-lines or the N-lines, depending on chromium concentration. This latter observation explains the difference in the results for the 2.1% and 0.94% samples in this temperature range. At low temperatures, for the 0.94% sample the predicted values for t_m for the N_1- and N_2-lines are, respectively, 2.0×10^{-3} and 1.4×10^{-3} s, while the observed values are 2.7×10^{-4} and 2.6×10^{-4} s. Since the predicted values were obtained using only the first term in the [] in equation (22), it was concluded that the discrepancy in observed and predicted values reflects the need for including the second term. This term is relevant when direct pumping of the pair systems by the exciting light is important.

These second case of interest is obtained when $A'_{21} > p_3$ and $n_2(0)/N_2(0) \gg \omega_{cr}/(A'_{21} - p_3)$. Then the second expression in equation (21) reduces to

$$n_2(t) = A\,\mathrm{e}^{-t/\tau_N} + B\,\mathrm{e}^{-t/\tau_R}\,, \tag{23}$$

where A and B are positive constants. This will give a double exponential decay as observed by Imbusch [75] for his more lightly doped samples.

Heber [87] recently examined the validity of using phenomenological rate equations to interpret fluorescence decay curves and energy transfer. He showed that the use of collective rate equations is justified when the probability of an ion being excited is independent of the position of the ion. This situation is realized when the excitation energy migrates rapidly among the ions. Since this is true for heavily doped ruby, he predicts that the use of collective rate equations is justified for heavily doped ruby. However, the time-resolved spectroscopy experiments performed by Heber on heavily doped ruby contradict this prediction. Using ruby laser excitation he was unable to observe any rise-time of the N_2-line fluorescence with equipment capable of 100 μs resolution. Collective rate equations predict $t_m \approx 1$ ms. He concludes from this that collective rate equations should not be used to describe these data.

These results are entirely consistent with the observations reported earlier by Powell et al. [5]. As mentioned previously, these workers attributed this discrepancy to the importance of direct pumping of the pairs by the exciting light, and thus the necessity of including the second term in (22) in predicting t_m. Heber neglects this term because he resonantly excites the R_1-line. However, as shown in Fig. 10e, there is a fourth-nearest neighbor pair level close to being at the same energy of the R_1-line and therefore capable of absorbing the R_1-line ruby excitation energy. For samples with 1% chromium concentration, the ratio of the number of fourth-nearest neighbor pairs to the number of isolated chromium ions is about 0.0625 [72], and to account for the observed discrepancy in the risetime it is only necessary to have $n_2(0)/N_2(0) \approx 0.0525$ in the second term of (22). Thus, although these are very rough estimates, it seems quite feasible that the risetime anomaly is due to direct pumping.

6.3 Temperature dependence of the multi-ion fluorescence band

Tolstoi et al. [59, 74, 86] reported results of measurements of the fluorescence intensity and decay time of the band near 7750 Å at several temperatures. They found that the decay time has a maximum value at about 300 °K and decreases at both higher and lower temperatures. The value of the decay time at this band was found to be different from those of the R- or N-lines at both 300 and 80 °K for a sample with 1.8% chromium.

The temperature dependence of the 7750 Å fluorescence band has also been investigated by Powell et al. [5], and the results of their lifetime measurements are shown in Fig. 19. Above approximately 400 °K the lifetime of this band coincides with that of the R- and N-lines for both of the samples investigated. Below this temperature a double exponential decay is observed in the sample with 0.94% chromium, with the longer lifetime equal to that of the R-lines. For the 2.1% chromium sample an initial rise in the fluorescence is observed with the tail of the fluorescence showing a decay time equal to that measured for the R-lines. Again the measured value of t_m is about a factor of 30 smaller than that predicted using only the first term in the [] in (22). This quantitative discrepancy is ascribed to the importance of direct pumping of the band, and the rest of these results can be satisfactorily explained by the cross-relaxation energy transfer model described by equations (20) to (23). It appears that the excited state population of the 7750 Å band is in thermal equilibrium with those of the single-ion and pair systems at high temperatures and, at low temperatures, the 7750 Å band is pumped through the single-ion system. The fact that the decay time of the R-lines, instead of that of the N-lines, appears in the decay pattern

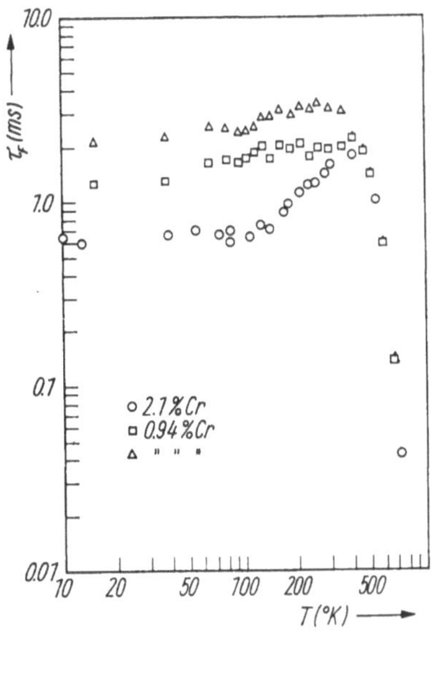

Fig. 19. Temperature dependence of the fluorescence decay time of the 7750 Å band in ruby with 2.1% and 0.94% chromium. Below about 300 °K in the 0.94% sample the fluorescence exhibits a double exponential decay. The decay times for both of these exponentials are shown in this figure (after [5])

of the 7750 Å band at low temperatures contradicts the conclusion of Gerlovin and Tolstoi [85] that the band is pumped directly by the pair systems.

6.4 Summary of energy transfer in heavily doped ruby

The model currently employed to explain energy transfer in heavily doped ruby is illustrated in Fig. 20. The external radiation creates an excited state in the single-ion system. Then the following statements can be made with regard to the transfer of this energy to other systems:

1. Above a critical chromium concentration of 0.4%, the excitation diffuses among the single ions in the lattice via a superexchange type of interaction, similar to a Frenkel exciton with a hopping rate of $k_h \approx 10^6\ s^{-1}$ for a 1% concentration. Below the critical concentration, the single-ion excitation is localized.

2. If the excitation reaches a single ion which is near to an ion pair, it can be transferred to the pair through an orbitally dependent exchange interaction with a rate constant of $k_t \approx 10^3\ s^{-1}$ for a 1% concentration at 77 °K.

3. The predominant mode of energy migration and transfer is through radiationless processes. (The effects of radiative processes will be discussed in the next section.)

4. At high temperatures the transfer of energy among all of the single-ion, pair, and multi-ion systems is efficient enough to cause the populations of their excited states to be in thermal equilibrium. At low temperatures the single-ion system pumps the pair and multi-ion system.

5. The fourth-nearest neighbor pair system is pumped more efficiently than the third-nearest neighbor pair system and the $2\bar{A}$ single-ion level pumps the pair systems more efficiently than the $\bar{E}$-level.

6. Some of the pair systems (especially fourth-nearest neighbors) have transitions which are nearly in resonance with the single-ion R-lines. This allows for direct pumping of these levels, although phonon-assisted energy transfer from single ions is the dominant means of exciting the pair systems.

7. Interaction with the Photon Field

This section discusses two aspects of the interaction of the fluorescence systems of heavily doped ruby with the photon field. The first of these involves an external beam of photons and its ability to produce laser action in the N-lines. The second deals with the effects of trapping fluorescence photons in the crystal.

Fig. 20. Model for energy transfer from single ions to pairs in heavily doped ruby (after [75]). ● represent Cr^{3+} ions, $h\nu$ is the energy of an external photon, and k_h and k_t are the hopping and energy transfer rates, respectively

7.1 Stimulated emission from the pair systems

The N_1- and N_2-lines of heavily doped ruby were the first transitions suggested as possible solid state optical laser lines [88, 89]. Schawlow pointed out that two conditions favor the establishment of a population inversion for these transitions. The first is that these lines terminate on upper components of the ground state manifolds and these levels can be thermally depopulated at low temperatures. The second is that the population of the excited state of the pairs is increased by energy transfer from single ions.

The first observations of stimulated emission at the N-lines were made concurrently by Schawlow and Devlin [90] in 0.5% ruby and by Wieder and Sarles [91] in 0.7% ruby. Above a pumping threshold of the same order as that required to obtain R-line laser operation in lightly doped ruby, they observed sharp increases in intensity, decreases in decay time, and a directional fluorescence emission at the N-lines. No stimulated emission was obtained at the R_1-line in heavily doped ruby. These experiments were carried out at 77 and 4.2 °K. It was found that laser action could occur simultaneously at the N_1- and N_2-lines. The threshold for the N_1-line was found to be slightly less than that for the N_2-line at 77 °K, and the situation was reversed at 4.2 °K. This may be due to the fact that at the higher temperature there is still some energy transfer from the N_2-system to the N_1-system, whereas at the lower temperatures this pumping is absent and, as shown in the previous section, the N_2-system is pumped more efficiently by the single ions than the N_1-system.

The fact that simultaneous laser action can be obtained in these two N-lines is additional proof that they come from different pair systems. This can only be observed at low temperatures where the various fluorescence systems are uncoupled. At higher temperatures the populations of their excited states are in thermal equilibrium and therefore it will be much harder to obtain a population inversion and, if stimulated emission were obtained, it should only occur at one line.

7.2 Effects of reabsorption in heavily doped ruby

In crystals a few millimeters thick such as those generally used in fluorescence investigations, it is possible for a photon emitted by one ion in the crystal to be absorbed by some other ion before escaping from the sample. This phenomenon is referred to as radiative re-absorption, self-absorption, or radiation trapping and generally lengthens the observed fluorescence lifetime and enhances the low-energy portion of the fluorescence spectrum with respect to the high-energy region.

Varsanyi et al. [92] showed that the fluorescence lifetime in lightly doped ruby can be lengthened by a factor of four over its intrinsic value due to radiation trapping. The effects of re-absorption on the fluorescence spectrum of heavily doped ruby have been investigated by Jekeli [93]. He observed self-reversal in the R-lines, but no intensity changes in the N-lines.

Powell et al. [54, 94] investigated re-absorption in heavily doped ruby at several temperatures and observed not only the self-reversal of the R-lines but also intensity and line-width changes in the N-lines. They divided the effects of re-absorption into two categories: those arising from the absorption of the radiation as it passes through the crystal, and those arising from the re-distribution and re-emission of the absorbed energy. In the first of these categories, selective absorption causes a spectral line to decrease in intensity, increase in halfwidth, and eventually exhibit self-reversal. In general, the effects in this category are greater at low temperatures and depend on the population of the lower level of the transition. The effects in the second category depend on such processes as vibronic transitions and energy transfer. These considerations, along with the knowledge of energy levels, vibronic spectra, and energy transfer processes, were used to explain all of the observed effects of re-absorption in the fluorescence spectra of heavily doped ruby.

A recent, interesting result which is not specifically related to re-absorption but rather to the initial absorption of light in heavily doped ruby, is the observed concentration dependence of the intensity of the photon echos in heavily doped ruby [95]. The intensity of the photon echos was observed to increase approximately as the square of the chromium concentration up to about 0.03 wt% Cr^{3+}, and then to decrease at higher concentrations. This can be attributed to the decrease in the dipolar relaxation time which occurs at high concentrations where the dipoles are strongly coupled. In interpreting these results, it was also necessary to account for the reduced absorption of the laser pulse due to the increased widths of the R-lines in heavily doped ruby [70].

8. Summary and Conclusions

The optical spectrum of heavily doped ruby is generally well explained. Spectral lines can be attributed to isolated single ions or exchange-coupled ion pairs. First- through fourth-nearest neighbor pairs have been identified and the ground state splittings explained using a phenomenological exchange theory. The single-ion and pair systems interact through an orbitally dependent exchange interaction which is efficient enough at high temperatures to cause the populations of their excited states to be in thermal equilibrium. At low temperatures the single-ion system pumps the pair systems.

The following areas could be the subject of further investigations:

1. A theoretical model for the pair coupling mechanism is needed which predicts the magnitude of the exchange interaction.
2. More theoretical work is needed to predict the splittings of the excited states of the pair systems.
3. The exact origin of the absorption pair band is still uncertain.
4. More work is needed in studying the fluorescence band near 7750 Å and its inteɪaction with the single-ion and pair systems.
5. The mechanism causing spin–lattice relaxation is still uncertain.
6. Energy transfer in heavily doped ruby could be better characterized, especially with respect to the migration in the single-ion system, the nature of the phonon-assisted transfer mechanism, and the fact that the pumping of the N-lines through the R_2-level is more efficient than the pumping through the R_1-level. An attempt should be made to verify the energy transfer theory proposed by Lyo [83].

References

[1] See, for example, D. L. WOOD, G. F. IMBUSCH, R. M. MACFARLAND, P. KISLIUK, and D. M. LARKIN, J. chem. Phys. **48**, 5255 (1968); A. M. TKACHUK and Z. N. ZONN, Optika i Spektroskopiya **26**, 587 (1969); Soviet Phys. — Optics and Spectroscopy **26**, 322 (1969).

[2] H. du Bois and G. J. Elias, Ann. Phys. (Germany) (Ser. 4) **27**, 233 (1908); **35**, 617 (1911).
C. E. Mendenhall and R. W. Wood, Phil. Mag. **30**, 316 (1915).
[3] O. Deutschbein, Ann. Phys. (Germany) (Ser. 5) **14**, 712 (1932); **20**, 828 (1934).
[4] S. F. Jacobs, Ph. D. Thesis, John Hopkins Univ., Baltimore 1957 (unpublished).
[5] R. C. Powell, B. DiBartolo, B. Birang, and C. S. Naiman, Phys. Rev. **155**, 296 (1967).
[6] R. S. Krishnan, Proc. Indian Acad. Sci. **A26**, 450 (1947).
[7] A. L. Schawlow, D. L. Wood, and A. M. Clogston, Phys. Rev. Letters **3**, 271 (1959).
[8] N. A. Tolstoi and A. P. Abramov, Optika i Spektroskopiya **14**, 691 (1963); Soviet Phys. — Optics and Spectroscopy **14**, 365 (1963).
[9] A. A. Kaplyanskii, A. K. Przhevuskii, and R. B. Rozenbaum, Fiz. tverd. Tela **10**, 2307 (1968); Soviet Phys. — Solid State **10**, 1864 (1969).
[10] A. M. Clogston, unpublished report.
[11] P. Kisliuk, A. L. Schawlow, and M. D. Sturge, in: Adv. Quantum Electronics, Vol. I, Ed. P. Grivet and N. Bloembergen, Columbia Univ. Press, New York 1964 (p. 725).
[12] E. A. Harris and J. Owen, Phys. Rev. Letters **11**, 9 (1963).
D. S. Rodbell, I. S. Jacobs, J. Owen, and E. A. Harris, Phys. Rev. Letters **11**, 10 (1963).
[13] P. W. Anderson, Solid State Phys. **14**, 99 (1963).
[14] M. J. Berggren, G. F. Imbusch, and P. L. Scott, Phys. Rev. **188**, 675 (1969).
[15] V. I. Cherepanov and A. E. Nikiforov, phys. stat. sol. **28**, 65 (1968); **33**, K151 (1969).
[16] A. E. Nikiforov, A. N. Men, and V. I. Cherepanov, Fiz. tverd. Tela **6**, 3288 (1964); Soviet Phys. — Solid State **6**, 2631 (1965).
[17] A. E. Nikiforov and V. I. Cherepanov, phys. stat. sol. **14**, 391 (1966).
[18] A. E. Nikiforov and V. I. Cherepanov, Optika i Spektroskopiya **14**, 737 (1967); Soviet Phys. — Optics and Spectroscopy **14**, 401 (1967).
[19] S. V. Vonsovskii, V. I. Cherepanov, A. N. Men, and A. E. Nikiforov, phys. stat. sol. **24**, 51 (1967).
[20] A. E. Nikiforov and V. I. Cherepanov, phys. stat. sol. (b) **44**, 269 (1971).
[21] P. Kisliuk, N. C. Chang, P. L. Scott, and M. H. L. Pryce, Phys. Rev. **184**, 367 (1969).
[22] M. H. L. Pryce, private communication.
[23] N. Laurance, E. C. McIrvine, and J. Lambe, J. Phys. Chem. Solids **23**, 515 (1962).
[24] B. N. Brockhouse, J. chem. Phys. **21**, 961 (1953).
[25] M. Tachiki and T. Nagamiya, J. Phys. Soc. Japan **13**, 452 (1958).
[26] J. B. Goodenough, Phys. Rev. **117**, 1442 (1960).
[27] W. P. Osmond, Proc. Phys. Soc. **79**, 394 (1962).
[28] P. Kisliuk and W. F. Krupke, unpublished report.
[29] L. Rimai, H. Statz, M. J. Weber, G. A. deMars, and G. F. Koster, Phys. Rev. Letters **4**, 125 (1960).
[30] H. Statz, L. Rimai, M. J. Weber, G. A. deMars, and G. F. Koster, J. appl. Phys., Suppl. **32**, 218S (1961).
[31] H. Statz, M. J. Weber, L. Rimai, G. A. deMars, and G. F. Koster, J. Phys. Soc. Japan **17**, Suppl. B-I, 430 (1962).
[32] J. Owen, J. appl. Phys., Suppl. **32**, 213S (1961).
[33] W. Gunsser, W. Hille, and A. Knappwost, Z. phys. Chem. **58**, 316 (1968).
[34] A. Jelenski, H. Szymczak, and J. Twarowski, in: Magnetic Resonance and Relaxation, Ed. R. Blinc, North-Holland Publ. Co., Amsterdam 1967 (p. 1205).
[35] A. E. Nikiforov and V. I. Cherepanov, Fiz. tverd. Tela **7**, 1162 (1965); Soviet Phys. — Solid State **7**, 931 (1965).
[36] Yu. L. Shelekhin, M. P. Votinov, and B. P. Berkovskii, Fiz. tverd. Tela **9**, 2119 (1967); Soviet Phys. — Solid State **9**, 1663 (1968); Fiz. tverd. Tela **8**, 589 (1966); Soviet Phys. — Solid State **8**, 469 (1966).

[37] G. F. IMBUSCH and M. B. GRAIFMAN, J. appl. Phys. **39**, 981 (1968).
[38] A. E. NIKIFOROV, Fiz. tverd. Tela **7**, 1248 (1965); Soviet Phys. — Solid State **7**, 1005 (1965).
[39] R. T. DALY, paper presented at the Opt. Soc. Amer. Meeting, Pittsburgh (Pa.), October 1961.
[40] P. KISLIUK and W. F. KRUPKE, Appl. Phys. Letters **3**, 215 (1963).
[41] P. KISLIUK and W. F. KRUPKE, J. appl. Phys. **36**, 1025 (1965).
[42] R. C. POWELL, B. DIBARTOLO, B. BIRANG, and C. S. NAIMAN, in: Optical Properties of Ions in Crystals, Ed. H. M. CROSSWHITE and H. W. MOOS, Interscience Publishers, New York 1967 (p. 207).
[43] L. F. MOLLENAUER and A. L. SCHAWLOW, Phys. Rev. **168**, 309 (1968).
[44] A. L. SCHAWLOW, J. appl. Phys., Suppl. **33**, 395 (1962).
[45] A. HADNI, Appl. Optics **2**, 977 (1963).
[46] J. F. MOSER, H. STEFFEN, and F. K. KNEUBUHL, Helv. phys. Acta **39**, 195 (1966).
[47] A. A. KAPLYANSKII and A. K. PRZHEVUSKII, Dokl. Akad. Nauk SSSR **142**, 313 (1962); Soviet Phys. — Doklady **7**, 37 (1962).
[48] A. A. KAPLYANSKII and A. K. PRZHEVUSKII, Proc. Internat. Conf. Luminescence, Budapest 1966 (p. 1380).
[49] A. A. KAPLYANSKII and A. K. PRZHEVUSKII, Fiz. tverd. Tela **9**, 257 (1967); Soviet Phys. — Solid State **9**, 190 (1967).
[50] G. C. SPITALE, Masters Degree Thesis, U.S.A.F. Inst. Technol., Wright-Patterson A. F. Base (Ohio) 1965.
[51] L. F. MOLLENAUER, Ph.D. Thesis, Stanford Univ., Palo Alto (Calif.) 1965.
[52] L. M. KANSKAYA and A. K. PRZHEVUSKII, Optika i Spektroskopiya **26**, 226 (1969); Soviet Phys. — Optics and Spectroscopy **26**, 121 (1969).
[53] A. A. KAPLYANSKII, V. N. MEDVEDEV, and A. K. PRZHEVUSKII, Zh. eksper. teor. Fiz., Pisma **5**, 427 (1967); Soviet Phys. — J. exper. theor. Phys., Letters **5**, 347 (1967).
[54] R. C. POWELL, Ph. D. Thesis, Arizona State Univ., Tempe (Ariz.) 1967.
[55] A. LINZ and R. E. NEWNHAM, Phys. Rev. **123**, 500 (1961).
[56] C. S. NAIMAN and A. LINZ, Proc. Symp. Optical Masers, Polytech. Inst. Brooklyn, 1963 (p. 369).
[57] F. GIRES and G. MAYER, J. Phys. Radium **22**, 832 (1961).
[58] F. VARSANYI and G. H. DIEKE, Phys. Rev. Letters **7**, 442 (1961).
[59] N. A. TOLSTOI, LIU SHUN-FU, and M. E. LAPIDUS, Optika i Spektroskopiya **13**, 242 (1961); Soviet Phys. — Optics and Spectroscopy **13**, 133 (1961).
[60] J. C. GILL, Nature (London) **190**, 619 (1961); Proc. Phys. Soc. **79**, 58 (1962).
[61] K. J. STANDLEY and R. A. VAUGHAN, Phys. Letters (Netherlands) **10**, 33 (1964); Phys. Rev. **139**, A1275 (1965).
[62] J. H. PACE, D. F. SIMPSON, and J. S. THORP, Proc. Phys. Soc. **76**, 697 (1960).
[63] W. B. MIMS and J. D. MCGEE, Phys. Rev. **119**, 1233 (1960).
[64] A. A. MANENKOV and A. M. PROKHOROV, Soviet Phys. — J. exper. theor. Phys. **11**, 527 (1960).
[65] J. C. GILL, in: Adv. Quantum Electronics, Ed. J. R. SINGER, Columbia Univ. Press, New York 1961 (p. 399).
[66] N. BLOEMBERGEN, S. SHAPIRO, P. S. PERSHAN, and J. O. ARTMAN, Phys. Rev. **114**, 445 (1959).
[67] M. J. WEBER, MIT Technical Memorandum T 309.
[68] A. KIEL, Phys. Rev. **120**, 137 (1960).
[69] H. HIRONO, J. Phys. Soc. Japan **16**, 766 (1961).
[70] R. C. POWELL, B. DIBARTOLO, B. BIRANG, and C. S. NAIMAN, J. appl. Phys. **37**, 4973 (1966).
[71] B. BIRANG, B. DIBARTOLO, and R. C. POWELL, J. appl. Phys. **38**, 5113 (1967).
[72] R. J. STRAIN, Ph.D. Thesis, Univ. of Illinois, Urbana 1963.
[73] I. WIEDER and L. R. SARLES, in: Adv. Quantum Electronics, Ed. J. R. SINGER, Columbia Univ. Press, New York 1961 (p. 214).

[74] N. A. Tolstoi and Liu Shun-Fu, Optika i Spektroskopiya **13**, 403 (1962); Soviet Phys. — Optics and Spectroscopy **13**, 224 (1962).
[75] G. F. Imbusch, Phys. Rev. **153**, 326 (1967).
[76] D. F. Nelson and M. D. Sturge, Phys. Rev. **137**, A1117 (1965).
[77] T. Förster, Ann. Phys. (Germany) (Ser. 6) **2**, 55 (1948).
[78] D. L. Dexter, J. chem. Phys. **21**, 836 (1953).
[79] R. Orbach, in: Optical Properties of Ions in Crystals, Ed. H. M. Crosswhite and H. W. Moos, Interscience Publishers, New York 1967 (p. 445).
[80] R. J. Birgeneau, J. chem. Phys. **50**, 4282 (1969).
[81] A. S. Moskvin and V. V. Dryzhinin, Optika i Spektroskopiya **29**, 899 (1970); Soviet Phys. — Optics and Spectroscopy **29**, 478 (1970).
[82] A. Szabo, Phys. Rev. Letters **25**, 924 (1970).
[83] S. K. Lyo, Phys. Rev. **B 3**, 3331 (1971).
[84] P. W. Anderson, Phys. Rev. **109**, 1492 (1958).
[85] I. Ya. Gerlovin and N. A. Tolstoi, Optika i Spektroskopiya **28**, 833 (1970); Soviet Phys. — Optics and Spectroscopy **28**, 451 (1970).
[86] N. A. Tolstoi and A. M. Tkachuk, Optika i Spektroskopiya **6**, 659 (1959); Soviet Phys. — Optics and Spectroscopy **6**, 427 (1959).
[87] J. Heber, phys. stat. sol. **42**, 497 (1970).
[88] A. L. Schawlow, in: Quantum Electronics, Ed. C. H. Townes, Columbia Univ. Press, New York 1960 (p. 553).
[89] A. L. Schawlow, in: Adv. Quantum Electronics, Ed. J. R. Singer, Columbia Univ. Press, New York 1962 (p. 50).
[90] A. L. Schawlow and G. E. Devlin, Phys. Rev. Letters **6**, 96 (1961).
[91] I. Wieder and L. R. Sarles, Phys. Letters (Netherlands) **6**, 95 (1961).
[92] F. Varsanyi, D. L. Wood, and A. L. Schawlow, Phys. Rev. Letters **3**, 544 (1959).
[93] W. Jekeli, J. Opt. Soc. Amer. **55**, 1442 (1965).
[94] R. C. Powell, B. DiBartolo, B. Birang, and C. S. Naiman, Bull. Amer. Phys. Soc. **11**, 483 (1966).
[95] A. Compaan and I. D. Abella, Phys. Rev. Letters **27**, 23 (1971).

(Received September 9, 1971)

Original Papers

phys. stat. sol. (a) **10**, 359 (1972)

Subject classification: 14.3 and 15; 22

Institute of Physics, University of Zagreb

Electrical Properties of Cubic $Cu_{2-x}Se$ Stable at Room Temperature

By

B. Mestnik, Z. Ogorelec, and Ž. Ambrožić

The electrical conductivity and thermoelectric power of non-stoichiometric cuprous selenide were investigated. The composition dependence of these properties was examined on samples with very great stoichiometric deviations, i.e. in the homogeneity range where $Cu_{2-x}Se$ is in the cubic phase at room temperature. This range was not covered by any previous examination. The validity of a simple model for self-dopation which gives a linear relation between the concentration of holes and stoichiometric deviations may be considered as the main result of the investigation. Many other relevant data about $Cu_{2-x}Se$ are given and the experimental and theoretical curves are compared and discussed.

Es wurde die elektrische Leitfähigkeit und Thermospannung von nichtstöchiometrischem Kupferselenid untersucht. Die Abhängigkeit dieser Eigenschaften von der Zusammensetzung wurde an Proben mit sehr großen stöchiometrischen Abweichungen, d. h. in dem Homogenitätsbereich, in dem $Cu_{2-x}Se$ bei Zimmertemperatur in der kubischen Phase vorliegt, untersucht. Dieser Bereich wurde bis jetzt nicht erforscht. Als Hauptergebnis der Untersuchung kann man die Gültigkeit eines einfachen Modells für die Selbstdotierung ansehen, das eine lineare Beziehung zwischen der Konzentration der Löcher und der stöchiometrischen Abweichungen ergibt. Es werden auch andere relevante Daten über $Cu_{2-x}Se$ angegeben und die experimentellen und theoretischen Kurven verglichen und diskutiert.

1. Introduction

Like other compounds of the $A^{I}_{2\pm x}B^{VI}$ group, cuprous selenide is a very suitable system for studying the influence of deviations from stoichiometry on the semiconducting or semimetallic properties. It has a wide homogeneity range and its composition can be easily varied. Measurements on this compound have been made before [1 to 3], but they were carried out at temperatures above 300 °C and close to the stoichiometric composition. The reason for this is very simple: there are certain cuprous halogenides (CuI, CuCl) which exhibit almost pure ionic conductivity in which above 300 °C only Cu ions take part. This allows the use of galvanic cells (e.g. Cu–CuI–$Cu_{2-x}Se$–Pt) for changing the composition of $Cu_{2-x}Se$. Below 300 °C this method cannot be applied.

Our investigation was concentrated on cubic $Cu_{2-x}Se$ with a composition between $Cu_{1.75}Se$ and $Cu_{1.82}Se$ which, according to the equilibrium diagram [4], is stable at room temperature. In order to vary the composition, we applied selective evaporation of Se [5].

2. Electrical Conductivity

Non-stoichiometric cuprous selenide was prepared by means of a direct synthesis of pure and precisely weighed components (5N, Fluka, Switzerland) in an evacuated and sealed quartz tube maintained at a temperature of 900 °C.

After the synthesis was completed (6 h), the tube was cooled to room temperature over a period of 12 h. The initial composition $(2 - x)$ was determined from the masses of Cu and Se used in the synthesis.

Electrical conductivity was measured on prismatic samples of cuprous selenide at temperatures ranging from room temperature to 170 °C. The measurements gave reproducible results. Since the composition of each sample can be varied by evaporating Se, both electrical conductivity and thermoelectric power were measured on one and the same sample, though each time with a different value of $(2 - x)$. The composition was varied from $Cu_{1.750}Se$ to $Cu_{1.827}Se$ because we wanted to include a homogeneity range in which the high-temperature or cubic phase is stable at room temperature. The equilibrium diagram of the Cu–Se system being uncertain in this composition range X-ray examinations proved to be necessary. The diffraction patterns, obtained at room temperature in a Nonius Guiner-de Wolf camera showed the existence of a cubic lattice at room temperature in the range of $2 - x \approx 1.75$ to 1.83. This is in accordance with Heyding's measurements.

The results of the measurements of electrical conductivity are shown in Fig. 1. As can be seen, there is a series of straight lines which show the dependence of electrical conductivity on temperature for different compositions. The experimental points between room temperature and 170 °C follow the equation $\sigma = \sigma_1 T^{-\alpha}$ where the average value of the exponent $\alpha = 1.4$ agrees with the theoretical value for pure phonon scattering. The values for σ_1 are between $3.07 \times 10^7\ \Omega^{-1}\ cm^{-1}$ ($Cu_{1.773}Se$) and $2.36 \times 10^7\ \Omega^{-1}\ cm^{-1}$ ($Cu_{1.827}Se$). By using the obtained results we can plot curves of the electrical conductivity versus composition at particular temperatures. The curves σ vs. $(2 - x)$ shown in Fig. 2 refer to three different temperatures (50, 100, and 150 °C). When the composition is between $1.773 \leqq 2 - x \leqq 1.827$ the experimental points lie on a straight line. Extrapolation of the lines for different temperatures to the

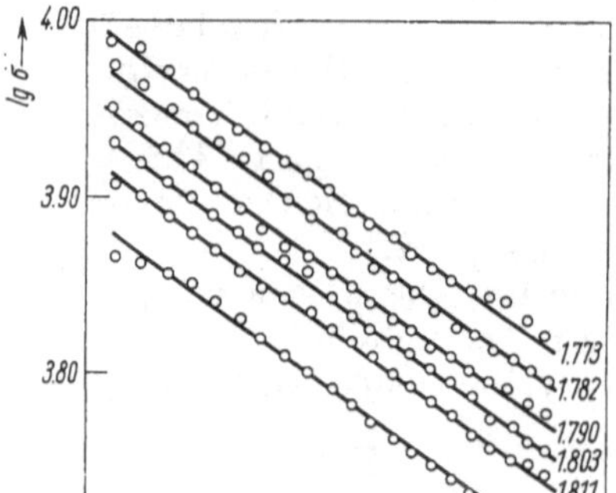

Fig. 1. Electrical conductivity of $Cu_{2-x}Se$ vs. temperature for different compositions $(2-x)$

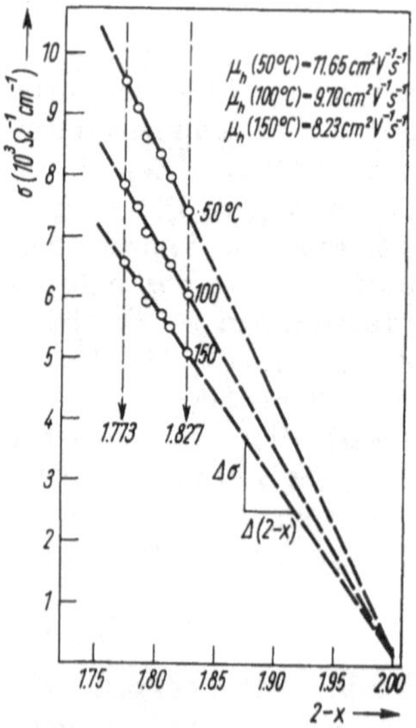

Fig. 2. Electrical conductivity of $Cu_{2-x}Se$ vs. composition for different temperatures

point of the ideal stoichiometric composition gives us the value of the residual conductivity 200 Ω^{-1} cm^{-1}. The linearity of the function σ vs. $(2 - x)$ suggests that a simple model for the self-doping of $Cu_{2-x}Se$, confirmed by earlier measurements at high temperatures [1 to 3], is valid also for the homogeneity range which we investigated. For, the electrical neutrality law allows the conclusion that each deficit Cu atom yields a hole in the valence band of the crystal. Since our experiment did not reveal any additional effects (e.g. dependence of hole mobility on composition), there must exist a linear relation between the concentration of holes and deviations from stoichiometry:

$$p = \frac{2\,A}{V_m} - \frac{(2-x)\,A}{V_m}\,, \tag{1}$$

where A is Avogadro's number, and V_m the molecular volume. Thus, the electrical conductivity $\sigma = e\,\mu_h\,p$ can be written as

$$\sigma = \frac{2\,A\,e\,\mu_h}{V_m} - \frac{(2-x)\,A\,e\,\mu_h}{V_m}\,, \tag{2}$$

where e is the electron charge and μ_h the mobility of holes. Since the concentration of holes is constant for a given composition, the conductivities σ change with temperature exclusively as a result of changing mobility μ_h. This enables us to determine the equation for the mobility of carriers as

$$\mu_h = B\,T^{-\alpha}\,, \tag{3}$$

where α is equal to 1.4 and B a constant. By using the diagram in Fig. 2, the mobility of the charge carriers can be evaluated by means of the following equation:

$$\mu_h = \frac{V_m}{A\,e}\,\frac{\Delta\sigma}{\Delta(2-x)}\,. \tag{4}$$

The calculated mobilities for three different temperatures are as follows:

$$\mu_h\,(50\ °C) = 11.65\ cm^2\,V^{-1}\,s^{-1}\,,$$
$$\mu_h\,(100\ °C) = 9.70\ cm^2\,V^{-1}\,s^{-1}\,,$$
$$\mu_h\,(150\ °C) = 8.23\ cm^2\,V^{-1}\,s^{-1}\,.$$

With the help of the known mobilities and by using equation (3), we obtain the constant B equal to 3.82×10^4 $cm^2\,V^{-1}\,s^{-1}\,deg^{1.4}$.

3. Thermoelectric Power

We expected a very small thermoelectric power α of non-stoichiometric cuprous selenide and a constancy of α over a relatively wide temperature range. Therefore, we measured α using a simple integral method: one end of a prismatic sample was held at 0 °C, while the temperature of the other end was changed. The thermoelectric voltage was measured between the ends. The measurements, which were carried out on one and the same sample each time with a different composition, gave a bundle of straight lines with varying slopes (Fig. 3). No difference between the heating and cooling cycles was observed. The slopes of the single lines in Fig. 3 give a constant thermoelectric power α_{sum}. However, in our experiment α_{sum} was related to copper. The absolute thermoelectric

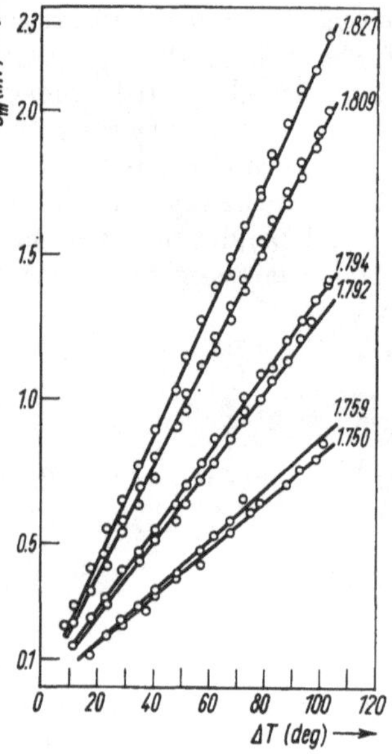

Fig. 3. Thermoelectric voltage U_{th} (measured against Cu), generated in $Cu_{2-x}Se$ samples. Reference temperature 0°C, heating and cooling rate 1 deg/min

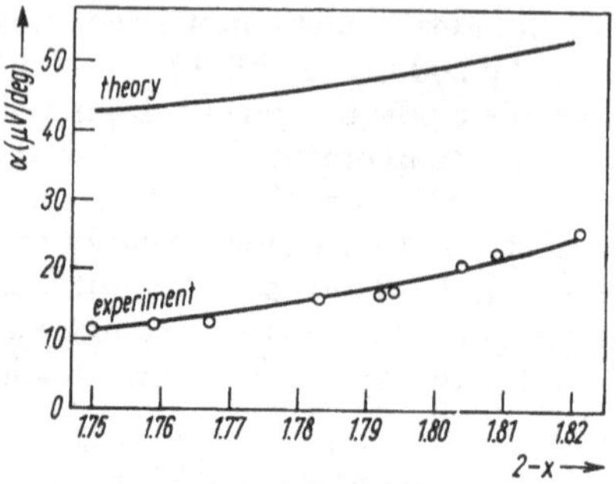

Fig. 4. Thermoelectric power of non-stoichiometric $Cu_{2-x}Se$ vs. compositions $(2-x)$

power (i.e. in relation to Pb) was obtained by adding $\alpha_{Cu} = +3.2\ \mu V/deg$. The dependence of $\alpha_{Cu_{2-x}Se}$ on the composition $(2 - x)$ is shown in Fig. 4.

The measured thermoelectric power can be compared directly with the theoretical calculations. These calculations can be made using the well-known equation for the thermoelectric power of p-type semiconductors:

$$\alpha = \frac{k}{e} (\delta - \eta) , \tag{5}$$

where k is the Boltzmann constant, η the reduced chemical potential of the holes $\left(\eta = (E_V - E_F)/kT\right)$, and δ a constant which, for a degenerate semiconductor with a wide forbidden energy gap, is given by

$$\delta = \frac{(r + 2)\, F_{r+1}(\eta)}{(r + 1)\, F_r(\eta)} . \tag{6}$$

$F_r(\eta)$ and $F_{r+1}(\eta)$ are Fermi-Dirac functions. r is the exponent in the relation $l \sim E^r$, where l is the mean free path of carriers and E their energy.

The exponent r depends on the scattering mechanism. In the case of scattering on thermal vibrations of the ionic lattice, the exponent r is equal to $1/2$ for temperatures below the Debye temperature. Using the general equation for the concentration of holes participating in the conductivity,

$$p = 4\pi \left(\frac{2\, m_h^* kT}{h^2}\right)^{3/2} F_{1/2}(\eta) , \tag{7}$$

and comparing relation (7) with equation (1), we obtain values for the Fermi-Dirac function $F_{1/2}(\eta)$. The effective mass of carriers m_h^* is equal to $4\,m$ [1]. The corresponding values for $F_{3/2}(\eta)$ and η were determined from the tables of

MacDougall and Stoner [6]. Using equation (5), we then calculate α, the thermoelectric power. The results of this calculation are given in Table 1 and shown in Fig. 4. As can be seen, the experimental curve follows the theoretical one,

Table 1

Change of certain parameters with changing composition

$2-x$	p (10^{21} cm^{-3})	$F_{1/2}(\eta)$	$F_{3/2}(\eta)$	η	α (μV/deg)
1.750	5.2516	20.7595	128.377	9.813	42.5688
1.760	5.0416	19.9293	120.198	9.545	43.6890
1.770	4.8315	19.0988	112.231	9.273	44.8954
1.780	4.6214	18.2683	104.488	8.996	46.2741
1.790	4.4114	17.4382	96.979	8.716	47.6529
1.800	4.2013	16.6077	89.704	8.430	49.2902
1.810	3.9912	15.7771	82.641	8.140	50.8412
1.820	3.7812	14.9470	75.828	7.844	52.6508

while the absolute values are considerably different. This difference may be due to the introduction of a wrong parameter into the calculations. We are not sure, for instance, that $m_h^* = 4\,m$ is the proper value for the effective mass of holes in the encountered range of compositions. Similar investigations on Ag_2S and Ag_2Se [7, 8] suggest that the effective mass may actually depend on deviations from stoichiometry. However, further investigations will have to be made before any definite conclusions can be drawn.

As can be seen from Table 1, for all compositions, the Fermi level is relatively deep in the valence band ($E_V - E_F > 0$). This means that $Cu_{2-x}Se$ must be regarded as a strongly degenerate semiconductor.

4. Conclusion

According to X-ray investigations the cubic $Cu_{2-x}Se$ phase has a homogeneity range at room temperature within the limits of 35.5 at% Se ($2-x = 1.83$) and 36.4 at% Se ($2-x = 1.75$) and is a p-type semiconductor with variable hole concentration.

The temperature dependence of the conductivity satisfies the law $\sigma = \sigma_1\,T^{-1.4}$ where σ_1 depends on the composition. The linearity of the function σ vs. $(2-x)$ indicates that deviations from stoichiometry dope the crystal in a very simple way, each deficit Cu atom yields a hole in the valence band.

The behaviour of the measured thermoelectric power is qualitatively in agreement with theoretical expectations. The disagreement in the absolute values may be caused by the dependence of the effective mass on the composition.

The thermoelectric power also indicated that non-stoichiometric $Cu_{2-x}Se$ can be regard as a strongly degenerate semiconductor for all the compositions examined.

Acknowledgement

The authors are indebted to Mr. A. Tonejc for excellent Debye-Scherrer films.

References

[1] G. LORENZ and C. WAGNER, J. chem. Phys. **26**, 1607 (1957).
[2] R. RONTIÉ, M. SUDRES, and J. MAHENC, J. Chim. phys. et Physico-Chim. biol. **64**, 1013 (1970).
[3] R. RONTIÉ, M. SUDRES, and J. MAHENC, J. electroanal. Chem. (Amsterdam) **25**, 489 (1970).
[4] R. D. HEYDING, Canad. J. Chem. **44**, 1233 (1966).
[5] B. ČELUSTKA and Z. OGORELEC, Croat. chem. Acta **41**, 73 (1969).
[6] J. MACDOUGALL and E. C. STONER, Phil. Trans. Roy. Soc. (London) **237A**, 350 (1938).
[7] S. MIYATANI, I. TOYOTA, and T. YONAGIHARA, J. Phys. Soc. Japan **23**, 35 (1967).
[8] S. MIYATANI, J. Phys. Soc. Japan **24**, 328 (1968).

(Received November 8, 1971)

phys. stat. sol. (a) **10**, 365 (1972)

Subject classification: 1.5 and 3; 22.1.3

Sektion Physik der Universität München

Preparation of Single Crystal Tellurium Films by Epitaxial Growth on Selenium

By

P. BAMMES, A. OTTO, and E. PETRI

Thin single crystal films of trigonal tellurium grow epitaxially from the vapour phase on the (10$\bar{1}$0) faces of trigonal selenium films. After vaporazing the selenium from the double layers selfsupporting tellurium films of a thickness below 1000 Å are obtained. The electron diffraction patterns are discussed.

Trigonales Tellur wächst in dünnen, einkristallinen Schichten aus der Dampfphase epitaxial auf (10$\bar{1}$0) Oberflächen von trigonalen Selenschichten. Nach dem Abdampfen des Selens von diesen Doppelschichten erhält man freitragende Tellurschichten mit einer Dicke unter 1000 Å. Die Elektronenbeugungsaufnahmen werden diskutiert.

1. Introduction

To prepare thin samples of single crystal tellurium, which are transparent in the electron microscope, various methods have been developed. Starting from bulk material etching [1 to 3], stripping off spangles [3], etching combined with electropolishing [4], and chemical polishing [5] gave transparent specimens. By evaporation onto single crystal NaCl [6] and by using the so-called sandwich method [7] single crystal tellurium films of a size of about 1 μm^2 were obtained.

For electron spectroscopy investigations [8] selfsupporting films of single crystal tellurium of a constant thickness less than 1000 Å within a diameter of at least 100 μm were needed. The methods mentioned above did not seem to fulfil these demands. The method of Griffiths and Sang [9] for preparing very thin single crystal selenium films of large size suggested an attempt to invert this method for the preparation of epitaxial tellurium films. Additionally a method had to be found to separate the fragile tellurium film from the selenium support without damage.

2. Preparation

We started with single crystal films of trigonal selenium of about 4000 Å thickness. These were prepared as Griffiths and Sang [9] described. Thin single crystal tellurium slices (about 0.2 mm) with large (10$\bar{1}$0) faces were freshly cleaved from bulk material. High purity selenium (99.999% Se) was evaporated in a vacuum of 10^{-5} Torr and deposited on the (10$\bar{1}$0) faces of the tellurium slices heated to 115 °C. The tellurium crystals, covered with a single crystal selenium layer, were brought onto a surface of 20% nitric acid. They were prevented from sinking by the surface tension and the tellurium was dissolved from below during about one hour. The selenium films remained on the surface of the nitric acid. The acid was then strongly diluted with water and the single crystal selenium layers were picked up on 50, respectively 100 μm diaphragms without supporting layer . The diaphragms were heated to 117 °C in a vacuum less than 10^{-5} Torr and then high-purity tellurium (99.99%) was evaporated onto the selenium films. Subsequently the diaphragms were heated to 160 °C at the same vacuum for one hour. In this way, the supporting selenium films were evaporated, and unsupported single crystal films of trigonal tellurium were obtained. This procedure was varied. At substrate temperatures of 100 and 122 °C during the evaporation of tellurium no single crystal tellurium films turned out. Instead of preparing the selenium films on diaphragms, as mentioned above, it was tried to prepare them on NaCl single crystals. However, this procedure led to a higher rate of damaged films.

3. Investigation of the Films

The transmission electron diffraction patterns of the selenium films used in the above described procedure (Fig. 1) had the same quality Griffiths' diffraction patterns showed. The electron diffraction patterns of the selenium–tellurium double layers (Fig. 2) show, that Griffiths' and Sang's [9] method of epitaxial growth of selenium on tellurium is well invertible. On the $(10\bar{1}0)$ face of selenium a tellurium layer of the same orientation arises. Fig. 2 shows the diffraction rings of polycrystalline tellurium beside the patterns of single crystal tellurium and selenium. This specimen was obtained after preparation of selenium on NaCl.

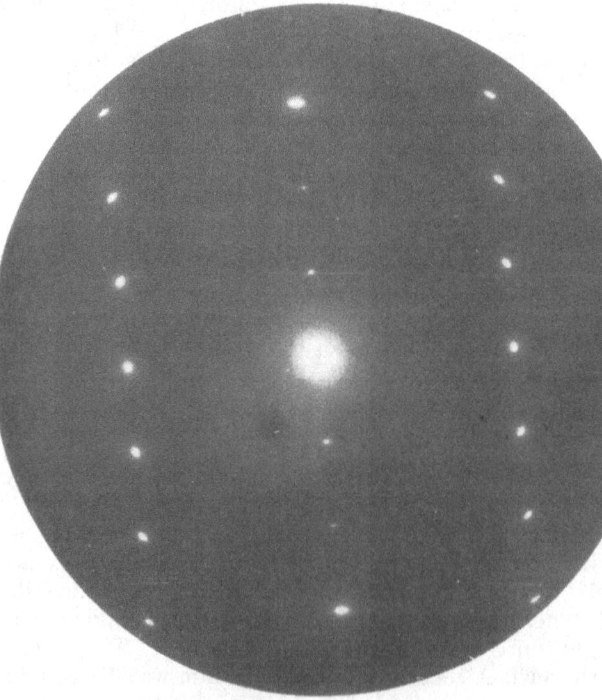

Fig. 1. Transmission diffraction pattern from a $(10\bar{1}0)$ faced selenium film at normal incidence of a 60 keV electron beam

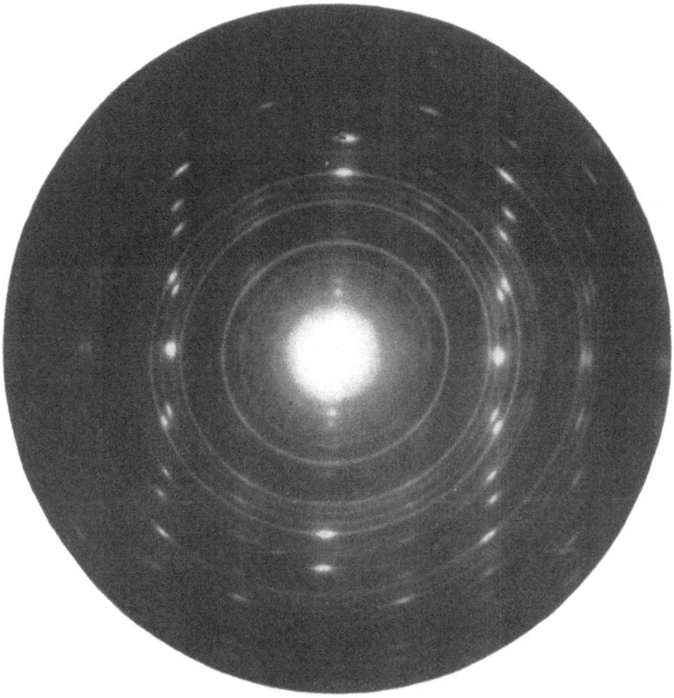

Fig. 2. Transmission diffraction pattern from a $(10\bar{1}0)$ faced double layer of selenium and tellurium at normal incidence of a 80 keV electron beam (see text)

The rings come from sections of the tellurium film, where no selenium film was between tellurium and NaCl. After separation of selenium by the first mentioned method in Section 2 the remaining films yielded diffraction patterns as shown in Fig. 3, proving that the films consisted of single crystal tellurium.

Checked in a light microscope, the tellurium films (less than 1000 Å thick) showed no cracks and were relatively smooth. Even electron microscopic magnification didn't make visible any structures within the diaphragm diameter of 100 μm.

The lattice constants calculated by use of the diffraction patterns are in good agreement with the known data. On the figures a weak distortion is observable, caused by the electron microscope. This effect was taken into account for the calculations. Beside the typical reflections of tellurium for the $10\bar{1}0$ reciprocal direction of the incident electron beam on Fig. 3 other reflections are to be seen:

a) in addition to the $(000n)$ reflections with $n = 3, 6, \ldots$ the diffraction patterns the forbidden (0001), (0002), . . . reflections (Fig. 2, 3, and 4),

b) among the expected rows of reflections $(000n)$, $(\bar{1}2\bar{1}n)$, $(\bar{2}4\bar{2}n)$, and $(1\bar{2}1n)$, $(2\bar{4}2n)$ further parallel rows of reflections (Fig. 3),

c) diffuse reflections of very low intensity attached to the strong reflections corresponding to a lattice with constants between tellurium and selenium. On Fig. 3 these reflections are hardly visible.

Furthermore the Se–Te double layers show additional reflections, which are marked by arrows in Fig. 4.

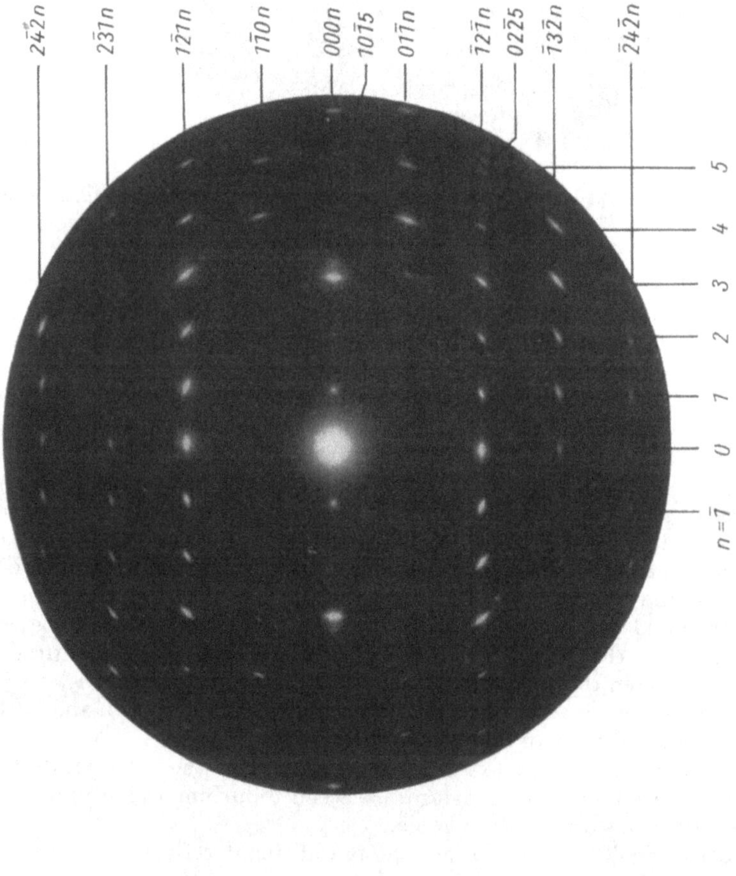

Fig. 3. Transmission diffraction pattern from a $(10\bar{1}0)$ faced tellurium film at normal incidence of a 80 keV electron beam

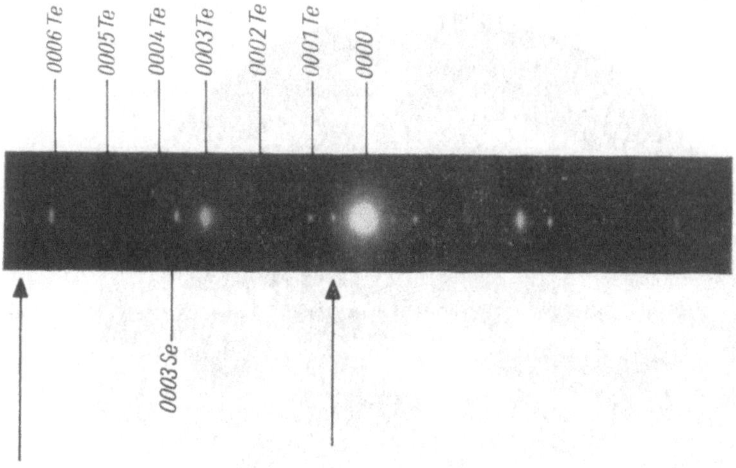

Fig. 4. Part of transmission diffraction pattern from a $(10\bar{1}0)$ faced double layer of tellurium and selenium at normal incidence of a 100 keV electron beam. Here the thickness of the selenium film is considerably less than that of the tellurium film

4. Discussion of the Additional Reflections

Concerning Fig. 3:

a) These reflections were explained by Griffiths and Sang [10] with elastic multiple scattering in case of selenium, and by Schüz [11] with dominating multiple scattering and particular systematic interaction in case of tellurium.

b) Even a slight bend of the film being extremely probable in the case of the used preparation explains the additional rows of reflections. Besides the expected section through the reciprocal lattice, additional, slightly inclined ($\approx 10°$) sections are obtained. These additional sections produce additional rows of reflections at certain distances from the (0000) reflection.

c) The satellites could refer to small regions of the lattice possessing lattice constants between those of selenium and tellurium. Because they appeared with rather different intensity at different samples, we suppose, a more intensive annealing might extinct them entirely. This question was not further investigated, because little volumes of differing properties in the films do not influence electron spectroscopic investigations [8].

Concerning Fig. 4:

These reflections were observed only at Se–Te double layers. They are easily interpreted as diffraction of the (0003) reflection of selenium (tellurium) in the tellurium (selenium) crystal.

Acknowledgements

The authors thank Prof. Rollwagen for enabling this work. Thanks are also due to W. Fellner for his help in the electron microscope investigations.

References

[1] L. C. Lovell, J. H. Wernick, and K. E. Benson, Acta metall. **6**, 716 (1958).
[2] J. S. Blakemore, J. W. Schultz, and K. C. Nomura, J. appl. Phys. **31**, 2226 (1960).
[3] P. Klein and K. Kleinhenz, Z. Naturf. **23a**, 530 (1968).
[4] H. Gottschalk, K. Kleinhenz, and E. Wagner, Z. Naturf. **25a**, 765 (1970).
[5] J. C. Doukhan, N. Doukhan, G. Saada, and B. Thomas, phys. stat. sol. **35**, 835 (1969).
[6] E. M. Hörl and J. Weiss, J. appl. Phys. **38**, 5132 (1967).
[7] E. Brüche and W. Schüz, Z. Phys. **199**, 135 (1967).
[8] P. Bammes, R. Klucker, E. E. Koch, and T. Tuomi, phys. stat. sol. (b) **49**, 561 (1972).
[9] C. H. Griffiths and H. Sang, Appl. Phys. Letters **11**, 118 (1967).
[10] C. H. Griffiths and H. Sang, Mat. Res. Bull. **2**, 515 (1967).
[11] W. Schüz, phys. stat. sol. **25**, 253 (1968).

(Received January 6, 1972)

phys. stat. sol. (a) **10**, 371 (1972)

Subject classification: 14.3.1 and 14.3.4; 22.9

Max-Planck-Institut für Physik und Astrophysik,
Abteilung Numerische Rechenmaschinen, München

Negative Resistance in Organic Monomolecular Layers Sandwiched between Metal Electrodes

By

K. H. Gundlach and J. Kadlec

Metal–insulator–metal sandwiches can exhibit low-frequency differential negative resistance (DNR) if they are placed in vacuum and formed by applying a few volts. This effect was till now established only for amorphous or polycrystalline insulators (mostly oxides). In attempts to explain the DNR phenomenon it is important to know whether it occurs also for other insulators and therefore monomolecular multilayers of cadmium arachidate are investigated. It is shown that such organic films can exhibit DNR very similar to that observed in amorphous and polycrystalline insulators. The striking difference, however, is that for monomolecular multilayers the DNR occurs at temperatures below about 225 °K, and not at room temperature as for the above-mentioned insulators. Moreover, multilayer sandwiches can be formed at 77 °K. The dependence of the DNR on the number of monomolecular layers is reported and the way in which electrode material and voltage polarity affect the forming process. The results will be discussed in relation to the various existing models proposed to explain the DNR phenomenon.

Metall–Isolator–Metall-Sandwichschichten können bei niedriger Frequenz negativen differentiellen Widerstand aufweisen, wenn die Proben ins Vakuum gebracht und durch Anlegen einer elektrischen Spannung von einigen Volt formiert werden. Dieser Effekt wurde bisher nur an amorphen oder polykristallinen Isolatorschichten, meistens Oxyden, nachgewiesen. Zur Erklärung der Erscheinung ist es wichtig zu wissen, ob sie auch bei anderen Isolatorschichten auftritt. Daher wurden monomolekulare Vielfachschichten aus Cadmiumarachidat untersucht. Es wird gezeigt, daß mit diesen organischen Schichten negativer Widerstand, ähnlich wie bei amorphen und polykristallinen Isolatoren, erzielt werden kann. Auffallende Unterschiede bestehen jedoch einmal darin, daß der negative Widerstand unterhalb der Temperatur von etwa 225 °K auftritt und nicht bei Zimmertemperatur wie im Falle amorpher oder polykristalliner Isolatorschichten. Außerdem können die genannten organischen Schichten bereits bei 77 °K formiert werden. Es wird über die Abhängigkeit des negativen Widerstandes von der Zahl der monomolekularen Schichten berichtet und über die Abhängigkeit des Formierungsprozesses von der Polarität der angelegten Spannung und dem Metall der Elektroden. Die Ergebnisse werden im Zusammenhang mit den vorliegenden Modellvorstellungen zur Erklärung des negativen Widerstandes diskutiert.

1. Introduction

A number of papers (see below) have reported on low-frequency voltage-controlled differential negative resistance (henceforth DNR) for metal–insulator–metal sandwich structures. Observations are generally made in a vacuum of 10^{-2} Torr or better; gases (e.g. He [1, 21], Ne [1, 21], H_2 [21], Ar [21]) can be used instead. The effect is not present in oxygen atmosphere. First sufficient voltage, in general 4 to 6 V, must be applied across the sandwich in order to "form" it [3]. A typical current (J)–voltage (U) characteristic obtained after forming is shown in Fig. 1.

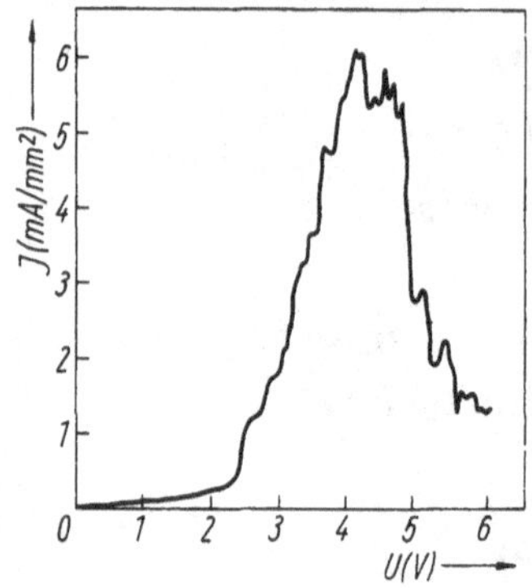

Fig. 1. Current–voltage characteristic of an Al–(Al oxide)–Au sandwich at room temperature showing DNR, Au positive. Oxide thickness was 170 Å

In the investigation of the effect, various types of sandwiches have been used: Al–(Al oxide)–Au [1, 2, 5, 7, 9, 10, 14, 18 to 25], Al–(Al oxide)–Al [20, 25, 26], Al–SiO–Au [1 to 3, 6, 8, 14, 15], Au–SiO–Au [30], Ta–Ta_2O_5–Au [1, 2], Zr–ZrO_2–Au [1, 2], Ti–TiO_2–Au [1], Au–ZnS–Au [23, 28], metal–silicone polymer–metal [11], Ni-collodium–Ag [12], Ag–CaF_2–Ag [12], and Ag–MgF_2–Ag [12, 13].

Al–(Al oxide)–metal junctions with top electrodes of Ag, Au, Cu, Co, Sn, In, Bi, Pb, Al, and Mg were investigated by Hickmott [20]. The insulators of all above-mentioned sandwiches were either amorphous or polycrystalline. Cleaved single crystals of mica, sandwiched between gold electrodes, did not show DNR [4]. It was therefore thought that DNR may be closely related to the structure of amorphous or polycrystalline insulators and hence limited to such materials.

In order to check this conjecture we have employed monomolecular layers of cadmium arachidate as insulator. This has the additional advantage to have an insulator whose thickness can be varied in discrete steps of 26 Å [17]. It will be shown that such layers can exhibit DNR very similar to that of oxides and other insulators. There are, however, characteristic differences between oxide and monolayer sandwiches, namely, the temperature at which forming is possible and the temperature range in which DNR can be established.

2. Previous Observations on Metal–Oxide–Metal Sandwiches

For comparative purposes we summarize here observations made for SiO and Al oxide sandwiches, which have been most extensively investigated. As already mentioned, Fig. 1 shows a typical *J–U* characteristic for an Al–(Al oxide)–Au sandwich exhibiting DNR. The oxide was produced by oxygen glow discharge as described in [29]. The measurement was performed at room temperature and at a pressure of about 10^{-3} Torr.

Where the temperature dependence of forming was investigated it is found that the sandwiches form only above a certain temperature T ($T \approx 220$ °K for Al–SiO–Au [15] and $T \approx 250$ °K for Al–(Al oxide)–Au[1])). For a given forming voltage the rate and degree of forming becomes larger at higher temperatures [15].

[1]) This value was found in our investigations on plasma oxidized Al–(Al oxide)–Au sandwiches (unpublished).

A sample formed (e.g. at room temperature) and then cooled to liquid nitrogen temperature exhibits DNR only for the rising part of the first voltage cycle [3, 22]. The disappearance of the DNR is practically permanent as long as the sample is held at $T \approx 77$ °K [3, 27]. The DNR obtained at room temperature becomes less pronounced with decreasing temperature and disappears at approximately 200 °K [14, 22].

It was found that Al–SiO–Au sandwiches can only be formed by biasing the Au electrode positively [15]. Al–SiO–Al sandwiches could not be formed [15]. Al–(Al oxide)–Au junctions form easily when Au is positive [7, 9, 10, 14, 22, 24, 25]. Hickmott, however, reported that he could form his Al–(Al oxide)–Au samples also for Al positive [20]. Likewise, Al–(Al oxide)–Al sandwiches were formed [20, 25, 26] and here again the forming was easier for top electrode positive [20].

3. Experimental Methods for Organic Monomolecular Layers

Monomolecular layers of arachidic acid (CH_3–$(CH_2)_{18}$–COOH) were spread on the surface of the water solution of a Blodgett-Langmuir trough. The hydrocarbon chains are nearly perpendicular to the surface with the carboxyl group in contact with the water solution containing Cd ions. Hence the arachidic acid will be to some extent, depending on the pH value of the solution, converted into cadmium arachidate. The thickness of such an organic monolayer[2]) is approximately 26 Å [17]. The detailed technique of spreading a monolayer on the surface of the water solution is described in several papers (e.g. [16, 17] and references therein).

To produce the sandwich structure a base electrode consisting of an Al stripe was first evaporated on a glass substrate. The procedure for depositing monolayers on it is as follows: When the substrate is dipped once into the solution and taken out again, it is covered by one monolayer. However, the second time that the substrate is put into the solution, a second monolayer is deposited and on taking the glass substrate out a third monolayer is transferred. Subsequent dippings behave in the same way and hence we have only produced organic insulators having odd numbers of monolayers. To complete the multilayer sandwich, top electrodes of Au or Al were evaporated. The area of a sample was about 1 mm². The samples were placed in a sealed brass box. It could be evacuated to 10^{-3} Torr and its temperature could be varied between 77 °K and room temperature. The J–U characteristics were plotted on a recorder, using a sawtooth voltage. For all characteristics shown in this paper the voltage increased at 0.3 V/s and decreased to zero within 0.02 s. It is known [3, 8, 27] that the J–U characteristics do not manifest DNR in the return trace if the voltage is removed in such short time. On the other hand, the fast decrease in voltage allows the observation of the so-called threshold voltage [3, 8, 27]. It is the voltage at which the junction changes from the high- to the low-impedance state. In Fig. 1 the threshold voltage is observed to be about 2.3 V.

Before reporting on DNR in multilayer sandwiches, we note that samples with base electrode of Au invariably showed short circuit. This observation, which was made for thickness from 1 to 11 monolayers, is in agreement with that of Handy and Scala [16] who did not find any insulating behaviour when

[2]) Throughout this paper the term monolayer designates a single organic monomolecular layer.

oxide-free base electrodes were used. There is experimental evidence that the adhesive forces between the monolayer and the gold surface is less strong than for oxide-covered metals, and it is thought that this leads to voids extending through the entire organic multilayers [16].

It is obvious from what is mentioned above and from the preparation method of the multilayer sandwiches, that there is an Al oxide film between the Al base electrode and the monolayers. The capacitance of a sandwich can be calculated on the basis of one monolayer alone, and since this agrees within a few per cent with measurements of the capacitance (cf. also [17]) it can be concluded that the Al oxide film is not thicker than 20 Å.

4. Negative Resistance for Organic Monomolecular Multilayers

DNR in Al–(Al oxide)–multilayer–Al (or Au) sandwiches can be observed as follows. The unformed sample is placed in a vacuum chamber at 10^{-3} Torr and cooled to liquid nitrogen temperature. Then a sawtooth voltage is applied across the junction, the top electrode being positive. As a rule, DNR is not observed in the first voltage cycle but appears in the second one; however, it disappears again after two to four further voltage cycles and the J–U characteristic becomes smooth. The latter state is shown at the bottom of Fig. 2, in this case for a sample having eleven monolayers and a top electrode of Al. Now the sandwich is warmed at about 5 deg/min and the J–U characteristic starts to change at about 150 °K. The most pronounced DNR (which means the peak-to-valley ratio of the current takes its maximum value) is observed at

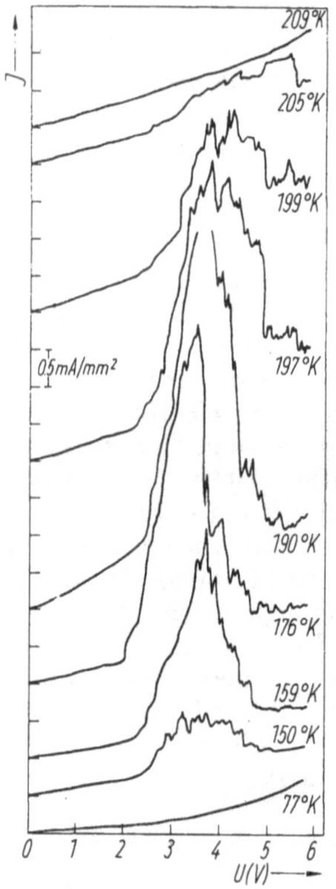

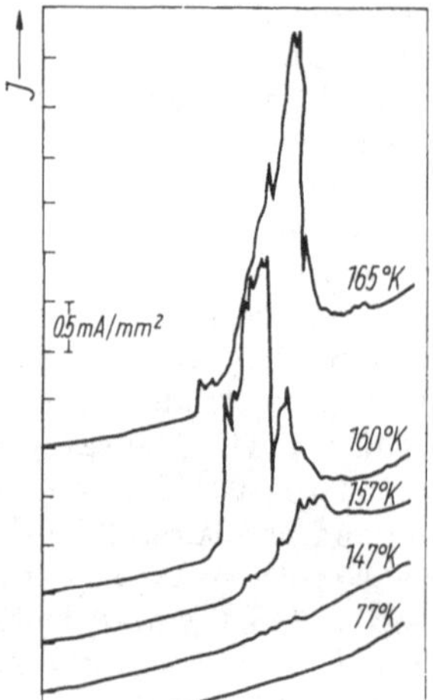

Fig. 3. Current–voltage characteristics of an Al–(Al oxide)–eleven monolayers–Au sandwich for increasing temperature, Au positive. For temperatures above about 170 °K monolayer sandwiches with top electrode of Au exhibit short circuit. The vertical scale is displaced for each curve

Fig. 2. Current–voltage characteristics of an Al–(Al oxide)–eleven monolayers–Al sandwich for increasing temperature, top electrode positive. Note the complete disappearance of the DNR above about 210 °K. The vertical scale is displaced for each curve for clarity about 190 °K and then the effect becomes smaller, disappearing above about 210 °K.

As shown in Fig. 3, similar results were obtained when the top electrode consisted of Au. In this case the junctions always exhibited short circuit above approximately 170 °K. If the sample is cooled again before it has become short circuit (e.g. from 165 °K) then the DNR disappears again at about 150 °K.

So far, the observations do not separate the temperature effects from purely time-dependent features, so we made investigations at various steady temperatures. Fig. 4 shows some results. In order to start from a standard state, we always cooled unformed samples to 100 °K and then applied the sawtooth voltage until the smooth J–U curve was obtained. Then the voltage was removed and the temperature raised. Now re-applying voltage at a constant temperature in the range

a) 110 to 150 °K produced DNR but not pronounced,
b) 150 to 190 °K gives better DNR than above,
c) 190 to 225 °K leads to pronounced DNR (see e.g. curve 4 in Fig. 4),
d) 225 to 300 °K causes an increase in conductivity with little or no DNR.

In all cases described above the DNR is observed only for the first few voltage cycles. This is illustrated by the curves 4 to 7 in Fig. 4. There it is also seen that the current maxima shift toward lower voltages with increasing number of voltage cycles until the DNR disappears. It can be reobtained as follows:

(i) Increasing the temperature in the range 100 to 225 °K (which means, e.g., from 140 to 150 °K); the larger the temperature increase, the better the DNR is developed.

(ii) Interrupting the voltage bias for one hour or more (see curve 8 in Fig. 4), In this case the DNR is more pronounced for longer resting times. Also in the observations (i) and (ii) the DNR disappears after a few voltage cycles.

We now list further observations made with multilayer sandwiches:

1. They can be formed by applying 5 to 6 V or more. Forming was possible only when the top electrode was biased positively. This applies for top electrodes of Au as well as Al.

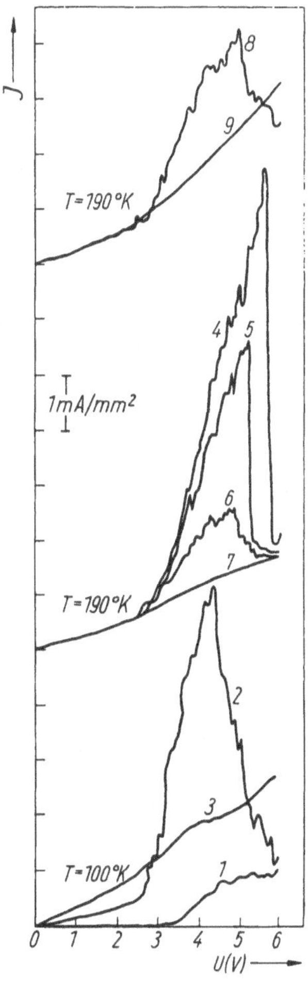

Fig. 4. Current–voltage characteristics of an Al–(Al oxide)–eleven monolayers–Al sandwich, top electrode positive. Curve 1: first voltage cycle of an unformed sample. Curve 2: second voltage cycle, the sample is now formed and exhibits DNR. Curve 3: smooth characteristic after a few voltage cycles. Sample temperature was 100 °K. Curves 4 to 7: first four voltage cycles after temperature increase from 100 to 190 °K. Curve 8: first voltage cycle after interrupting junction bias for 3 h, at 190 °K. Curve 9: smooth curve obtained after some more voltage cycles at 190 °K

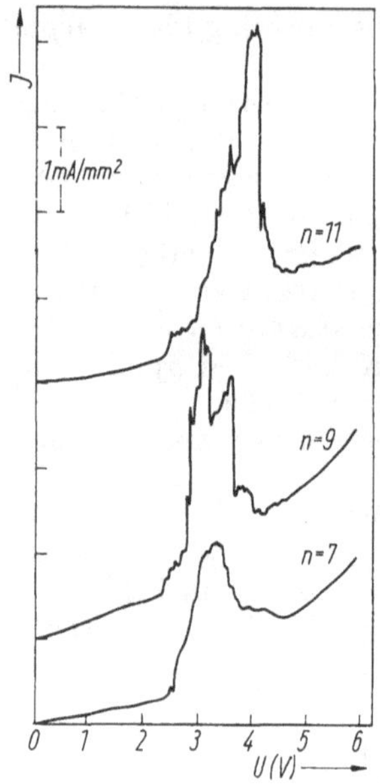

Fig. 5. Dependence of the current–voltage characteristic on the number of monolayers for Al–(Al oxide)–n monolayers–Au sandwiches at $T \approx 170$ °K with Au positive

2. After forming, the DNR is observed for both polarities of voltage bias. This provides a good method to make sure whether a sample is formed or not: cool it e.g. to 77 °K and have no voltage applied for at least one hour; then, biasing the base electrode positive, a formed sample exhibits DNR in the first voltage cycle.

3. The threshold voltage lies between 2 and 3 V.

4. DNR becomes more pronounced with increasing number of monolayers. This is exemplified for Al–(Al oxide)–multilayers–Au samples in Fig. 5. For a total of 35 samples with top electrodes of Al we found the following average values for the peak-to-valley ratio: 4.2 for eleven monolayers, 2.1 for nine monolayers, and 1.45 for seven monolayers. For sandwiches having less than five monolayers no DNR could be obtained.

5. For a given number of monolayers it was found that lower dc conductivity at 6 V was associated with more pronounced DNR.

6. DNR disappears when the ambient oxygen pressure exceeds 0.1 to 1 Torr.

Before discussing experimental results in detail, we wish to summarize arguments showing that the DNR in Al–(Al oxide)–multilayer–metal samples is either an effect of the organic monolayers and the Al oxide film, or perhaps an effect of the organic monolayers alone, but does not purely arise from the Al oxide film. In Section 3, we deduced that the Al oxide film is not thicker than about 20 Å; DNR has not in fact been observed for such thin Al oxide films no matter how they were prepared. When we omitted the monolayer from the trough and repeated our manufacturing process, the resulting "dummy" sandwiches showed no DNR; in fact they either were short-circuited or had very low resistance. Furthermore, the forming process and the observation of DNR exhibit quite different temperature dependence in the two types of sandwiches (cf. Section 2 and results given in this section). Finally, we have observed that the DNR becomes more pronounced with increasing number of monolayers.

5. Discussion

There are about five distinct models proposed to explain DNR [27, 28]. Some of them interpret also forming, switching, memory effects, dead time, the voltage for the peak current in the DNR characteristic, and others. However, no model accounts fully for the numerous observations demonstrated particularly for oxide sandwiches. It seems that the situation becomes more complicated considering also our results on multilayer sandwiches because they exhibit some properties not identical with those of oxide sandwiches. Nevertheless, a few suggestions and comments can be made.

Most of the models proposed to explain forming involve motion of metal ions through the insulator [27, 28]. The organic monomolecular layers consist of long hydrocarbon chains perpendicular to the film plane of the insulator [17]. The binding forces between contiguous chains are very weak. Hence the mobility of impurity ions (here metal ions) in the direction perpendicular to the insulator film plane is expected to be considerably higher than in oxides which do not have such an anisotropy in binding forces. This may be the reason why multilayer sandwiches can be formed at 77 °K while silicium oxide [15] and aluminium oxide[3]) sandwiches do not for mbelow about 220 and 250 °K, respectively. The observation that multilayer sandwiches cannot be formed (at 77 °K) when the base electrode is positive is explained by this suggestion on the basis of the thin oxide film (cf. the end of Section 3) inhibiting the injection of metal ions from the base electrode into the organic layers.

For monomolecular multilayers the existence of DNR could be explained in principle by nearly all of the existing models, as in the case of amorphous and polycrystalline insulators. No single theory, however, explains the detailed behaviour. As described in Section 4, the DNR for multilayer sandwiches appears below 225 °K and occurs, at constant sample temperature, only for the first or the first few voltage cycles. The latter observation is probably related to the so-called dead time discussed in detail by several authors. In the filamentary model of Dearnaley et al. [27] the dead time is a measure of the time between filament breaking and reaching a state from which it may reform. The transition between these two states is tentatively interpreted as relaxation of space charge polarization. In the energy band model of Simmons and Verderber [3, 8] the dead time is related to electron diffusion through the insulator and is simply given by $(\nu P)^{-1}$, where P is the phonon-assisted tunneling probability and ν is the attempt-to-escape frequency. Both models predict strong increase of the dead time with decreasing temperature. So it is not surprising that multilayer sandwiches show a dead time in the range of hours because they exhibit DNR at temperatures below 225 °K. But the situation is more complicated. Bernard et al. [10] have reported that the dead time of Al–(Al oxide)–Au sandwiches increases irreversibly from seconds to hours by warming the samples up to 500 °K for a peroid of one hour. (The dead time was measured at room temperature.) This observation cannot be explained by the existing theories.

The voltage U_p for the current peak in the J–U characteristic has tentatively be related to positions of impurity levels in the forbidden band of the insulator (see e.g. [27]). Furthermore, from Hickmott's results [1, 6] the empirical correlations $U_p \sim \sqrt{E_g}$ and $U_p \approx -0.36\sqrt{\varepsilon_r} + 4$ can be found.[4]) Here E_g and ε_r are the energy gap and the relative dielectric constant, respectively, of the insulator. (We note that ε_r and E_g are not independent [27].) For monomolecular layers of cadmium arachidate [17] $\varepsilon_r = 2.4$. Hence, using the second of the above correlations, one obtains $U_p \approx 3.45$. This value is, within experimental error, in agreement with experiment for measurements where the multilayer sandwich was warmed steadily (cf. Fig. 2 and 3). But in measurements at constant temperatures (Fig. 4) higher values of U_p are observed. Hence U_p can apparently not be related simply to parameters of band models which would

[3]) Own investigations, cf. footnote [1]).

[4]) Hickmott [6] proposed instead of $U_p \approx -0.36\sqrt{\varepsilon_r} + 4$ the correlation $U_p^2 \approx 10.3 - 0.18\,\varepsilon_r$. We found from Fig. 14 in [1] that the former correlation agrees better with experimental data than the latter one.

be expected to be constant for sandwiches of the same material. This suggestion is confirmed by a recent work of Sutherland [28]. He reported for Au–ZnS–Au sandwiches that U_p can be increased drastically by applying voltage in the region of the current minimum of the J–U characteristic for a period of a few minutes. This means, the voltage U_p may be changed for a given sample.

To account for the above-mentioned observations on ZnS sandwiches Sutherland [28] has proposed a filamentary model similar to that of Dearnaley et al. [27]. In these models the DNR is thought to arise from local breaking of thin conducting filaments by Joule heating. The voltage U_p is related to the distribution of filament resistances and this in turn depends on spacing of positive centres in the conducting filaments. It is difficult to say whether such models would apply also to monomolecular multilayers. At present we see no way to explain the disappearence of DNR above a certain temperature (cf. e.g. Fig. 2). It is also not immediately obvious why the DNR becomes more pronounced with increasing number of monolayers.

In conclusion, we have shown that low-frequency differential negative resistance, observed in recent years for a wide range of amorphous and polycrystalline insulators, occurs also when organic monomolecular layers are used as insulator. This experimental evidence strongly suggests that the phenomena under consideration are to a large extent independent of specific properties of the insulator material. Consequently, theories involving sophisticated energy band models (e.g. different impurity levels within the insulator forbidden band) are unlikely to be correct. This hypothesis seems to be supported by recent observation that the dead time and the voltage for the current peak can be changed for a given sample.

Acknowledgements

We would like to acknowledge the fruitful discussions with Prof. H. Billing and Dr. J. Bernard. Thanks are due to Miss Ch. Degendorfer for careful preparation of the sandwich structures.

References

[1] T. W. Hickmott, J. appl. Phys. **33**, 2669 (1962).
[2] T. W. Hickmott, J. appl. Phys. **34**, 1569 (1963).
[3] J. G. Simmons and R. R. Verderber, Proc. Roy. Soc. **A301**, 77 (1967).
[4] L. Eckertová and J. Jiskra, Czech. J. Phys. **B18**, 131 (1968).
[5] L. Eckertová, phys. stat. sol. **18**, 3 (1966) and references therein.
[6] T. W. Hickmott, J. appl. Phys. **35**, 2679 (1964).
[7] J. Bernard, G. Delacote, and Y. Mentalecheta, phys. stat. sol. **31**, 315 (1969).
[8] J. G. Simmons and R. R. Verderber, Radio and Electronic Engineer **34**, 81 (1967).
[9] J. Emmer, phys. stat. sol. (a) **3**, K191 (1970).
[10] J. Bernard, R. Haug, Y. Mentalecheta, and Y. Tregouet, phys. stat. sol. (a) **6**, K127 (1971).
[11] H. T. Mann, J. appl. Phys. **35**, 2173 (1964).
[12] T. Lewowski, S. Sendecki, and B. Sujak, Acta phys. Polon. **28**, 343 (1965).
[13] R. A. Cola, J. G. Simmons, and R. R. Verderber, Proc. National Aerospace Electronics Conf., 1964 (p. 118).
[14] T. W. Hickmott, J. appl. Phys. **36**, 1885 (1965).
[15] R. R. Verderber, J. G. Simmons, and B. Eales, Phil. Mag. **16**, 1049 (1967).
[16] R. M. Handy and L. C. Scala, J. Electrochem. Soc. **113**, 109 (1966).

[17] K. H. Drexhage and H. Kuhn, in: Proc. Internat. Symp. Basic Problems in Thin Film Physics, Clausthal/Göttingen 1965, Ed. H. Mayer and R. Niedermayer, Vandenhoeck and Ruprecht, Göttingen 1966 (p. 339).
[18] G. S. Kreynina, L. N. Selivanov, and T. J. Shumskaya, Radio Engng. Electron. Phys. **5**, 219 (1960).
[19] S. R. Pollack, W. O. Freitag, and C. F. Morris, Electr. Chem. Tech. **1**, 96 (1963).
[20] T. W. Hickmott, J. appl. Phys. **35**, 2118 (1964).
[21] J. Bernard, M. Decker, and Y. Mentalecheta, C. R. Acad. Sci. (France) **270**, 1419 (1970).
[22] Y. Kumagai, K. Nakajima, Y. Shibata, and Y. Koike, Japan. J. appl. Phys. **32**, 586 (1963).
Y. Kumagai, S. Nishimatsu, and Y. Shibata, Japan. J. appl. Phys. **33**, 649 (1964).
[23] R. R. Sutherland, K. O. Legg, and R. A. Collins, Thin Solid Films **6**, R39 (1970).
[24] L. Eckertová and R. Hrach, in: III. Czech. Conf. Electronics and Vacuum Physics Transactions, Prague 1965, Academia, Prague 1967 (p. 313).
[25] R. Uzan, A. Roger, and A. Cachard, Le Vide **127**, 38 (1967).
[26] J. A. Roger, P. Thevenard, C. H. S. Dupuy, and R. Uzan, in: IV. Czech. Conf. Electronics and Vacuum Physics, Prague 1968 (p. 353).
[27] G. Dearnaley, A. M. Stonehame, and D. V. Morgan, Rep. Progr. Phys. **33**, 1129 (1970).
[28] R. R. Sutherland, J. Phys. D (appl. Phys.) **4**, 468 (1971).
[29] K. H. Gundlach and J. Hölzl, Surface Sci. **27**, 125 (1971).
[30] R. A. Collins and R. D. Gould, Solid State Electronics **14**, 805 (1971).

(Received January 10, 1972)

phys. stat. sol. (a) **10**, 381 (1972)

Subject classification: 10.2; 22.1.2

Department of Metallurgy, University of Oxford

Dislocation Contrast in X-Ray Topographs of Very Thin Crystals

By

B. K. TANNER

The minimum thickness for visibility of dislocations in X-ray topographs has been measured for several reflections. The fraction of an extinction distance at which this image loss occurs is not constant but is dependent on reflection, and varies from about 0.4 for low-order reflections to about 0.15 for high-order reflections. An explanation of this effect is given and shown to be in qualitative agreement with experiment. Image loss in thin crystals in the Bragg case is demonstrated.

L'épaisseur d'un échantillon au-dessous de laquelle une dislocation n'est pas visible par topographie aux rayons X a été mesurée dans le cas de plusieurs réflections. Cette épaisseur dépend de la réflection utilisée et est comprise entre 0,4 fois la distance d'extinction dans le cas des réflections d'ordre bas et 0,15 fois la distance d'extinction dans le cas des réflections d'ordre élevé. Une explication de ce phénomène est proposée, et les résultats théoriques sont en accord qualitatif avec les mesures experimentales. La disparition de l'image dans le cas des topographies par réflection (cas de Bragg) est aussi mise en évidence.

1. Introduction

One of the major disadvantages of the technique of X-ray topography [1, 2] is that, as a result of the large image widths, only in specimens with dislocation density less than about 10^5 lines/cm^2 are individual dislocations visible. In recent years, as crystal growth techniques have improved, many crystals have been grown nearly dislocation-free. However, as deformation begins, the dislocation density rises very rapidly and to make a complete study of the early deformation stages of many metal systems, it would appear necessary to use X-ray topography in conjunction with high-voltage electron microscopy. As it would be advantageous to use the same specimen in the two instruments, it is important that the contrast of defects of X-ray topographs of very thin crystals be examined.

It has been argued [3] that in crystals thinner than about 1/3 to 1/6 of an extinction distance, ξ_g, direct images of dislocations should cease to be visible in the Laue case. While it was experimentally demonstrated that dislocations did become invisible in very thin specimens, no figure was given for the thickness at which invisibility occurred. Attempts to feather down the edges of MgO and iron–silicon alloy crystals chemically gave only a rough figure of about 1/3 of an extinction distance [4]. In this paper, the thickness at which invisibility occurs is measured to an accuracy of about 1 μm for several reflections using different radiations.

2. Theoretical Argument

In a projection topograph, the diffracted beam intensity reaching the film is equal to the integrated intensity and independent of the shape of the wave front, provided that the integration is carried out across the whole of the rocking curve [5]. The integrated plane wave reflecting power for zero ab-

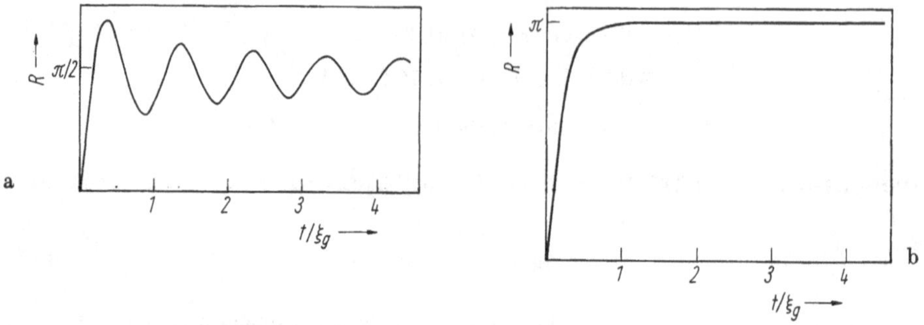

Fig. 1. Reflecting power as a function of crystal thickness for plane polarised radiation and no absorption. a) Laue case; b) Bragg case

sorption has been calculated, [6] and this is shown in Fig. 1 for the Laue and Bragg cases with plane polarised X-rays.

The direct image in X-ray topographs is considered to be formed when the lattice planes around a defect are misoriented by about the width of the perfect crystal rocking curve, [7]. Here, X-rays from parts of the source not exactly satisfying the Bragg condition in the perfect crystal are reflected. Suppose this region of 'bad' material is of thickness Δ. In general, Δ/t will be considerably less than unity and as the X-rays diffracted in the 'bad' region are not diffracted by the perfect crystal, the 'bad' region diffracts kinematically. The intensity from the defect is thus proportional to Δ/t. Two cases will now be considered:

a) Crystal is thick, i.e. $t \gg \xi_g$.

b) Crystal is thin, i.e. $t \ll \xi_g$.

The following is a summary, and extension to the Bragg case of the argument in reference [3].

2.1 *Laue case — zero absorption*

Inspection of Fig. 1a shows that in a thick crystal, the intensity oscillates about a mean value. The gradient of the intensity vs. thickness curve at any point is always less than at zero thickness. Thus, the intensity lost from reducing the crystal thickness from t to $t - \Delta$ is always less than that reflected from the bad region of thickness Δ. Direct images are therefore always more intense than the background. However, in thin crystals less than about $\xi_g/8$, the intensity is proportional to the crystal thickness. The net effect from a 'bad' layer of thickness Δ and the perfect material of thickness $t - \Delta$, is to produce the same intensity as that from the perfect crystal t. Dislocations then cease to be visible and only large misorientations, such as scratches, where no X-rays from the source can satisfy the Bragg condition in the misoriented material are detectable.

2.2 *Bragg case — zero absorption*

In the Bragg case no oscillations are observed, Fig. 1b. The intensity reflected is independent of crystal thickness for a thick crystal, and the intensity reflected from a crystal of thickness t will be equal to that from a crystal of thickness $t - \Delta$. Direct images in the Bragg case are always more intense than background. Again, in the thin crystal of thickness less than about $\xi_g/8$, the intensity in the perfect region becomes proportional to the thickness. Dislocations are not then expected to be visible in very thin crystals in the Bragg case.

3. Experimental Methods and Results

A polished slice of float-grown silicon of orientation (111) and initially about 200 μm thick was mounted at a low angle on a glass slide and mechanically polished until the thin end gave appreciable transmission of white light. A short chemical polish left the specimen free from elastic strain and varying in thickness from about 5 to 100 μm. Fig. 2a, b show a repeat of the Penning and Goemans experiment, except that a wedge-shaped specimen is used here rather than a flat one. It is immediately obvious that defects appearing using $CuK\alpha_1$ radiation ($\xi_g = 15$ μm) are invisible with $AgK\alpha_1$ radiation ($\xi_g = 47$ μm), both topographs being taken with the same 220 reflection in transmission.

When polarisation is taken into account, it is possible to measure quite accurately the thickness at which visibility ceases. Inspection of the AgKα topograph indicates which dislocation is just visible, and the thickness at this point may be determined from the position of the relevant dislocation with respect to the thickness fringes in the CuKα topograph. It is important to ensure that one is starting at the first fringe, and this may be checked by reference to longer extinction distance reflections. Including errors in estimating the positions of the maxima and minima and variations in the wedge angle, the measured accuracy is about 1 μm in the thickness. Utilising this method, the fraction of the σ-polarisation extinction distance at which visibility ceases has been measured for several reflections and the results are shown in Table 1.

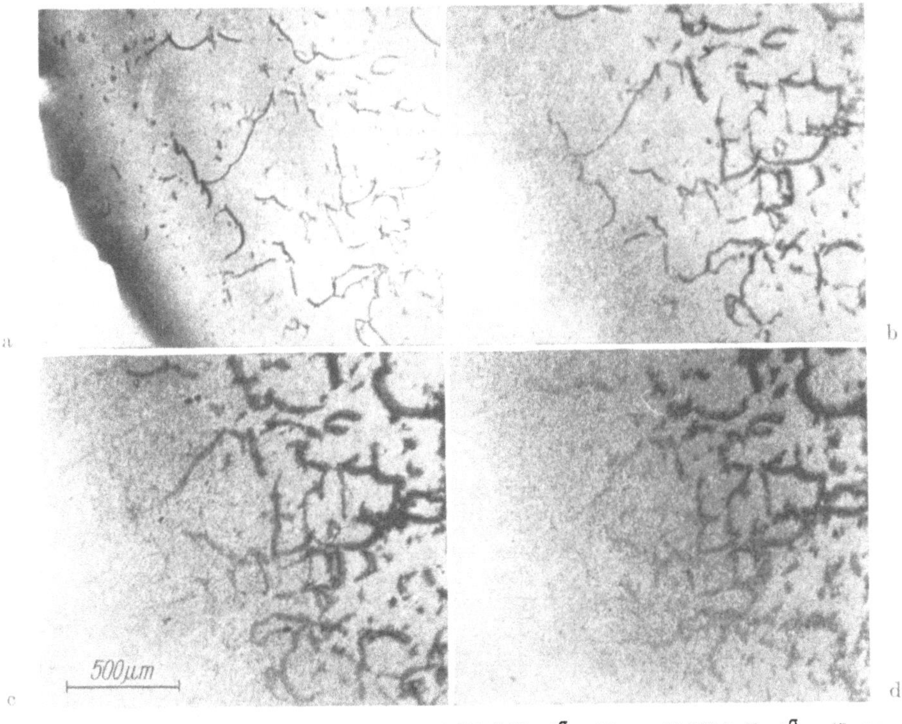

Fig. 2. Loss of image in thin crystal in Laue case. a) 220 CuKα $\xi_g^\sigma = 15$ μm; b) 220 AgKα $\xi_g^\sigma = 47$ μm; c) 440 AgKα $\xi_g^\sigma = 71$ μm; d) 555 MoKα $\xi_g^\sigma = 128$ μm

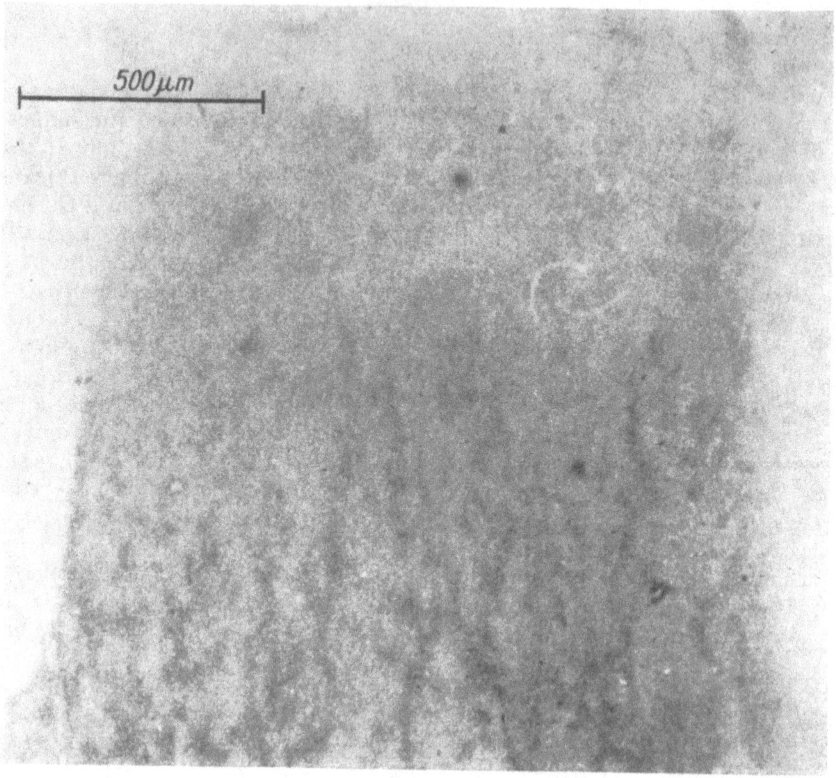

Fig. 3. Loss of image in Bragg case. AgKα 555 reflection, thin area at top of photograph

Table 1
Position of invisibility — Laue case

radiation	reflection	thickness for image loss (μm)	t/ξ_g^σ
AgKα	$\bar{2}\bar{2}0$	19	0.41
	$20\bar{2}$	17	0.37
	$\bar{1}11$	17	0.32
	$\bar{1}\bar{1}1$	18	0.34
	111	20	0.38
	$\bar{3}\bar{3}3$	20	0.25
	$\bar{4}\bar{4}0$	19	0.26
MoKα	$\bar{2}\bar{2}0$	15	0.42
	$\bar{2}\bar{4}2$	18.5	0.39
	$\bar{3}\bar{1}1$	21	0.38
	$\bar{5}\bar{5}5$	20	0.16

Fig. 3 demonstrates the loss of image contrast in the thin crystal in the Bragg case. Shown is a 555 surface reflection topograph using AgKα radiation and the thin end of the wedge is at the top of the photograph.

4. Discussion

It is clear from both, Fig. 2 and from Table 1, that a general figure for the fraction of an extinction distance at which image loss occurs cannot be given. A definite decrease is observed as one goes to higher order reflections, which cannot be explained either by experimental error or polarisation effects. However, an explanation may be sought in the following way:

Consider the effect of increasing the thickness of misoriented material, Δ, for a constant thickness of crystal, t, where the intensity is no longer a linear function of t.

The visibility is defined as

$$V = \frac{I_\Delta + I_{t-\Delta} - I_t}{I_\Delta + I_{t-\Delta} + I_t}\,.$$

Then, if $\Delta/t \ll 1$,

$$I_{t-\Delta} + I_\Delta < \Phi\, t\,,$$

where Φ is the gradient of the intensity–thickness curve at zero thickness.

If $\Delta/t \approx 1$,

$$I_{t-\Delta} + I_\Delta = \Phi\, t\,,$$

as the perfect crystal of thickness $t - \Delta$, above and below the defect diffracts kinematically. Thus the visibility is greater for $\Delta/t \approx 1$ than for $\Delta/t \ll 1$, at the same value of t. Defects are thus expected to remain visible in thinner material in the case where $\Delta/t \approx 1$ than where $\Delta/t \ll 1$.

The thickness of misoriented material which contributes to the image will be dependent on the perfect crystal rocking curve width, and for simplicity, will be considered inversely proportional to it. Then,

$$\Delta \sim g\, \xi_g\,,$$

where $\boldsymbol{g}$ is the reciprocal lattice vector. Expressing the crystal thickness t as a fraction k_t of the extinction distance gives

$$\frac{\Delta}{t} \sim \frac{g}{k_t}\,.$$

Thus, as g increases, the visibility at a given value of k_t increases. A plot of t/ξ_g against g, where t is the thickness for image loss, shows considerable scatter. This is not surprising, as the thickness of misoriented material will be dependent upon the projection which the dislocation strain field makes in the X-ray incidence plane. However, restriction to the 111 reflections, which have similar

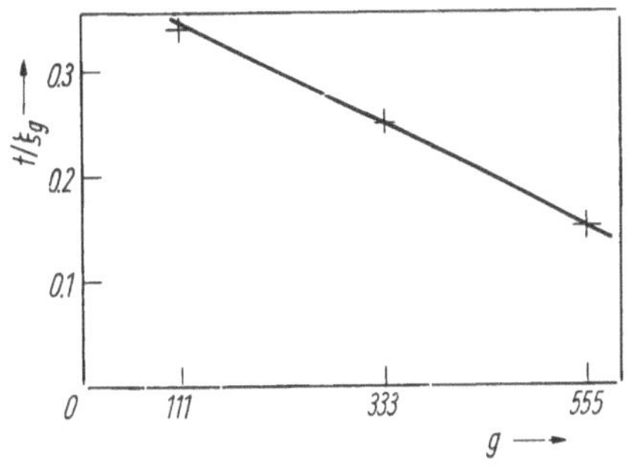

Fig. 4. Fraction of extinction distance at which loss occurs as a function of reciprocal lattice vector $\boldsymbol{g}$ for the 111 reflections

projections of misoriented material, yields a monotonic decrease of t/ξ_g with increasing g. This effect, seen in Fig. 4, is in good qualitative agreement with the above argument.

5. Conclusion

The fraction of an extinction distance at which image loss occurs is dependent upon the reflection used. In low-order reflections the minimum thickness at which good dislocation contrast is observed is about 0.4 ξ_g. This value falls to about 0.15 ξ_g for higher order reflections.

Acknowledgements

The author wishes to thank Prof. P. B. Hirsch (F.R.S.) for laboratory facilities, and Dr. C. J. Humphreys for criticising the manuscript, and his interest and encouragement throughout this study. The assistance of Mr. H. R. Pettit in polishing the specimen, and Mr. R. Ogden of Plessey (U.K.) in providing the silicon, is gratefully recorded. Financial support from the Science Research Council and Oriel College, Oxford is acknowledged.

References

[1] A. R. LANG, J. appl. Phys. **29**, 597 (1958).
[2] A. R. LANG, Acta cryst. **12**, 249 (1959).
[3] P. PENNING and A. H. GOEMANS, Phil. Mag. **18**, 297 (1968).
[4] A. R. LANG, Modern Diffraction and Imaging Techniques in Material Science, North-Holland Publ. Co., 1970 (p. 407).
[5] S. TAKAGI, J. Phys. Soc. Japan **26**, 1239 (1969).
[6] W. H. ZACHARIASEN, Theory of X-Ray Diffraction in Crystals, Wiley, 1945.
[7] A. AUTHIER, Adv. X-Ray Analysis **10**, 9 (1967).

(Received December 28, 1971)

phys. stat. sol. (a) **10**, 387 (1972)

Subject classification: 1.5 and 3; 14.3.1; 16; 22.4.1

Union Carbide European Research Associates, S.A., Bruxelles

La recristallisation des films évaporés de sulfure de cadmium dopés à l'argent

Par

C. HÉRINCKX[1]), W. DE SUTTER[2]), A. FOURDEUX[3]) et N. TERAO[3])

La recristallisation des couches de sulfure de cadmium évaporées, qui présente les caractéristiques d'une recristallisation secondaire, est fortement influencée par la présence de quantités minimes d'impuretés à des concentrations de dopage du semi-conducteur (4×10^{-3} at%). Le rôle activateur ou inhibiteur de ces impuretés correspond au rôle accepteur ou donneur des centres qu'elles créent. La recristallisation ne dépend pas de la formation d'une seconde phase riche en impuretés. Des micrographies électroniques du front de recristallisation ont été obtenues et montrent que la région recristallisée ne contient pas d'inhomogénéités mais bien quelques sous-grains légèrement désorientés. Une explication est avancée d'après laquelle le moteur de la recristallisation serait l'augmentation d'énergie de surface des grains due à la recombinaison des lacunes de soufre ionisées et des impuretés acceptrices.

The recrystallization of evaporated cadmium sulphide films, which has the salient features of a secondary recrystallization, is influenced by the presence of minute amounts of impurities of the same order of magnitude as the doping levels of monocrystalline CdS ($4 \times \times 10^{-3}$ at%). The role of the impurities as activators or inhibitors of recrystallization is related to the type of ionized centres created (acceptor or donor centres). The recrystallization does not depend on the segregation of a second phase containing the impurities. Transmission electron micrographs of the recrystallization front show that the recrystallized region is free of inhomogeneities but contains some slightly disoriented subgrains. An explanation is presented which attributes the driving force of the recrystallization to an increase of the surface energy of the grains due to the surface recombination of ionized sulphur vacancies and acceptor impurities.

1. Introduction

Les films évaporés de sulfure de cadmium se présentent sous la forme de microcristaux dont la composition, qui s'écarte généralement du rapport stoechiométrique, varie en fonction des conditions de préparation.

La mobilité des porteurs de charge dans ces films est considérablement réduite par rapport à celle des monocristaux en raison de leur structure désordonnée et des nombreux joints de grains qui diffusent les électrons. Cette réduction de la mobilité constitue un inconvénient majeur pour les applications des films évaporés de CdS dans les domaines de la photoconductivité et de la conversion d'énergie lumineuse en énergie électrique.

Plusieurs méthodes destinées à accroître, grâce à divers traitements suivant l'évaporation, la taille des cristaux dans les couches de CdS ont été décrites et étudiées par divers auteurs [1 à 4]. Toutes consistent à induire un grossissement

[1]) Université de l'Etat à Mons, Faculté des Sciences, 7000 Mons, Belgique.
[2]) Vrije Universiteit Brussel, 1640 St. Genesius-Rode, Belgique.
[3]) Université Catholique de Louvain, Unité de métallurgie physique, 3030 Heverlee, Belgique.

des grains par un traitement thermique, généralement précédé d'une addition d'impuretés bien choisies (argent ou cuivre). Par contre, le rôle inhibiteur de la recristallisation joué par un trop grand excès d'un des constituants (S ou Cd) a été observé ainsi que l'influence de certaines impuretés (oxygène, indium) qui réduisent la taille des domaines cristallins obtenus après le traitement thermique [2, 3].

Kahle et Berger [4] ont rangé le phénomène de recristallisation des couches de CdS dans la catégorie des recristallisations secondaires, ce qui est incontestablement correct mais ne fournit pas une explication détaillée du mécanisme de cette recristallisation. En particulier le rôle des impuretés activatrices ou inhibitrices de la croissance cristalline n'est pas complètement éclairci. Le but du présent travail est d'apporter quelques résultats expérimentaux et des interprétations de nature à préciser davantage l'effet du dopage à l'argent sur la recristallisation secondaire du sulfure de cadmium. Ces interprétations seront basées sur un dosage précis de l'argent, ainsi que des mesures électriques et des observations de la texture et de la structure des couches par rayons X et microscopie électronique.

2. Préparation

Le sulfure de cadmium utilisé pour la préparation des couches est un produit contenant quelques p.p.m. d'impuretés qui est dégazé à chaud avant utilisation. L'évaporation est faite sous vide à partir d'un creuset d'alumine chauffé à 800 °C par rayonnement dans une enceinte en quartz. Cette enceinte de forme cylindrique est presque complètement obturée dans sa région centrale par une pièce en graphite à laquelle sont fixées les pastilles (d'un diamètre de 14 mm) servant de substrats pour les couches. Ces pastilles sont en verre dont le coefficient de dilatation thermique est adapté à celui du CdS. Elles sont protégées pendant la période de chauffage du creuset par un masque en graphite solidaire du porte-substrat et actionné de l'extérieur de l'enceinte à vide par un système magnétique. Le porte-substrat est chauffé par rayonnement à l'aide d'un four situé à l'extérieur de l'enceinte en quartz et dont la température peut varier entre l'ambiante et 400 °C.

Pendant l'évaporation, un équilibre dynamique s'établit entre la pression de vapeur du CdS dans la région de l'enceinte comprise entre le creuset et le porte-substrat et la pression (10^{-6} Torr) maintenue par le groupe de pompage au-dessus du porte-substrat. La pression d'équilibre établie de la sorte maintient constante la composition de la vapeur de CdS et assure la condensation au niveau des substrats d'une couche de 10 μm en un temps variant entre quelques minutes et une heure.

3. Expériences de recristallisation

3.1 Introduction

Dans toutes les expériences décrites dans cette section un activateur de la recristallisation a été utilisé. L'agent activateur a été déposé sur les couches de CdS par évaporation sous vide et le phénomène de recristallisation a été suivi au cours du traitement thermique des échantillons suivant le procédé classique [1 à 4] d'observation microscopique entre polariseurs croisés.

Sauf mention contraire, c'est l'argent qui a été utilisé généralement comme activateur.

Les couches activées de CdS, chauffées dans une atmosphère inerte (argon sec, azote purifié), recristallisent à une température comprise entre 350 et 400 °C. Lorsque toute la surface d'un échantillon n'est pas recouverte d'activateur, la recristallisation ne se produit que dans la région recouverte (tout au moins aux températures de recuit utilisées ici, qui ne dépassent jamais 550 °C). Lorsque la surface du film de CdS présente des défauts (griffes ou inclusions d'impuretés) il arrive que ceux-ci servent de centres de nucléation de la recristallisation [2], mais en l'absence de tels défauts ces centres se trouvent toujours à la périphérie du dépôt d'activateur.

Les deux types habituels de croissance cristalline en présence d'activateur ont été observés, c'est-à-dire des domaines dans lesquels l'orientation de l'axe *c* varie d'un point à l'autre et dont la forme pennée rappelle les sphérulites, ou des grands grains plats contenant de nombreux sous-grains et à l'intérieur desquels l'orientation de l'axe *c* est à peu près constante [2, 4]. La morphologie des couches recristallisées dépend des conditions de l'expérience (mode de préparation, type et quantité d'activateur, traitement thermique, etc.). Les effets de ces différents facteurs ont été étudiés et sont décrits ci-dessous.

3.2 Expériences

3.2.1 Température de la source

Une série de couches ont été préparées en faisant varier la température de la source de vapeur (creuset) et en maintenant constante la température des substrats (200 °C). Les temps d'évaporation ont été ajustés de manière à condenser des épaisseurs voisines de 10 μm (ces temps variaient de deux heures à 700 °C à cinq minutes à 1000 °C). Après l'addition d'activateur, toutes les couches se sont recristallisées entre 350 et 400 °C sans présenter de différences appréciables quant à la vitesse de recristallisation ou la morphologie des échantillons. Dans le mode d'évaporation en enceinte quasi étanche utilisé ici, la température de la source n'affecte donc que la dynamique de la condensation du CdS.

3.2.2 Température du substrat

La source étant portée à 800 °C, les substrats ont été maintenus pendant l'évaporation du CdS et au cours d'expériences successives à 25, 100, 200 et 300 °C. Un effet marqué de la température du substrat sur la stoechiométrie des couches de CdS a été observé. Il était rendu visible par la couleur des dépôts passant du brun foncé au jaune clair. Dans les couches évaporées sur des supports à 25 et 100°C la recristallisation n'a pas pu être déclenchée en raison de l'excès de cadmium. Pour des températures de substrat de 200 et 300°C, la couche était de teinte orange et est passée au jaune foncé caractéristique des monocristaux de CdS au cours de la recristallisation qui s'est faite sans difficulté. Un essai d'évaporation a été mené à 400°C, mais dans ce cas la vitesse de condensation du CdS était très faible et la recristallisation n'a pu être obtenue.

3.2.3 Ecarts à la stoechiométrie

Afin de mieux cerner l'influence de la composition, les couches évaporées de CdS furent placées sous vide en ampoules de quartz scellées et furent portées à 500 °C pendant 48 h en présence de vapeur saturante d'un des constituants. Après un tel traitement dans la vapeur de soufre pur (pression saturante 1,5 atm), les couches devinrent jaune clair et la recristallisation fut totalement inhibée.

En présence de vapeur saturante de cadmium (0,013 atm) les couches conservèrent leur teinte orange et la recristallisation se produisit, donnant naissance à des cristaux un peu plus désordonnés qu'en l'absence de traitement. Au cours de cette expérience, la composition des couches s'est vraisemblablement fort peu modifiée.

Une troisième série d'échantillons fut traitée de la même façon sous vide, mais sans addition dans l'ampoule d'un des constituants. Les couches devinrent jaune clair et la recristallisation fut inhibée. Ce résultat s'explique par une perte de cadmium qui confère aux échantillons recuits sous vide des propriétés analogues à celles obtenues sous pression de soufre. Un excès de Cd est donc nécessaire à la recristallisation.

3.2.4 Activateurs et inhibiteurs — Effets de compensation

On sait [1 à 3, 5] que le cuivre peut servir d'activateur. Cette propriété a été confirmée par nos expériences mais les cristaux furent plus désordonnés que ceux activés à l'argent. Le plomb n'active pas la recristallisation, mais ne l'inhibe pas: après un recuit à 600 °C en présence d'un excès de plomb évaporé sur le CdS, une activation à l'argent a joué son rôle normal et les couches ont pu être recristallisées.

L'indium, par contre, est un inhibiteur: des échantillons où l'on avait dissous une quantité d'indium supérieure à 0,15 at% n'ont pu être recristallisés, quelle que soit la quantité d'argent utilisée comme activateur. Pour des concentrations d'indium de l'ordre de 0,08 at%, l'effet inhibiteur de cette impureté préalablement dissoute dans les échantillons a pu être compensé par l'adjonction d'une quantité saturante d'argent qui a permis la recristallisation dans les conditions normales. Dans une série d'échantillons où l'on avait fait diffuser environ 0,03 at% d'indium, l'adjonction d'une quantité équivalente d'argent (en at%) n'a pas permis la recristallisation, alors que l'évaporation ultérieure, sur les mêmes échantillons, d'une nouvelle dose de 0,03 at% d'argent a assuré une recristallisation normale. On voit ici les rôles opposés de l'indium et de l'argent qui, à des concentrations faibles, se compensent mutuellement.

3.2.5 Concentrations efficaces d'activateur

On sait déjà que jusqu'à la limite de solubilité de l'argent dans le CdS, ce métal se comporte comme un activateur. D'autre part, puisqu'il conserve ce rôle à des concentrations très faibles, il était utile de rechercher le minimum efficace d'argent permettant la recristallisation. Une quantité de 5 mg d'argent a donc été évaporée, au travers d'un masque tournant, sur un support rectangulaire en verre situé à 20 cm de la source d'argent et préalablement recouvert d'un film de 6 µm de CdS. Le masque tournant, placé très près de la surface du CdS, était taillé de manière à assurer une décroissance logarithmique de la quantité d'argent déposée le long de l'échantillon. Après recuit à 400 °C pendant une heure, la limite de la zone recristallisée a été repérée et correspondait à 0,004 at% d'Ag. Cette concentration ne représentant que moins d'un Å d'Ag en moyenne déposé sur la surface du CdS, une expérience de contrôle a été réalisée en activant une couche de CdS avec la concentration minimum efficace ($\approx$ 0,004 at%). Un film de 6 µm de CdS a été placé à 20 cm d'une source contenant 0,5 mg d'Ag. Après évaporation de l'Ag, la couche de CdS a été chauffée et la recristallisation a commencé à 525 °C, en un point de l'échantillon, et s'est propagée lentement, transformant en 150 min toute la surface de l'échantillon en un seul domaine à structure „pennée".

3.2.6 Diffusion de l'activateur

Sur une couche de CdS a été évaporée une quantité sursaturante d'Ag au travers d'un masque dégageant une zone circulaire au centre de l'échantillon. Après chauffage à 350 °C pendant 30 min, le CdS situé sous le dépôt d'Ag était recristallisé sous la forme „pennée". Poursuivant alors le traitement thermique à 500 °C, une lente progression du front de recristallisation a été observée, à une vitesse de l'ordre de 40 µm/h. Cette progression radiale est évidemment liée à la diffusion de l'activateur parallèlement au plan de la couche. Il est à remarquer (Fig. 1) qu'il n'y a pas de corrélation entre les orientations des axes *c* des cristaux formés au cours de cette diffusion de l'argent et celles des domaines initiaux. Ces cristaux affectent d'ailleurs l'autre forme de recristallisation des couches de CdS (celle en grains plats) avec une certaine périodicité radiale. Ils forment des domaines annulaires d'une largeur de l'ordre de 0,2 mm qui se développent tangentiellement, c'est-à-dire perpendiculairement aux rayons des anneaux. Ce mode de croissance tangentielle, qui a été reproduit maintes fois au cours de ces expériences de diffusion d'activateur, est typique de la recristallisation en grains plats. Son origine mérite d'être étudiée plus à fond. Il est vraisemblable que cette croissance tangentielle est liée à la diffusion relativement rapide des lacunes de soufre qui précède la diffusion radiale, plus lente, des atomes d'argent et crée des accumulations de défauts qui s'opposent à la croissance cristalline dans la direction perpendiculaire au front de croissance.

3.2.7 Epaisseur critique

Des couches d'épaisseurs variables ont été recouvertes d'une quantité sursaturante d'Ag puis recuites à 500 °C dans l'azote pour provoquer la recristallisation. Lorsque leur épaisseur était inférieure à 1 µm, les couches de CdS ne recristallisaient pas. Lorsqu'elle dépassait 2 µm, elles recristallisaient toujours. Entre ces deux limites, la recristallisation était aléatoire. Par un traitement de recuit progressif, on peut accroître de manière très appréciable les dimensions des domaines cristallins et réduire le nombre de sous-grains.

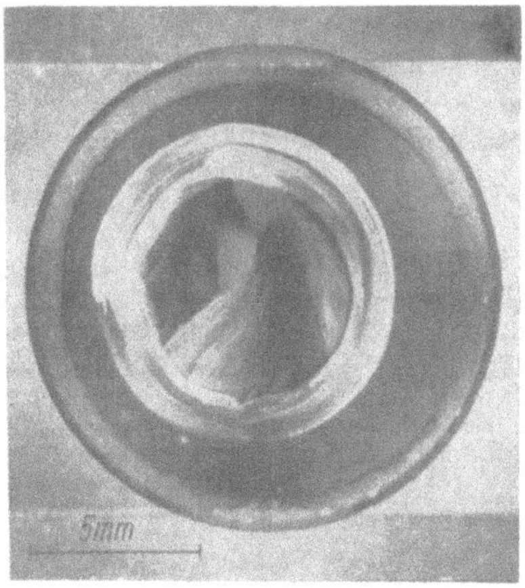

Fig. 1. Progression de la recristallisation d'une couche de CdS par diffusion à l'activateur (Ag) à la périphérie de la zone activée

Par ailleurs, des estimations de la concentration minimum efficace d'activateur ont montré que cette concentration augmente lorsque l'épaisseur des couches décroît. La mesure de la concentration minimum efficace d'activateur dans les couches de 2 ou 3 μm d'épaisseur est toutefois très délicate et les résultats relativement imprécis.

3.3 Morphologie et structure

3.3.1 *Diffraction des rayons X et des électrons*

Les couches de CdS, évaporées sur des substrats à 200 °C, qui ont été étudiées dans le présent travail ont toutes une texture analogue à celle décrite dans la littérature [3, 5] (microcristaux de 0,3 à 1 μm de largeur, dont les axes *c* sont perpendiculaires au support). Cette texture a été confirmée par diffraction de rayons X et par diffraction d'électrons sur des échantillons amincis en faisant varier l'inclinaison de la couche par rapport au faisceau d'électrons.

3.3.2 *Microscopie électronique*

Les épaisseurs des couches étudiées ici étant comprises entre 2 et 10 μm, il a fallu amincir les échantillons afin de pouvoir les examiner par transmission à l'aide d'un microscope travaillant sous 100 kV. La technique d'amincissement a été décrite en détail par ailleurs [6] et consiste en une série de polissages mécaniques et électrolytiques effectués sur les deux faces des couches détachées de leur support. Les échantillons observés sont donc situés à peu près à mi-épaisseur des couches.

La Fig. 2 montre l'aspect des couches évaporées, avant l'activation à l'argent. Elles sont constituées de petits cristaux bien définis dont la taille moyenne, de 0,5 à 1 μm, croît avec l'épaisseur de la couche. A l'intérieur des grains, on observe de nombreuses dislocations et quelques parois de polygonisation. La structure des couches est fibreuse, l'axe *c* des petits cristaux étant quasi perpendiculaire à la surface des échantillons.

Afin d'observer l'effet de l'addition d'activateur sur la structure des couches avant la recristallisation, un dépôt d'argent a été évaporé sur la moitié de la surface d'un échantillon, l'autre moitié étant mise à l'abri des impuretés. La Fig. 3 montre l'aspect de la moitié activée de l'échantillon après chauffage dans

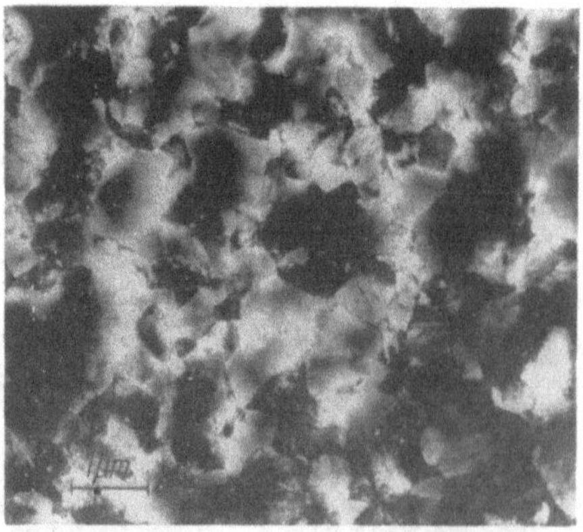

Fig. 2. Aspect d'une couche amincie de CdS telle qu'évaporée — épaisseur de l'échantillon de départ 4,4 μm

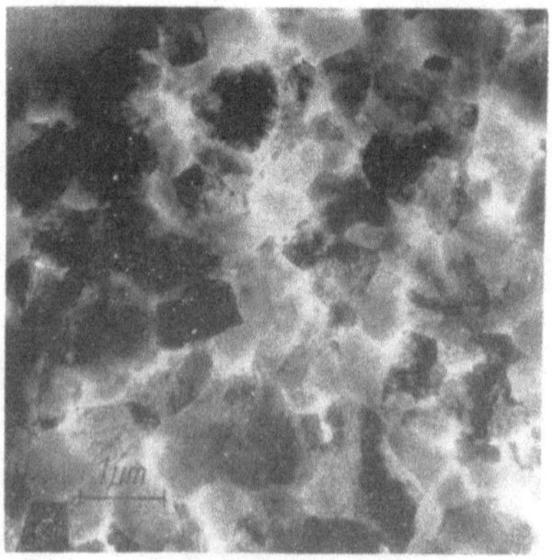

Fig. 3. Même échantillon que celui de la Fig. 2 sur lequel un excès d'argent a été évaporé, puis diffusé pendant 4 h à 300 °C sous azote

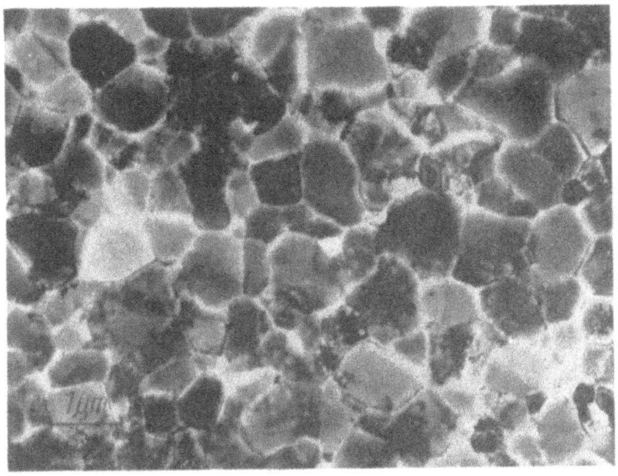

Fig. 4. Même échantillon que celui des Fig. 2 et 3 après chauffage sous azote pendant 4 h à 300 °C

l'azote pendant 4 h à 300 °C. A la suite de ce traitement, l'argent a diffusé dans la couche, mais aucun précipité d'argent ou d'un composé d'argent n'a été observé en microscopie par transmission. La structure révélée par la Fig. 3 est voisine de celle de la Fig. 2 (sans Ag) et présente la même dispersion de la taille des grains. Il n'y a donc pas eu de migration des joints de grains, leur mouvement étant vraisemblablement inhibé par la présence des impuretés (Ag) dissoutes.

Par contre, la moitié non activée du même échantillon, qui a évidemment subi le même traitement thermique que la région de la Fig. 3, présente une taille des grains beaucoup plus uniforme (Fig. 4) indiquant qu'un mouvement général des joints a eu lieu.

Des couches sur lesquelles un excès d'Ag a été évaporé ont été examinées après recristallisation partielle. La structure des régions non recristallisée et recristallisée d'un même échantillon a ainsi pu être observée, ainsi que la morphologie du front de recristallisation et de son voisinage immédiat.

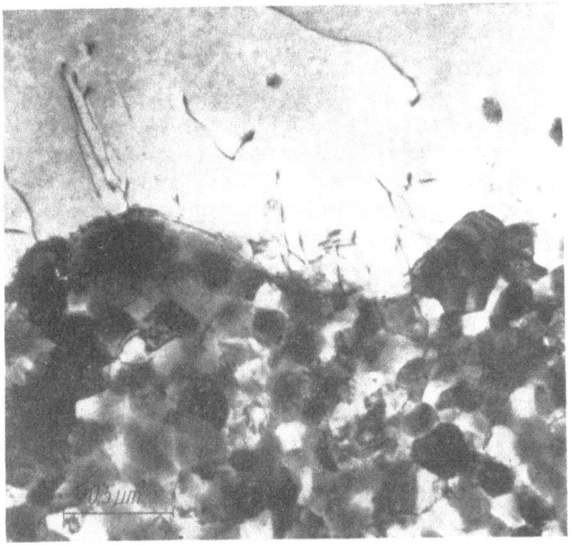

Fig. 5. Aspect du front de recristallisation d'une couche de CdS de 3,5 µm d'épaisseur sur laquelle un excès d'argent a été déposé et qui a été recuite dans l'azote à 400 °C

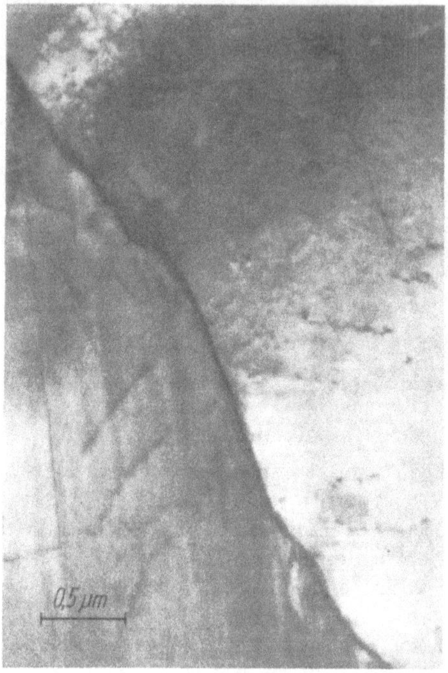

Fig. 6. Micrographie électronique par transmission montrant un sous-joint de polygonisation, des dislocations et des défauts d'empilement dans une région recristallisée d'une couche de CdS dopée à l'argent

La Fig. 5 illustre cette observation et montre que l'aspect du front de recristallisation est caractéristique d'une recristallisation secondaire (développement d'un très grand grain au détriment d'un ensemble de microcristaux présentant une texture). Aucun précipité d'argent ou de composé d'argent n'a été observé dans le volume des échantillons prélevés dans les régions recristallisée et non recristallisée. Cette dernière présente une structure en petits cristaux avec une distribution non uniforme des grains, voisine de celle obtenue avant activation. La région recristallisée est constituée de grands cristaux d'une taille allant de plusieurs dizaines de microns à quelques millimètres. Ces grands cristaux renferment de nombreux sous joints de polygonisation ainsi que quelques dislocations et défauts d'empilement (Fig. 6).

4. Conductibilité électrique et rôle des impuretés

Les mesures des propriétés électriques du CdS ont généralement pour but d'évaluer la mobilité des porteurs de charge ainsi que la densité et la répartition des niveaux donneurs et accepteurs. Dans les couches évaporées, la mobilité des électrons (porteurs majoritaires) est fortement influencée par la structure du film de CdS, la grosseur des grains et surtout leur orientation cristallographique par rapport au support [5].

Les mesures électriques rapportées ci-dessous fournissent des informations sur le processus de diffusion de l'argent précédant la recristallisation, ainsi que sur le rôle de ce métal en tant qu'agent dopant du CdS microcristallin. Quelques tentatives ont également été faites pour observer les variations de conductibilité au cours de la recristallisation, mais toutes ont échoué en raison de la destruction des contacts électriques due aux déplacements de matière au cours du grossissement des grains.

Ces mesures ont été faites sur des couches évaporées dans les conditions décrites ci-dessus (c'est-à-dire à 800 °C sur des supports chauffés à 200 °C). Afin d'éviter les effets des impuretés provenant de l'atmosphère ambiante (notamment l'oxygène), les conductibilités électriques ont été mesurées sous vide (10^{-6} Torr) dans l'appareil d'évaporation de l'argent et juste après la dite évaporation. Pour assurer les contacts électriques, des essais ont été réalisés à l'aide de divers matériaux (or, platine, carbone, molybdène). L'indium, souvent utilisé pour établir des contacts ohmiques sur le CdS, était exclu en raison de sa diffusion aisée dans les couches et de son rôle d'impureté donatrice qui l'oppose à celui de l'argent. Les meilleurs résultats ont été obtenus à l'aide d'électrodes faites de platine, déposé par pulvérisation cathodique sur le support avant l'évaporation des films de CdS, dont l'épaisseur était comprise entre 3 et 7 μm.

Les mesures des conductibilités électriques dans le plan des couches ($\sigma_{||}$) ont été faites au travers d'un interstice de 20 μm entre les électrodes de platine. L'argent était déposé sur le CdS sous forme d'une couche très mince, mesurée par le glissement de fréquence d'oscillation d'une lame de quartz, et correspondant à la quantité d'argent tout juste suffisante pour induire la recristallisation du CdS.

Des mesures de la conductibilité dans la direction perpendiculaire au plan des couches ($\sigma_{\perp}$) ont également été réalisées. Dans ce cas, l'écart entre les électrodes de platine était de l'ordre de 1 mm et une couche d'argent épaisse (plusieurs μm) évaporée sur le CdS servait à la fois d'agent dopant et de contre-électrode. L'avantage d'une telle disposition réside dans sa symétrie, la surface de la couche de CdS étant partagée en deux moitiés mises en série par l'intermédiaire de la couche d'argent et parcourues en sens inverse par le courant.

Quoique les propriétés électriques des films évaporés de CdS varient de plusieurs ordres de grandeur en fonction des conditions de préparation, et notamment des écarts à la stoechiométrie résultant du processus d'évaporation–condensation [5], les lots d'échantillons étudiés ici étaient relativement homogènes, avec une conductibilité à 20 °C de l'ordre de 10^{-3} Ω^{-1} cm^{-1}.

Avant la diffusion de l'argent la photoconductivité est assez faible et les deux composantes $\sigma_{||}$ et $\sigma_{\perp}$ sont pratiquement égales. La conductibilité électrique des couches étudiées ici est donc isotrope. Pourtant la distribution des axes c des microcristaux est très anisotrope, leur dispersion autour de la normale à la couche n'étant que de quelques degrés (cf. plus haut). On ne devrait donc pas s'attendre à un effet considérable sur la conductibilité d'un changement d'orientation des microcristaux par rapport au support. Or, un accroissement de mobilité des électrons a été observé [5] lors de la réorientation — avant grossissement important — des grains dans les couches de CdS traitées à 250 °C.

Cet accroissement de mobilité doit, semble-t-il, être imputé à une meilleure coordination des orientations respectives des grains voisins ou à un changement de structure ou de composition des joints entre microcristaux.

Des tentatives ont été faites pour déterminer la mobilité de Hall des couches utilisées pour les mesures de σ. Dans l'obscurité, l'effet Hall était noyé dans un bruit de fond relativement considérable provenant des fluctuations locales de densité de porteurs au voisinage des défauts de réseau très nombreux dans les couches de CdS. La mobilité y étant faible, l'équilibre thermodynamique entre les pièges à électrons et la bande de conduction n'y est pas constamment assuré et des charges spatiales s'y créent et s'y annulent en un processus aléatoire. Un éclairage intense permet d'améliorer le rapport signal/bruit, et dans ces conditions une mobilité de 0,2 à 1 cm^2/Vs a été mesurée pour les couches non recristallisées. Après recristallisation ces couches présentaient des mobilités 10 à 50 fois plus élevées mais fort dépendantes de l'illumination.[4])

La diffusion de l'Ag a pu être suivie par des mesures de conductibilité au cours du traitement thermique effectué par paliers de 25 en 25 degrés. Jusqu'à

---

[4]) Remarquons qu'on peut obtenir des valeurs de mobilité de Hall plus reproductibles sur des couches présentant un écart à la stoechiométrie (excès de Cd) qui les rend plus conductrices mais perturbe la diffusion de l'Ag et la recristallisation.

150 °C la résistance dans l'obscurité des échantillons recouverts d'Ag évaporé ne change pratiquement pas. Ce faible coefficient de température indique que la densité de porteurs „intrinsèques" varie fort peu avec T. Par contre, à 175 °C $\sigma_{||}$ diminue lentement avec une constante de temps de l'ordre de 7 à 8 h. Cette évolution de la résistance est d'autant plus rapide que la température est élevée et provient de la diffusion de l'argent dans la couche de CdS. A 250 °C, une constante de temps de l'ordre de 2 h a été observée. La valeur d'équilibre de $\sigma_{||}$ est dix fois plus faible que la conductibilité initiale, ce qui confirme le rôle d'accepteur de l'argent dans le CdS, semi-conducteur de type n. Remarquons que pour les échantillons servant à l'observation des variations de $\sigma_{||}$, la quantité minimum d'argent provoquant la recristallisation a été utilisée, ce qui correspond à 8×10^{17} atomes d'argent par cm^3. Par ailleurs, la densité initiale d'électrons libres mesurée par effet Hall sous illumination intense dans les couches non dopées est de l'ordre de $10^{16}/cm^3$. Il y a donc compensation des centres ionisés jouant le rôle d'accepteurs (Ag) constitués par des atomes d'argent en position réticulaire de cadmium et un nombre quasi équivalent de centres donneurs (lacunes de soufre ionisées V_S^+) suivant un mécanisme bien connu dans le cas des composés semi-conducteurs IIB–VIA [7].

D'autres mesures, faites dans la direction transverse, n'ont montré aucun effet de l'argent en dessous de 250 °C, et à cette température une diminution de $\sigma_{\perp}$ plus lente et moins marquée que celle de $\sigma_{||}$. La valeur d'équilibre de $\sigma_{\perp}$ est deux fois plus faible que la conductibilité initiale. Aucune explication n'a été trouvée pour cette différence de propriétés parallèles et transverses. Il faut toutefois remarquer que $\sigma_{\perp}$ est mesurée avec un très large excès d'argent qui donne à la fin du processus de diffusion du métal un CdS saturé en argent. La concentration en métal (Cu, Ag, Au) correspondant à la saturation des monocristaux de CdS a été déterminée par Dreeben [8] qui l'évalue au maximum à 2×10^{-2}% en poids, ce qui correspond dans le cas de l'argent à $2{,}7 \times 10^{-2}$ at%, soit $5{,}4 \times 10^{18}$ atomes d'argent par cm^3. La compensation des centres ionisés accepteurs et donneurs dans le cas des échantillons utilisés pour les mesures de $\sigma_{\perp}$ se fait donc à une autre échelle que dans le cas des mesures de $\sigma_{||}$.

A des températures dépassant 300 °C, la diffusion de l'argent s'opère en quelques minutes et le phénomène n'a pu être suivi convenablement par des mesures électriques en raison de l'inertie thermique du support de l'échantillon et des défauts de contact thermique dûs au fait que les couches de CdS sont placées sous vide. Il est à noter que les mesures électriques indiquent que la diffusion de l'argent dans le volume de la couche précède toujours la recristallisation de celle-ci.

5. Discussion

La recristallisation des couches évaporées de sulfure de cadmium est un phénomène favorisé par la présence de certaines impuretés (Cu, Ag). Depuis la découverte par Gilles et Van Cakenberghe [1] de cette recristallisation en grands domaines, le rôle activateur de ces impuretés a été attribué à une diminution de l'énergie d'activation du processus de croissance cristalline [3, 4].

Indépendamment du fait que l'introduction d'impuretés a généralement pour effet de bloquer au contraire le développement cristallin, on voit mal comment des domaines recristallisés très étendus (de l'ordre de plusieurs dizaines de mm^2) pourraient se développer dans des couches où une énergie d'activation réduite par la présence des impuretés devrait multiplier les centres de nucléation. Ces centres sont au contraire moins nombreux dans les couches activées que dans celles qui ne le sont pas [4].

Addis résout cette contradiction en supposant la formation d'une phase riche en argent où les atomes seraient plus mobiles que dans le CdS sans impuretés et par conséquent l'énergie d'activation de la recristallisation plus faible. Par diffusion des atomes d'argent dans les microcristaux, cette seconde phase serait concentrée au niveau du front de recristallisation. La progression de ce front s'accompagnerait du dépôt d'un précipité d'argent dans la zone recristallisée qui contiendrait donc des inclusions de métal localisées préférentiellement aux joints de grains.

Cette hypothèse d'un mouvement des atomes d'argent qui se déplaceraient en même temps que le front de recristallisation suppose que les atomes de métal acquièrent au voisinage de ce front une vitesse de diffusion considérable. En effet, au cours de la recristallisation à 400 °C, le front se déplace à une vitesse de l'ordre de 1 cm/h, alors que la vitesse de diffusion du cuivre (fort semblable à celle de l'argent) mesurée par Somorjai et Lyon [9] dans les cristaux de CdS est, à 700 °C, de l'ordre de 0,05 cm/h. D'autre part, l'expérience décrite ci-dessus de recristallisation à la périphérie d'une zone recouverte d'argent en excès montrait une vitesse de progression du front, à 500 °C, de 0,004 cm/h. Dans cette expérience, le déplacement du front était conditionné par la vitesse de diffusion de l'argent dans la couche de CdS, qui est donc un processus plusieurs centaines de fois plus lent que la recristallisation. Quoique l'on ne puisse pas écarter l'hypothèse d'un accroissement de la mobilité des atomes de métal parallèlement au front de croissance et dans une petite zone entourant celui-ci, il faut pour le moins considérer avec circonspection la possibilité d'une augmentation de cette mobilité de trois ordres de grandeur.

Quoi qu'il en soit, les expériences décrites dans les sections précédentes montrent que la formation d'un précipité consécutif à l'introduction d'impuretés ne joue pas de rôle déterminant dans le processus de recristallisation. En effet, la recristallisation peut être induite par l'addition de quantités minimes d'argent (ou de cuivre), dans des proportions six fois moindres que celles qui correspondent à la limite de solubilité du métal dans le CdS. De plus, les micrographies électroniques ne montrent dans ce cas aucune phase secondaire au niveau du front de recristallisation ni aucun précipité dans la zone recristallisée.

L'argent et l'indium dissous dans les couches de CdS dans de faibles concentrations jouent des rôles inverses en tant qu'agents dopants du semi-conducteur, et en même temps comme activateur et inhibiteur de la recristallisation. En généralisant quelque peu, on peut déduire de ceci que l'origine du rôle des impuretés dans le mécanisme de recristallisation est électronique et non pas structurel. Une impureté acceptrice active la recristallisation, tandis qu'une impureté donatrice l'inhibe. Ces deux effets se compensent mutuellement dans les limites de la solubilité de ces impuretés dans le CdS monocristallin. Toute addition d'activateur en concentration plus élevée qu'un optimum situé entre le minimum révélé par les expériences décrites ci-dessus (soit 4×10^{-3} at%) et la limite de solubilité (soit $2{,}7 \times 10^{-2}$ at% dans les cristaux) a pour effet d'augmenter l'énergie d'activation de la recristallisation. Cela explique que dans des couches exemptes de griffes ou autres défauts, la germination et la croissance des grains commencent toujours à la périphérie d'un dépôt abondant de métal, dans la région où le dopage optimum est assuré. Le précipité observé par certains auteurs [5, 10] dans les régions où a été évaporé un large excès de métal est vraisemblablement analogue aux inclusions observées dans les monocristaux de CdS [8]. Dans ces grands monocristaux, des inclusions décorent les défauts de réseau lorsque la limite de solubilité des impuretés (Ag, Cu, Au) est dépassée ou même lorsque ces monocristaux contiennent un trop large excès d'un des constituants (Cd).

Le fait que l'épaisseur de la couche de CdS exerce une influence sur la quantité relative d'activateur (plus élevée pour les couches d'épaisseur inférieure à 4 μm) et sur la cinétique de la recristallisation traduit l'importance des états de surface, qui modifient l'équilibre des charges spatiales et compensent les centres accepteurs. Pour des épaisseurs inférieures à 1 μm, la recristallisation est totalement inhibée.

Ayant reconnu le rôle électronique des impuretés, il reste à comprendre quel est le moteur de la recristallisation et le mécanisme de celle-ci.

Kahle et Berger [4] ont mis en évidence le fait qu'il s'agit d'une recristallisation secondaire dont des exemples sont bien connus dans le cas des métaux en feuilles minces [11]. Ce qui constitue toutefois l'originalité du processus étudié ici c'est le fait qu'il soit activé par l'adjonction d'impuretés acceptrices. Si dans cette recristallisation secondaire une modification de l'énergie d'activation ne constitue pas l'effet majeur des impuretés, il faut rechercher celui-ci dans une modification de l'énergie libre de surface des grains qui composent la couche évaporée de CdS.

Les couches de sulfure de cadmium obtenues par évaporation du composé stoechiométrique étant toujours de type n, elles contiennent une densité appréciable de lacunes de soufre ionisées V_S^+, en plus éventuellement d'un certain nombre d'atomes de cadmium en position interstitielle qui sont susceptibles de jouer le rôle d'impuretés donatrices. Etant donné la grande densité de défauts structurels représentés par les joints de grains, on peut s'attendre à ce que les lacunes de soufre se trouvent en plus grand nombre à la surface des grains où la concentration des porteurs libres est donc assez élevée. Lors de l'introduction d'impuretés acceptrices (Ag^-), celles-ci se recombinent avec les lacunes V_S^+ à la surface des grains, où leur densité est suffisamment grande pour que cette recombinaison directe soit possible. Le résultat est une diminution très nette de la densité de porteurs libres à la périphérie des grains, et par conséquent une augmentation de l'énergie de surface de ces grains. Remarquons ici qu'une telle augmentation d'énergie résultant d'une recombinaison en surface a été observée au cours du dopage au cuivre de monocristaux de CdS, avec pour conséquence une diminution de la tension de vapeur de ces cristaux [9].

Cette augmentation d'énergie de surface bloque la recristallisation primaire normalement observée pour les couches de CdS non dopées et accroît d'autant le gain d'énergie résultant de la recristallisation secondaire. C'est donc probablement l'accroissement d'énergie libre de surface des grains qui sert de gâchette au mécanisme de la recristallisation secondaire, conférant à celle-ci sa soudaineté et limitant sa nucléation à un nombre extrêmement réduit de centres distincts.

Remerciements

Les auteurs sont redevables à Mme. G. Heckmus de la préparation des échantillons amincis pour la microscopie électronique et des clichés, et à Mr. W. Breemans de la réalisation des mesures électriques. Ils remercient le Dr. H. Hatwell pour les nombreuses et fructueuses discussions concernant la matière du présent travail.

Bibliographie

[1] J. M. Gilles et J. Van Cakenberghe, Nature (London) **182**, 862 (1958); Solid State Phys. in Electronics and Telecommunications, Vol. 2, Part 2, Academic Press, New York 1960 (p. 900).
[2] A. Vecht et A. Apling, phys. stat. sol. **3**, 1238 (1963).
[3] R. R. Addiss, Trans. X. Nat. Vacuum Symp., MacMillan Co., New York 1963 (p. 354).
[4] W. Kahle et H. Berger, phys. stat. sol. (a) **2**, 717 (1970).
[5] J. Dresner et F. V. Shallcross, Solid State Electronics **5**, 205 (1962); J. appl. Phys. **34**, 2390 (1963).
[6] A. Fourdeux, G. Heckmus et A. Malbrancke, Union Carbide E. R. A. Techn. Rep. 19/71.
[7] Voir par exemple F. A. Kröger, The Chemistry of Imperfect Crystals, North-Holland Publ. Co., Amsterdam 1964.
[8] A. Dreeben, J. Electrochem. Soc. **111**, 174 (1964).
[9] G. A. Somorjai et H. B. Lyon, J. chem. Phys. **43**, 1456 (1965)
[10] H. Hatwell, communication privée.
[11] Voir par exemple J. W. Christian, The Theory of Transformations in Metals and Alloys, Pergamon Press, 1965 (p. 737).

(Received December 1, 1971)

phys. stat. sol. (a) **10**, 401 (1972)

Subject classification: 19; 6; 18.1; 22.9

Laboratoire de Radioélectricité, Université de Rennes

Paramagnetic Relaxation of Dy^{3+} Diluted in YES Observation of the Direct Process

By

B. Hagene

Measurements of the relaxation time on a crystal of 15% Dy^{3+} in YES have been carried out with the magnetic field parallel to the crystal axis. An Orbach process is observed at the highest temperatures, but the experimental data recorded at low fields do not fit the Orbach relation which takes into account the magnetic interactions. At low temperatures a process is observed which in terms of temperature or magnetic field offers well the features of a direct process. The discussion of the experimental conditions shows that the misorientation accountable for this phenomenon can only be due to a lattice distortion around the paramagnetic ions.

Nous avons effectué des mesures du temps de relaxation de Dy^{3+} dilué à 15% dans un cristal de YES, le champ magnétique étant parallèle à l'axe du cristal. Nous observons un processus Orbach aux plus hautes températures, mais les points expérimentaux relevés en champ faible sont mal représentés par la relation d'Orbach qui tient cependant compte des interactions magnétiques. Il apparaît à basse température un mécanisme qui présente bien, en fonction de la température et du champ magnétique, les caractères d'un processus direct. La discussion des conditions expérimentales montre que le défaut d'orientation responsable de ce phénomène ne peut être dû qu'à une distorsion du réseau autour des ions paramagnétiques.

1. Introduction

Various authors tried to check the theoretical prevision of Orbach [1] about the non-existence of a direct relaxation process in dysprosium ethylsulfate when the magnetic field is parallel to the crystallographic axis. It is indeed possible to show [2] that the relaxation time of this salt is described by a general relation of the form

$$\tau_1^{-1} = 1.7\times 10^{-11}\, H^4\, T \sin^2\theta \cos^2\theta + \\ + 1.7\times 10^7\, e^{-23/T} + 1.6\times 10^7\, e^{-29/T} + 1.4\times 10^{-5}\, T^9 \quad (1)$$

the second Orbach term and the Raman term being negligible at about 2 °K.

There are, however, few contradictions between the results of the experimental works performed at $\theta = 0$ on concentrated salts. Cooke et al. [3] do not detect any direct process, opposite to Hudson and Mangum [4] although they work in very similar conditions; van den Broek and van der Marel [2] observe the direct process on polycrystalline powders, but a study on a single crystal leads to similar results; Kalbfleisch [5] detects only the Orbach process, but Kump [6] at lower temperatures finds a law of the form $\tau_l^{-1} \sim T^1$, notwithstanding an accurate orientation of the crystal.

We have taken up this problem by a study of Dy^{3+} ions diluted in yttrium ethylsulfate, assuming that the cross relaxation phenomena in the concentrated salt may disturb the measurements. First a short study has been carried out on a undiluted DyES sample as a comparison with previous results.

2. Experimental Procedures and Samples

The relaxation time is measured by the magneto-optic method described elsewhere [5, 7]. It is particularly suitable for measurements of τ_1 in dependence on the magnetic field, especially low fields; it is sensitive enough in the case of DyES to allow the study of diluted salts; lastly it is the method used by Kalbfleisch and Kump.

Our experiments have been made on two types of samples: (I) crystals of undiluted DyES obtained from 99.9% purity dysprosium oxide (main impurities: Ho_2O_3, Gd_2O_3, and Tb_4O_7), and a sample of the same quality kindly handed over by the Kammerlingh-Onnes Laboratory in Leiden, (II) crystals of DyES diluted to 15% in YES prepared from 99.99% yttrium oxide (main impurities: Tb, Dy). Crystals were carefully grown from a saturated aqueous solution, about 13 °C (growth rate: 1 cm³/month). The optical quality of the samples is excellent. Slabs about 4 mm thick have been cut normal to the crystal axis (within ± 2°).

3. Undiluted Salt

Fig. 1 shows the temperature dependence of τ_1 for four values of the magnetic field. For $H > 1$ kOe the experimental curves fit well with functions of the form

$$\tau_1^{-1} = A\,T + B\,e^{-23/T}\,.$$

However the coefficient A is nearly independent of H, contrary to our expectation for the direct process. This is corroborated by the curves in Fig. 2 and 3 showing the magnetic field dependence of τ_1 at two different temperatures. If we try to fit our low field measurements to the law of Brons-van Vleck (or the law of Orbach simplified by $H_{\mathrm{dd}}^2 \gg H_{\mathrm{hf}}^2$)

$$\tau_1^{-1} = \tau_{10}^{-1}\,(b_{\mathrm{dd}} + p\,C\,H^2)\,(b_{\mathrm{dd}} + C\,H^2)^{-1}$$

we find at $T = 2.1$ °C: $p \approx 0.15$ $b_{\mathrm{dd}}/C = (300)^2$ Oe², i.e. values near those determined on powder samples by van den Broek and van der Marel. However, p decreases quickly when T decreases, which is inconsistent with the theoretical prediction.

It is likely that the maximum of τ_1 at about $H = 5$ kOe conspicuous at 2.11 °K (Fig. 2) can be ascribed to a phonon-bottleneck coming from an acoustical mis-

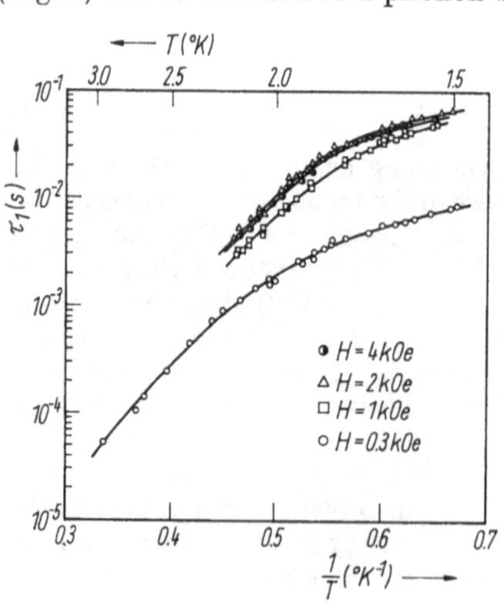

Fig. 1. 99.9% DyES. Temperature dependence of τ_1 for several magnetic field intensities

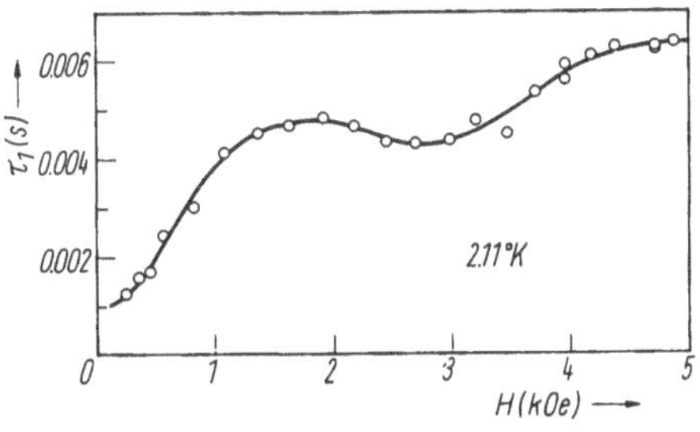

Fig. 2. 99.9% DyES. Magnetic field dependence of τ_1 at $T = 2.11$ °K

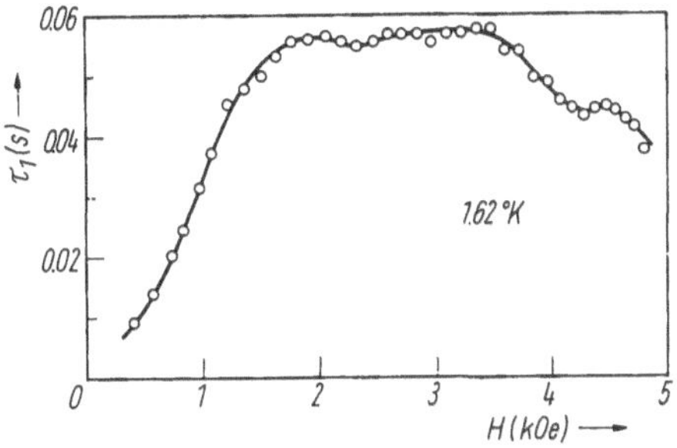

Fig. 3. 99.9% DyES. Magnetic field dependence of τ_1 at $T = 1.62$ °K

match between the crystal and the bath: there is a boundary thermic resistance similar to that observed on CeES by Glättli [8] and on HoES by Hagene [9]. The maximum of τ_1 agrees roughly with the maximum of the Schottky specific heat of the spin system.

All these results are in good agreement with the observations of Kump, except the order of magnitude of A which we find at least ten times larger.

4. Diluted Salts

For the highest values of the field and in the highest temperature range, the curves $\lg \tau_1 = f(1/T)$ in Fig. 4 show a rectilinear portion corresponding to the Orbach process, and are well described by an expression of the form $\tau_1^{-1} = B \exp(-\Delta/kT)$ with $\Delta/k = (23 \pm 1)$ °K.

At low temperatures, a more efficient process appears especially the more the higher the field is. Assuming that it concerns a direct process of the form

$$\tau_1^{-1} = \xi \, H^5 \coth \frac{g_{\parallel} \, \beta \, H}{2 \, kT}$$

(the approximation $g_{\parallel} \, \beta \, H \ll 2 kT$ has not been retained on account of the large values of $g_{\parallel}$ and H) the different coefficients could be fitted to describe exactly

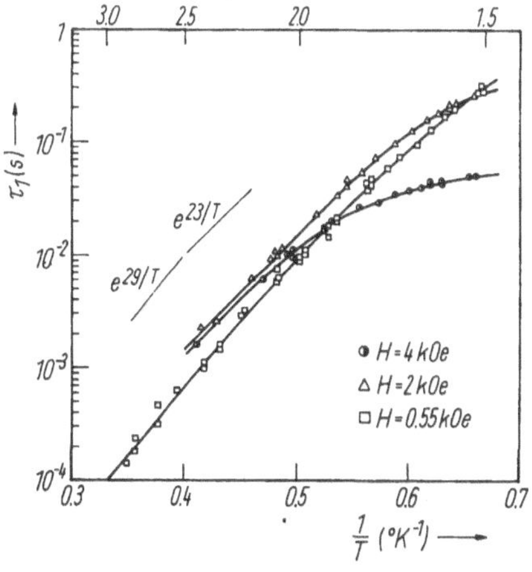

Fig. 4. $[Dy_{0.15}Y_{0.85}]$ES. Temperature dependence of τ_1 for several magnetic field intensities. The curves for $H = 2$ and 4 kOe are calculated and fitted to the experimental points (see equation (2))

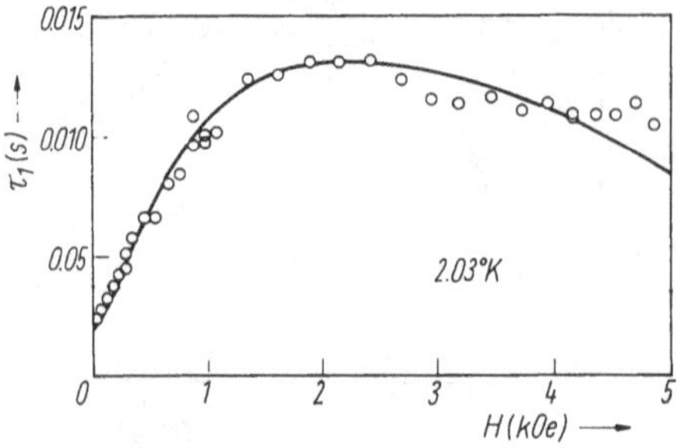

Fig. 5. $[Dy_{0.15}Y_{0.85}]ES$. Magnetic field dependence of τ_1 at $T = 2.03$ °K. The curve is calculated and fitted to the low field experimental points

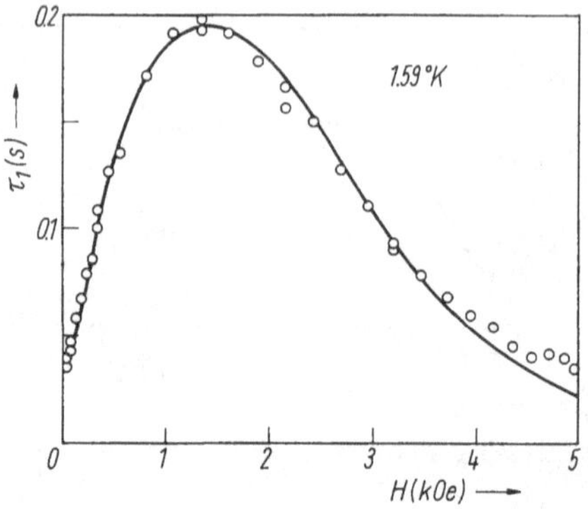

Fig. 6. $[Dy_{0.15}Y_{0.85}]ES$. Magnetic field dependence of τ_1 at $T = 1.59$ °K. The curve is calculated and fitted to the experimental points (see equation (4))

the experimental data with the expressions

$$\left.\begin{aligned} \tau_1^{-1} &= 0.96 \coth(0.73/T) + 0.63\times 10^7\, e^{-23/T} \qquad H = 2\text{ kOe}\,, \\ \tau_1^{-1} &= 13 \coth(1.45/T) + 0.63\times 10^7\, e^{-23/T} \qquad H = 4\text{ kOe}\,. \end{aligned}\right\} \tag{2}$$

These measurements were completed by a study of the magnetic field dependence at two different temperatures (Fig. 5 and 6): (I) $T = 2.03$ °K where, even in a high field, the Orbach process prevails, (II) $T = 1.59$ °K where, at $H =$ $= 2$ kOe, the direct process is already important.

The coefficient $B = 0.63\times 10^7$ in the expressions (2) is near that theoretically calculated (see equation (1)) and it does not seem to vary with the magnetic field above 2 kOe. In low fields when the direct term is small, we tried to explain the magnetic field dependence of τ_1 with the Orbach expression [1] for the salts with $g_\perp = 0$ which is the case for DyES:

$$\begin{aligned} \frac{1}{\tau_1} &= \frac{1}{\tau_{1\infty}} \frac{H^2 + H_{hf}^2 + H_{dd}^2}{H^2 + H_{hf}^2 + \frac{1}{2} H_{dd}^2} \\ &= \frac{1}{\tau_{1\infty}} \frac{H^2 + \alpha}{H^2 + \beta} = \frac{1}{\tau_{10}} \frac{q H^2 + \beta}{H^2 + \beta}\,, \end{aligned} \tag{3}$$

with $q = \tau_{10}/\tau_{1\infty} = \beta/\alpha$. Thus we determine $q = 0.185$ with good accuracy. In the case $T = 1.59$ °K, the values found for α and β lead to aberrant values of H_{dd}^2 and H_{hf}^2: $H_{hf}^2 < 0$ and $H_{dd}^2/2 = b_{dd}/C = (430)^2$ Oe^2, whereas we expect $b_{dd}/C \approx (120)^2$ Oe^2 [9]. On the other hand α seems to vary with temperature. Adding the term corresponding to the direct process calculated by Huber [10] the results is not modified, the relaxation time in zero field predicted for the direct process is very long, about one second. Therefore, if equation (3) fits very well with our experimental results, it does not seem that in the existing state of the theory it is possible to specify correctly how the various interactions interfere in α and β.

In high enough fields the first term in (2) accounts for the decrease of τ_1 when H increases. ξ may be also determined from the results of Fig. 6 by the expression

$$\frac{1}{\tau_1} = \xi\, H^5 \coth \frac{g_\parallel \beta H}{2\, kT} + \frac{1}{\tau_{10}} \frac{q H^2 + \beta}{H^2 + \beta}\,, \tag{4}$$

where the various parameters take the following values:

$$T = 1.59\ ^\circ\mathrm{K}\,, \qquad g_{\parallel} = 10.8\,, \qquad p = 0.185\,,$$
$$\xi = 1.7 \times 10^{-17}\,, \qquad \beta = (206)^2\ \mathrm{Oe}^2\,, \qquad \tau_{10} = 25\ \mathrm{ms}\,.$$

The experimental points above 3.5 kOe deviate from the curve.

The determination of ξ from the results achieved at 2.03 °K is evidently very indefinite. Assuming for ξ the same value as for $T = 1.59$ °K, equation (4) describes accurately our measurements (curve in Fig. 5) when $H < 2.5$ kOe and with

$$\beta = (250)^2\ \mathrm{Oe}^2\,, \qquad \tau_{10} = 2.3\ \mathrm{ms}\,.$$

5. Discussion

Referring to the direct term in equation (1) calculated by Orbach in the approximation $g_{\parallel}\,\beta\,H \ll 2\,kT$, we obtain for $H = 2$ kOe

$$\tau_1^{-1} \approx 270 \sin^2\theta \cos^2\theta\; T\,.$$

The comparison with our experimental relation shows that an angle of $\theta \approx 4°$ is sufficient to justify our observations on the diluted salt. However, this angle is superior to the misorientation of the crystal axis in the magnetic field. On the other hand, the crystal rotation in its holder and the exchange of the sample do not affect appreciably the measured relaxation time although it must vary as $\sin^2\theta$ about $\theta = 0$. Therefore we can conclude that it is really a direct process we observe even when θ is very small.

One may think that there is a distortion of the lattice around the paramagnetic ions, involving a small modification in the symmetry of the electric field built up by the nine water molecules. Kump ascribes his observations on undiluted DyES to such distortions; we note also that Gill's measurements lead to a non-zero value of $g_{\parallel}$ [11]. The origin of the distortion could be the fixation of supplementary water molecules during the very slow crystal growth (but a density measurement has given within $1^0/_{00}$ the value determined from the structure by X-rays by Ketelaar [12]), or the difference between the Dy^{3+} and Y^{3+} ionic radii, which is 6%. It would be necessary to try to give evidence of such deformations, e.g. by the study of the line-shape in EPR.

The relaxation process observed at the lowest temperatures on the concentrated salt by Kump and Hagene cannot be ascribed to a simple direct process as in the case of the diluted salt. The decrease of the relaxation time passing from the 15% diluted salt to the concentrated salt is spectacular, particularly in low field (τ_1 is divided by about eight). These may have several reasons:

a) The increase of the magnetic dipole interaction which is expressed by a very strong decrease of τ_1 in zero external field, when the relaxation occurs in the presence of the internal field only.

b) The broadening of the lines of the EPR spectrum and thus of the band of phonons able to interact with the spins.

c) A better efficiency of the cross-relaxation processes, particularly through foreign ions with a high relaxation rate. It would be then necessary to carry out a study on undiluted DyES similar to that described elsewhere on NdES [9, 13]: measurement of τ_1 in a high purity salt, study of the part of foreign ions added in a small ratio.

6. Conclusion

As in the case of NdES our experiments show that it is very critical to interpret the measurements of relaxation time in concentrated salts as long as we are not certain that the presence of impurities does not deeply disturb the behaviour of the ion studied. Even a small dilution (15%) allows, at least in the case of DyES, to observe only the usual processes of spin–lattice relaxation. On the other hand, the existence of the direct process seems certain even when the crystal axis is roughly parallel to the field. Lastly the discrepancy between the theoretical expression of the field dependence of the relaxation time for the Orbach process and our experimental data cannot proceed from a lack of experimental accuracy.

References

[1] R. ORBACH, Proc. Roy. Soc. **A264**, 485 (1961).
[2] J. VAN DEN BROEK and L. C. VAN DER MAREL, Physica (Utrecht) **30**, 565 (1964).
[3] A. H. COOKE, C. B. P. FINN, B. W. MANGUM, and R. ORBACH, J. Phys. Soc. Japan, Suppl. BI **17**, 462 (1962).
[4] R. P. HUDSON and B. W. MANGUM, Proc. XI. Coll. Ampère, Eindhoven 1962.
[5] H. KALBFLEISCH, Z. Phys. **181**, 13 (1964).
[6] U. KUMP, phys. stat. sol. **34**, 691 (1969).
[7] B. HAGENE, Rev. Phys. Appl. **2**, 270 (1967).
[8] H. GLÄTTLI, Canad. J. Phys. **46**, 103 (1968).
[9] B. HAGENE, Thèse, University Rennes 1971.
[10] D. HUBER, Phys. Rev. **131**, 190 (1963).
[11] J. C. GILL, Proc. Phys. Soc. **82**, 1066 (1963).
[12] J. A. A. KETELAAR, Physica (Utrecht) **4**, 619 (1937).
[13] B. HAGENE, phys. stat. sol. (a) **9**, 643 (1972).

(Received December 28, 1971)

phys. stat. sol. (a) 10, 407 (1972)

Subject classification: 14.4.1; 22.8.1

Institut für Angewandte Physik der Universität Münster

Contribution of Domain Wall Motion to the Permittivity of Rochelle Salt

By

P. GURK

The contribution ε_w of the domain walls to the dielectric constant of Rochelle salt (sodium potassium tartrate) within the frequency region of 0.02 Hz and 50 kHz is studied. The dispersions of ε_w as well as their dependence on time are related to three electrical after effects already known, and which are characterized by the following activation energies: $W_1 = 0.51$ eV, $W_2 = 0.66$ eV, and $W_3 = 1.4$ eV. A simplified model of the domain wall and its motion is proposed. Based on this model the time constants τ_i of the after effects and the thickness δ of the domain walls can be determined by measuring ε_w. In the ferroelectric range of temperature it follows for δ: 500 Å $< \delta <$ 2500 Å, and for the time constants 2×10^{-7} s $< \tau_1 < 5 \times 10^{-6}$ s and 5×10^{-4} s $< \tau_2 < 4 \times 10^{-2}$ s.

Es wird der Beitrag ε_w der Domänenwände zur Dielektrizitätskonstante bei Seignettesalz (Kalium–Natrium–Tartrat) im Frequenzbereich zwischen 0,02 Hz und 50 kHz untersucht. Die hier gefundenen Dispersionen von ε_w sowie ihre Zeitabhängigkeit werden mit drei bereits bekannten elektrischen Nachwirkungserscheinungen in Zusammenhang gebracht, welche durch folgende Aktivierungsenergien charakterisiert werden können: $W_1 = 0{,}51$ eV, $W_2 = 0{,}66$ eV und $W_3 = 1{,}4$ eV. Es wird ein einfaches Modell der Domänenwand sowie ihrer Bewegung angegeben. Aus den Messungen können auf Grund dieses Modells die Zeitkonstanten τ_i der Nachwirkungen sowie die Domänenwanddicke δ bestimmt werden. Im ferroelektrischen Temperaturbereich ergibt sich für δ: 500 Å $< \delta <$ 2500 Å und für die Zeitkonstanten 2×10^{-7} s $< \tau_1 < 5 \times 10^{-6}$ s bzw. 5×10^{-4} s $< \tau_2 < 4 \times 10^{-2}$ s.

1. Introduction

The initial mechanical free dielectric constant ε of a ferroelectric multi-domain crystal consists of several contributions. The contribution ε_w, caused by domain walls, is added to the dielectric constant ε^0 of the single-domain sample. This wall contribution has already been investigated in detail in some other ferroelectric substances [1 to 4]. Corresponding studies in Rochelle salt (RS), particularly at very low frequencies, have not become known so far.

Therefore the wall contribution ε_w in RS within the frequency region of 0.02 Hz to 50 kHz shall be studied in this paper.

Applying a square-wave voltage optical and electrical studies of the dynamics of the domain walls in RS show that the domain walls move mainly sideways [5]. In what follows it is assumed that this sidewise wall motion is exclusively responsible for the contribution ε_w measured when small sinusoidal fields are applied.

Caloric effects (difference between adiabatic and isothermal dielectric constant) which can cause a dispersion in the very low frequency region are disregarded here because they are too small compared with the dispersions to be discussed.

2. Model for the Domain Wall Motion

Before discussing the dynamics of domain walls in detail something has to be said about the after-effects which are known to influence strongly the dielectric behaviour of RS in the parelectric as well as in the ferroelectric phase. The motion of the domain walls in particular is determined decisively by these after-effects.

Three after-effects have been investigated in detail [6]. Phenomenologically each of them can be described quantitatively by an internal electric field $E_e(t)$ equivalent to an external field. The dependence of $E_e(t)$ on time obeys the following differential equation:

$$\tau_i \dot{E}_{ei}(t) = \frac{a_i}{\varepsilon_0} D(t) - E_{ei}(t); \qquad i = 1, 2, 3 \tag{1}$$

($D(t)$ dielectric displacement, the time constants τ_i as well as the constants a_i depend on temperature T, $\tau_i \sim \exp(W_i/kT)$, W_i activation energy, k Boltzmann constant; cf. Fig. 6).

As the three after-effects are independent of each other the fields $E_{ei}(t)$ can be added, if D is given.

In the ferroelectric state D can be substituted by the spontaneous polarization P without any considerable error. Therefore:

$$\tau_i \dot{E}_{ei}(t) = \frac{a_i}{\varepsilon_0} P(t) - E_{ei}(t) . \tag{1a}$$

With P constant the equilibrium value of $E_{ei}(t)$ is given by

$$E^0_{ei} = \frac{a_i}{\varepsilon_0} P . \tag{2}$$

For the following these assumptions are made:

1. The sample contains N flat domain walls which divide the whole sample in $(N + 1)$ parallel plates. These plates have the same direction as the ferroelectric axis (Fig. 1).
2. Under the influence of an electric field each wall moves as a whole in the direction of the x-axis, which is perpendicular to the wall plane. The motion does not change the polarization profile of the wall.
3. The displacement of the wall does not exceed its thickness.
4. Each wall obeys the same equation of motion.
5. In the middle of the wall at x_w (defined by $P(x_w) = 0$) the sum of all electric fields vanishes ($E_a + \sum_i E_{ei}(x_w) = 0$; E_a external field).

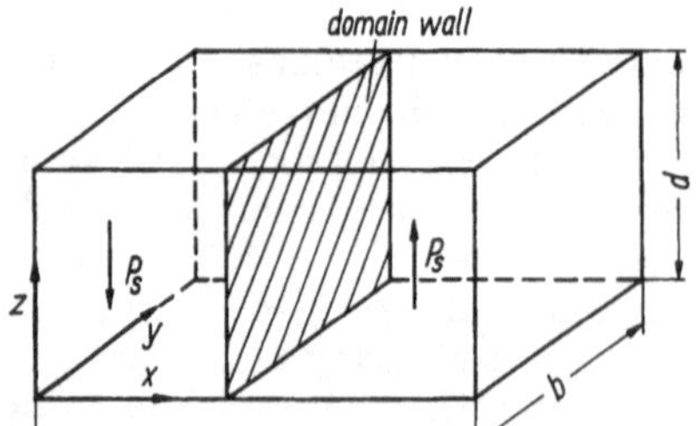

Fig. 1. Schematic diagram of the sample containing N domain walls. Only one wall is represented

6. Equation (1a) holds also within the wall region.

With regard to assumption 4 the problem is reduced to the study of the motion of one single wall.

A simple presentation of the relation between the wall displacement and the external field can be obtained by determining E_a as a function of a given wall displacement $x_w(t)$, that is of a given change of the average sample polarization $\Delta P(t) = -2 P_s n x_w(t)$; ($n = N/l$; cf. Fig. 1). Therefore we look for the time function of the field $E_a(t)$ which has to be applied to the electrodes in order to realize the given wall displacement $x_w(t)$.

To simplify further a linear polarization profile is assumed (the rest position of the wall is $x = x_w = 0$):

$$P(x) = P' x \quad \text{for} \quad |x| \leq \delta/2 , \tag{3}$$
$$|P(x)| = P_s \quad \text{for} \quad |x| \geq \delta/2 .$$

Here the wall thickness δ has been introduced so that $P' = 2 P_s/\delta$ (cf. Fig. 2)

The motion of a wall is given by:

$$x_w = \hat{x}_w e^{j \omega t} . \tag{4}$$

The amplitude $\hat{x}_w$ is chosen so small that $\hat{x}_w \leq \delta/4$. Then one gets:

$$P(x, t) = P' (x - x_w) \quad \text{for} \quad |x| \leq \delta/4 . \tag{5}$$

Considering first of all only one after-effect (index 1 e.g.) the internal field E_{e1} in the middle of the wall [12] is

$$E_{e1}(x_w) = \frac{a_1}{\varepsilon_0} P' \left(1 - \frac{1}{1 + j \omega \tau_1}\right) x_w . \tag{6}$$

A corresponding term results for the second after-effect (index 2).

The relaxation frequencies $f_1 = (2 \pi \tau_1)^{-1}$ and $f_2 = (2 \pi \tau_2)^{-1}$ of these two after-effects lie within the range of the applied measurement frequencies. So in the region of the wall the internal fields E_{e1} or E_{e2} are periodically built up and reduced to some extent. The time constant τ_3, however, is so large that the field E_{e3} does not change during many periods. By adequate treatment of the sample [12] the field E_{e3} vanishes practically at the very beginning of a measuring sequence ($t_e = 0$; compare Section 3). Afterwards it gradually grows to its static final value which is reached only several hours later. During this evolution a linear local variation for E_{e3} is assumed in the vicinity of the rest position of the wall

$$E_{e3}(x, t_e) = E_3'(t_e)\, x \tag{7}$$

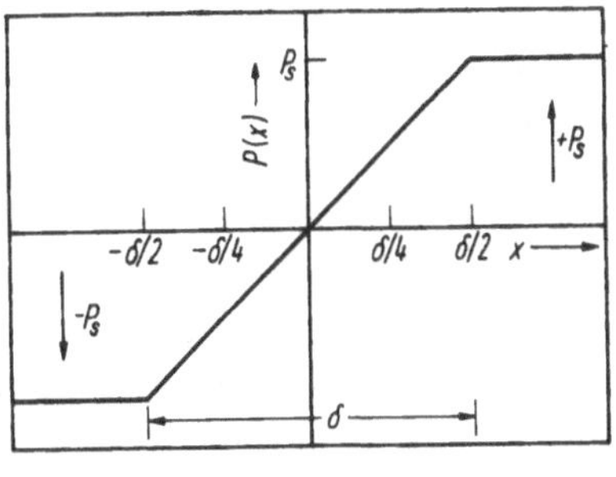

Fig. 2. Assumed dependence of polarization on the coordinate x within the region of a domain wall

with $E_3'(t_e)$ growing monotonously during the evolution time t_e. For the middle of the wall it follows:

$$E_{e3}(x_w, t_e) = E_3'(t_e)\, x_w\,. \tag{7a}$$

Defining the dielectric modulus $\beta_w = (\varepsilon_w)^{-1}$,

$$\varepsilon_0\, E_a = \beta_w(\omega)\, \Delta P = -\beta_w(\omega)\, 2\, P_s\, n\, x_w(t) \tag{8}$$

holds for a sample containing n walls per unit length.

According to the assumption 5 it follows from equation (6) and (7a):

$$\beta_w(\omega) = \frac{1}{2\, P_s n}\left[\varepsilon_0\, E_3'(t_e) + a_1\, P'\left(1 - \frac{1}{1 + j\,\omega\,\tau_1}\right) + \right.$$
$$\left. + a_2\, P'\left(1 - \frac{1}{1 + j\,\omega\,\tau_2}\right)\right]. \tag{9}$$

Equation (9) describes two relaxation processes. As the time constants τ_1 and τ_2 differ by orders of magnitude, two separate semi-circles are obtained in the β_w-plane. As only $E_3'(t_e)$ depends on time the semi-circles shift along the positive β_w'-axis with t_e increasing. Their diameters and their frequency distribution remain unchanged.

Internal friction and restoring force are characteristics of a relaxational behaviour. In the case of the relaxation of the domain walls described above the restoring force is given by the field E_{e3} and the friction by the periodical change of the fields E_{e1} and E_{e2}.

3. Experimental Procedure

To measure the complex dielectric constant ε, a low-frequency bridge was used. It allows to determine capacitances up to 6 nF and conductances between 10^{-6} and $10^{-14}\ \Omega^{-1}$ within the frequency range of 0.01 to 1000 Hz. For frequencies above 1 kHz a commercial capacitance bridge (General Radio Type 1615 A) was used.

The temperature stability during the measurement was typically better than $\pm$ 0.001 deg. A relative humidity of 15% was provided.

The experiments were carried out as follows: The crystal was stored at least for one day at a temperature of about 25 °C (above the upper Curie-point) while a strong ac-field (50 Hz) was applied. In this field the crystal was slowly cooled down to the measuring temperature. The amplitude of the field was chosen at least twice the coercive field. After the measuring temperature had been reached, the amplitude of the ac-field was continuously diminished to zero. Thereby a defined initial state of the sample was guaranteed. This point of time is to be defined as the zero mark of the evolution time t_e. As a result of the strong ac-field the internal field E_{e3} of the slow after-effect is zero at the beginning of the measuring sequence: $E_{e3}(x, t_e = 0) = 0$.

The amplitude of the measuring field was about 1 V/cm. It was chosen so small that a linear electric response was observed.

In order to receive the wall contribution ε_w from the measured dielectric constant $\varepsilon = \varepsilon^0 + \varepsilon_w$, the value ε^0 of the single domain crystal had to be found, where ε^0 should be the mechanical free constant.

Fig. 3. Dispersion of ε_w at various stages of evolution time t_e. The centres of the circles are marked by (×). A further dispersion at frequencies above 1 kHz is indicated by the small semi-circles. $T = -16.52$ °C, $\hat{E}_a = 1.2$V cm^{-1}

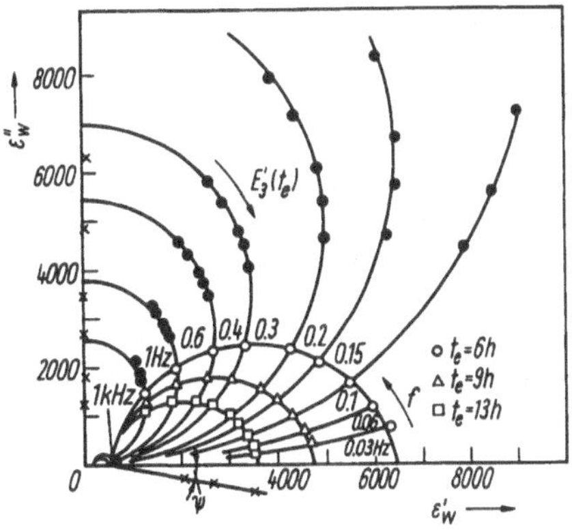

In the ferroelectric phase one gets

$$(\varepsilon_0 \varepsilon^0)^{-1} = -2A = \left.\frac{dE}{dD}\right|_{D_s = D(E=0)}, \tag{10}$$

assuming that the relation between the field and the dielectric displacement D is given by

$$E = A\,D + B\,D^3\,. \tag{11}$$

The value A was obtained by analysing the hysteresis loop, as it has been described by Müser [7], ε^0 is regarded as a real value.

4. Results

In Fig. 3 ε_w is plotted as a function of the evolution time t_e and the frequency. A strong dispersion is observed. It is seen that values for given t_e as well as the points measured at constant frequency lie on circles. The centres of the latter are situated approximately on the imaginary axis. With increasing t_e points belonging to the same measuring frequency move towards the origin.

To measure the frequency dependence of ε_w (about ten measuring points), an interval $\Delta t_e \approx 30$ min is necessary. From Fig. 3 it is obvious that the shift of ε_w in time is so slow that the measurements performed during Δt_e may be

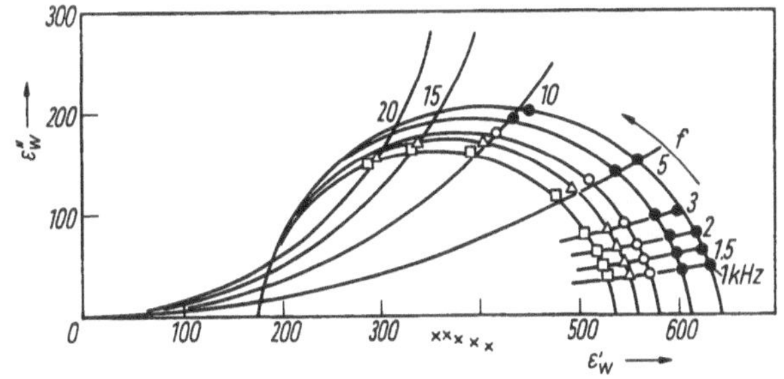

Fig. 4. High-frequency dispersion of ε_w at $T = -16.52$ °C. The results are taken from the measurement of Fig. 3 but represented in an enlarged scale

regarded as being performed simultaneously, if $\Delta t_e/t_e \ll 1$, and therefore the corresponding points in the ε_w-plane can be connected. As we see in Fig. 3 these points form relaxation circles of the Cole–Cole type. Here the decrease of the relaxation strength with increasing t_e is obvious.

At higher frequencies the dispersion described above is followed by a second one as indicated in Fig. 3. It is small in comparison to the first one. In Fig. 4 this high-frequency dispersion is plotted in an enlarged scale. The diagram is analogous to Fig. 3.

At higher temperatures all the after-effects proceed more rapidly. The same holds for the time dependence of ε_w at constant frequency (circles above the ε_w''-axis), and the relaxation frequencies of the dispersions measured at constant t_e (circles above the ε_w'-axis) are shifted to higher values with increasing temperature.

In order to compare the experimental results with the theoretical ones (Section 2) ε_w has been transformed into $\beta_w = (\varepsilon_w)^{-1}$. As can be seen from Fig. 5 the result is in exact accordance with the description by means of equation (9).

According to equation (9) the time constants τ_1 and τ_2 can be achieved immediately from the frequencies at the tops of the two relaxation circles. In the case of the high-frequency dispersion, however, we measured only below the relaxation frequency. The latter was computed from the measured values. Here a correction has to be considered to eliminate the error caused by piezoelectric resonance vibrations of the sample. In Fig. 6 the τ-values obtained thus are plotted as full points. The straight lines for τ_1 and τ_2 are determined by the τ-values which are measured very precisely in the parelectric temperature ranges. For the activation energies of the three relaxation processes considered one gets: $W_1 = 0.51$ eV, $W_2 = 0.66$ eV, and $W_3 = 1.4$ eV.

Furthermore it is possible to obtain the domain wall thickness δ from the diameter d_i of the semicircles shown in Fig. 5. According to equation (9) the

Fig. 6. Temperature dependence of the timeconstants τ_1, τ_2, and τ_3 of the after effects which influence the domain wall motion

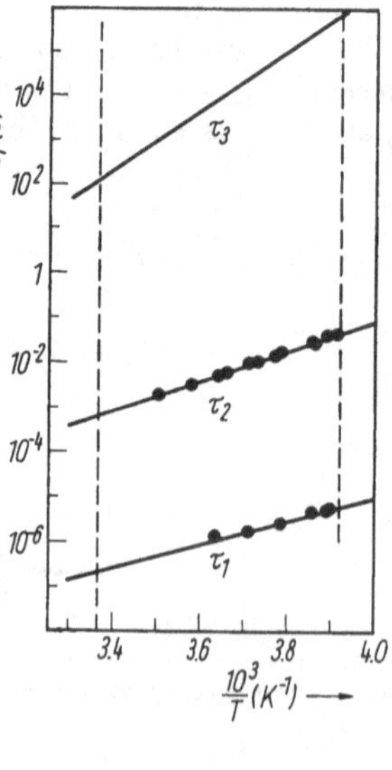

Fig. 5. Dispersion of $\beta_w = \varepsilon_w^{-1}$ at various stages of t_e. The high-frequency relaxation circle is plotted only for one t_e. $T = -15.92$ °C

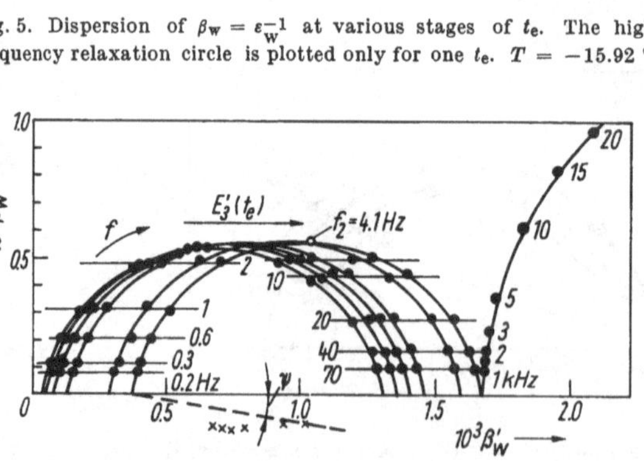

diameters d_i are given by:

$$d_i = \frac{a_i P'}{2 P_s n} = \frac{a_i}{\delta n}. \tag{12}$$

To compute δ, the values of the domain wall density n and the constants a_i of the after-effects must be known. n is found by counting the domains which were made visible by a polarization microscope. Here the same conditions as in the measurement of ε_w were set up (rel. humidity, treatment of the sample etc.). The constant a_1 can be calculated if the relative humidity and the temperature are given [8]. At temperatures above 0 °C, however, the measurement of the dispersion of ε_w, which belongs to this very fast after-effect, leads to considerable difficulties as high frequencies are needed. They stimulate the above mentioned mechanical resonance vibrations of the sample which invalidate the results entirely.

On the other hand the low-frequency relaxation circle (index 2) has been measured in the whole ferroelectric region. The temperature dependence of the after-effect constant a_2, however, is not known precisely. Here the a_2-values accurately measured in the two parelectric temperature ranges have been interpolated within the ferroelectric range assuming a temperature dependence $a_2 \sim \exp(-c/kT)$. The domain wall thickness δ which has been found by analysing this relaxation circle according to equation (12) is represented in Fig. 7. Remarkable and in accordance with theoretical considerations [9] is the strong increase of δ when the Curie-points are approached. The wall thicknesses found here are considerably larger than the values given by Zhirnov [9] and by Mitsui and Furuichi [10]. On the other hand the wall thickness determined by Petroff [11] by means of X-ray topographical investigations are closer to the values given here. The δ-values measured in the two samples shown in Fig. 7 differ considerably. It is assumed that this discrepancy is due to the imperfect knowledge of a_2.

For some temperatures the thickness δ has been computed by means of the high-frequency circle (index 1). Considering the limited measurement accuracy, these δ-values agree with the corresponding δ-values stated in Fig. 7 (upper curve). The experiments show that the dynamics of the domain walls in Rochelle salt are essentially determined by internal fields which are due to

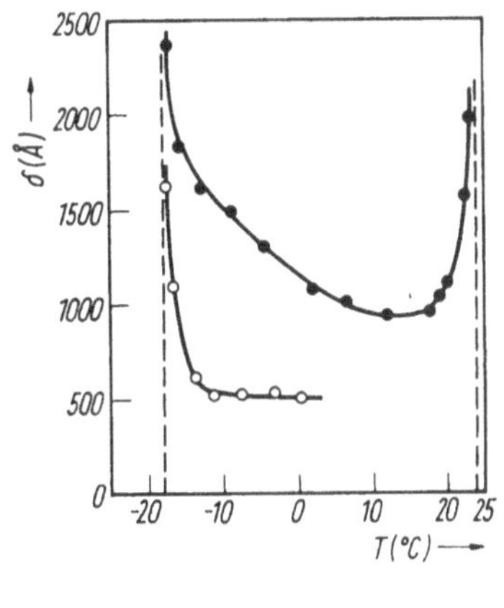

Fig. 7. Domain wall thickness, δ, obtained by analysing the dispersion of ε_w of two samples

imperfections. The analysis demonstrates, however, that the assumed structure of the wall, i.e. its thickness, is essentially independent of the special loss mechanism and can be regarded as a property of the ideal crystal.

Acknowledgements

In conclusion I would like to express my gratitude to Prof. Dr. Dr. h. c. Dr. h. c. H. Bittel for his interest in this work and for supplying the experimental equipment. Also I am indebted to Prof. Dr. H.-G. Unruh for many helpful and stimulating discussions.

References

[1] J. FOUSEK, Proc. EMF, Saarbrücken 1969 (p. 147).
[2] P. BORNAREL and J. LAJZEROWICZ, J. appl. Phys. **39**, 4339 (1968).
[3] J. FOUSEK and V. JANOUŠEK, phys. stat. sol. **13**, 195 (1966).
[4] P. LAUGINIE and F. GILLETTA, Proc. EMF, Saarbrücken 1969 (p. 219).
[5] H. BITTEL, H. E. MÜSER, and G. BERNDES, Forschungsbericht des Landes NRW, Nr. 1955, Westdeutscher Verlag Köln u. Opladen 1968.
[6] H.-G. UNRUH, Z. angew. Phys. **16**, 315 (1963).
[7] H. E. MÜSER, Z. angew. Phys. **12**, 300 (1960).
[8] H.-G. UNRUH and E. SAILER, Z. Phys. **224**, 45 (1969).
[9] V. A. ZHIRNOV, Soviet Phys. — J. exper. theor. Phys. 8, 822 (1959).
[10] T. MITSUI and J. FURUICHI, Phys. Rev. **90**, 193 (1953).
[11] J. F. PETROFF, phys. stat. sol. **31**, 285 (1969).
[12] P. GURK, Dissertation, Münster 1971.

(Received January 27, 1972)

phys. stat. sol. (a) 10, 415 (1972)

Subject classification: 8 and 15; 14.3; 22.7

Instituut voor Lage Temperaturen en Technische Fysika,
Laboratorium voor Vaste-Stof-Fysika en Magnetisme, University of Leuven

Thermoelectric Properties of p-Type Bi_2Te_3–Sb_2Te_3–Sb_2Se_3 Alloys and n-Type Bi_2Te_3–Bi_2Se_3 Alloys in the Temperature Range 300 to 600 K[1]

By

M. IMAMUDDIN and A. DUPRE

A conventional absolute method was used for the measurement of the thermal conductivity of pressed and sintered p-type Bi_2Te_3–Sb_2Te_3–Sb_2Se_3 alloys and n-type Bi_2Te_3–Bi_2Se_3 alloys in the temperature range of 300 to 600 K. The measurements were made in directions parallel and perpendicular to the pressing direction. Measurements of the Seebeck coefficient and the electrical resistivity for these alloys were performed between 300 and 575 K in the pressing direction only. The maximum values of the figure of merit for the p- and n-type samples are 17×10^{-4} and 16×10^{-4} K^{-1}, respectively. It is concluded that a thermoelectric generator, having these alloys for its p- and n-branches will attain its highest efficiency in the temperature range between 370 and 480 K.

Die thermoelektrischen Eigenschaften von gepreßten und gesinterten p-leitenden Bi_2Te_3–Sb_2Te_3–Sb_2Se_3 und n-leitenden Bi_2Te_3–Bi_2Se_3 Legierungen wurden im Temperaturbereich 300 bis 600 K mit einer konventionellen Methode untersucht. Diese Messungen wurden parallel und senkrecht zur Preßrichtung durchgeführt. Der Seebeckkoeffizient und der elektrische Widerstand dieser Legierungen wurden im Temperaturbereich 300 bis 575 K nur in Preßrichtung gemessen. Die Maximalwerte der Leistungsfaktoren von p- und n-leitenden Proben betragen 17×10^{-4} bzw. 16×10^{-4} K^{-1}. Daraus folgt, daß ein thermoelektrischer Generator mit diesen Legierungen als p- bzw. n-Zweig seine höchste Leistungsfähigkeit im Temperaturbereich 370 bis 480 K erreicht.

1. Introduction

The theoretical considerations of Ioffe [1] suggested that solid solution alloying could improve the thermoelectric figure of merit ($Z = \alpha^2/(\varrho K)$) of semiconductors, where impurity scattering is predominant, by decreasing the lattice thermal conductivity (K_L) without affecting the Seebeck coefficient (α) and the electrical resistivity (ϱ).

This concept led to considerable research in recent years on the thermoelectric properties of solid solution alloys of compound semiconductors and, in particular, those in the generalized systems, $(Bi, Sb)_2(Te, Se)_3$ and $(Bi)_2(Te, Se)_3$.

However, most of this work was limited to a room temperature study. In some cases the alloys were doped, in others they were not; in some the material was sintered,in others it was a single crystal. For reasons such as these the results have not always been comparable.

The thermoelectrical and physical properties of thermoelectric materials which are prepared by the powder metallurgical method, are dependent on

[1]) An account of this work forms part of a D.Sc. thesis submitted by M. Imamuddin to the University of Leuven, Belgium.

a large number of factors [2, 3] such as the chemical composition, the degree of purity, the quantity of the starting powder, the shape and the size of the grains, the surface layer on the grains, the method of powder preparation.

Inhomogeneity of chemical composition is important in determining the thermoelectric properties of a material in that it always leads to a lower figure of merit for the material than does homogeneity. Inhomogeneity enhances the thermal conduction by "circulating currents". These many factors lead to the fact that even only with one particular composition a large number of investigations are needed to get a final explanation of the measured properties.

The present study of the thermoelectric properties of p-type Bi_2Te_3–Sb_2Te_3–Sb_2Se_3 alloys doped with Pb and n-type Bi_2Te_3–Bi_2Se_3 alloys doped with i. SbI_3 and ii. AgI were made over the temperature range from 300 to 600 K to see their applicability in a thermoelectric generator to be operated at high temperature.

2. Preparation of Ingots

The pressed powder samples for this investigation were supplied by the metallurgie Hoboken (Belgium). The p- and n-type alloys have the following compositions:

p-type: 73 mol% Sb_2Te_3, 24 mol% Bi_2Te_3, 3 mol% Sb_2Se_3, 0.1 or less wt% of Pb (sample No. HP 28).

n-type: 80 mol% Bi_2Te_3, 20 mol% Bi_2Se_3 with excess of Te and doped with i. SbI_3 (sample No. HP 36), ii. AgI (sample No. HP 37).

The alloys were prepared by melting together the proper proportions of the respective elements and dopants in sealed, evacuated quartz tubes. The purity of the basic elements is 99.99% for Bi, Se, Sb and 99.999% for Te. The melts were heated and agitated, then removed from the furnace and allowed to cool rapidly. The alloys were then mechanically broken, grinded and pressed unidirectionally and subsequently annealed at a temperature of 450 °C. Annealing produced bonding and homogenization by diffusion, recrystallization and graingrowth. Metallographic studies revealed that the blocks contained little or no oxides and that they are nearly without porosity. As far as the grain size is concerned, it should be noted that the unidirectional stress broadens the grains somewhat in the direction perpendicular to the stress. Starting with powders of < 50 mesh, individual grains with maximum dimensions of $0.04 \times \times 0.7$ mm were observed.

All samples which were used for thermal conductivity measurements were cut from these ingots in a manner such that the heat-flow path was a) parallel, b) perpendicular to the pressing direction. Samples used for measurements of α and ϱ were cut from the ingot with the long axis parallel to the pressing direction.

3. Methods of Measurements

A conventional steady-state absolute method was used for the measurement of the thermal conductivity (K) of pressed and sintered p- and n-type alloys in the temperature range 300 to 600 K. In each case the measurements were done parallel and perpendicular to the pressing direction of the samples. The thermal conductivity was measured by determining the temperature difference across the sample of known dimensions with a known amount of heat flowing through the sample. The temperature difference over the sample was measured by a calibrated vacuplus–vacuminus (NiCr–Ni) thermocouple. New formulae were deduced and used for the calculation of the heat loss due to radiation. The contact resistance of the sample with the gradient heater and the heat sink was negligible. The measurements of the Seebeck coefficient and electrical resistivity for these p- and n-type alloys in the pressing direction were done in the temperature range of 300 to 575 K by the usual methods.

4. Experimental Results

The results of the measurements of the thermal conductivity of pressed powder p-type alloys of (mol%) composition $(Bi_2Te_3)_{24}(Sb_2Te_3)_{73}(Sb_2Se_3)_3$ doped with Pb and n-type alloys of composition $(Bi_2Te_3)_{80}(Bi_2Se_3)_{20}$ doped with i. SbI_3, and ii. AgI over the temperature range of measurements are plotted as a function of $1/T$ in Fig. 1, 2, and 3. The $K - K_e$ values (Fig. 1, 2, and 3) are obtained from the measured total thermal conductivity by using the relation

$$K = K_L + K_e + K_{amb}, \tag{1}$$

where K_L, K_e, and K_{amb} are respectively the lattice, electronic, and ambipolar contribution to the thermal conductivity. The electronic thermal conductivity, K_e, is calculated from the measured values of the electrical resistivity by using the nondegenerate expression

$$K_e = \left(\frac{k}{e}\right)^2 \sigma T \left(\lambda + \frac{5}{2}\right), \tag{2}$$

where $\sigma = 1/\varrho$ is the electrical conductivity and λ is a constant for each scattering process. We have assumed acoustic lattice scattering ($\lambda = -1/2$). The nondegenerate expression is justified from the work of Birkholz and Haacke [4], Wendler and Schreiner [5], and many others. The electrical resistivity data of the samples also support the nondegeneracy. The results of the measurements of the Seebeck coefficient, and of the electrical resistivity for both p- and n-type alloys in the pressing direction are shown in Fig. 4. The electrical resistivity

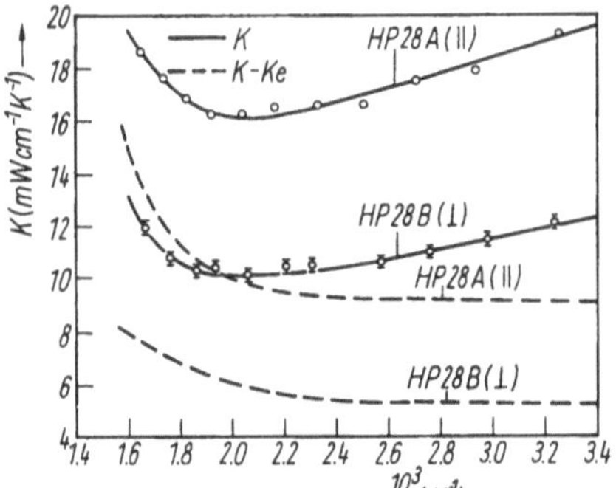

Fig. 1. Plot of thermal conductivity K and $K - K_e$ (where $K_e = 2\,(k/e)^2\,\sigma\,T$) against reciprocal absolute temperature for p-type $Bi_{0.5}Sb_{1.5}Se_{0.09}$ alloys doped with 0.1 wt% or less of Pb. The symbols ○ and Φ represent the results in the pressing and perpendicular to the pressing direction, respectively

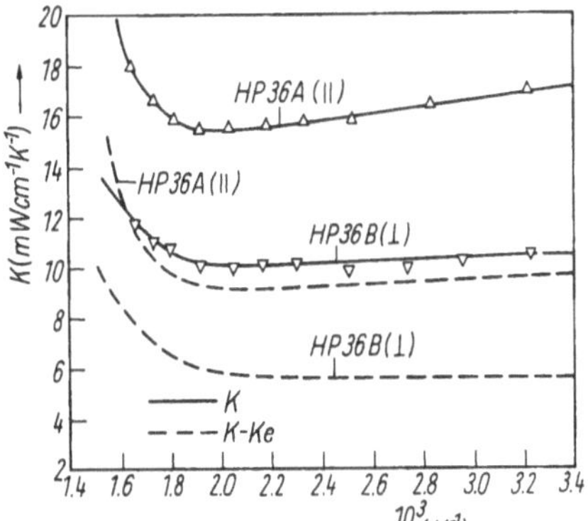

Fig. 2. Plot of thermal conductivity K and $K - K_e$ (where $K_e = 2\,(k/e)^2\,\sigma\,T$) against reciprocal absolute temperature for n-type alloys of composition $Bi_2Te_{2.41}Se_{0.59}$ with excess of Te and doped with SbI_3. The symbols △ and ▽ represent the results in the pressing and perpendicular to the pressing direction, respectively

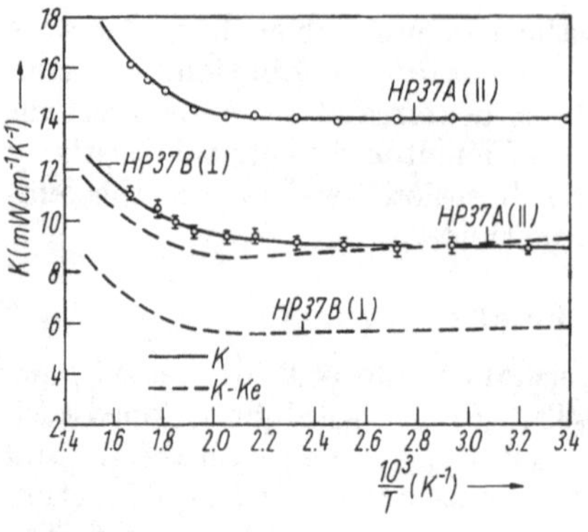

Fig. 3. Plot of thermal conductivity K and $K - K_e$ (where $K_e = 2\,(k/e)^2\,\sigma\,T$) against reciprocal absolute temperature with excess of Te in n-type alloys of composition $Bi_2Te_{2.41}Se_{0.59}$ doped with AgI. The point symbols have the same meaning as in Fig. 1

anisotropy ratio in these alloys is about 1.5. The temperature dependence of the figure of merit Z in both p- and n-type alloys in the pressing direction is shown in Fig. 5. The Z value is calculated from the measured values of α, ϱ, and K by

$$Z = \frac{\alpha^2}{\varrho\,K}\,. \tag{3}$$

Results of measurements of the thermoelectric properties for both p- and n-type alloys are tabulated in Table 1. The results shown in the table do not represent all of the data recorded.

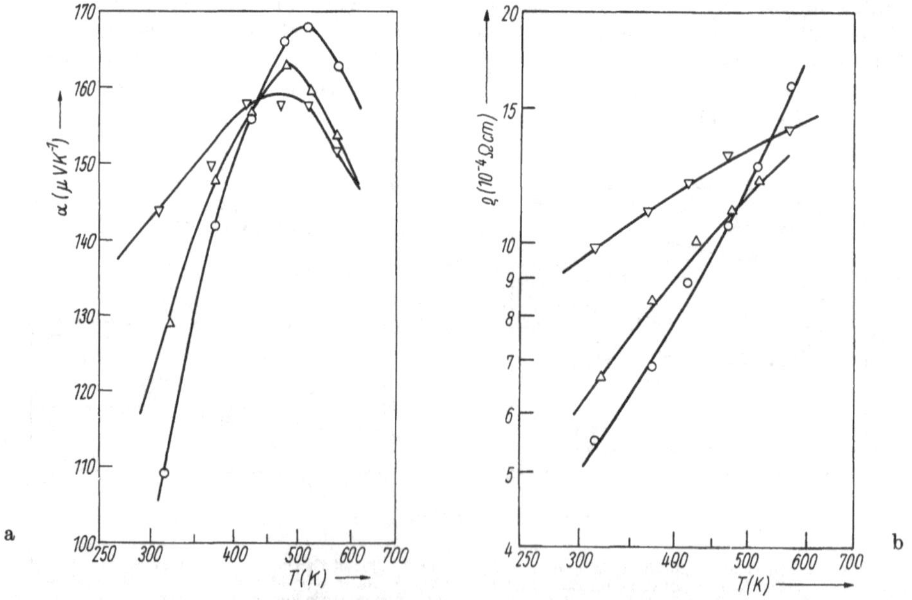

Fig. 4. Temperature dependence of Seebeck coefficient (α) and electrical resistivity (ϱ) in p-type $Bi_{0.5}Sb_{1.5}Se_{0.09}$ alloy doped with Pb and in n-type $Bi_2Te_{2.41}Se_{0.59}$ alloys doped with a) SbI_3 and b) AgI_3. The point symbols have the following meaning: ○ = sample No. HP 28A (||); △ = sample No. HP 36A (||); and □ = sample No. HP 37A (||)

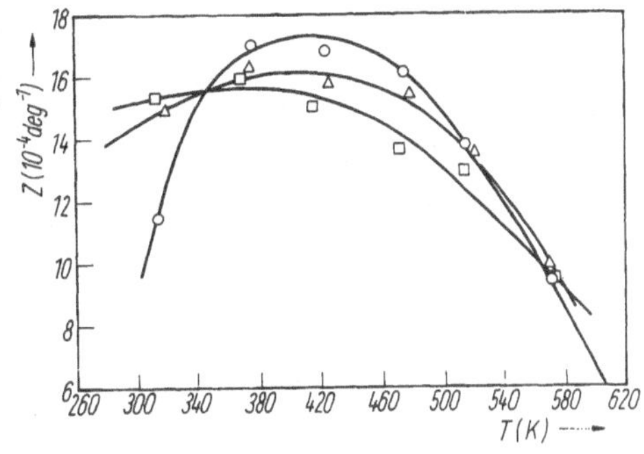

Fig. 5. Temperature dependence of figure of merit (Z) in p-type $Bi_{0.5}Sb_{1.5}Se_{0.09}$ alloy doped with Pb and in n-type $Bi_2Te_{2.41}Se_{0.59}$ alloys doped with i. SbI_3 and ii. AgI. The point symbols have the same meaning as in Fig. 4

Table 1

Thermoelectric properties of p- and n-type alloys in the pressing direction only

sample No.	T (K)	α (μV K^{-1})	ϱ (10^{-4} Ωcm)	α^2/ϱ (μV^2/K^2 Ωcm)	K (10^{-3} W/cmK)	Z (10^{-4} K^{-1})
HP 28A(‖)	315	109	5.5	21.59	18.80	11.49
	377	142	6.8	29.44	17.30	17.02
	426	156	8.8	27.64	16.60	16.65
	476	166	10.5	26.23	16.30	16.09
	517	168	12.5	22.58	16.40	13.77
HP 36A(‖)	321	129	6.7	24.83	16.70	14.87
	376	148	8.3	26.24	16.10	16.30
	428	157	10.0	24.65	15.70	15.70
	480	163	11.0	24.15	15.50	15.58
	522	159	12.0	21.06	15.60	13.50
	571	149	13.5	16.45	16.60	9.91
HP 37A(‖)	314	145	9.8	21.44	14.00	15.31
	369	150	11.0	22.50	14.00	16.07
	418	158	12.0	20.80	14.00	14.86
	472	158	13.0	19.20	14.05	13.59
	517	158	13.0	19.20	14.50	13.04
	573	152	14.0	16.50	15.70	9.46

5. Analysis and Discussion

5.1 Temperature dependence of thermal conductivity

The literature contains divergent data on the thermal conductivity of $(Sb_2Te_3)_x(Bi_2Te_3)_y(Sb_2Se_3)_{1-x-y}$ and $(Bi_2Te_3)_x(Bi_2Se_3)_{1-x}$ systems. This can be explained by the different carrier concentrations in the samples studied, whose effect on K_L is not always taken into account. Sometimes [6] the value of K_L is given without giving the corresponding value of α. In addition in some papers, for example in [4], the data on the lattice thermal conductivity of $Bi_2Te_{3-x}Se_x$ solid solutions with low selenium content are given without considering intrinsic conductivity and are thus increased. Another cause of discrepancy among various data on the lattice thermal conductivity, particularly for the Bi_2Te_3–Sb_2Te_3–Sb_2Se_3 and Bi_2Te_3–Bi_2Se_3 systems, is the proper choice of the Lorentz number that is used to calculate the electronic thermal conductivity.

Below the intrinsic range, i.e. when there is no ambipolar contribution to the thermal conductivity, the lattice thermal conductivity is determined by taking the difference between the measured total thermal conductivity and the electronic thermal conductivity calculated from the measured electrical conductivity. The Lorentz number L which relates the electronic thermal conductivity to the electrical conductivity is a function of the type of carrier scattering and the position of the Fermi level. In these systems the electronic thermal conductivity is a major part of the total thermal conductivity, so the differences in the Lorentz number can lead to large differences in the calculated conductivity.

5.1.1 *p-type (Bi, Sb)$_2$(Te, Se)$_3$ alloys*

In face of all the uncertainties the results of our measurements of the thermal conductivity of p-type alloys $(Sb_2Te_3)_{73}(Bi_2Te_3)_{24}(Sb_2Se_3)_3$ doped with 0.1 wt% or less of Pb are in good agreement with other published values of the thermal conductivity. Fig. 1 shows the temperature dependence of the thermal conductivity of the p-type pseudoternary alloys over the temperature range of measurements. It is well known that the thermal conductivity of pressed powders along the pressing direction is greater than the conductivity perpendicular to the pressing direction. The thermal conductivity anisotropy ratio in the two principle directions for these pseudoternary alloys is 1.60 at room temperature which is smaller than the value of 2 for Bi_2Te_3. This anisotropy ratio of our alloys is almost constant over the whole temperature range of measurements. The total thermal conductivity (Fig. 1) decreases with increasing temperature as one would expect from the consideration of the electronic contribution to the thermal conductivity. This contribution decreases as the temperature increases. However, beyond a temperature of 485 K the thermal conductivity is seen to rise again. This general trend is also exhibited by the near intrinsic samples investigated by several other workers [7, 8]. Fig. 1 shows also the temperature dependence of $K - K_e$. Proportionality of $K - K_e$ to $1/T$ is not found. This deviation may be attributed to an ambipolar contribution arising from the diffusion of electron–hole pairs down the temperature gradient with the onset of intrinsic conduction. The ambipolar contribution is seen more clearly by the rise of $K - K_e$ at higher temperatures. Testardi et al. [9] and Schwartz [10] found a similar rise and deviation from proportionality of $K - K_e$ to $1/T$. It is clear from Fig. 1 that there is, in any case, an appreciable ambipolar component in the $K - K_e$ values of all the samples. The onset of intrinsic behaviour in sample No. HP 28A above 400 K is also evident through a decrease in the slope of the Seebeck coefficient–temperature curve in Fig. 4a.

To avoid this deleterious ambipolar effect, the materials have to be doped well above the intrinsic conduction. However, with the small band gaps of the compound (Bi, Sb)$_2$ (Te, Se)$_3$ this degree of doping results in degeneracy and, consequently, lowering of the figure of merit. The requirement of large band gap with respect to KT, therefore, is obvious.

The ambipolar contribution to the total thermal conductivity at room temperature in the present alloys is considered negligible and therefore, $K - K_e$ may be taken as the lattice thermal conductivity of the samples at room temperature.

The room temperature value of 9.15 mW cm^{-1}K^{-1} of the lattice thermal conductivity of the present measurement in the pressing direction (sample No. HP 28A) is in good agreement with the results for other alloys of similar composition. Probably our results can be best compared with those of Wendler and Schreiner [5]. They have reported a value of 8.76 mW $cm^{-1}K^{-1}$ for a sintered and pressed p-type alloy of composition $(Sb_2Te_3)_{74}(Bi_2Te_3)_{26}$ doped with 0.01 wt% of Pb. A comparison of the room temperature value of K_L of the present work with those reported in the literature is shown in Table 2.

However, it should be noted that all the above mentioned measurements (with the exception of those of Wendler and Schreiner [5]) were made either on zone refined or Bridgman-grown polycrystalline materials while those used in the present measurements are sintered and pressed powders.

Therefore their work cannot be compared directly with ours.

There is no available published work on alloys of similar composition in the higher temperature range. Hence we cannot compare our high-temperature results. In order to calculate the ambipolar contribution to the total thermal conductivity we must know the electron and hole conductivities and the energy band gaps of the alloys.

Table 2

Lattice thermal conductivity of p-type alloys at room temperature (K_L is expressed in mW $cm^{-1}K^{-1}$)

composition	impurity	K_L	reference
$(Sb_2Te_3)_{73}(Bi_2Te_3)_{24}(Sb_2Se_3)_3$	0.1 wt% Pb	9.15	present work
$(Sb_2Te_3)_{74}(Bi_2Te_3)_{25}$	0.01 wt% Pb	8.76	Wendler and Schreiner [5]
$(Sb_2Te_3)_{75}(Bi_2Te_3)_{25}$	excess of Te	9.0	Champness et al.[7]
$(Sb_2Te_3)_{72}(Bi_2Te_3)_{25}(Sb_2Se_3)_3$	Te	9.6	Yim et al. [11]
$(Sb_2Te_3)_{74}(Bi_2Te_3)_{22}(Bi_2Se_3)_4$	Te	10.0	Gramberg and Plust [8]
$(Sb_2Te_3)_{74}(Bi_2Te_3)_{26}$	I	9.5	Goldsmid [12]
$(Sb_2Te_3)_{73}(Bi_2Te_3)_{27}$	undoped	10.7	Testardi et al. [9]
$(Sb_2Te_3)_{72}(Bi_2Te_3)_{28}$	excess of Bi	5.0	Rosi et al. [13]

5.1.2 *n-type* $(Bi)_2(Te, Se)_3$ *alloys*

The total thermal conductivity (Fig. 2 and 3) for the n-type alloys of molar composition $(Bi_2Te_3)_{80}(Bi_2Se_3)_{20}$ with excess of tellurium and doped with i. SbI_3 and ii. AgI decreases with increasing temperature and reaches a minimum at about 480 K for the samples doped with SbI_3 and at about 420 K for the samples doped with AgI and it rises again after these minima. This is the general trend for the near intrinsic samples. The thermal conductivity anisotropy ratios in the two principal directions for n-type alloys doped with i. SbI_3 and ii. AgI are about 1.6 and 1.5, respectively. These ratios are almost constant over the whole temperature range. Fig. 2 and 3 show the temperature dependence of K and $K - K_e$ for the n-type alloys. Again no proportionality of $K - K_e$ to $1/T$ is found in these alloys. $K - K_e$ shows a minimum at about 480 K for the SbI_3 doped alloys and at about 420 K for the AgI doped alloy. Beyond these minima $K - K_e$ increases sharply with the increase of temperature. This deviation of $K - K_e$ may again be attributed to the ambipolar contribution. This contribution is more evident at high temperatures. The onset of intrinsic behaviour above 480 K for SbI_3 doped samples and above 420 K for AgI doped samples is also evident from the deviation of the almost straight lines of Fig. 4a at high temperatures where a plot of α against $\ln T$ is shown.

The room temperature values of $K - K_e$ of 9.50 mW $cm^{-1}K^{-1}$ and 9.25 mW $cm^{-1}K^{-1}$ in the pressing direction for samples doped with SbI_3 and AgI, respectively, are in good agreement with other published values of the lattice thermal conductivity. Wendler and Schreiner [5] have reported a value of 8.40 mW $cm^{-1}K^{-1}$ for a sintered and pressed n-type alloy of composition $(Bi_2Te_3)_{80}(Bi_2Se_3)_{20}$ doped with 0.075 wt% of CuBr.

Gordiakova et al. [14] have reported a value of 8.00 mW $cm^{-1}K^{-1}$ for iodine doped n-type alloys of the above mentioned composition. These alloys were sintered and pressed. Penn [15] studied the thermoelectric properties of n-type alloys and their relationship to those of pressed powder samples. He found a value of 8.5 mW $cm^{-1}K^{-1}$ at room temperature for a pressed n-type alloy of composition $(Bi_2Te_3)_{85}(Bi_2Se_3)_{15}$. The slightly larger values of our samples are believed to be due to the ambipolar contribution. It also suggests a Lorentz number slightly greater than 2. Some other workers have also studied n-type alloys of similar composition but different methods of preparation and different dopings make any comparison of results hazardous. However the reported values of the lattice thermal conductivity of their samples are similar to ours. For example, Rosi et al. [13] have reported a value of 9.0 mW $cm^{-1}K^{-1}$, Goldsmid [12] a value of 10.80 mW $cm^{-1}K^{-1}$, Champness et al. [16] a value of 12.5 mW $cm^{-1}K^{-1}$, and Birkholz [17] a value of 12.0 mW $cm^{-1}K^{-1}$ for n-type alloys of composition $(Bi_2Te_3)_{80}(Bi_2Se_3)_{20}$.

As in the case of p-type alloys, a comparison of high-temperature results of n-type alloys is impossible because of a lack of data. To make a more complete analysis for both p-type and n-type alloys one should

1. measure the Hall coefficient and Hall mobility,
2. measure the energy band gap, and
3. know the carrier concentrations.

5.2 Temperature dependence of α, ϱ, and Z of doped pseudoternary and pseudobinary alloys

Fig. 4a shows the variation of the Seebeck coefficient α with $\ln T$ over the temperature range of measurements for the pseudoternary Bi_2Te_3–Sb_2Te_3–Sb_2Se_3 alloys and pseudobinary Bi_2Te_3–Bi_2Se_3 alloys. Goldsmid [18] showed that the Seebeck coefficient of a nondegenerate extrinsic semiconductor having spherical or ellipsoidal bands should vary linearly with $\ln T$ in a temperature range where the carrier concentration remains constant. The plot of the Seebeck coefficient α against $\ln T$, according to this theory, should give a straight line of slope $3\,k/2\,e$ (or 129 μV/K). Fig. 4a, in fact, shows almost straight lines over a limited range. The slopes of the n-type samples doped with i. SbI_3 and ii. AgI are respectively 119 and 70 μV/K while the slope for the p-type sample is 170 μV/K. These compare favourably with the theoretical value of 129 μV/K since such factors as a rise in the density of states effective mass with temperature, a warped energy surface, and a change in the energy dependence of the relaxation time can affect these values. The curvature of the lines at high temperature is due to the onset of intrinsic conduction.

If the carrier concentration in the samples was constant around room temperature, the plot of the resistivity ϱ against absolute temperature in Fig. 4b may be regarded as indicating the inverse of the temperature variation of the mobility [18]. The average slopes of the lines give the dependence of ϱ on temperature. The average slope, in the present case, of the p-type alloy is 1.6 and for the n-type alloys it is about 1.4. For nondegenerate semiconductors with acoustic lattice scattering the mobility is expected to vary as $T^{-1.5}$ and hence the resistivity should vary as $T^{1.5}$. Since the observed temperature dependence is almost consistent with the theoretical prediction it would appear that the lattice scattering, as is assumed in our calculations for K_e, is the dominant scattering mechanism in our samples. For comparison, in pseudoternary Bi_2Te_3–Sb_2Te_3–Sb_2Se_3 alloys the reported values are 1.2 for n-type and 1.6 for p-type [11] in the range 150 to 300 K.

Comparison of the maximum values of α^2/ϱ for our samples with the values obtained by other workers shows that the values of α^2/ϱ obtained in the present work are substantially lower than the data of other workers. This can be explained by the high sensitivity of the properties of such alloys on the conditions of processing, on structure of materials, on crystallization or growth condition. Therefore, it seems that the present samples are lacking perfectness in some of these criteria.

In most materials, the figure of merit is a function of temperature and this temperature dependence may differ widely for different materials. In the present case (Fig. 5) the Z value attains a maximum of 17×10^{-4} K^{-1} for the p-type alloys and 16×10^{-4} K^{-1} for the n-type alloys at about 370 K and it remains relatively constant over the temperature range 370 to 480 K. Therefore, a thermoelectric generator employing these alloys for its n- and p-branches, will have its highest efficiency or coefficient of performance over this range. It would appear that the decrease of Z after this range is due in some measure to the ambipolar thermal conductivity. It is also apparent from Fig. 5 that the SbI_3 doped n-type alloys are more suitable than AgI doped n-type alloys.

The Z values in our samples at room temperature are inferior to those reported in the literature. This is probably due in some measure to the heavy doping of the alloys to increase the figure of merit at a high temperature, since heavy doping raises the temperature at which the material becomes intrinsic.

6. Conclusions

The agreement with the theory of our results of the lattice thermal conductivity is qualitative. To make a more complete analysis of the thermal conductivity results of the present work one has to separate the ambipolar contribution from the total thermal conductivity. Therefore one should measure the Hall coefficient and the energy band gap of these alloys and one should also have knowledge of the carrier concentrations of the samples.

The Z values for all of our samples are relatively constant over a temperature range of 370 to 480 K and after this range the figure of merit decreases rapidly with increasing temperature due in some measure to the ambipolar thermal conductivity. Therefore, a thermoelectric generator employing these alloys for its n- and p-branches, will have its highest efficiency over a temperature range of 370 to 480 K. It is also apparent that the SbI_3 doped n-type alloys are more suitable as thermoelectric materials than AgI doped n-type alloys.

Acknowledgements

The authors would like to acknowledge the assistance of Mrs. C. Cerstelotte in the experimental work and wish to express their gratitude to the Metallurgie Hoboken, Olen (Belgium) for supplying the samples and the data of Seebeck coefficient and electrical resistivity. One of the authors (M.I.) is grateful to the Belgian Ministry of Education and Culture for a grant.

References

[1] A. F. Ioffe, Canad. J. Phys. **B34**, 1342 (1956).
[2] H. Schreiner and F. Wendler, Z. Metallk. **52**, 218 (1961).
[3] H. Schreiner, Österr. Chem. Z. **63**, 37 (1962).
[4] U. Birkholz and G. Haacke, Z. Naturf. **17a**, 161 (1962).
[5] F. Wendler and H. Z. Schreiner, Z. Metallk. **56**, 853 (1965).
[6] O. Beckman, P. Bergvall, and T. Tripathi, Ark. Fys. **28**, 215 (1964).
[7] C. H. Champness, P. T. Chiang, and P. Parekh, Canad. J. Phys. **43**, 653 (1965).
[8] G. von Gramberg and H. G. Plust, Helv. phys. Acta **36**, 237 (1963).
[9] L. R. Testardi, J. N. Bierly, and F. J. Donahoe, J. Phys. Chem. Solids **23**, 1209 (1962).
[10] R. J. Schwartz, J. appl. Phys. **38**, 2865 (1967).
[11] W. M. Yim, E. V. Fitzke, and F. D. Rosi, J. Mat. Sci. **1**, 52 (1966).
[12] H. J. Goldsmid, J. appl. Phys. **32**, 2198 (1961).
[13] F. D. Rosi, B. Abeles, and R. V. Jensen, J. Phys. Chem. Solids **10**, 191 (1959).
[14] G. N. Gordiakova, G. V. Kokosh, and S. S. Sinani, Soviet Phys. — J. tech. Phys. **3**, 1 (1958).
[15] A. W. Penn, Advanced Energy Conversion **7**, 257 (1968).
[16] C. H. Champness, W. B. Muir, and P. T. Chiang, Canad. J. Phys. **45**, 3611 (1967).
[17] U. von Birkholz, Z. Naturf. **13a**, 780 (1958).
[18] H. J. Goldsmid, Proc. Phys. Soc. **71**, 633 (1958).

(Received January 28, 1972)

phys. stat. sol. (a) **10**, 425 (1972)

Subject classification: **14.2**

Zentralinstitut für Festkörperphysik und Werkstofforschung der Deutschen Akademie der Wissenschaften zu Berlin, Dresden

Magnetization and Size Effect of Hard Superconductors

By

A. HANDSTEIN

For cylindrical samples of hard superconductors formulae basing upon the theory of Irie and Yamafuji are deduced and used for a calculation of the irreversible portion of magnetization. It is possible to represent the irreversible magnetization in a simple reduced form. The values of maximum magnetization of specimens of a Nb–64 at% Ti alloy with different thickness and annealing fit to the straight line $-4\pi\,\overline{M}_{max} = 0.45_8\,H_{max} + 31.6$. $H_{c1} = 58$ Oe and the reversible magnetization curve of this alloy were determined. By these the irreversible portion of experimental magnetization curves is determined and compared with the calculations. The causes of deviations are discussed.

Auf der Grundlage der Theorie von Irie und Yamafuji wird ein Formalismus für die Berechnung des irreversiblen Anteils der Magnetisierung zylindrischer Proben harter Supraleiter abgeleitet. Die irreversible Magnetisierung kann in einer vereinfachten reduzierten Form dargestellt werden. Die Werte der maximalen Magnetisierung von Proben aus Nb–64 At% Ti liegen auf der Geraden $-4\pi\,\overline{M}_{max} = 0{,}45_8\,H_{max} + 31{,}6$. Für diese Legierung wurden $H_{c1} = 58$ Oe und die reversible Magnetisierungskurve bestimmt. Damit wird der irreversible Anteil der experimentellen Magnetisierungskurven bestimmt und mit den Rechnungen verglichen. Die Ursachen von Abweichungen werden diskutiert.

1. Introduction

The hard superconductors have magnetization curves differing much from those of ideal type II superconductors (Fig. 1). The fluxoids penetrating into the bulk sample at applied fields $H > H_{c1}$ interact with the defects in the crystal lattice of hard superconductors. Thereby in the bulk material a gradient of flux density is formed. The concept of the critical state introduced by Bean [1] and Kim et al. [2] was the basis for a far extending comprehension of the behaviour of hard superconductors. Above all it enabled to understand the size effect, that means the dependence of magnetization of the sample thickness.

The irreversible portion $-4\pi\,M^{irr}$ causes a rise of magnetization at increasing applied fields $H > H_{c1}$. In this range a fluxoid front penetrates into the bulk sample and attains the centre at the penetration field H_p (branch 1 and 2, respectively). After passing through a maximum $-4\pi\,M^{irr}$ decreases because the pinning force is commonly lowered by the increasing flux density. At the upper critical field H_{c2} the bulk sample turns into the resistive state. A following reduction of the applied field causes the fluxoid gradient to be formed in reversed direction (branch 3). At $H = 0$ the sample retains the remanent magnetization $4\pi\,M(0)$.

Several models exist in which the relation between the gradient of flux density $\partial B/\partial r$ and the induction B was fitted to the models by adjusting the parameters

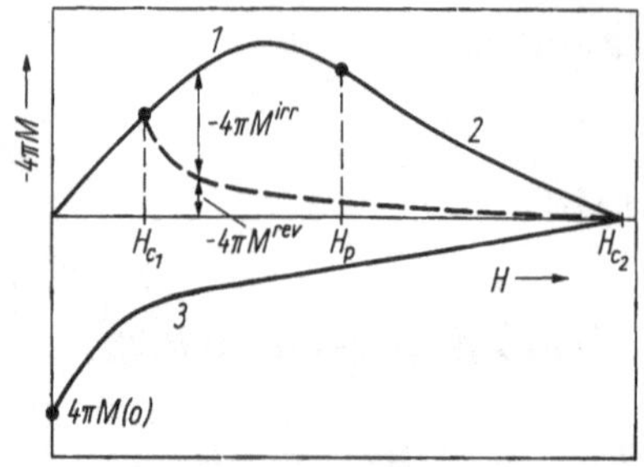

Fig. 1. Magnetization curve of a hard superconductor (schematically)

[1 to 5]. On account of its general formulation the phenomenological expression of Irie and Yamafuji [5] is especially suitable for a description of the irreversible behaviour of magnetization. For that reason it is used for the calculation in the case of cylindrical samples in a longitudinal applied field.

2. Theoretical Considerations

For the gradient of flux density we obtain from [5] the relation

$$\frac{\partial B^2}{\partial r} \pm \alpha\, B^\gamma = 0\,. \tag{1}$$

The sign is defined by the branch of the magnetization curve. The parameter α describes the intensity of the interaction between pinning centres and fluxoids. The parameter γ depends on the density and the distribution of pinning centres and its value varies within $0 \leqq \gamma \leqq 1$. Equation (1) is only valid in the bulk sample and describes the irreversible behaviour. Therefore the values to be calculated are marked by the subscript "irr". First of all the difference $H - B_0$ of applied field H and the induction B_0 inside the sample near to the surface is neglected, because in hard superconductors these values are mostly small. By integration of equation (1) we obtain the behaviour of the induction $B = B(r, H)$ on the single branches of the magnetization curve. Substitution of $B = B(r, H)$ into Maxwell's equation

$$-4\,\pi\,\overline{M} = H - \frac{1}{\pi\, R^2\, l} \int\limits_0^R B(r, H)\, 2\,\pi\, r\, l\, \mathrm{d}r \tag{2}$$

yields the irreversible portions of the mean magnetization of a cylindrical sample of radius R (length $l \gg R$).

2.1 Branch 1: $0 \leqq H \leqq H_p^{\mathrm{irr}}$

$$\left.\begin{aligned} B(r, H) &= 0\,, && 0 \leqq r \leqq r_1\,,\\ B(r, H) &= \left[H^{2-\gamma} - \frac{2-\gamma}{2}\,\alpha\,(R - r)\right]^{1/(2-\gamma)}, && r_1 \leqq r \leqq R\,. \end{aligned}\right\} \tag{3}$$

The radius r_1 is found from $B(r_1, H) = 0$:

$$r_1 = R - \frac{2}{(2-\gamma)\,\alpha}\, H^{2-\gamma}\,. \tag{4}$$

On the condition $r_1 = 0$ we obtain the penetration field

$$H_p^{irr} = \left(\frac{2-\gamma}{2} \alpha R\right)^{1/(2-\gamma)} . \tag{5}$$

Substitution of equation (3) into (2) yields

$$-4\pi \overline{M}^{irr} = H\left[1 - \frac{4}{\alpha R^2 (3-\gamma)}\left(R - \frac{2}{\alpha(5-2\gamma)} H^{2-\gamma}\right) H^{2-\gamma}\right]. \tag{6}$$

On branch 1 there is the maximum of the magnetization curve following from the condition $d(-4\pi \overline{M}^{irr})/dH = 0$:

$$\left.\begin{aligned} H_{max}^{irr} &= \left(\frac{\alpha R}{4}\right)^{1/(2-\gamma)} \left[(3-\gamma)\left(1 - \sqrt{\frac{1-\gamma}{3-\gamma}}\right)\right]^{1/(2-\gamma)}, \\ H_{max}^{irr} &= \left(\frac{\alpha R}{4}\right)^{1/(2-\gamma)} f_2(\gamma)^{1/(2-\gamma)}, \end{aligned}\right\} \tag{7}$$

and

$$-4\pi \overline{M}_{max}^{irr} = H_{max}^{irr} \frac{2-\gamma}{5-2\gamma}\left(1 + \sqrt{\frac{1-\gamma}{3-\gamma}}\right) = H_{max}^{irr} f_1(\gamma) . \tag{8}$$

According to equation (8) the ratio $(-4\pi \overline{M}_{max}^{irr})/H_{max}^{irr}$ depends on γ only. At the penetration field the magnetization amounts to

$$-4\pi \overline{M}_p^{irr} = \frac{1}{5-2\gamma} H_p^{irr} . \tag{9}$$

2.2 Branch 2: $H_p^{irr} \leqq H < H_{c2}$

$$B(r, H) = \left[H^{2-\gamma} - \frac{2-\gamma}{2} \alpha (R - r)\right]^{1/(2-\gamma)}, \qquad 0 \leqq r \leqq R . \tag{10}$$

Substituting of equation (10) into (2) we obtain

$$-4\pi \overline{M}^{irr} = H + \frac{8}{\alpha^2 R^2 (3-\gamma)(5-2\gamma)} \times$$

$$\times \left[H^{5-2\gamma} - \left(H^{2-\gamma} - \frac{2-\gamma}{2} \alpha R\right)^{\frac{5-2\gamma}{2-\gamma}}\right] - \frac{4}{\alpha R (3-\gamma)} H^{3-\gamma} . \tag{11}$$

Apart from H_p^{irr} there are not any marked points on this branch. The calculation is not valid for $H \to H_{c2}$ because here the magnetization values do not tend towards zero.

2.3 Branch 3: $0 \leqq H < H_{c2}$

$$B(r, H) = \left[H^{2-\gamma} + \frac{2-\gamma}{2} \alpha (R - r)\right]^{1/(2-\gamma)}, \qquad 0 \leqq r \leqq R . \tag{12}$$

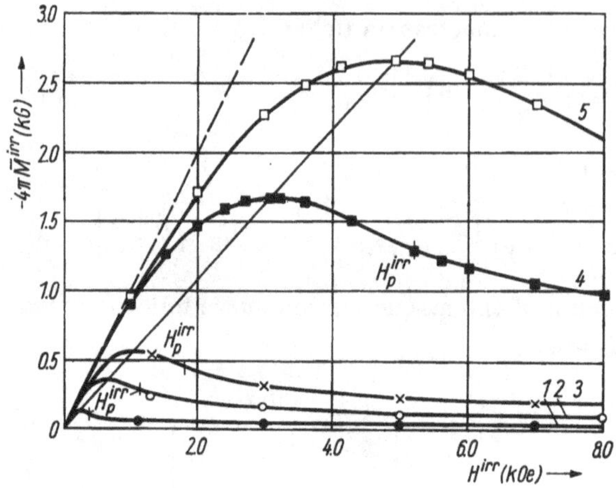

Fig. 2. Calculated magnetization curves for variable values of α and constant values $\gamma = 0.5$ and $R = 10^{-2}$ cm.
(1) $\alpha = 1 \times 10^6\ G^{1.5}\ cm^{-1}$;
(2) $\alpha = 5 \times 10^6\ G^{1.5}\ cm^{-1}$;
(3) $\alpha = 1 \times 10^7\ G^{1.5}\ cm^{-1}$;
(4) $\alpha = 5 \times 10^7\ G^{1.5}\ cm^{-1}$;
(5) $\alpha = 1 \times 10^8\ G^{1.5}\ cm^{-1}$

Substituting of equation (12) into (2) yields

$$4\pi M^{irr} = \frac{8}{\alpha^2 R^2 (3-\gamma)(5-2\gamma)} \times$$

$$\times \left[\left(H^{2-\gamma} + \frac{2-\gamma}{2}\,\alpha R \right)^{\frac{5-2\gamma}{2-\gamma}} - H^{5-2\gamma} \right] - \frac{4}{\alpha R (3-\gamma)} H^{3-\gamma} - H\,. \quad (13)$$

At the field strength $H = 0$ the sample retains the remanent magnetization

$$4\pi \overline{M}(0) = H_p^{irr} \frac{2(2-\gamma)^2}{(3-\gamma)(5-2\gamma)} = H_p^{irr} f_3(\gamma)\,. \quad (14)$$

These equations were numerically evaluated by a computer. For a constant value γ a family of magnetization curves is obtained. The values of the maximum magnetization fit a straight line independent of α or R. For $\gamma = 0.5$ and $R = 10^{-2}$ cm this behaviour was represented for different values α in Fig. 2. The similarity of the magnetization curves in Fig. 2 causes us to try a represen-

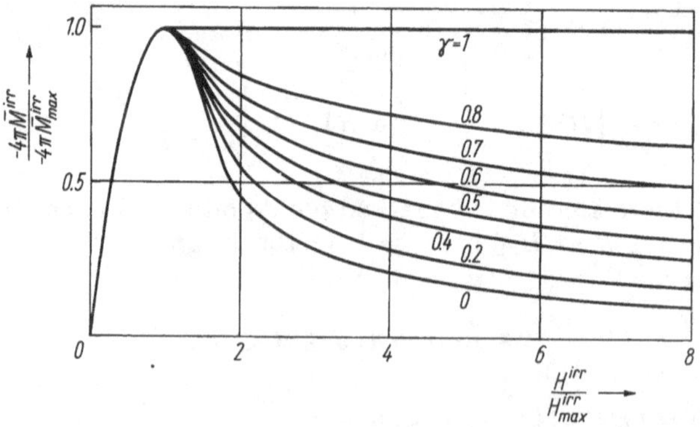

Fig. 3. Reduced magnetization curves $(-4\pi \overline{M}^{irr})/(-4\pi \overline{M}^{irr}_{max}) = f(H^{irr}/H^{irr}_{max})$ for $0 \leq \gamma \leq 1$

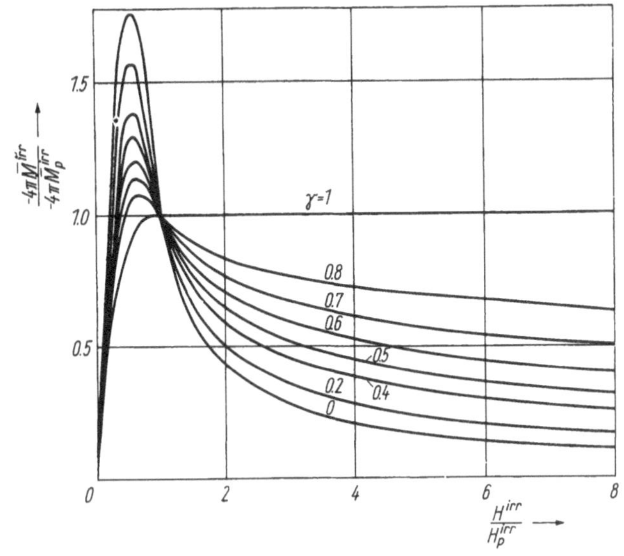

Fig. 4. Reduced magnetization curves $(-4\pi\bar{M}^{irr})/(-4\pi\bar{M}_p^{irr}) = f(H^{irr}/H_p^{irr})$ for $0 \leq \gamma \leq 1$

tation of the curves in a simple form analogous to the principle of corresponding states. The reduction of the curves with the values at the maximum and at the penetration field shows this to be possible (Fig. 3 and 4, respectively). In both cases we obtain a family of curves the behaviour of which is only determined by the parameter γ. In Fig. 3 at a field strength $H^{irr}/H_{max}^{irr} < 1$ the splitting of the curves is not distinct because the single curves are close together. In the equations (3) to (11) α and R exist only as the product (α R). Consequently they have always the same exponent. A doubling of R at a constant value α has formally the same influence as a doubling of α at constant R, etc. This behaviour is interesting with regard to the experimental results.

3. Preparation of Samples and Experimental Technique

The specimens were prepared from a Nb–64 at% Ti alloy. Pieces of wire were taken out at different stages of wire drawing. Thus each specimen has a different strength of cold deformation. The diameters amounted to $D = 0.16$, 0.24, 0.32, 0.40, and 0.48 mm. In order to enlarge the number of different specimens they were annealed at 400 °C for increasing and at 600 °C for decreasing magnetization values, respectively. The heat treatment was carried out for one hour.

The external magnetic field was supplied by a superconducting solenoid. The magnetization was measured by a ballistic method at 4.2 K. An 11000 turn pick-up coil and a bundle of 3 to 20 wires according to their diameter were used to obtain sufficiently large deflections of the galvanometer. The length of each sample was 19 mm. The demagnetization factor was neglected because of the relative error of magnetization measurements of about 6%.

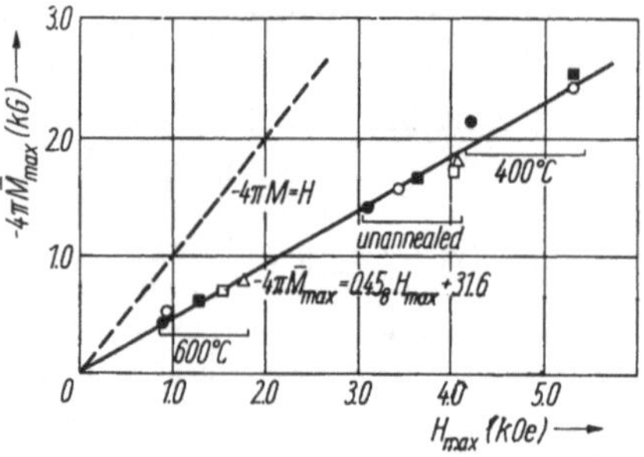

Fig. 5. Maximum magnetization $-4\pi \bar{M}_{max}$ vs. field strength H_{max}.
◉ $D = 0.16$ mm, unannealed or 400 °C, or 600 °C;
○ $D = 0.24$ mm, unannealed or 400 °C, or 600 °C;
■ $D = 0.32$ mm, unannealed or 400 °C, or 600 °C;
□ $D = 0.40$ mm, unannealed or 600 °C;
△ $D = 0.48$ mm, unannealed or 600 °C

4. Experimental Results

4.1 Measurements of magnetization

The magnetization curves of the various specimens form a family of curves similar to those of Fig. 2 or 8. The maximum values of magnetization $-4\pi \bar{M}_{max}$ fit the straight line ($-4\pi \bar{M}_{max}$ and H_{max} in G)

$$-4\pi \bar{M}_{max} = 0.45_8 H_{max} + 31.6 \tag{15}$$

obtained by a special curve fitting technique. In Fig. 5 the values of 13 samples were plotted against H_{max}. It is seen that the values $-4\pi \bar{M}_{max}$ of non-annealed specimens with different diameters fit the same straight line as those of the specimens with the same diameter and various defect structures owing to annealing. In these experiments the size effect is superimposed to the differently strong pinning effect of the defect structures obtained by wire drawing and annealing. The magnetization values of samples with a diameter $D \geqq 0.40$ mm annealed at 400 °C were lowered by flux jumps. These could not be taken into consideration.

4.2 Determination of H_{c1} and of the reversible magnetization curve

In papers [6, 7] it was demonstrated that the straight line of maximum magnetization can be used for a determination of the lower critical field H_{c1}. We extrapolate the values $-4\pi \bar{M}_{max}$ of the straight line (15) to that of an ideal sample of a type II superconductor. At the intersection point of (15) with the straight line $-4\pi M = H$ of ideal diamagnetism we obtain the lower critical field $H_{c1} = 58$ Oe for the Nb–64 at% Ti alloy. This value corresponds well to the results of other Nb–Ti alloys (Fig. 6).

The upper critical field $H_{c2}(4.2\text{ K}) = 103$ kOe was found from $H_{c2}(1.2\text{ K}) = 145$ kOe [10] and the temperature dependence given by Coffey [11]. The thermodynamic critical field $H_c(4.2\text{ K}) = 1.20$ kOe was estimated from the equation

$$H_{c1} = \frac{H_c^2}{H_{c2}} \left(\ln \frac{H_{c2}}{\sqrt{2}\, H_c} + 0.08 \right) \tag{16}$$

obtained by elimination of $\varkappa$ from Abrikosov's solution of the Ginzburg-Landau equation (reference [12]). $\varkappa \approx 61$ was estimated from equation

$$H_{c2} = \sqrt{2}\, \varkappa\, H_c\,. \tag{17}$$

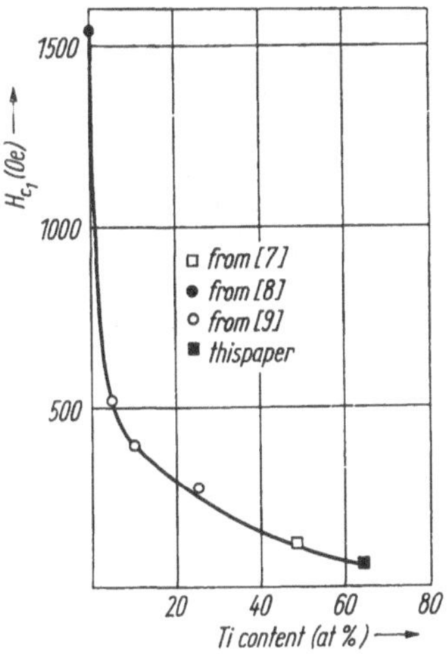

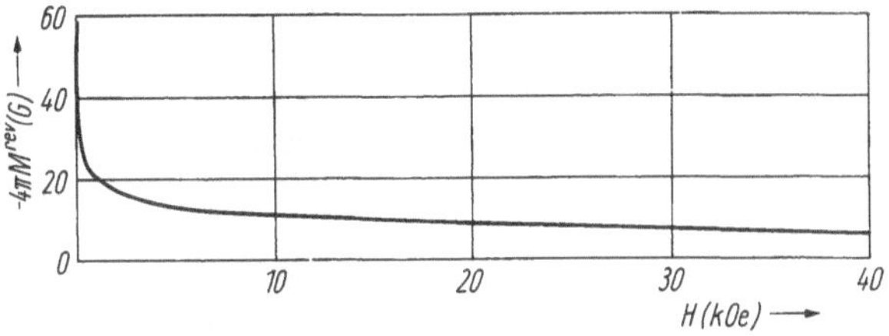

Fig. 7. Reversible magnetization curve for $\varkappa \approx 61$, $H_{c1} = 58$ Oe, and $H_{c2} = 103$ kOe at 4.2 K

Fig. 6. Lower critical field H_{c1} of Nb–Ti alloys vs. Ti content at 4.2 K

The values H_{c1}, H_{c2}, and $\varkappa$ permit to determine the reversible magnetization curve according to the calculations of Koppe and Willebrand [13]. The diagram of the reversible magnetization curves of these authors was extrapolated to $\varkappa \sqrt{2} \approx 86$. Then the reversible curve of Nb–64 at% Ti was calculated (Fig. 7).

4.3 Determination of the irreversible portions of magnetization

Considering exactly the distribution of the fluxoid density inside a hard superconductor on branches 1 or 2, we obtain in the bulk sample a magnetic field

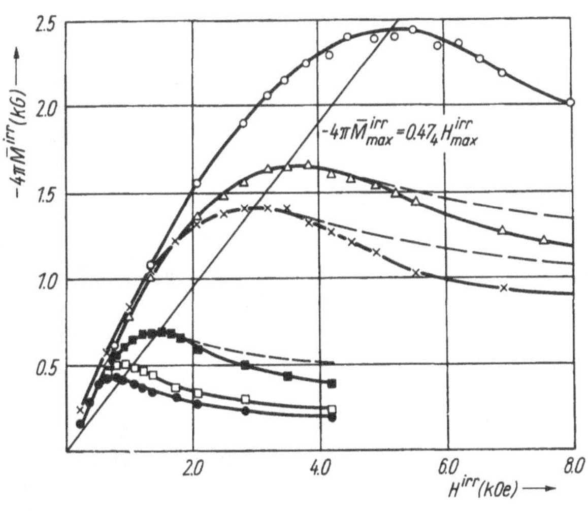

Fig. 8. Irreversible portion of experimental magnetization curves. Dashed lines calculated ($\gamma = 0.76$). ○ $D = 0.24$ mm, 400 °C; △ $D = 0.32$ mm, unannealed; × $D = 0.16$ mm, unannealed; ■ $D = 0.40$ mm, 600 °C; □ $D = 0.24$ mm, 600 °C; ● $D = 0.16$ mm, 600 °C

Table 1

Characteristic values of the samples (values at the maximum and at $H = 0$ from the experiment, H_p^{irr} calculated, $\gamma = 0.76$)

sample No.	1	2	3	4	5	6	7	8	11	12	13	14	15
annealing (°C)	—	—	—	—	—	400	400	400	600	600	600	600	600
diameter D (mm)	0.16	0.24	0.32	0.40	0.48	0.16	0.24	0.32	0.16	0.24	0.32	0.40	0.48
$-4\pi \overline{M}_{max}^{irr}$ (G)	1410	1560	1660	1745	1770	2145	2440	2550	425	508	610	690	770
H_{max}^{irr} (Oe)	3000	3450	3590	3830	4000	4130	5150	5200	770	880	1220	1450	1610
H_p^{irr} (Oe)	4452	5023	5285	5598	5765	6451	7673	7884	1243	1456	1868	2167	2410
$\bar{\alpha}$ (10^6 G$^{1.24}$ cm^{-1})	6.74_2	5.22_0	4.16_9	3.58_2	3.09_6	10.67_8	8.82_7	6.84_6	1.38_6	1.12_4	1.14_8	1.10_4	1.05_0
$4\pi \overline{M}(0)$ (G)	1334	614	1798	1748	1922	2530	2472	2430	497	582	730	853	947

strength H reduced by an amount depending on a diamagnetic screening field. According to Fig. 7 in thin samples or those with a small irreversible amount of magnetization the portion of reversible magnetization certainly has an influence. Therefore the experimental magnetization values were corrected by the values from Fig. 7. The irreversible portion of several magnetization curves is shown in Fig. 8. The maximum values of the curves fit the straight line

$$-4\pi \overline{M}_{max}^{irr} = 0.47_4\, H_{max}^{irr} \tag{18}$$

intersecting the origin by definition. According to equation (8) the parameter $\gamma = 0.76$ results from $f_1(\gamma) = 0.47_4$. $\gamma = 0.76$ describes the irreversible behaviour of all samples investigated at low fields. The pinning strength $\bar{\alpha}$ and the magnetization curves can be determined by adapting the corrected experimental curves to equations (7) and (8), respectively.

It is seen from Table 1 that the values of the pinning strength $\bar{\alpha}$ increase with increasing degree of cold deformation and decreasing diameter. This influence of cold deformation has been known already for a long time [14]. After annealing the sequence of $\bar{\alpha}$ remains. Their difference increases after annealing at 400 °C, because the fibre structure [15] and the quantity of nuclei for precipitation processes are strongly influenced by the degree of cold deformation [16]. After annealing at 600 °C $\bar{\alpha}$ alters merely by about 24% over the whole range.

The size effect appears well in samples 1 and 8. Both have almost the same value $\bar{\alpha}$, By doubling the diameter D $-4\pi M_{max}^{irr}$, H_{max}^{irr}, and $\overline{H}_p^{irr}$ increase by about 77%.

The remanent magnetization of several samples (cf. Table 1) is falsified by flux jumps. $4\pi \overline{M}(0)$ can be calculated from equation (14) with $f_3(\gamma) = 0.395$. According to the data of Table 1 a good agreement exists only for samples 1 to 5 with bad superconducting properties and sample 6.

5. Discussion

At higher fields $H > H_{max}$ the three calculated curves in Fig. 8 (dashed lines) show the calculated magnetization values to be larger than the experimental ones. We obtain this fact at the other samples, too. There are mainly two causes for this. The phenomenological theory fails at higher fluxoid densities, since is does not take into consideration the drop of the order parameter ψ and, therefore, in the theory the gradient $\partial B/\partial r$ does not tend towards zero at $H \to H_{c2}$. For that reason several authors [17 to 20] introduced a correction factor

$$\left(1 - \frac{H}{H_{c2}}\right)^n \tag{19}$$

with $1 \leqq n \leqq 2$. This factor takes into consideration the real physical behaviour in the range $H_{c1} \ll H \leqq H_{c2}$. We obtained a better agreement of calculated curves with experimental ones likewise by use of equation (19).

Further on adapting the theory to the experiment was realized at the maximum of magnetization curves. Here the front of flux lines has not still attained the centre of the sample. The parameters α and γ obtained at the maximum describe only the behaviour of the irreversible magnetization of the external range of the cylindrical sample, because the mean magnetization likewise regards essentially the behaviour of this range. Here the pinning force density is supposed to be greater than in the interior of the sample. This can be caused by the heterogeneous distribution of the fibre texture over the cross section of the sample and by atoms possibly diffused into the external range. Unfortunately the penetration field H_p cannot be taken directly from the magnetization curves in order to realize a fit at H_p. It turned out that the values of γ obtained by the attempt to fit H_p^{irr} from the equations (5) and (9) iteratively to the experimental values are $\gamma < 0.76$. According to [5] the parameters $\gamma < 0.76$ belong to a lower mean density of pinning centres. It is seen in Fig. 3 and 4 that the magnetization curves are more rigorously decreasing for low γ-values than for larger ones at increasing applied fields.

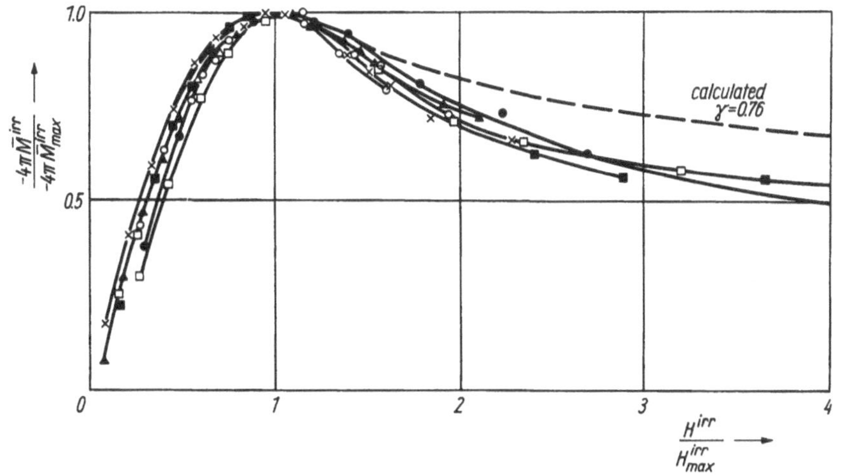

Fig. 9. Reduced magnetization curves $(-4\pi\bar{M}^{irr})/(-4\pi\bar{M}^{irr}_{max}) = f(H^{irr}/H^{irr}_{max})$, Nb–64 at% Ti. Dashed lines calculated ($\gamma = 0.76$). ○ $D = 0.24$ mm, 400 °C; ▲ $D = 0.32$ mm, unannealed; × $D = 0.16$ mm, unannealed; ■ $D = 0.40$ mm, 600 °C; □ $D = 0.24$ mm, 600 °C; ● $D = 0.16$ mm, 600 °C

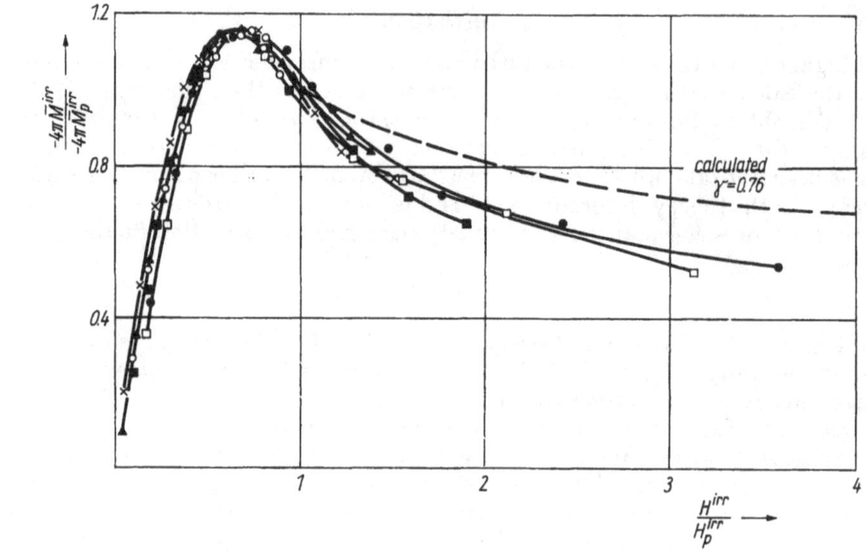

Fig. 10. Reduced magnetization curves $(-4\pi \overline{M}^{irr})/(-4\pi \overline{M}_p^{irr}) = f(H^{irr}/H_p^{irr})$, Nb–64 at% Ti (marks as in Fig. 9)

In Fig. 9 and 10 the experimental magnetization curves are plotted in a reduced form. The calculated curves deviate from the experimental ones for the reasons mentioned above. In spite of several deviations the reduced experimental curves agree well but they overlap at some points. This fact is supposed to be caused by a different dependence of the pinning forces of various defect types on the magnetic field strength and above all by instrumental errors. The agreement of the reduced curves confirms, however, the validity of the equations from Section 2 at variations of α and R. Because of the various types and the complicated distribution of defects the behaviour of the irreversible magnetization cannot consequently be described by only one parameter γ at different applied fields.

The phenomenological expression of Irie and Yamafuji [5] describes well the irreversible magnetization of hard superconductors. Campbell et al. investigated the magnetization of an eutectic Pb–Bi alloy [21]. The maximum values of magnetization curves of variously annealed and variously thick samples

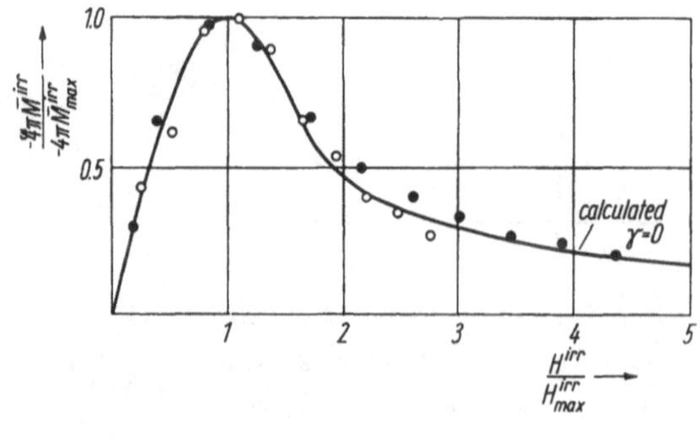

Fig. 11. Reduced magnetization curves $(-4\pi \overline{M}^{irr})/(-4\pi \overline{M}_{max}^{irr}) = f(H^{irr}/H_{max}^{irr})$ of cylindrical samples made of Pb–56 at% Bi (diameter 2 mm; magnetization values from [21]). ○ 20 °C 1 h; ● 114 °C, 0.5 h

fitted a straight line with the slope $m = 0.63 = f_1(\gamma)$ (see equation (8)). The model used by the authors for the calculations corresponds to the case $\gamma = 0$ with a small density of pinning centres. The magnetization values of two samples taken from [21] and the calculated curve for $\gamma = 0$ in the reduced form are shown in Fig. 11. The main reason for the good agreement is that the defects causing the irreversible behaviour are non-superconducting Bi particles distributed homogeneously over the cross section. In a recent paper Pfeiffer et al. applied the theory of Irie and Yamafuji to the calculation of magnetization curves of Nb–V samples. They obtained a good agreement for $\gamma = 0.2$ [22] which again confirmed the general validity.

Acknowledgements

It is a pleasure to acknowledge helpful discussions with Prof. F. Lange. Thanks are also due to Mrs. R. Findeisen for her experimental assistance.

References

[1] C. P. Bean, Phys. Letters (Netherlands) **8**, 250 (1962).
[2] Y. B. Kim, C. F. Hempstead, and A. R. Strnad, Phys. Rev. **129**, 528 (1963).
[3] J. Silcox and R. W. Rollins, Appl. Phys. Letters **2**, 231 (1963).
[4] K. Yasukōchi, T. Ogasawara, N. Usui, and S. Ushio, J. Phys. Soc. Japan **19**, 1649 (1964).
[5] F. Irie and K. Yamafuji, J. Phys. Soc. Japan **23**, 255 (1967).
[6] A. F. Prekul and N. V. Volkenštein, Fiz. Metallov i Metallovedenie **23**, 741 (1967).
[7] A. Handstein, phys. stat. sol. (a) **4**, K157 (1971).
[8] D. Eckert, private communication.
[9] M. S. Lubell and R. H. Kernohan, J. Phys. Chem. Solids **32**, 1531 (1971).
[10] T. G. Berlincourt and R. R. Hake, Phys. Rev. **131**, 140 (1963).
[11] H. T. Coffey, Phys. Rev. **166**, 447 (1968).
[12] J. Lowell, Cryogenics **5**, 185 (1965).
[13] H. Koppe and J. Willebrand, J. low-Temp. Phys. **2**, 499 (1970).
[14] J. P. McEvoy, Jr. and R. F. Decell, J. appl. Phys. **35**, 982 (1964).
[15] C. Baker, J. Mat. Sci. **5**, 40 (1970).
[16] U. Zwicker, R. Löhberg, and W. Heller, Z. Metallk. **61**, 836 (1970).
[17] H. Voigt, Z. Phys. **213**, 119 (1968).
[18] K. Dichtel and H. Koppe, J. low-Temp. Phys. **1**, 137 (1969).
[19] W. A. Fietz and W. W. Webb, Phys, Rev. **178**, 657 (1969).
[20] E. W. Urban, J. appl. Phys, **42**, 115 (1971).
[21] A. M. Campbell, J. E. Evetts, and D. Dew-Hughes, Phil. Mag. **18**, 313 (1968).
[22] K. Pfeiffer, R. Papp, and H. W. Weber, phys. stat. sol. (b) **46**, K77 (1971).

(Received January 10, 1972)

phys. stat. sol. (a) **10**, 437 (1972)

Subject classification: 13; 21.1; 22.5.2

Institute for Solid State Physics, University of Tokyo (a), and Horiba Ltd. (b), Kyoto

Compton Profile Measurements by Use of Solid State Detector

By

T. FUKAMACHI (a), S. HOSOYA (a), Y. HOSOKAWA (b), and H. HIRATA (b)

It is practically impossible to measure the Compton profile of any heavy material, because its strong absorption very much reduces the intensity. This difficulty has been overcome by use of a non-dispersive solid state detector (SSD) together with γ-rays or high energy X-rays, because the effective intensity in this case becomes 10^2 to 10^4 times stronger than with the conventional dispersive crystal analyser. As for the energy resolution, the ratio of the profile width to the resolution of an SSD can be made comparable with the conventional case by choosing a suitable radioactive γ-ray source. The preliminary measurements have already shown very promising results and suggest various applications feasible only by this innovation.

La mesure du profil de Compton était presque impossible pour le matériel relativement lourd à cause de l'absorption sensible qui affaiblit l'intensité. La difficulté peut être surmontée par l'emploi d'un détecteur à solide (SSD) nondispersif et des rayons X à haute énergie ou rayons γ, où l'intensité effective sera 10^2 á 10^4 fois plus fort par rapport de celle à réalisée avec un analysateur à cristal ordinaire. À propos de la résolution d'énergie, le rapport de la largeur de profil à la résolution d'énergie du SSD peut être comparable à celui des cas conventionel. Les mesures préliminaire indiquent que les diverses applications de cette technique serant prospères.

1. Introduction

The measurement of the Compton profile has been used for studying electron states through obtaining the electron momentum distribution. In such a study, however, it has so far been practically impossible to collect the data precise enough for a detailed analysis on relatively heavy materials [1, 2], because the intensity is very much reduced due to the absorption by the specimen itself and the analysing crystal and also due to the necessary angular limitation by a Soller slit system. Therefore, an extremely powerful X-ray generator has been wanted for extending this kind of study, for instance, to the first transition metals.

On the other hand, the use of a non-dispersive solid state detector (SSD) in the field of fluorescence X-ray spectroscopy has already drastically shortened the measuring time and very much simplified the operation [3]. In X-ray microanalysis, too, an SSD has already widely been used [3]. In the present report, it is pointed out that the above difficulty in the Compton measurements has nearly perfectly been overcome by the use of an SSD followed by a multichannel pulse height analyser. When the same X-ray source is used, the effective intensity increases by a factor of about 10^2 to 10^4 by this method. Such a big change may not be attained even by the use of a very powerful X-ray generator conceivable at present.

It is true that the SSD has less resolution than the crystal analyser in a relatively low-energy region, but the former becomes at least comparable with the latter as the energy becomes higher, as shown in Section 2. Thus γ-rays from suitable radioactive isotopes rather than high-energy X-rays are preferable, and the relatively small intensity of γ-rays is covered by the smaller absorption. This tendency is further reinforced by other factors; the higher the energy used is, the wider the Compton profile becomes, and in general the S–N ratio and the stability are more easily improved in SSD, and the error accompanying the deconvolution procedures because γ-rays have nothing like the doublet structure.

In Section 3 preliminary measurements are described using both X-rays and γ-rays as well as instruments readily accessible. The better results will be obtained, if proper instruments can be used. Moreover, the present system may become more preferable by possible future improvements of SSD such as mentioned in Section 4. This innovation, which realized the drastic shortening of the time necessary for measurements, enables us to carry out various kinds of researches so far impossible concerning the Compton effect. These possibilities will be described briefly in Section 5.

2. Resolution of Analysers Relative to Compton Profile Width

In the case of dispersive energy analysis, the resolution can be improved simply by the use of a higher order reflection at the expense of intensity, while this cannot be done in the case of an SSD. In the low-energy region, an SSD is generally far lower in resolution than a crystal analyser. However, as the energy becomes higher, the Compton profile becomes wider, and the energy resolution decreases more rapidly in a crystal analyser than in SSD. Thus an SSD can become even superior to a crystal in resolution relative to the Compton profile width, when the energy reaches a certain value. This situation will be described in the following section.

2.1 Crude approximation

A crude approximation holds in the region of 5 to 30 keV X-rays used for ordinary diffraction work. When a free electron scatters an X-quantum, the energy loss of the latter is calculated from energy and momentum conservations:

$$\Delta E = \hbar\omega_0 - \hbar\omega = \frac{p^2 - p_0^2}{2\,m} \approx \frac{8\,\pi^2\,\hbar^2}{m}\,\frac{\sin^2\theta_c}{\lambda_0^2} + \frac{4\,\pi\,\hbar\,q}{m}\,\frac{\sin\theta_c}{\lambda_0}, \tag{1}$$

where ω_0 and ω are the angular frequencies of the incident and scattered X-rays, respectively, m is the electron mass, and $2\,\theta_c$ is the scattering angle. $q = -\,p_0 \cos\varphi$, where $\boldsymbol{p}_0$ is the momentum of an electron before scattering and φ is the angle between $\boldsymbol{p}_0$ and the scattering vector. Formula (1) is valid within the approximation of $\omega \approx \omega_0$. The first term of (1) is the loss of X-ray energy scattered by an electron at rest. This energy loss is expressed by

$$E'_{\mathrm{Cpt}} = \frac{2\,(\hbar\omega_0)^2}{m\,c^2}\sin^2\theta_c \sim (\hbar\omega_0)^2\,. \tag{2}$$

This is shown by the curve (a′) in Fig. 1 for $\sin\theta_c = 1$. The second term in the last expression of (1) expresses the Doppler broadening due to the finite velocity

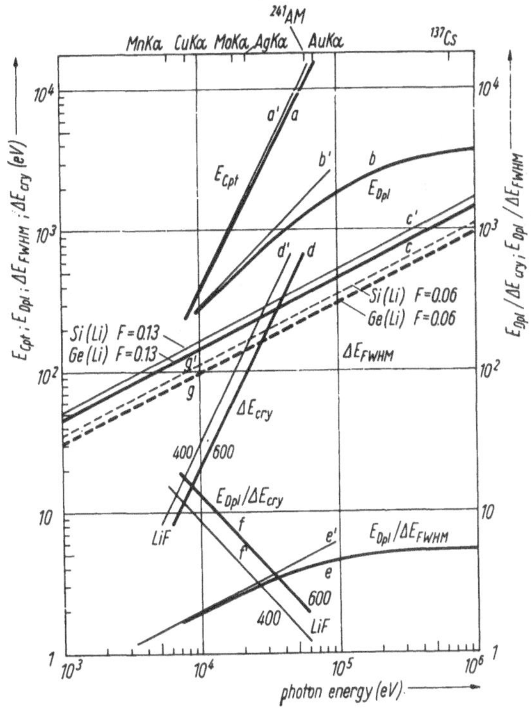

Fig. 1. The energy dependence of the Compton shift (E_{Cpt}), the Compton profile width (E_{Dpl}), full width at half-maximum of an SSD (ΔE_{FWHM}), and resolution of an analysing crystal (ΔE_{cry}). The details are described in the text

of an electron. The broadening of the Compton profile is expressed by

$$E'_{Dpl} = \frac{2}{m\,c} \sin\theta_c\, \Delta q\, \hbar\omega_0 \sim \hbar\omega_0\,. \tag{3}$$

This is shown by the curve (b') in Fig. 1 for $\sin\theta_c = 1$ in a typical case, $\Delta q = 2$ atomic units (A.U.)

The full width at half-maximum (FWHM) of the intrinsic resolution of an SSD [4] is given by

$$\Delta E_{FWHM} = 2.355\,(F\,\varepsilon\,\hbar\omega_0)^{1/2} \sim (\hbar\omega_0)^{1/2}\,, \tag{4}$$

where F is the Fano factor, ε is the averaged energy necessary to produce a hole–electron pair. The value of ε at 77 °K is 3.92 eV for a Si(Li) detector and 2.98 eV for a Ge(Li) [4]. The value of ΔE_{FWHM} for a real detector is larger than that given by (4) especially in the energy region of several keV. This is due to effects such as dark current and thermal noise in the detector and various noises in the pre-amplifier input circuit with a field effect transistor (FET), as mentioned later. However, in the energy region of 20 keV, the relation (4) is a fairly good approximation. The curve (c) in Fig. 1 shows the value of ΔE_{FWHM} of a Ge(Li) detector with $F = 0.13$ [3, 5]. In the case of a crystal analyser, the energy resolution may be given corresponding to the angular divergence of a Soller slit, which is usually about 7′. This resolution is also shown (curves (d) and (d′)) for the 600 and 400 reflection of a LiF crystal analyser, respectively. Thus the resolution is given by

$$\Delta E_{\text{cry}} = \frac{\Delta\lambda}{2\pi c \hbar} (\hbar\omega_0)^2 \sim (\hbar\omega_0)^2 . \tag{5}$$

In the Compton profile measurement, the important value is the ratio of E_{Dpl} to ΔE_{FWHM}. As shown by the curve (e') in Fig. 1, the value of $E_{\text{Dpl}}/\Delta E_{\text{FWHM}}$ for Ge(Li) is 2 at 10 keV and 5 at 60 keV. A more accurate estimation for the above-mentioned values is made in the following section.

2.2 *Relativistic approximation*

In the case of high-energy X-rays, the energy transferred to an electron becomes larger and the approximation $\omega \approx \omega_0$ becomes less valid. In the relativistic approximation, it has been well known that the momentum component along the polar axis z is given by

$$z = -p_0 \cos\varphi = q\left[1 + \frac{\Delta\lambda}{\lambda_0} + \left(\frac{\Delta\lambda}{\lambda_0}\right)^2 (4 \sin^2\theta_c)^{-1}\right]^{-1/2}, \tag{6}$$

where

$$q = \frac{m c}{2\lambda_0 \sin\theta_c}\left[\lambda - \lambda_0 - \frac{4\pi\hbar}{m c}\sin^2\theta_c\right]. \tag{7}$$

$\Delta\lambda = \lambda - \lambda_0$, and λ_0 and λ are the wavelengths of the incident and scattered X-rays, respectively. For $p_0 = 0$, therefore, $q = 0$ and λ takes the value

$$\lambda = \lambda_0 + \frac{4\pi\hbar}{m c}\sin^2\theta_c . \tag{8}$$

With this value, an exact value of E'_{Cpt} is obtained:

$$E_{\text{Cpt}} = \frac{\hbar\omega\left[\frac{4\pi\hbar}{m c}\right]\sin^2\theta_c}{\lambda_0 + \left[\frac{4\pi\hbar}{m c}\right]\sin^2\theta_c} . \tag{9}$$

Curve (a) in Fig. 1 represents equation (9) with $\sin\theta_c = 1$. It is seen there, that curve (a') does not deviate very much from curve (a) if the energy is lower than 100 keV. Equation (9) shows that most of the X-ray energy is transferred to an electron in the region where $\lambda_0 \ll [4\pi\hbar/(m c)]\sin^2\theta_c$. This value corresponds to the X-ray energy of 256 keV if the scattering angle is high.

Next, the Doppler broadening E_{Dpl} should be estimated. In (6), the broadening of z is taken to be $\Delta z = z_1 - z_2 = 2$ A.U., where z_1 and z_2 are the high- and low-energy values at the half-maximum, respectively. If the values of λ corresponding to z_1 and z_2 are expressed by λ_1 and λ_2, respectively, E_{Dpl} is given by

$$E_{\text{Dpl}} = 2\pi c \hbar \frac{\lambda_2 - \lambda_1}{\lambda_2 \lambda_1} . \tag{10}$$

Curve (b) in Fig. 1 was calculated from (10) by using the values, λ_1 and λ_2, calculated from the extrema of λ satisfying (6). Curve (b) is not very different from curve (b′) up to the MoKα–AgKα region. Curve (e) in Fig. 1 shows the most important figure of merit $E_{Dpl}/\Delta E_{FWHM}$, where the value ΔE_{FWHM} should be chosen corresponding to the energy value at each position of the Compton profile.

2.3 Comparison between SSD and a crystal analyser

The curves (e) and (f) in Fig. 1 clearly show the relative resolutions of an SSD and a crystal analyser. Around the energy region of 30 keV, these two analysers are comparable in resolution. However, this resolution is inferior to that of the widely used measurement with MoKα and the 400 reflection of LiF. At least this resolution can be attained also by using an SSD, if the radiation with 300 keV or even higher energy such as γ-rays from ^{137}Cs is used. For certain studies, the present method using an SSD together with AgKα or AuKα radiation may be preferred at the expense of the resolution, when a very strong intensity is an essential merit.

3. Examples of Measurements

3.1 With a usual X-ray generator

As was shown in Section 2, the high-energy X-rays are preferable for an SSD, and the Kα radiation, at least AgKα should be used. The usual X-ray generator for diffraction work is too low in voltage to be used efficiently for AgKα or AuKα radiations. The current standard generator for fluorescence X-ray spectroscopy with 100 kV can efficiently be used for at least AgKα, while a 200 kV generator would be desirable for AuKα. In the present preliminary measurements, a sealed-off tube with an Ag target could be used only at 45 kV. The available SSD was Si(Li) with a resolution of 260 eV at MnKα. In Fig. 2, curve (a) shows the energy distribution of the X-rays from the tube and curve (b) that of Ru-filtered X-rays. In the first case, the tube current was suppressed to such a small value that it could not be read by a milliammeter. The continuous X-rays are indicated relatively stronger at the low-energy

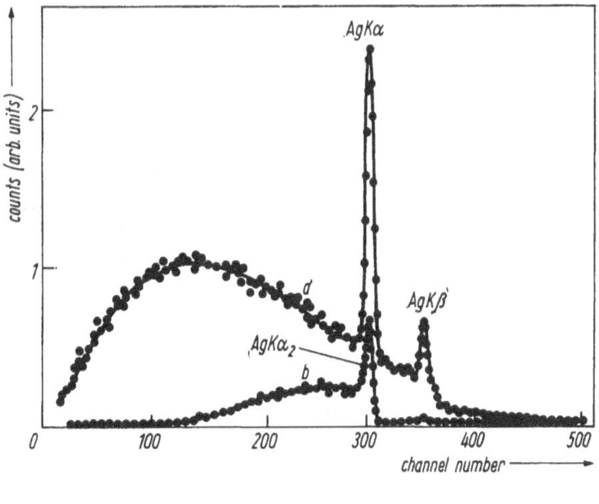

Fig. 2. The spectrum of X-rays from a tube with an Ag target. The energy analysis was carried out by an Si(Li) SSD directly (curve a) and after Ru filtering (curve b)

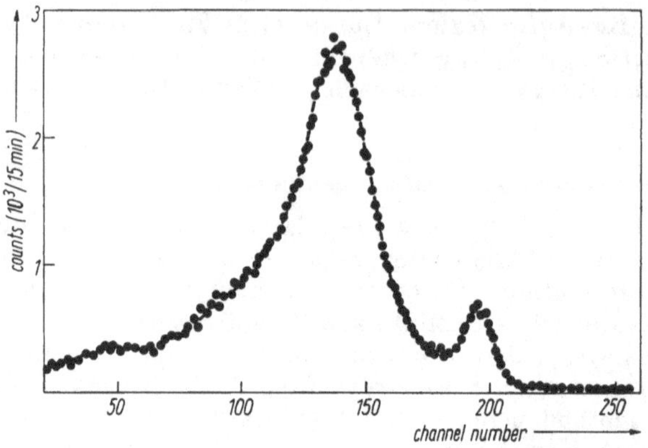

Fig. 3. The Compton profile of a LiF single crystal measured with 22 keV $AgK\alpha_2$ at the energy division of 25.3 eV/channel. An X-ray generator was operated at 56 keV and 12 mA. The scattering angle $2\theta_c$ is 130°. Other details are described in the text

side than at the other side, because the efficiency of the SSD is higher for the lower-energy X-rays. In curve (b), the background at the low-energy side is relatively high.

In a more recent demonstrative measurement, the Compton profile of a LiF single crystal taken with AgKα is shown in Fig. 3. In this case, the incident beam passed through a Soller slit of 7′ divergence, then through a Ru filter, and was reflected by a pyrolytic graphite monochromator. The used SSD was Si(Li) with a resolution of 166 eV at MnKα. As shown in Fig. 3, the intensity proved to be still very strong and the background was very low.

In any case, unwanted X-rays in the incident beam should be suppressed to such an extent that they cannot make the effective dead time longer and the linearity range narrower. It may be also desirable to use a suitable filter which is less absorbing for the Compton scattered X-rays and more absorbing both for the fluorescence X-rays from the sample and for the background in the low-energy region. In the case of a Ni sample, for instance, it proved to be very helpful to put a Ru filter in front of the SSD.

With the above-mentioned SSD, the energy of AgKα is not high enough to give good results, as Fig. 1 indicates. High-energy radiation such as AuKα is favourable for the study of relatively heavy materials so far hardly studied because of the strong absorption. Here only AgKα and AuKα radiations were mentioned, because the α_1-cut filter is available for each of them [6].

3.2 With a radioactive source

γ-rays with energy higher than 1.02 MeV cause the creation of electron–positron pairs with more or less non-negligible probability and it makes the Compton effect less intensive [7]. Therefore, suitable radioactive sources are for example, ^{57}Co (136, 122, 14.4 keV), ^{109}Cd (87.7 keV), ^{241}Am (59.6, 26.4 keV), and ^{137}Cs (662 keV). However, there may be studies for which, for instance, ^{22}Na (1.277 MeV) is preferable. In this case, the pair creation may not prevent the necessary analysis.

In the present preliminary work, the 59.6 keV radiation from ^{241}Am has been used. The corresponding energy distribution is shown in Fig. 4. As expected, the S–N ratio is very good. However, the background level is somewhat different on the low- and high-energy sides. This is apparently due to the Compton

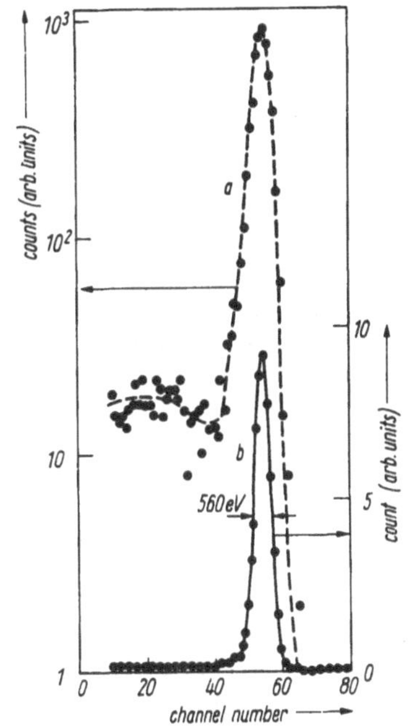

Fig. 4. The spectrum of 59.6 keV γ-rays from a ^{241}Am source. The analysis was done by SSD and plotted in logarithmic (curve a) and linear (curve b) scales. The Compton scattering due to the source itself is appreciable only in curve a

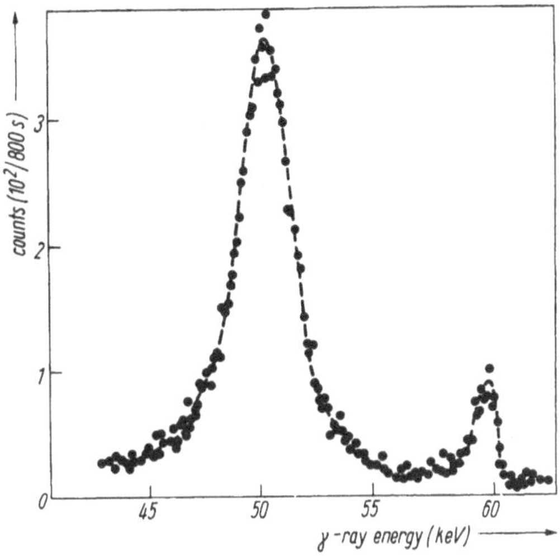

Fig. 5. The Compton profile of an Ni powder sample by use of 59.6 keV γ-rays from ^{241}Am. The measurement was carried out with the scattering angle of (128 ± 10)° and with an energy division of 112 eV/channel

scattering inside the radioisotope source itself. According to the present measurements, the Compton profile intensity with the RI source is smaller by a factor of 10^{-2} to 10^{-4} than the intensity with a usual X-ray tube. However, the RI source, which is very stable and has little noise, proved to be very useful for the measurement even of relatively heavy materials. Namely Fig. 5 shows the Compton profile of a Ni powder sample measured during only 800 s. The 20 mCi ^{241}Am source used which is readily accessible had the shape of a ring, although a stronger source with a more suitable shape is commercially available. Besides, a far more efficient Ge(Li) detector [8] with resolution high enough is also available, although only a Si(Li) detector with moderate resolution could be used in the present work. Therefore, a counting rate by probably 10 to 10^2 times higher may be easily obtained, if a more intense RI source and a more sensitive Ge(Li) are used. As mentioned above, the deconvolution procedure can be done more reliably when an SSD is used. By the way, before deconvolution, a correction should be made for the energy dependence of the sensitivity of the SSD used.

4. The Possible Highest Resolution of SSD

The resolution is usually defined at the energy of MnKα. According to this definition, the present minimum value of ΔE_{FWHM} is 153 eV for a Si(Li) detector [3] and 162 eV for a Ge(Li) [8]. In formula (4), the values of ε at 77 °K are

2.98 and 3.92 eV for Ge(Li) nad Si(Li), respectively [4]. The ΔE_{FWHM} resulting with $F = 0.13$ is shown in Fig. 1, curves (c) and (c′) for Ge(Li) and Si(Li), respectively. However, equation (4) is valid only in the high-energy region, because, in the low-energy region, the noise due to the pre-amplifier circuit is dominant and (4) should be replaced by

$$\Delta E^{tr}_{FWHM} = [\Delta E^2_{amp} + \{2.355\,(F\,\varepsilon\,\hbar\omega_0)^{1/2}\}^2]^{1/2}\,. \tag{11}$$

As concerns the sensitivity, Ge(Li) is superior to Si(Li) for energies higher than about 20 keV, although Si(Li) is to a small extent superior to Ge(Li) for the energy range between 11.1 and about 20 keV because of the GeK absorption edge [8]. Although the variety of available detectors, especially Ge(Li), are so far limited in size and resolution, the Fano factor of an SSD can in principle be made smaller by choosing appropriate size and structure, and can be made as small as 0.06 [5]. The resolution curve for this case is plotted in Fig. 1 as (g) and (g′) for Ge(Li) and Si(Li), respectively. An SSD corresponding to curve (g), if available, gives a curve better than (e). Then the resolution becomes better than 165 eV, which is the value obtained by the 400 reflection from an analysing crystal for AgKα radiation.

It is also useful to make the value of ε smaller in order to improve the resolution, although this contradicts the endeavour to make an SSD usable at room temperature. Another factor of making ε effectively smaller is to improve the perfection of the detector crystal. Anyway, some improvement in resolution may be expected in future.

5. Future Aspects of Compton Measurements

As was described above, the necessary time for the Compton measurements has been drastically shortened while a good resolution is secured. This innovation enables us to carry out various measurements so far practically impossible. Some of them are listed here.

5.1 Study of anisotropy of the momentum distribution of 3d electrons

As it was shown above, a Ni sample was easily measured even using 59.6 keV γ-rays from ^{241}Am. Therefore, it is possible to detect the anisotropy of the momentum distribution of 3d electrons, for instance, in vanadium pointed out by Weiss [9]. In general, highly absorbing heavy materials may be studied, especially when γ-ray sources like 87.7 keV from ^{109}Cd or 122 keV from ^{57}Co are used.

5.2 Compton measurement on a single crystal satisfying the Bragg condition

As was studied [10] experimentally and theoretically, the intensity of the Compton scattering decreases when a perfect crystal reflects the incident X-ray beam. It is theoretically predicted [11] that the Compton profile is affected by the existing diffraction wave. This fact has already been nearly confirmed in our laboratory (T. F. and S. H.) for a graphite sample using a dispersive analyser, though it is a mosaic crystal rather than a perfect crystal. The peak of a Compton profile at least is shifted when the incident X-rays are reflected. Such a measurement will far more easily and precisely be carried out by the use of an SSD. Incidentally, the experimental results in this case will reflect the averaged state of crystal imperfections.

5.3 Measurements of the binding effect in relatively heavy materials

The binding effect or Raman scattering has very weak intensity and has so far been found at least in C [12 to 14], Li [13, 14], Be [13, 14], and B [13 to 15] by a dispersive crystal analyser. For the heavier atoms, the binding energy is separated from the thermal diffuse scattering (TDS) peak far enough to be detected even by an SSD. Besides this kind of measurement may drastically be advanced using the following coincidence method.

5.4 Measurement of the Compton scattered X-rays in coincidence with fluorescence light or X-rays

The atom excited by the Compton scattering regains its stable state by emitting a fluorescence radiation. Therefore, at least in principle, the Compton profile contributed from one of various electrons can, to some extent seperately, be measured by the use of a coincidence technique. Apart from an SSD, a similar experimental technique has been used [16 to 18].

Two of the authors (T. F. and S. H.) have more recently succeeded in preliminary coincidence measurements with fluorescence X-rays, as will separately be published.

5.5 Measurement of Compton profile in coincidence with Compton-scattered electrons

In the high-energy region, the Compton scattered electrons cannot change very much the momentum direction when they are emitted from the crystal, although they are scattered in the crystal. If the Compton scattering is measured by an SSD in coincidence with electrons coming out in each direction and having each energy value, then the more specified momentum distribution of electrons may be obtained instead of the linear momentum distribution.

5.6 Measurement of orientation distribution of electron spins in crystals

Each photon in the usual characteristic radiation from a conventional X-ray tube is linearly polarized before observed, namely it consists of a pair of inverse circular polarizations. In the closed K-shell, electrons are degenerate in energy with antiparallel spins. If this radiation is subjected to Compton scattering by a ferromagnetic material, then partially circular polarized X-rays may be obtained. If this beam is scattered by a magnetic material, after being monochromated by a crystal, and then analysed by an SSD, the information concerning orientation distribution of electron spins in this material may be, in principle, obtained. If the nuclei at β-decay are kept under controlled conditions, the emitted γ-rays can be circularly polarized. This γ-ray beam may be suitable as a source for this kind of experiment [19, 20].

Acknowledgements

The authors are pleased to acknowledge that one of the motives for carrying out the present work was the discussion between one of the authors (T. F.) and Prof. M. Tokonami, University of Osaka. They are also thankful to Mr. K. Iwai, Sophia University, for elucidating discussion about spin polarization described in the last part of the present paper.

References

[1] M. COOPER and B. G. WILLIAMS, Phil. Mag. **17**, 1079 (1968).
[2] R. J. WEISS, Phys. Rev. **24**, 883 (1970).
[3] R. S. FRANKEL and D. W. AITKEN, Appl. Spectroscopy **24**, 557 (1970).
[4] D. W. AITKEN, IEEE Trans. Nuclear Sci. **15**, No. 3, 10 (1968).
[5] H. R. ZULLIGER and D. W. AITKEN, IEEE Trans. Nuclear Sci. **17**, No. 3, 187 (1970).
[6] S. HOSOYA, Japan. J. appl. Phys. **7**, 1 (1968).
[7] C. M. DAVISSON, in: α-, β-, and γ-Ray Spectroscopy, Vol. 1, Ed. K. SIEGBAHN, North-Holland Publ. Co., 1965 (p. 69).
[8] ORTEC Technical Data, 8000 Series Ge(Li) Low Energy Photon Spectrometer, May 1971; ORTEC Inc., 100 Midland Road, Oak Ridge, Tenn. 37830.
[9] R. J. WEISS, X-Ray Determination of Electron Distribution, Appendix 5B, Wiley, New York 1966.
[10] S. ANNAKA, S. KIKUTA, and K. KOHRA, J. Phys. Soc. Japan **21**, 1559 (1966).
[11] H. SANO, K. OHTAKA, and Y. H. OHTSUKI, J. Phys. Soc. Japan **27**, 1254 (1969).
[12] M. COOPER and J. A. LEAKE, Phil. Mag. **13**, 603 (1966).
[13] T. SUZUKI, J. Phys. Soc. Japan **22**, 1139 (1967).
[14] T. SUZUKI, T. KISHIMOTO, T. KAJI, and Ts. SUZUKI, J. Phys. Soc. Japan **29**, 730 (1970).
[15] N. G. ALEXANDROPOULOS and G. G. COHEN, Phys. Rev. **187**, 455 (1969).
[16] J. W. MOTZ and G. MISSONI, Phys. Rev. **124**, 1458 (1961).
[17] Z. SUJKOWSKI and B. NAGEL, Ark. Fys. (Sweden) **20**, 323 (1961).
[18] S. SHIMIZU, Y. NAKAYAMA, and T. MUKOYAMA, Phys. Rev. **140**, A806 (1965).
[19] R. M. STEFFEN and H. FRAUENFELDER, see reference [7], Vol. 2 (p. 1453).
[20] P. M. PLATZMAN and N. TZAR, Phys. Rev. **B2**, 3556 (1970).

(Received December 28, 1971)

phys. stat. sol. (a) **10**, 447 (1972)

Subject classification: 16; 13.4; 14.3.4; 22.8

Istituto di Fisica dell' Università, Parma (a) and Escuela de Fisica de la Universidad de La Habana (b)

Electric Field Variation of Photoconductivity Quenching Due to Hot Electron Capture in $ZnIn_2S_4$

By

N. Romeo (a) and O. Vigil (b)

The study of the photoconductivity quenching as a function of the electric field in the ternary compound $ZnIn_2S_4$ has given evidence of a level localized at about 1.77 eV above the valence band which is responsible for the capture of hot electrons and for the N-type negative resistance. By a rough estimate of the quenching compared with the experimental it was possible to determine the potential barrier offered by the centre to the electrons, for a fixed illumination intensity of the gap energy light upon the crystal.

L'étude du quenching de la photoconducibilité en fonction du champ électrique en monocristaux de $ZnIn_2S_4$ nous a donné la possibilité de trouver un niveau localisé au voisinage de 1,77 eV au-dessus du sommet de la bande de valence qui est responsable pour la capture des électrons chauds et ainsi de la resistance negative type N. Un'estimation approximée du quenching comparée avec les résultats experimentaux a donné la possibilité de déterminer la barrière de potentiel que le niveau présente aux électrons, pour une intensité donnée de lumière de frequence correspondente à l'énergie de la bande interdite.

1. Introduction

$ZnIn_2S_4$ ternary compounds, if they are illuminated by gap energy light, present an N-type negative resistance. This effect has been explained as due to electron capture by a centre with repulsive potential barrier, though direct evidence of such an interpretation has not been given [1]. The purpose of this work is the study of the photoconductivity quenching as a function of the electric field. This study has enabled us to identify a level in the forbidden gap which is responsible both for quenching and for quenching variation produced by the electric field. This fact leads us to conclude that this level is the level with the repulsive potential barrier which captures the electrons at high electric fields and that it is responsible for the N-type negative resistance. The comparison between an approximate calculation of the quenching as a function of the electric field and the experimental results enables us to determine the potential barrier that the centre presents for electrons at a fixed illumination intensity of the crystal by gap energy light.

2. Experimental Results

When monocrystals present a N-type negative resistance due to electron capture by centres with a repulsive potential barrier, the capture can be studied using the method of Pratt and Ridley [2].

It consists in recording on the same sample the dc I–U characteristic and a pulse I–U characteristic with pulses shorter than the capture-time of the hot electrons. The pulse characteristic shows only the mobility variation as a function of the electric field while the dc characteristic shows both the mobility variation as a function of the electric field and the capture variation as a function of the electric field. A comparison between the two characteristics gives the possibility of studying the capture directly. This method has, on the other hand, the disadvantage that the pulse current to be measured is low in comparison with the capacity effect that masks the measurement in insulating crystals such as those we have used. We have studied the electron capture in $ZnIn_2S_4$ crystals by observing the photoconductivity quenching variation as a function of the electric field.

The crystals studied have a resistivity of about 10^{10} Ωcm, in the dark, and a mobility of about 50 $cm^2\ V^{-1}\ s^{-1}$ as measured by photo Hall effect. All the measurements have been made at room temperature. The contacts have been made by indium evaporated under high vacuum.

First of all we have made a study of the photoconductivity quenching as a function of the wavelength of the secondary light, at low electric fields. The results are presented in Fig. 1; the quenching has a maximum at about 7000 Å (1.77 eV). Then we have studied quenching as a function of the electric field, using as secondary light a light source filtered by a narrow band interferometric filter centered at 7000 Å. The results are presented in Fig. 2 where we show the I–U characteristics, when the crystal is illuminated by primary light whose intensity is equal to the intensity used for studying the quenching variation as a function of the electric field. This is necessary because, as reported in [1], the electric field at which the capture begins depends on the illumination intensity of the crystals. This fact means that the potential barrier of the centre to electrons is a function of the illumination intensity. For this reason, in our experiments, we have used a fixed illumination intensity corresponding to a power of 6×10^{-4} W/cm² absorbed by the sample.

The results of Fig. 2 are significant because, at the same electric field of about 900 V/cm and the light intensity used, the I–U characteristic becomes sublinear and the photoconductivity quenching begins to decrease.

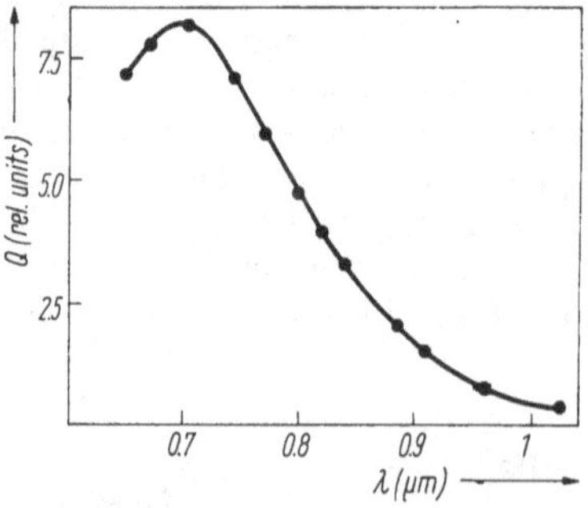

Fig. 1. Quenching of the photoconductivity as a function of the wavelength of the secondary light

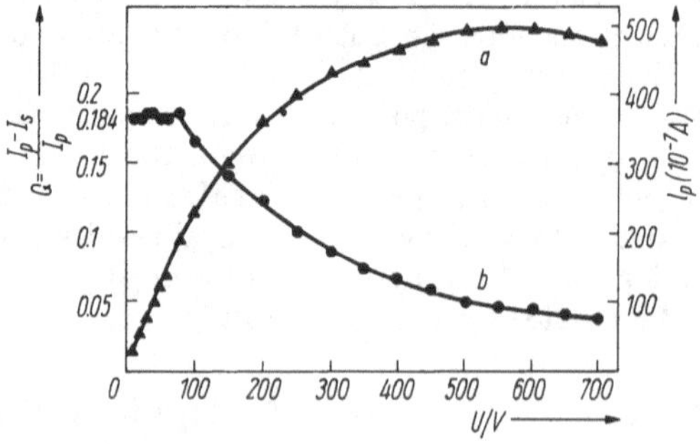

Fig. 2. (a) I–U characteristic of a $ZnIn_2S_4$ monocrystal, illuminated by gap energy light. The contacts have been made by indium. The electrode distance is 0.088 cm. (b) Quenching variation as a function of the voltage applied to the same crystal

3. Interpretation and Discussion of Results

The photoconductivity quenching is given by

$$Q = \frac{I_p - I_s}{I_p}, \tag{1}$$

where I_p is the primary photocurrent, i.e. the current which flows through the crystal when illuminated only by energy gap light and I_s is the secondary current, i.e. the current we have when the crystal is illuminated both by energy gap light and by lower energy light. If we denote by Q_0 the quenching at low electric fields, with no capture by the repulsive centre, we have

$$Q_0 = \frac{\mu_p n_0 - \mu_s (n_0 - n_1)}{\mu_p n_0}, \tag{2}$$

where μ_p and μ_s are the electron mobilities when the crystal is illuminated by primary light and by primary and secondary light, respectively, n_0 is the number of electrons in the conduction band when we have only the primary light and n_1 is the number of electrons recombining when the crystal is illuminated by both light sources. As reported by Bube and MacDonald [3], μ_s may be different from μ_p for two reasons: a) the secondary light may change the occupation of charged scattering centres and b) the secondary light may cause the minority carriers to play an important role in the conductivity process. In our case, as we work at room temperature, the effect of the change in the occupation of charged centres might be negligible and, besides, also the contribution to the conductivity of the holes set free by the secondary light might be negligible, if we work with a weak intensity of secondary light.

In any case, we have made photo-Hall effect measurements at low electric fields, both by primary light and by primary and secondary light together, with light intensities suitable for studies of the quenching effect as a function of electric field. In both cases, we found for the crystal the same value of mobility, i.e., if there are differences between μ_s and μ_p, they are within the experimental error.

If we consider $\mu_p \approx \mu_s$ at low electric fields, the quenching may be written as

$$Q_0 = \frac{n_1}{n_0}. \tag{3}$$

For the calculation of n_1 we use a kinetics of the type

$$\frac{dn}{dt} = \Phi - A\, n\, p, \tag{4}$$

where Φ is the flux of primary light absorbed by the crystal and $A\, n\, p$ is the band-to-band recombination rate. In (4) we neglect all capture types different from the band-to-band recombination. At equilibrium we have

$$\Phi = A\, n_0\, p_0, \tag{5}$$

where n_0 and p_0 are the equilibrium numbers of electrons and holes, respectively. In the presence of secondary light, the expression (5) becomes

$$\Phi = A\, (n_0 - n_1)\, p, \tag{6}$$

where p is the number of holes in the valence band with the secondary light and n_1 is the number of electrons recombining. From (5) and (6) we have

$$n_1 = n_0 \left(1 - \frac{p_0}{p}\right);$$

from this expression:

$$Q_0 = 1 - \frac{p_0}{p}. \tag{7}$$

For the calculation of p we use the kinetics:

$$\frac{\mathrm{d}p}{\mathrm{d}t} = \Phi_1 - C\,[N - (N_0 - p)]\,p\,, \tag{8}$$

where Φ_1 is the flux of the secondary light and $C\,[N - (N_0 - p)]$ is the recapture rate of the holes by the centre. N is the total number of centres and N_0 the number of centres occupied by holes, the number of centres occupied by electrons is, at the beginning, $(N - N_0)$. At the equilibrium, if we neglect p compared with N_0, considering weak intensities of the secondary light, we obtain:

$$\Phi_1 = C\,(N - N_0)\,p\,, \tag{9}$$

$$p = \frac{\Phi_1}{C\,(N - N_0)}. \tag{10}$$

This equation together with equation (7), gives

$$1 - Q_0 = \frac{C\,p_0}{\Phi_1}(N - N_0)\,. \tag{11}$$

At high electric fields, in (11) the number of centres occupied by holes N_0 is varied by Δn, the number of electrons captured by the centre, i.e. N_0 must be replaced by $N_0 + \Delta n$. At high electric fields, therefore, we have

$$1 - Q = \frac{C\,p_0}{\Phi_1}(N - N_0 - \Delta n)\,, \tag{12}$$

where Q is the quenching at high electric fields. Combining (11) and (12) one obtains

$$\frac{Q_0 - Q}{1 - Q_0} = -\frac{\Delta n}{N - N_0}. \tag{13}$$

If we use for Δn the value reported by Ridley and Watkins [4]:

$$\Delta n = -\,n_0\,(1 - \gamma)\,,$$

where γ is given by $\gamma = B_0/B$ (B_0 is the rate for the capture by the centre at low fields and B at high fields), one obtains

$$\frac{Q_0 - Q}{1 - Q_0} = \frac{n_0}{N - N_0}(1 - \gamma)\,. \tag{14}$$

The expression (14) may be written in the following way. At highest electric fields, γ is low in comparison with unity and Q in comparison with Q_0, so that

$$\frac{n_0}{N - N_0} = \frac{Q_0}{1 - Q_0} . \tag{15}$$

Replacing (15) in (14) we have to a good approximation

$$Q = Q_0 \gamma . \tag{16}$$

For fitting experimental results obtained with (16), we have supposed, as suggested in [4], that γ is given by

$$\gamma = \frac{B_0}{B} = \frac{e^{-U_0/KT}}{e^{-U_0/KT_e}} , \tag{17}$$

where U_0 is the potential barrier offered to the electrons by the centre, K is the Boltzmann constant, T is the lattice temperature and T_e the electron temperature. By means of (17) expression (16) becomes:

$$Q = \frac{Q_0 \exp\left(\frac{U_0}{KT_e}\right)}{\exp\left(\frac{U_0}{KT}\right)} . \tag{18}$$

To compare the expression (18) with the experimental data, it is necessary to know T_e as a function of the electric field. In crystals like $ZnIn_2S_4$, where one can expect that optical modes are as important as acoustic modes in determining T_e as a function of the electric field, it would be correct to introduce into (18) a formula for T_e which considers both acoustic and optical modes. Since, in this way, we should arrive at extremely complex formulae which would be difficult to be compared with experimental data, we try to make an experimental comparison by using the very simple formulae by Gunn [5] where optical modes are neglected. We consider that the electron temperature is a different function of electric field for intermediate and high electric fields. Therefore equation (18) may be written in two ways,

$$\frac{Q}{Q_0} = \exp\left[-\frac{U_0}{KT}\left(1 - \frac{1}{1 + \alpha E^2}\right)\right] \tag{19}$$

for intermediate electric fields assuming [5]:

$$T_e = T(1 + \alpha E^2) ,$$

where α is a parameter depending on the scattering type, and

$$\frac{Q}{Q_0} = \exp\left[-\frac{U_0}{KT}\frac{T_e - T}{T_e}\right] = \exp\left[-\frac{U_0}{KT}\left(\frac{E - E_s}{E}\right)\right] \tag{20}$$

for high electric fields, where we have supposed $T_e \propto E$; E_s is the value of the electric field at which the quenching begins to decrease.

In order to compare (20) with our experimental data, we show in Fig. 3 quenching as a function of $(E - E_s)/E$ in a semilogarithmic scale. For high

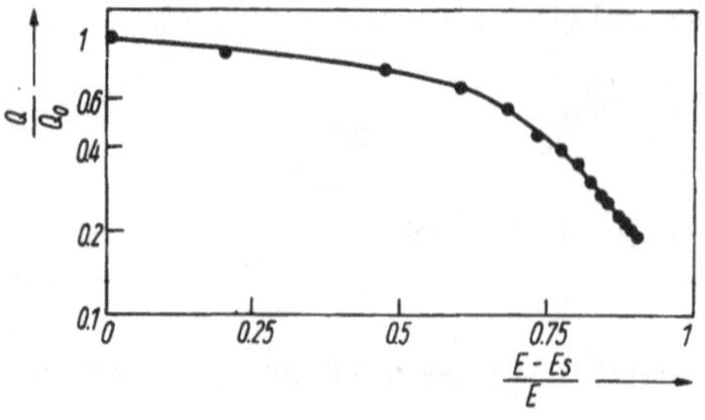

Fig. 3. Semilogarithmic curve of the ratio between the quenching at high field and the quenching at low fields as a function of $(E - E_s)/E$, where E is the electric field and E_s the threshold electric field at which the quenching begins to vary

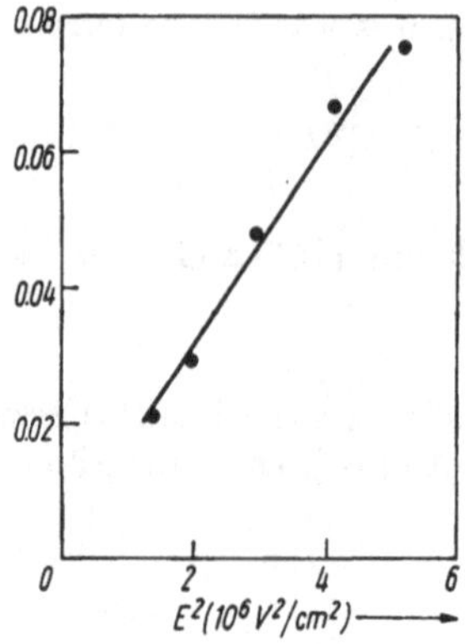

Fig. 4. Plot of the expression

$$\left(\frac{1}{1 + \frac{\ln Q/Q_0}{U_0/KT}} - 1\right)$$

as a function of E^2 (see text)

electric fields, one obtains a straight line from the slope of which the potential barrier U_0 is calculated to be 0.14 eV, at an intensity of primary light corresponding to a power of about 6×10^{-4} W/cm² absorbed by the crystal. This value of U_0 is an acceptable result.

To see if (19) fits the first part of the curve, we have using the value of 0.14 eV for U_0 in (19) plotted in Fig. 4 the expression

$$\left(\frac{1}{1 + \frac{\ln Q/Q_0}{U_0/KT}} - 1\right)$$

as a function of E^2.

This gives a nearly straight line from which one can obtain the value of $\alpha \approx \approx 1.5 \times 10^{-8}$ cm²/V². In our case if one considers that the scattering is due only to lattice vibrations because the measurements have been made at room temperature, we take for α the value [5] $\alpha = (\frac{3}{8})\,(\mu_0/c^2)$, where μ_0 is the mobility at low fields and c the sound velocity in the medium.

Taking for μ_0 the value $\mu_0 = 50$ cm²V⁻¹s⁻¹ one can obtain then $c \approx 2 \times 10^5$ cm/s. Also this result is acceptable. Notwithstanding we have used approximate formulae for T_c as a function of the electric field and we have entirely neglected optical modes, we observe some agreement with experimental data. This fact could indicate that optical modes are not excited at the temperature at which we have worked and so acoustic modes are prevalent. In any case, we think that a further study is necessary especially about the type of scattering as a function of the temperature. With our work, we wish to point out that in the case where one has electron capture by a repulsive centre which is responsible for quenching, the study of the quenching as a function of the electric field may become a powerful means not only for determining the capture itself and the centre localization in the forbidden gap but also for determining some parameters which enter the study of the capture.

Acknowledgements

We wish to thank Prof. F. Bassani for useful suggestions and Dr. C. Paorici for having grown the monocrystals. We are, besides, very grateful to Prof. A. Levialdi who, with his interest, has allowed the collaboration between the Physics Institute of Parma and the Physics School of La Habana, giving the possibility for this work.

References

[1] S. Mora, C. Paorici, and N. Romeo, J. appl. Phys. **42**, 2061 (1971).
[2] R. G. Pratt and B. K. Ridley, Proc. Phys. Soc. **81**. 996 (1963).
[3] R. H. Bube and H. E. MacDonald, Phys. Rev. **121**, 473 (1961).
[4] B. K. Ridley and T. B. Watkins, J. Phys. Chem. Solids **22**, 156 (1961).
[5] J. B. Gunn, Progr. in Semicond. **2**, 213 (1957).

(Received February 1, 1972)

phys. stat. sol. (a) **10**, 455 (1972)

Subject classification: 16 and 20.1; 11; 22.6

*Laboratoire de Spectroscopie et d'Optique du Corps Solide
Groupe de Recherche du C.N.R.S., Strasbourg*

Near Infrared Optical and Photoelectric Properties of Cu_2O

I. Infrared Absorption in Cu_2O after Annealing in Oxygen or He^+ Particles Bombardments

By

B. PREVOT, C. CARABATOS, and M. SIESKIND

The effects on the infrared absorption of annealing polycristalline samples of Cu_2O under oxygen atmosphere are reported. The absorption near the fundamental edge is characterized by several absorption bands with peaks at 0.65, 0.76, 0.88, 0.95, 1.10 μm and the strongest at 1.28 μm. These bands appear simultaneously and their magnitude depends upon the oxygen pressure during the annealing with a $(p_{O_2})^{2/3}$ law. In the same region of the spectrum, the effects of irradiation by α-particles from 1 to 3 MeV on analogous samples are studied also. Weaker bands appear at the same wavelengths. From comparison the approximate number of defects responsible for the absorption: 10^{17} cm^{-3} are evaluated. The discussion of the experimental results together with those of Zouaghi (see Part II) obtained by photoconductivity measurements will be presented in Part III.

On a étudié le spectre infrarouge de Cu_2O polycristallin recuit sous pression d'oxygène. Le spectre est caractérisé par plusieurs bandes d'absorption situées à 0,65, 0,76, 0,88, 0,95, 1,10 μm et la plus intense à 1,28 μm. Ces bandes d'absorption apparaissent simultanément et leur intensité croît avec la pression d'oxygène durant le recuit suivant une loi en $(p_{O_2})^{2/3}$. Nous avons également étudié, dans la même région spectrale, les effets optiques dus à l'irradiation d'échantillons de Cu_2O par des particules α de 1 à 3 MeV. Des bandes apparaissent aux mêmes longueurs d'onde que précédemment, mais légèrement moins intenses. De la comparaison de ces deux effets, nous avons pu déduire la densité approximative des défauts introduits et responsables des bandes observées: soit 10^{17} cm^{-3}. La discussion de ces résultats, ainsi que ceux obtenus par Zouaghi (Partie II) en photoconductivité, sera donnée dans un article commun (Partie III).

1. Introduction

Many experimental and theoretical papers have been devoted to the optical properties of cuprous oxide (Cu_2O). Three types of transitions giving rise to an optical absorption may be distinguished: fundamental and excitonic transitions related to the band structure [1], transitions due to the lattice vibrations [2], and transitions correlated with defects, particularly with non-stoichiometry defects of this compound.

Since Wagner's work [3], such defects of stoichiometry probably introduced during the preparation, are well known; Cu_2O is typically a non-stoichiometric material. For this reason, many electrical, photoelectrical, or optical properties, e.g. light emission in the near infrared, were often interpreted by assuming the existence of energy levels within the forbidden gap and connected to the excess of oxygen in the crystal.

In a previous paper, Schwab et al. [4] have reported qualitative results of the annealing effect under oxygen pressure on Cu_2O. New absorption bands appear in the near infrared. Their intensities increase with the annealing oxygen pressure.

The purpose of the present paper is to correlate quantitatively those optical properties Cu_2O with defects introduced in a the crystal by two ways:

– a chemical process, i.e. the annealing under different oxygen pressures;

– a physical process, such as bombardment with charged particles. In both cases, formation of copper vacancies among others is expected in a first approximation. The second procedure will provide an order of magnitude of the number of defects responsible for the observed bands.

In Section 2 of this paper, the method of preparation of our samples is described and the experimental procedure is shortly given. Section 3 is devoted to the statement of the optical results obtained with annealed Cu_2O samples. In Section 4, irradiation techniques and optical results are presented for irradiated samples pre-annealed or not. In Section 5, we discuss some numerical results deduced from irradiation experiments. In part II, Zouaghi will report results obtained from photoconductivity measurements and in part III, discussions and conclusions will be presented.

2. Samples and Experimental Procedure

Thin polycrystalline plates of Cu_2O are prepared by a preliminary oxidation of 'Johnson-Matthey' (London) spectroscopically "pure" copper-100 μm thickness – following the method described by Grun [5] or by Schwab et al. [4]. This oxidation is performed at 1050 °C and under 250 Torr oxygen for all samples during about 10 min. The samples are then annealed during 6 h at the same temperature but under different oxygen pressures; three types of annealed samples have been selected: those prepared under 50, 100, and 250 Torr of oxygen. The cooling is obtained by evacuating the sample in the cold part of the enclosure while vacuum is produced in the furnace. Starting from copper plates of thickness 100 μm, one obtains about 140 μm thick Cu_2O oxygen doped samples. Measurements of transmission and reflection curves have been performed on a Perkin Elmer spectrophotometer (Model 112 G) by a point by point method with intervals of 200 Å. Experiments at liquid nitrogen temperature have also been made using a metallic cryostat allowing both transmission (or reflection) and reference signal measurements.

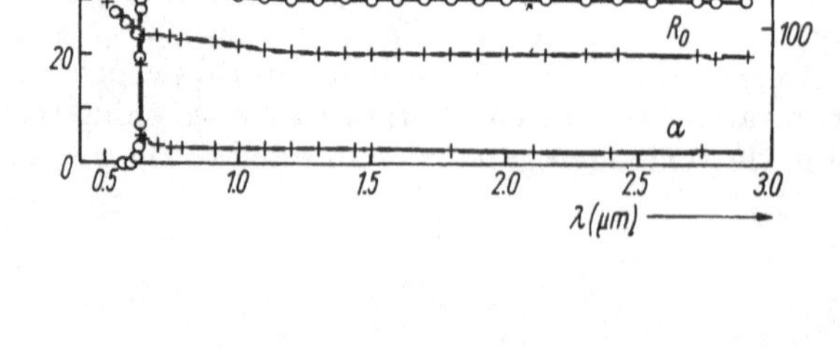

Fig. 1. Transmission (T) and reflection (R) curves measured (○) on a 100 μm thick parallel face reference sample 12 C2 at T = 293 °K; absorption (α) and reflection coefficient (R_0) of an air – Cu_2O dioptra (+) calculated from these points with the help of equations (1) and (2) are also plotted

3. Infrared Absorption Results on Annealed Samples

Generally, the surface of annealed samples presents strongly diffusing irregularities which hinder accurate reflection measurements. Therefore, the reflectivity has been determined on a reference non-annealed Cu_2O sample polished mechanically. The reflection R and transmission T curves are given on Fig. 1. The absorption α and reflection R_0 coefficients of an air-cuprite dioptra have been also plotted versus wavelength; both have been calculated with the help of the following relations:

$$T = \frac{(1 - R_0)^2 \, e^{-\alpha d}}{(1 - R_0^2 \, e^{-2\alpha d})}, \tag{1}$$

$$R = R_0 (1 + T \, e^{-\alpha d}), \tag{2}$$

where d is the thickness of the sample (21). On Fig. 2 transmission curves at room temperature and at 77 °K are shown for the 5 A sample which was annealed under 250 Torr oxygen during 6 h.

By a comparison with the transmission of the 12C2 reference sample (Fig. 1), one can deduce the following observations. The transmission is decreased in the region of the near infrared and new absorption bands are introduced between the fundamental edge and 2.0 μm.

At room temperature, only a step of absorption is observed in the region of 1 μm, but at 77 °K, the bands appear clearly at 1.28, 1.10, 0.95, 0.88, 0.76 and 0.65 μm. On some of the samples, weaker bands are found at 1.38 μm and at 1.68 μm. As usually, the fundamental absorption edge slightly shifts to higher energies at liquid nitrogen temperature.

If we assume that the reflection coefficient R_0 of an annealed sample is equal to that of the reference sample (12C2) ($R_0 = 20\%$, referring to Fig. 1) and take into account that R_0 does not vary appreciably with temperature, the variation of the absorption coefficient versus wavelength may be calculated for samples annealed under different pressures. In Fig. 3, we have plotted the absorption

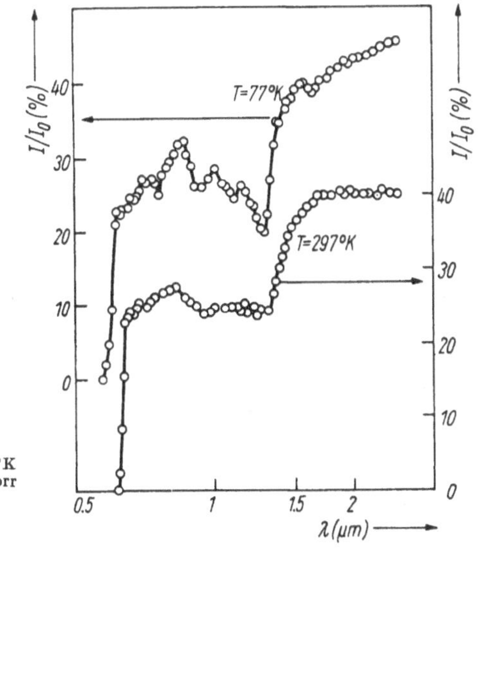

Fig. 2. Transmission curves at 297 °K and at 77 °K concerning a Cu_2O sample 5A annealed under 250 Torr oxygen during 6 h at 1050 °C

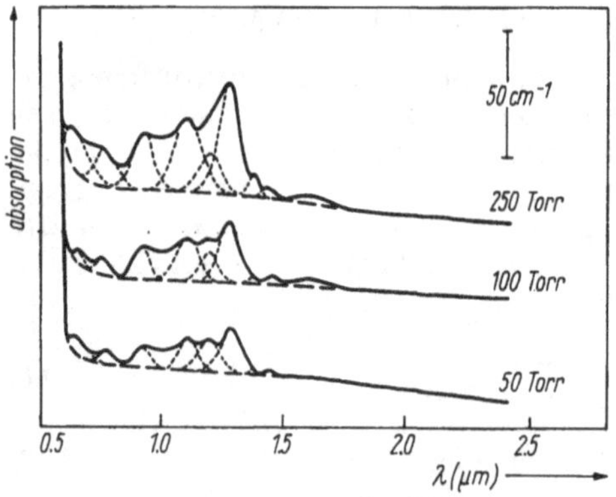

Fig. 3. Comparative variation of the absorption coefficient after annealing under different oxygen pressures. The vertical bar (50 cm^{-1}) gives an idea of the intensity of the absorption

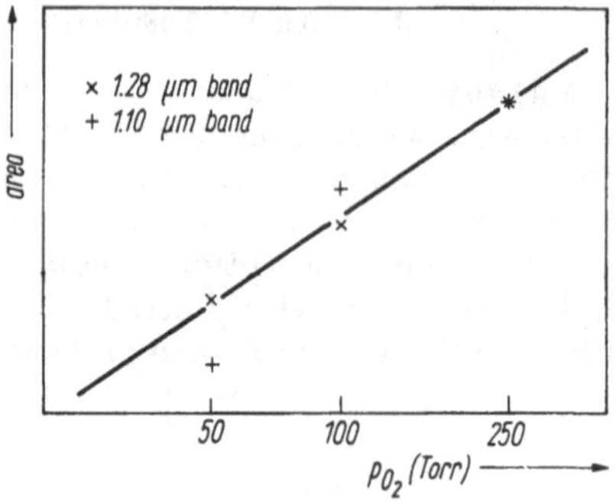

Fig. 4. Variation of the intensity of two strongest bands with oxygen pressure during the annealing in lg – lg coordinates. The slope is about 2/3

coefficient for three samples annealed under 50, 100, and 250 Torr oxygen; for clarity, we have graphically decomposed the spectra into constitutive bands. The band at 0.65 μm is not quite visible because it is superimposed to the fundamental edge. The intensity of the observed bands increases with the annealing oxygen pressure $p(O_2)$; we have tentatively evaluated the areas of the two strongest bands at 1.28 μm and at 1.10 μm for the three samples.

Fig. 4 shows the variation of the areas versus the oxygen pressure (double lg scale); the slope of the line is about 2/3.

On the other hand, the transmission of the samples apart from the bands at longer wavelengths is found to decrease with increasing $p(O_2)$.

4. Irradiation of the Samples — Optical Absorption Results

As the non-stoichiometry of Cu_2O is expected to be due to copper vacancies, it was interesting to compare the previous results with the absorption of samples presenting a lack of Cu atoms created by α-particles irradiation. A Van de Graaff[1]) accelerator was used to submit samples of Cu_2O, annealed or not, to He^+ particles (hereafter He^+ particles will be noted α-particles). The highest energy of the particles was about 3 MeV. The current intensity of the beam and the focussing on the sample were regulated. The irradiation enclosure in which the bombardment took place, was electrically isolated and served as a Faraday cup. The total number of incident particles (integrated flux) was obtained on a current integrator. Bombardments have been realized at room or at low temperature using a sample holder connected with a nitrogen dewar. Although the temperature of the samples was not measured during irradiation, it was evident that high beam irradiations considerably heated the crystal producing alterations. Therefore it was necessary to operate with a low flux during a long time to allow heat dissipation.

---

[1]) All bombardments were realized in the Low Energy Laboratory of the Nuclear Center of Strasbourg.

Fig. 5 shows transmission spectra of initially as stoichiometric as possible Cu_2O before and after irradiation with α-particles of different energies. The structures are more pronounced with increasing energy and the transmission minima are located at the same wavelengths as for annealed samples. With particles of energies lower than 1 MeV, no optical effect was observed. In addition, we have performed a great number of bombardments by varying the electrical intensity of the beam (between 0.1 μA and 2.0 μA) and the integrated flux (between 10 μC and 4000 μC). It follows that the dose parameter N_s, which gives the total number of particles colliding the sample under normal incidence on a normalized area of 1 cm² during the experiment, lies between 10^{15} and 10^{17} α-particles per cm². This upper limit dose points to defect saturation effects with clustering, accompanying prolonged irradiation.

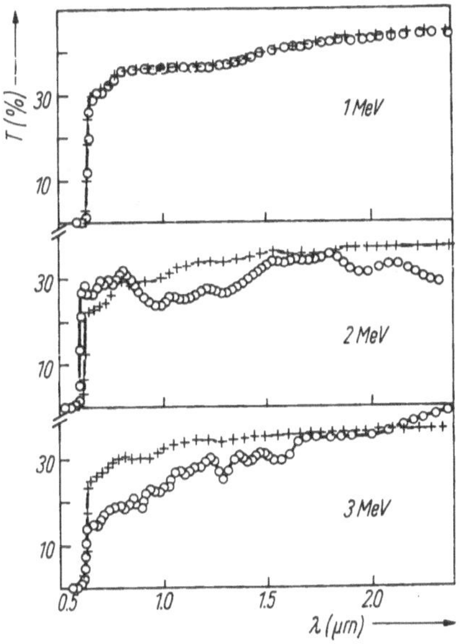

Fig. 5. Effect of the energy of the α-particles on the transmission. (+) and (○) are transmissions measured at 77 °K before and after α-bombardment of equal dose-parameter, respectively

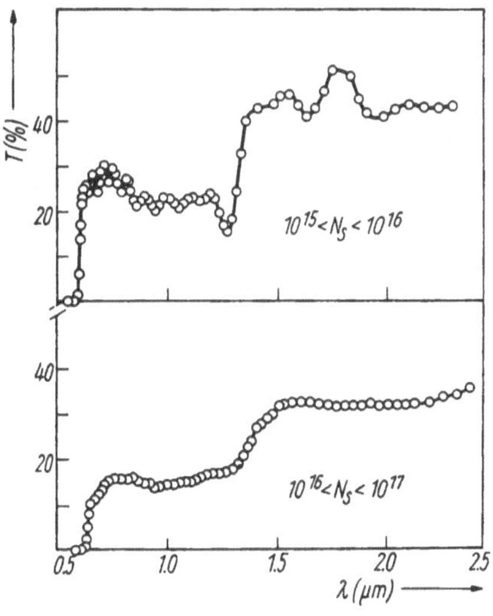

Fig. 6. Effect of the dose-parameter N_s on the transmission curve of a preannealed sample, $T = 77$ °K

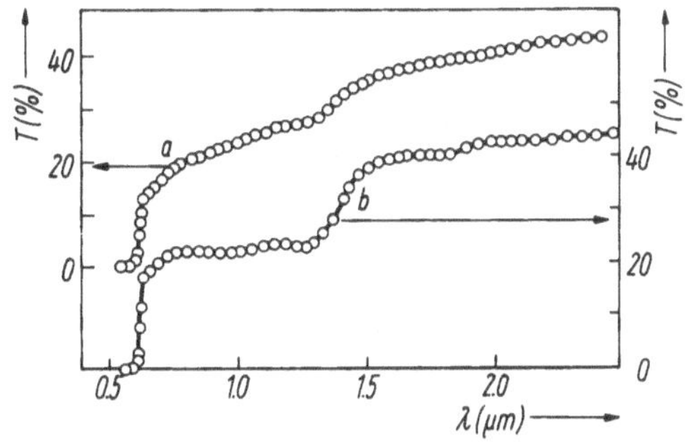

Fig. 7. Temperature effect on sample D 42-4 when bombarded strongly. (a) 77 °K, after annealing and α-irradiation; (b) 293 °K after annealing only. The disappearance of structure between 0.7 and 1.4 μm proves that the sample has again a better stoichiometric structure

Annealed samples, showing a transmission similar to that of Fig. 2, were also irradiated. No new bands were observed, but two different effects appeared for equal time exposures: for dose parameter N_s ranging about 10^{15} and 10^{16} αcm^{-2}, the previous bands were enhanced; on the contrary, for N_s greater than 10^{16} αcm^{-2}, the overall transmission was reduced and the bands broadened. Both cases are represented on Fig. 6. Fig. 7 shows that α-particles can also bleach out the bands due to annealing; the lower curve gives the transmission at room temperature for an annealed sample before irradiation, whereas the upper curve gives the transmission at low temperature for the same sample but after irradiation. Not only the structure of Fig. 2 disappears, but in addition we observe an increase of the transmission in a region where the bands were located previously. This effect probably involves a thermal oxygen degasing of the samples related to local annealing; for N_s greater than 10^{16} αcm^{-2}, the exposure time was not sufficient to allow heat dissipation because of the low thermal conductivity of Cu_2O.

5. Estimation of the Number of Defects Due to Irradiation

Informations about stopping cross section and range of penetration of α-particles in Cu_2O are not found in the literature. Gobeli [6] reported values of the stopping cross section for α-particles in atomic copper and oxygen:

$$\varepsilon(\mathrm{Cu})_{\mathrm{at}} = 66 \times 10^{-15}\ \mathrm{eV\ cm^2\ atom^{-1}}$$

$$\varepsilon(\mathrm{O})_{\mathrm{at}} = 30 \times 10^{-15}\ \mathrm{eV\ cm^2\ atom^{-1}}\,.$$

More suggestive are the following values: $N(\mathrm{Cu}) \cdot \varepsilon(\mathrm{Cu})$ at $= 5.5 \times 10^{-3}$ dyn, $N(\mathrm{O}) \cdot \varepsilon(\mathrm{O})$ at $= 1.25 \times 10^{-3}$ dyn if we take into account the number of each sort of atoms in the Cu_2O unit cell.

Further Baroody [7] considered in detail the interaction of irradiating particles with a diatomic substance. Following his treatment, we conclude that a collision with a copper atom is about five times more probable than with an oxygen atom, therefore we have neglected the latter one.

The range of He^+ particles in Cu_2O is expected to be very short. Referring to Whaling [8], in pure copper the penetration depth X of 3 MeV α-particles is of the order of 6 μm. The Bragg Kleeman's empirical formula, quoted by Blanc [9] allows more accurately for the constitution of the stopping material.

$$X/X_0 = 3.4 \times 10^4\ M^{1/2}/d \tag{3}$$

gives X by the knowledge of X_0, the penetration depth of He^+ particles in air; M is the reduced mass and d the density of the stopping material. We determine $X = 6.4$ μm with

$$M_{\mathrm{Cu_2O}} = \frac{143}{3}\,, \quad d_{\mathrm{Cu_2O}} = 6.17\ \mathrm{g\ cm^{-3}}\,, \quad \text{and} \quad X_0(3\ \mathrm{MeV}) \approx 1.66\ \mathrm{cm}\ [6]\,.$$

The description of radiation effects and the counting of defects created by irradiation with neutrons, charged particles, or γ-rays was developed extensively during the last two decades, namely by the following authors: Seitz and Koehler [10], Dieners and Vineyard [11], Snyder and Neufeld [12], Billington and Crawford [13], Kinchin and Pease [14], and Leibfried [15].

For an estimation of the defect concentration, we choose to use a simple formalism rather than a complicated one, considering the small number of known parameters for Cu_2O. Following Seitz [10] we have taken

$$N_d = N_A \, \sigma_d(E) \, \varphi \, \bar{\nu}(E_p) \, , \tag{4}$$

where N_A is the number of target atoms per unit volume (5.2×10^{22} atoms cm^{-3}),

$$\sigma_d(E) \approx \frac{\pi b^2}{4} \frac{E_p(\mathrm{Max})}{E_d} \tag{5}$$

the total displacement cross section as given by [13] in which E_p(Max) is the maximum of energy transfer in an elastic collision, E_d the Wigner energy (25 eV for Cu) and $b = Z_1 Z_2 e^2/E$ the classical distance of closest approach. φ is the integrated flux, usually of the order of 5×10^{15} particles cm^{-2}, and $\bar{\nu}(E_p)$ the number of displacements resulting from a simple primary event. Following the treatment of (14) equivalent to those of [12]and [16], we have taken

$$\bar{\nu}(E_p) = \frac{1}{2} \ln \left[\frac{E_p(\mathrm{Max})}{E_d} \right]. \tag{6}$$

Table 1 summarizes the values of N_d calculated for two energies.

Table 1

E (MeV)	σ_d (barns)	$\bar{\nu}$	N_d (10^{17} cm^{-3})
2	76.5	6.05	0.8
3	51	6.25	1.2

The knowledge of the defect concentration N_d makes it possible to use Kravets's formula [17] which gives the "experimental" oscillator strength

$$f_{\mathrm{exp}} = \frac{m_0 c^2}{\pi N_d e^2} \int n \, \alpha \, d\nu' \, . \tag{7}$$

m_0 and e are mass and charge of the electron at rest, respectively, c is the light velocity, n is the refractive index (constant and equal to 2.72 in our spectral region: see R_0 on Fig. 1), and α is the absorption coefficient plotted versus ν', the wave number. Choosing the band at 1.28 μm, we get an approximate value

$$f_{\mathrm{exp}} \approx 5 \times 10^{-3}$$

which is greater than f_{exp} for second class excitonic transitions spectra (1) but lower than f_{exp} of colour centres, for example in alkali halides.

6. Conclusions

This paper is concerned with two experimental results. First, we have observed new absorption bands in the near infrared; their connection with oxygen excess in the Cu_2O lattice is established. These bands are more pronounced at low temperature with peaks at 0.65, 0.76, 0.88, 0.95, 1.10, 1.28 μm, and a weaker one at 1.63 μm. An evaluation of the magnitude of the more important bands, correlated to the annealing oxygen pressure gives a $(p_{O_2})^{2/3}$ law. By submitting equivalent samples to irradiation with charged particles, we have observed the same bands. Therefore, we have used two processes which give rise to the same effects. Annealed and irradiated samples show the same structures, enhanced at integrated flux less than 10^{16} αcm^{-2}; for higher values, the bands disappear. This latter effect involves probably a thermal oxygen degasing related to local annealing of the sample. The exposure time did not allow heat dissipation.

We estimated quantities concerning the interaction of charged particles with Cu_2O such as, displacement cross section $\sigma_d(E)$, number of displacements resulting from a single primary event $\bar{\nu}(E)$, and the density of defects $N_d(E)$ due to irradiation. Although the determinations are rough, it is plausible to consider the range of $N_d \approx 10^{17} cm^{-3}$ as a good approach.

Some questions about irradiations were not yet solved, in particular the distribution of the defects in the sample. As the range of α-particles is short compared to the thickness of our samples (6 μm against 140 μm), one can suppose that the defects are not created in the whole crystal and therefore explain the weakness of the bands observed after irradiation only. Another detail is to be discussed: what happens to the α-particles after their slowing-down in the crystal? Barnes and Mazey [18], using electron microscope techniques, have examined point defect clusters produced in copper foils after α-particles bombardments. In both foils, this through which α particles had passed and that in which they came to rest, they observed dislocation loops ($r \approx 200$ Å) and a background of smaller dots ($r \approx 20$ Å).[2]) By annealing at about 350 °C, the dislocation loops and smaller dots grew to form dislocation tangles and eventually small helium bubbles ($r \approx 40$ Å) in the same number as previously. This means that the vacancies produced by irradiation provide some space for the helium atoms to form bubbles. This must be the case for our irradiation experiments. The same effect may happen with oxygen in the annealed samples.

At last, we may mention the work of Lisitsa et al. [20] who also studied the absorption and photoconductivity of Cu_2O with various content of oxygen before and after neutron irradiation. Before irradiation, they found no clearly expressed maxima in the absorption spectrum. However, neutron irradiation gave rise to three intensive bands at 1.2 μm, 1.95 μm, and 2.2 μm, respectively.

We also have investigated the effects of fast neutrons (≈ 1 MeV) on the absorption of Cu_2O in the range between 0.65 μm and 3.5 μm; but in our experiments no modification of the spectrum was observed.

Acknowledgements

The authors are indebted to Mrs. D. Magnac, Professor of Physics, for having kindly allowed to use the accelerator for production of He^+ particles and to Mr. S. Nikitine, Professor of Physics, for many discussions, to Dr. C. Schwab, Mrs. Robino, and G. Schwalbach for sample preparation.

---

[2]) Song et al. [19] also dealt with such aggregations of defects, involving dislocation loops of $r \approx 150$ Å, to give an explanation of fine structure in the excitonic series of Cu_2O and their displacement towards higher energies after neutron irradiations.

References

[1] S. Nikitine, Progress in Semiconductors **6**, 233 (1962).
[2] M. O'Keeffe, J. chem. Phys. **39**, 1789 (1963).
C. Noguet, C. Schwab, C. Sennett, M. Sieskind, and C. Viel, J. Phys. (Paris), **26**, 317 (1965).
C. Carabatos, A. Diffine, and M. Sieskind, J. Physique **29**, 529 (1968).
C. Carabatos, phys. stat. sol. **37**, 773 (1970).
[3] C. Wagner and H. Hammen, Z. Phys. Chem. **40**, 197 (1938).
[4] C. Schwab, A. Diffine, G. Koehl, C. Carabatos, M. Sieskind, P. Calme, and C. Viel, J. Physique **29**, 317 (1968).
[5] J. B. Grun, Thèse, Revue d'Optique **41**, 439 (1962).
[6] G. W. Gobeli, Phys. Rev. **103**, 275 (1956).
[7] E. M. Baroody, Phys. Rev. **112**, 1571 (1958).
[8] W. Whaling, Handbuch der Physik **34**, 193 (1953).
[9] D. Blanc, Les Radioéléments, Masson Paris 1966.
[10] F. Seitz and J. S. Koehler, Solid State Phys., Vol. 2, Academic Press, Inc., New York 1956.
[11] G. J. Dieners and G. H. Vineyard, Radiation Effects in Solids, Interscience Publishers, New York (1957).
[12] W. S. Snyder and J. Neufeld, Phys. Rev. **103**, 862 (1956); **97**, 1636 (1955); **99**, 1326 (1955).
[13] D. S. Billington and J. M. Crawford, Jr., Radiation damage in Solids, Princeton University Press, New Jersey 1961.
[14] G. H. Kinchin and R. S. Pease, Rep. Progr. Phys. **18**, 1 (1955).
[15] G. Leibfried, Bestrahlungseffekte in Festkörpern, Teubner, Stuttgart 1965.
[16] W. A. Harrison and F. Seitz, Phys. Rev. **98**, 1530 (1955).
[17] T. P. Kravets, Izv. vuzov. Moskva. Inz. Uch. **2**, 6 (1912).
B. S. Neporent and T. P. Kravets, Trud. Fiz. Akad. Nauk SSSR 1959 (p.308).
[18] R. S. Barnes and D. J. Mazey, Phil. Mag. **5**, 1247 (1960).
[19] K. S. Song, S. Nikitine, J. B. Grun, and M. Grosmann, C. R. Acad. Sci. (France) **259**, 1087 (1964).
[20] M. P. Lisitsa and G. A. Kholodar, Ukr. fiz. Zh. **12**, 1151 (1966).
[21] S. J. Fray, A. R. Goodwin, F. A. Johnson, and J. E. Quarrington, J. sci. Instrum. **40**, 387 (1963).

(Received October 29, 1971)

phys. stat. sol. (a) **10**, 465 (1972)

Subject classification: 11; 22; 22.6

Europäisches Institut für Transurane, Karlsruhe

Properties of Fission Spikes in UO_2 and UC Due to Electronic Stopping Power

By

H. BLANK

For the electronic stopping power of fission fragments in UO_2 and UC a distinction is made between the initial energy distribution of width r_i around the path of the fission fragment and a limiting temperature distribution with width r'_i. A value of r'_i can be fixed in a way which allows to compare different substances, here UO_2 and UC. The real temperature distribution depends upon the ratio $n = r_i/r'_i$. Using transmission electron microscopy, the ratio n is found to be close to one for U_4O_9 and the same is concluded for UO_2. This initial temperature distribution is used to calculate the thermoelastic stress field around a fission spike in UO_2. A rough estimate for UC shows that in this case $n \gg 1$, this means that the thermoelastic stress field in UC is of no practical importance, like in metals.

Beim elektronischen Bremsvermögen eines Spaltproduktes in UO_2 und UC kann man zwischen einer Anfangsenergieverteilung der Breite r_i um die Spaltspur und einer begrenzenden Anfangstemperaturverteilung der Breite r'_i unterscheiden. Ein Wert von r'_i kann derart festgelegt werden, daß man verschiedene Substanzen, hier UO_2 und UC, vergleichen kann. Die wahre Temperaturverteilung hängt vom Verhältnis $n = r_i/r'_i$ ab. Mit Hilfe der Elektronenmikroskopie läßt sich für U_4O_9 ein Verhältnis $n = 1$ abschätzen; dies sollte näherungsweise auch für UO_2 gelten. Mit dieser Anfangstemperaturverteilung läßt sich das thermoelastische Spannungsfeld um eine Spaltspur in UO_2 berechnen. Eine grobe Abschätzung für UC zeigt, daß dort $n \gg 1$, d. h. in diesem Fall hat das thermoelastische Spannungsfeld wie bei Metallen keine Bedeutung.

1. Introduction

About 80% of the energy liberated in a nuclear fission event is given to the two fission fragments as kinetic energy

$$2\,E_{kin} = m_1\,v_1^2 + m_2\,v_2^2\,.$$

By interaction of these fast moving ions with (i) the crystal electrons and (ii) the crystal atoms of the fuel matrix, this kinetic energy is transformed into heat, the primary form of fission energy. In spite of the fact that at present all nuclear power production is based on this concept, the fundamental physical processes, transformation of kinetic energy into lattice vibrations and damage to the crystal lattice of the nuclear fuel, have received relatively little attention. A book on fission damage in crystals published recently even does not mention fission damage in nuclear fuels like UO_2, UC, or UN at all.

A closer look to the properties of fission spikes in nuclear fuels reveals that still large gaps exist in our present knowledge. This paper draws attention to a problem, which to the knowledge of the author has not yet been treated, thermoelastic stresses due to fission spikes.

2. Estimation of Electronic Stopping Power for Fission Fragments in UO_2 and UC

According to Thomson and Lindhard [3] roughly about 3% of the kinetic energy of the median light fission fragment and about 6% of the median heavy fission fragment are expended in atomic collisions in UO_2 and UC, and the balance of 97 and 94% respectively of the initial fragment energy is given to the electronic system of the crystal.

From the work of Bohr, especially Chapters 3.5, 4.4, and 5.3 of reference [1], one may deduce an electronic stopping formula valid for heavy substances:

$$-\frac{dE}{dR} \approx N\, 1.7 \cdot 4\pi Z_1^{2/3} Z_2^{1/3} a_0 e^2 \frac{v}{v_0}, \tag{1}$$

where N number of atoms with atomic number Z_2 per cm^3 of stopping substance, Z_1 atomic number of fission fragments, v fragment velocity, $v_0 = e^2/\hbar$, and $a_0 = \hbar^2/m\, e^2$.

Since in equation (1) the influence of oxygen in UO_2 and of carbon or nitrogen in UC and UN cannot be properly accounted for the formula of Lindhard et al. [2] is used here. The much used Bethe-Bloch formula overestimates the influence of oxygen and underestimates the influence of uranium in comparison to equation (2):

$$-\frac{dE}{dR} = N\, 8\pi a_0 e^2 \frac{Z_1^{7/6}}{[1 + (Z_1/Z_2)^{2/3}]^{3/2}} \frac{v}{v_0}. \tag{2}$$

Both formulae give at any instant a linear relation between electronic stopping power $(dE/dR)_e$ and the path R travelled by the fission fragment. The "electronic range" R_e of the fission fragment (which is only slightly larger than the true range including atomic collisions) is given by

$$R_e = \frac{2}{k} \frac{v_0}{v'} E', \tag{3}$$

where E' and v' are the initial energy and velocity of the fragment, and the constant k is easily derived from equation (2).

In compounds with constituents j the total electronic stopping power $(dE/dR)_e^{tot}$ is

$$\left(\frac{dE}{dR}\right)_e^{tot} = \sum_j \left(\frac{dE}{dR}\right)_e^{(j)} \tag{4}$$

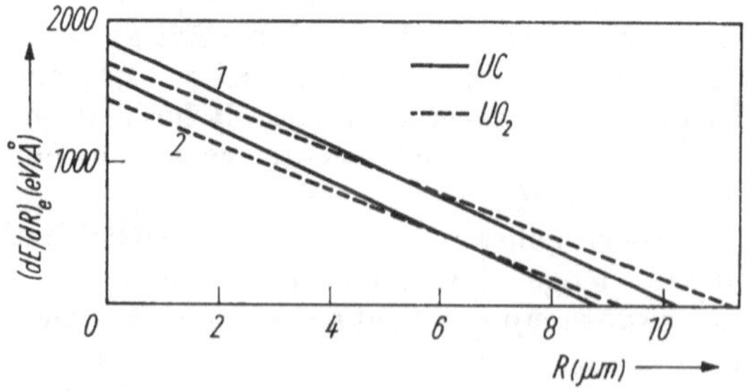

Fig. 1. Electronic stopping power of the median light (1) and the median heavy (2) fission fragment calculated for UO_2 and UC according to equations (2) and (4)

and the total electronic range R_e^{tot} is given by

$$\frac{1}{R_e^{tot}} = \sum_j \frac{1}{R_e^{(j)}}. \quad (5)$$

The electronic stopping powers and ranges of the median light and median heavy fission fragments in UO_2 and UC according to equations (2) to (5) are shown in Fig. 1.

3. Observation of Fission Tracks by Transmission Electron Microscopy (TEM)

Though under nuclear reactor conditions fission spikes are only created at high crystal temperatures (500 °C $\leqq T \leqq$ 2600 °C in oxides and at about 500 °C $\lesssim T \lesssim$ 1500 °C in carbides and nitrides), many details important at high temperatures can be learned from their properties at low temperatures.

Let us regard crystals of simple cubic structures with reasonably high melting temperatures. As examples for metals, semiconductors, and insulators one may take Pt, GaAs, and $BaTiO_3$. The crystal be at a temperature $T \approx 0.3\ T_m$, i.e. at a temperature where vacancies are not mobile. If in such a crystal a fission fragment is slowed down, under certain circumstances its track may be revealed by TEM as a straight line either by absorption contrast or by diffraction contrast.

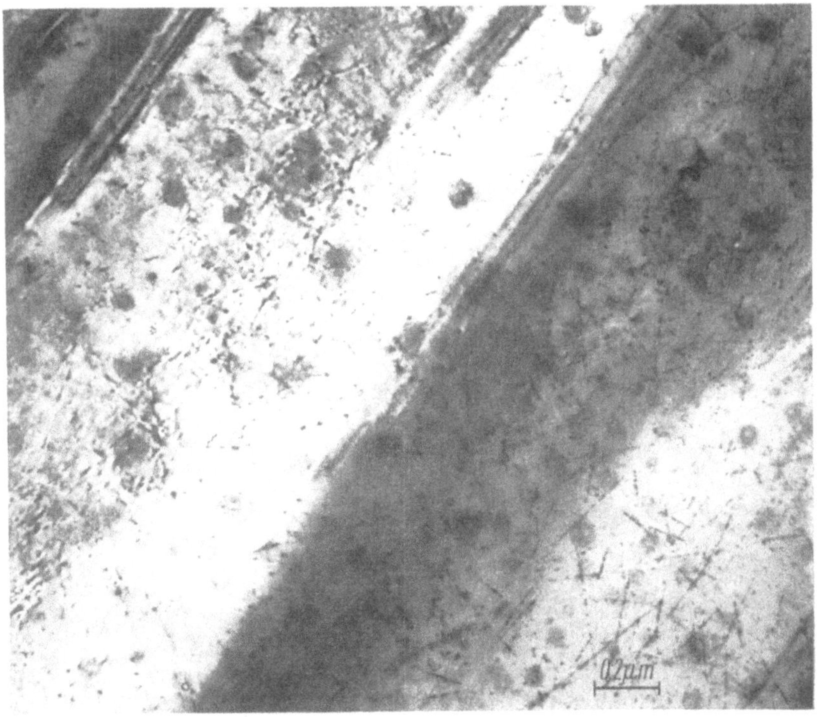

Fig. 2. Fission fragment tracks in $BaTiO_3$ observed by TEM. Depending on the diffraction conditions used, the tracks show either diffraction contrast (upper left) or absorption contrast (lower right). (Picture taken by the author during a stay as guest scientist at the solid state physics department of S.C.K.–C.E.N. Mol.)

In the first case the crystal structure has been completely or partly disordered around the path of the fragment, in the second case a permanent displacement field of cylindrical symmetry around the fragment path has been created.

Fission tracks are clearly produced in $BaTiO_3$, see Fig. 2, but no tracks are observed in bulk metal and semiconductors. Regarding its electrical properties, UO_2 must be expected to show a behaviour intermediate between a semiconductor and an insulator.

In UO_2 irradiated at temperatures $T \approx 100$ °C tracks visible by diffraction contrast form only, if they lie at distances $l \lesssim 200$ Å close to the crystal surface [5, 7]. The contrast is produced by a radial displacement field around the core of the track [6]. This displacement field must have been formed when the instantaneous lattice strain induced by the energy $\mathrm{d}E/\mathrm{d}R$ was able to relax versus the nearby crystal surface.

Within the bulk UO_2 crystal no damage visible by TEM is observed. This indicates that

(i) no permanent lattice strain is produced, and

(ii) no true melting and quench re-solidification of the crystal lattice can have occurred around the fission track.

If, however, the UO_2 contains large precipitates of U_4O_9, see Fig. 3, fission tracks are revealed by absorption contrast in U_4O_9. The images of the tracks are formed by locally disordering the otherwise ordered oxygen superlattice, thus destroying the diffraction properties of the ordered structure. Tracks with

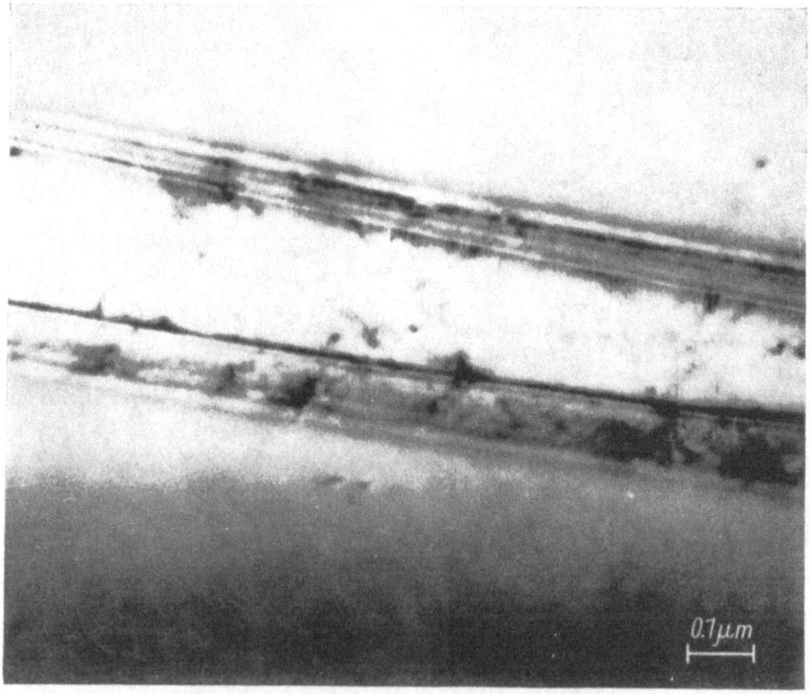

Fig. 3. Fission tracks in U_4O_9. The disordered oxygen superlattice is revealed by absorption contrast. Fission tracks are not produced in the UO_2 lattice adjacent to the U_4O_9 needle. (Picture taken during the author's stay at S.C.K.-C.E.N. Mol.)

slowly varying widths $2\,r_d$ in the range

$$60\ \text{Å} \lesssim 2\,r_d \lesssim 120\ \text{Å} \tag{6}$$

are observed, however again, no sharp boundary is found as would be expected from a molten zone around the core of the track.

Assuming that the oxygen sublattice in U_4O_9 has been disordered by the temperature pulse emitted from the fission track, an estimate on the temperature–time history around the core of the track in U_4O_9, in our case at a mean crystal temperature $T \approx 400$ °K, can be made in the following way:

U_4O_9 transforms to UO_{2+x} at $T \geqq 1150$ °C and this disordered state can easily be quenched to $T < 1150$ °C. Disorder will be produced already if the excess oxygen moves at least one atomic distance d.

From the mean square displacement

$$\overline{x}^2 = 6\,D\,\tau \tag{7}$$

one may derive a relation

$$\tau = \frac{d^2}{6\,\nu_D}\,e^{H_m/RT} \tag{7a}$$

if one puts $\overline{x}^2 = d^2$, and $D = d^2\,\nu_D \exp(-H_m/RT)$, which defines a limiting condition for disordering by thermal activation with respect to time τ and temperature T at the observed track radii r_d of relation (6). $H_m = 26.5$ kcal/mol is the activation energy of migration for oxygen in UO_{2+x} and ν_D is the Debye frequency. Both relations (6) and (7a) will be confronted in Section 4.2 with the laws of heat flow from a line source of energy along the path of the fission fragment as shown in Fig. 1.

4. Determination of the Initial Temperature Distribution for Linear Heat Flow around a Fission Spike

4.1 Initial temperature distribution for heat conduction

The energy $(dE/dR)\,\Delta R$ along a section ΔR of the fission track will be concentrated at first, i.e. 10^{-15} s after the passage of the fragment, within a radius $r_0 \lesssim 10$ Å around the track. Then, this energy will spread radially and finally will transform into heat.

Hence, after a certain time t', which we take as $t = 0$, there will be an initial temperature distribution around the track which may be characterized conveniently by the radius r_i of its steepest temperature gradient and which defines the starting condition for linear heat flow.

The value of r_i depends sensibly upon the mechanisms by which $(dE/dR)\,\Delta R$ is transformed into heat, and on comparing UO_2 with UC one should expect that $r_i\,(UO_2) \ll r_i\,(UC)$.

Taking dE/dR constant along a certain increment ΔR of track length the temperature distribution $T(r, t)$ around that increment is given by

$$T(r, t) = \Delta T(r, t) + \overline{T}\,, \tag{8}$$

where $\overline{T}$ is the mean crystal temperature and for $\Delta T(r, t)$ we use as a simple approximation (i.e. the initial distribution is assumed to be of Gaussian type)

$$\Delta T(r, t) = \frac{Q}{4\,\pi\,k\,(t + t_0)} \exp\left[-\frac{r^2}{4\,k\,(t + t_0)}\right] \tag{9}$$

with

$$Q = \frac{\mathrm{d}E/\mathrm{d}R}{C_p\,\varrho}\,, \tag{9a}$$

where C_p, ϱ, and k are heat capacity, density, and thermal diffusivity, respectively, of the crystal in question.

The parameter t_0 defines the radius r_i explained above by the relation

$$r_i = (2\,k\,t_0)^{1/2}\,. \tag{9b}$$

The problem is now to find reliable values of r_i for UO_2 and UC. To this end we regard another characteristic radius r_i', independent of the details of energy transformation in the fission spike, which describes just the maximum possible linear heat flow in the crystal with a heat distribution of cylindrical symmetry.

This limiting heat flow is determined by the maximum possible temperature gradient and can be estimated from the requirement, that the difference in thermal energy between neighbouring atoms in the lattice must be sensibly less than $k'T$ (k' Boltzmann constant) even near the melting temperature, hence

$$k'(\mathrm{grad}\ T)_{\max} = \frac{\gamma\,k'T}{d}\,; \quad \gamma \ll 1\,. \tag{10}$$

Taking for T the melting point and $\gamma \approx 0.025$ we get for both UO_2 and UC

$$(\mathrm{grad}\ T)_{\max} \approx 2 \times 10^9\ \mathrm{deg\ cm^{-1}}\,. \tag{10a}$$

On the other hand we have from equation (9)

$$(\mathrm{grad}\ T)_{\max} = \frac{Q\,\mathrm{e}^{-0.5}}{2\,\pi\,(2\,kt_0')^{3/2}}\,. \tag{11}$$

Identifying the left-hand side of equation (11) with (10a) yields in analogy to (9b)

$$r_i' = (2\,kt_0')^{1/2}\,. \tag{11a}$$

At a value $\mathrm{d}E/\mathrm{d}R = 1000$ eV/Å of the electronic stopping power, for instance, we get for t_0', r_i', and $\Delta T'(0, 0)$ the values shown in Table 2, which indicate that the limiting temperature distribution is highly peaked in all three crystals UO_2, U_4O_9, and UC.

The crystal properties used in these calculations are collected in Table 1.

We can now proceed to compare the initial temperature distribution around a fission spike characterized by r_i in equation (9b) with the limiting possible distribution in the crystal according to (11) and (11a).

Qualitatively three cases are possible:

$$r_i = n\,r_i' \quad \text{with} \quad n \gtreqless 1\,. \tag{12}$$

For $n = 1$ the energy distribution around the fission spike transforms directly into the limiting distribution in the crystal with the values of Table 2

$$t_0 \approx t_0'\,; \quad \Delta T(0, 0) = \Delta T'(0, 0)\,. \tag{12a}$$

For $n > 1$ the energy $\mathrm{d}E/\mathrm{d}R$ has been spread quicker and wider in the first moment of the life of the spike than by hypothetical heat conduction, thus

Table 1

Values of properties of UO_2 and UC

	T (°C)	UO_2	U_4O_9	UC
C_p (cal/mol deg) heat capacity (Ws/g deg)	1000	22 0.341	23.5 0.363	18 0.301
ϱ (g/cm³) density	20	10.97	11.33	13.63
k (cm²/s) diffusivity $4k$	1000	6.68×10^{-3} 2.67×10^{-2}	3.65×10^{-3} 1.46×10^{-2}	4.87×10^{-2} 0.195
K (W/cm deg) thermal conductivity	1000	2.5×10^{-2}	1.50×10^{-2}	0.20
ϱ_{el} (Ωcm) electrical resistivity	20 1000 1200 1900	10^3 5 2.2 0.27		5.7×10^{-5} 1.6×10^{-4} 1.84×10^{-4} 2.7×10^{-4}
ν_D (s⁻¹) Debye frequency		1.6×10^{13}		
ν Poisson ratio		0.317		0.286
E (dyn/cm²) Young modulus	20 1400 2400	2.23×10^{12} 1.9×10^{12} $\approx 1.57 \times 10^{12}$		2.25×10^{12} 2.00×10^{12}
α (deg⁻¹) thermal expansion	20 1400 2400	10.54×10^{-6} 10.57×10^{-6} 11.00×10^{-6}		$\approx 10^{-5}$
$(dE/dR)_{eff}$ (eV/Å) (Ws/cm)	— —	1000 1.6×10^{-8}		

Table 2

Initial conditions of heat distribution in UO_2 and UC for $(dE/dR)_{eff} = 1000$ eV/Å

	t_0' (s)	r_i' (Å)	$\Delta T'(0, 0)$ (°C)	Q (cm² deg)
UO_2	2.62×10^{-11}	59	1950	4.28×10^{-9}
UC	3.36×10^{-12}	57.2	1895	3.90×10^{-9}
U_4O_9	4.48×10^{-11}	57.1	1890	3.89×10^{-9}

leading to a value

$$t_0 = \frac{n^2 r_i'^2}{2k} = n^2 t_0'; \; n > 1 . \tag{12b}$$

This makes the initial temperature distribution around the fission spike wider and less peaked. The third case $n < 1$ is of no interest here.

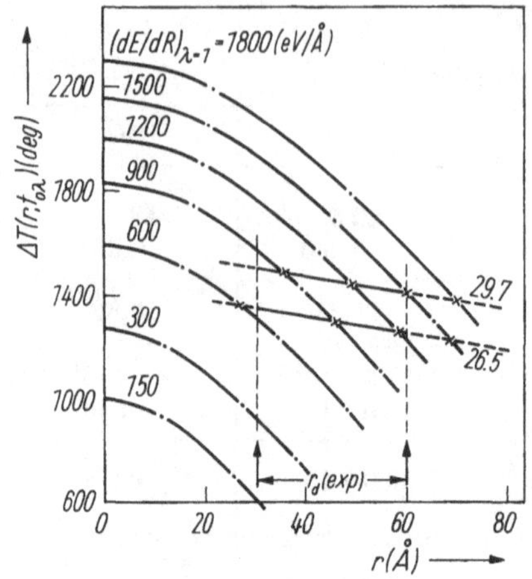

Fig. 4. Initial temperature distributions $\Delta T(r, t_{0\lambda})$ with $n = 1$ at seven values ($\lambda = 1, 2, \ldots 7$) of (dE/dR) in units of eV/Å along the range of a fission spike in U_4O_9; see also Fig. 1. The crosses indicate temperatures of disordering $T_\lambda - \bar{T}$ calculated from equation (7a) by putting $t_{0\lambda} = \tau_\lambda$. Two activation energies 26.5 and 29.7 kcal/mol were used in (7a) for the migration of oxygen. The observed range of track widths deduced from Fig. 3 is indicated by arrows and allows roughly to correlate r_d with a certain range of dE/dR, see text

4.2 Initial temperature distribution in U_4O_9 and UO_2

We are now able to estimate the initial temperature distribution around a fission spike in U_4O_9 by combining the results of Sections 3 and 4.1. In Fig. 4 the limiting temperature distributions for seven values $(dE/dR)_\lambda$ along the range of the fission fragment have been drawn using calculated values of $t'_{0\lambda}$ by equations (11a) and (12) and putting $t_0 = t'_{0\lambda}$ in equation (9). The case $n = 1$ is applied as a first guess; it turns out by chance to be the correct one for U_4O_9, see below.

If we interpret now the time constant $t'_{0\lambda}$ for each value of $(dE/dR)_\lambda$ as a time of disordering τ_λ, we get by equation (7a) the corresponding temperature of disordering T_λ,

$$T_\lambda = \frac{H_m}{R \ln (6\, t'_{0\lambda}\, \nu_D/d^2)} \,.$$

This yields a value for $r_{d\lambda}$, the radius of disordering at the value $(dE/dR)_\lambda$ of the electronic stopping power, if one identifies the temperature difference $T_\lambda - \bar{T}$ on the corresponding limiting temperature distribution $\Delta T(r_i\, t'_{0\lambda})$, hence

$$T_\lambda - \bar{T} = \Delta T(r_{d\lambda}, t'_{0\lambda}) \,.$$

The graphically obtained values of $r_{d\lambda}$ are connected by the straight line labelled 26.5 kcal/mol in Fig. 4. The two arrows at $r = 30$ and 60 Å, see relation (6), indicate that with the numerical values used the experimental width of the fission tracks in U_4O_9 can be correlated with an effective electronic stopping power in the range 600 eV/Å $\lesssim (dE/dR)_{eff} \lesssim$ 1200 eV/Å. If one assumes that $r_d =$ = 60 Å corresponds to the beginning of the track, this would mean that about 30% of the electronic stopping power does not transform into heat around the track but is carried away instantaneously by X-rays and elastic waves under the assumption that $n = 1$. However, this point needs further investigation.

In order to show the influence of a possible error in the value of the activation energy of migration used $r_{d\lambda}$ values for $H_m = 29.7$ kcal/mol are also shown in Fig. 4. If this H_m value would apply instead of 26.5 kcal/mol the observed track width would correspond to about 800 eV/Å $\lesssim (dE/dR)_{eff} \lesssim$ 1500 eV/Å.

Another error is introduced into the correlation between track width r_d, effective value $(dE/dR)_{eff}$, and corresponding location R along the track by the error in the exact value of r_i, i.e. of n. Replacing, for instance, $n = 1$ by $n = \sqrt{2}$ in equation (12b) means reducing the peak temperatures along the path of the track to such an extent, that practically no thermal disorder of the oxygen superstructure in U_4O_9 would be possible. Hence, small variations in n give already drastic changes in $\Delta T(r, t_0)$.

This discussion shows that the initial temperature distribution around a fission track in U_4O_9 depends sensibly on parameters not yet known with exact precision, but that the distribution shown in Fig. 4 with $n = 1$ nevertheless is a very useful approximation.

Because of the very similar properties of UO_2 and U_4O_9 one may safely conclude that also in the case of UO_2 it is appropriate to define the temperature distribution equation (9) by relation (12a), i.e. with $n \approx 1$.

4.3 Initial temperature distribution in UC and UN

Here no calibration by experiment is available at present, but since both UC and UN are metallic conductors, it is possible to make a rough order of magnitude estimate for the width r_i of equation (9b), even if the energy band calculations [11 to 13] have not yet yielded good values of the Fermi energy due to difficulties in describing the 5f electrons [12].

Highly excited electrons of energy K in the range 40 eV $\leqq K \leqq$ 100 eV [8, 9], which are emitted from the core of the fission track in the first 10^{-15} s, at first should loose their energy by plasmon excitation with a mean free path l_{pl} of the order of 10 Å [14] and characteristic energy losses $E_{pl} \approx 15$ eV.

Afterwards the plasmons should decay into single excited electrons with energies 5 to 10 eV above the Fermi level. These electrons will share their energy by electron–electron collisions with conduction electrons and also should induce interband transitions.

Using the free-electron model with $r_s \lesssim 2$ [14][1]), the mean free path l_e of these excited electrons will depend on energy approximately as [14, 15]

$$l_e = l_0 E^{-\lambda}; \qquad l_0 \approx 10^3 \text{ Å}, \qquad \lambda = 1.5 \text{ to } 1.7 \tag{13}$$

as long as $E \gg kT$. In our case, we may take $T \approx 1500$ °K, i.e. $kT = 0.13$ eV.

In view of the uncertainties involved and of the rough approximation needed, the following arguments lead to a guess of r_i: The energy K of the primary electrons will be used at first in a volume with radius $r_i \approx 40$ Å around the fission fragment path for plasmon excitation within a time of the order of 10^{-15} s.

This volume will then emit electrons in the energy range 5 eV $\lesssim E \lesssim$ 10 eV with mean free paths given by (13) loosing large amounts of their initial energy in each electron–electron collision. Since the life time τ_e at each energy E is of the order $\tau_e \approx h/E$ and only a few collisions are necessary for one excited electron to degrade its energy from, say, 10 to 1 eV, the whole procedure will need a time $t' < 10^{-13}$ s. According to relation (13) electrons with $E \approx 1.5$ eV will have $l_e \approx 500$ Å, this means that we can expect at $t' \approx 10^{12}$ s $r_i' > 500$ A, or roughly $r_i \approx 1000$ Å.

[1]) Assuming a Fermi energy of about 10 eV for UC gives $r_s \lesssim 2$, similar to the case of Al treated by Quinn.

From Table 2 we have $r_i' \approx 60$ Å for UC, hence we get by equation (12) $n \approx 16$ and by (12b) $t_0 > 100\, t_0'$. This results in peak temperatures of $\Delta T(0, t_0) <$ < 20 deg which are negligible when compared with values in U_4O_9 and UO_2.

5. Thermoelastic Properties of Fission Spikes

The initial temperature distribution obtained in Section 4.2 can now be used to describe the thermoelastic stress field of a fission spike in UO_2 at a mean crystal temperature $\overline{T}$.

According to Table 2 thermal conduction starts at an "age of the spike" $t_0 \approx 3 \times 10^{-11}$ s with $r_i \approx 60$ Å for $(\mathrm{d}E/\mathrm{d}R)_{\mathrm{eff}} = 1000$ eV/Å. The crystal volume within this radius will have undergone thermal expansion at that time and will have caused elastic compression waves travelling in radial direction with the velocity of the longitudinal component of elastic waves in the solid given by

$$v_l = \left[\frac{2\,\mu\,(1-\nu)}{\varrho\,(1-2\,\nu)}\right]^{1/2}. \tag{14}$$

Using the elastic constants of UO_2 at 1000 °C gives $v_l = 4.5 \times 10^5$ cm/s, which is in any case larger than the velocity $\mathrm{d}r/\mathrm{d}t = v_T$,

$$v_T = \frac{2\,k}{r}, \tag{14a}$$

of the maximum of the temperature pulse described by equation (9) at times $t_0 + t$, with $0 \leqq t$. Therefore, even for small r and $t = 0$ in equation (9) the lattice will have its proper equilibrium thermal expansion due to the temperature distribution $\Delta T(r, t)$ of equations (8) and (9). Hence, one can use this distribution in spite of its initial rapid change with time in order to calculate the associated thermoelastic stress field.

An energy balance at time $t = 0$ in equation (9) can be written

$$\frac{\mathrm{d}E}{\mathrm{d}R} - \frac{\mathrm{d}W}{\mathrm{d}R} = \left(\frac{\mathrm{d}E}{\mathrm{d}R}\right)_{\mathrm{eff}}, \tag{15}$$

$$\left(\frac{\mathrm{d}E}{\mathrm{d}R}\right)_{\mathrm{eff}} \Delta R = E_T + E_\sigma\,, \tag{15a}$$

where W is the energy carried off by elastic waves and X-rays, E_T is the thermal energy stored in the temperature distribution equation (9) and E_σ is the elastic energy stored in the thermoelastic stress field.

Along an increment ΔR of the fission fragment path one can regard $(\mathrm{d}E/\mathrm{d}R)_{\mathrm{eff}}$ as constant and calculate the thermoelastic stresses with cylindrical symmetry [10]. By using the case of plane strain we get the strain and stress tensors

$$\varepsilon = \begin{pmatrix} \varepsilon_{rr} & 0 & 0 \\ 0 & \varepsilon_{\varphi\varphi} & 0 \\ 0 & 0 & 0 \end{pmatrix}, \qquad \sigma = \begin{pmatrix} \sigma_{rr} & 0 & 0 \\ 0 & \sigma_{\varphi\varphi} & 0 \\ 0 & 0 & \sigma_{zz} \end{pmatrix} \tag{16), (17}$$

with the following expressions for the strain components:

$$\varepsilon_{rr} = \alpha' \, T_0 \left[e^{-x} - \frac{1 - e^{-x}}{2x} \right], \tag{16a}$$

$$\varepsilon_{\varphi\varphi} = \alpha' \, T_0 \, \frac{1 - e^{-x}}{2x}. \tag{16b}$$

The meaning of α', T_0, and x is as follows:

$$\alpha' = \frac{1 + \nu}{1 - \nu} \alpha , \tag{16c}$$

$$T_0 = \frac{Q}{4 \pi k (t + t_0)} \quad \text{with} \quad Q = \frac{(\mathrm{d}E/\mathrm{d}R)_{\mathrm{eff}}}{\varrho \, C_p}, \tag{16d}$$

$$x = \frac{r^2}{4 k (t + t_0)}. \tag{16e}$$

The stress components are

$$\sigma_{rr} = -2 \mu \alpha' \, T_0 \left[\frac{1 - e^{-x}}{2x} \right], \tag{17a}$$

$$\sigma_{\varphi\varphi} = 2 \mu \alpha' \, T_0 \left[\frac{1 - e^{-x}}{2x} - e^{-x} \right], \tag{17b}$$

$$\sigma_{zz} = -2 \mu \alpha' \, T_0 \, e^{-x}. \tag{17c}$$

If one wants to calculate the elastic energy density one must substract the thermal expansion from the thermoelastic strains.

$$\varepsilon_{rr}^{(e)} = \varepsilon_{rr} - \alpha \, T , \tag{18a}$$

$$\varepsilon_{\varphi\varphi}^{(e)} = \varepsilon_{\varphi\varphi} - \alpha \, T , \tag{18b}$$

and the elastic energy density $w = 1/2 \, \sigma_{ij} \, \varepsilon_{ij}$ can be used to calculate the energy E_σ of equation (15a). It turns out to be at most of the order of 1% of E_T. More important for practical purposes, however, is the magnitude and behaviour in time and space of the stress components equations (17).

The functions σ_{rr} and $\sigma_{\varphi\varphi}$ normalized to $2 \mu \alpha' \, T_0 = 1$ have been plotted in Fig. 5 as function of x. Comparing (17) with (16) shows that the same curves can also be used for $\varepsilon_{\varphi\varphi}$ and ε_{rr}, respectively, if the sign is properly accounted for.

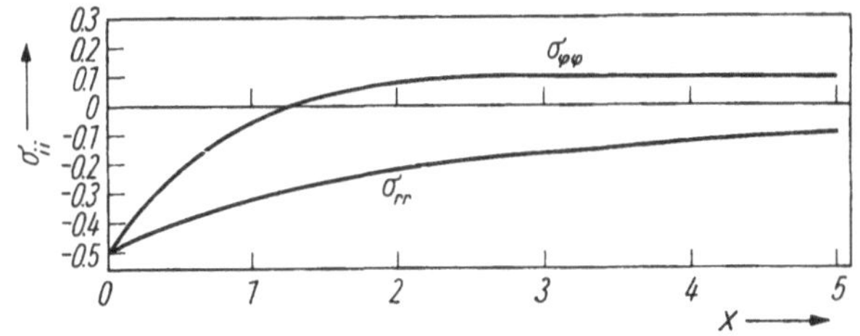

Fig. 5. Plot of normalized stress components σ_{rr} and $\sigma_{\varphi\varphi}$ vs. $x = r^2/4 \, k(t + t_0)$, see equation (16e). The scale factor for the stresses is $2 \mu \alpha' \, T_0$, see equations (16) and (17)

By choosing a value of t in equations (16d) and (16e) one can calculate $\sigma_{ii}(r)$ by using these curves. They show that σ_{rr} is a compressive stress for all radii whereas $\sigma_{\varphi\varphi}$ changes from compressive to tensile stress at $x_0 = 1.26$ and has a very flat tensile maximum near $x_m = 3$.

For large values of x both σ_{rr} and $\sigma_{\varphi\varphi}$ show an asymptotic behaviour:

$$-\sigma_{rr} \approx \sigma_{\varphi\varphi} \approx \frac{2\,\mu\,\alpha'\,T_0}{2\,x} \sim \frac{1}{r^2}. \tag{19}$$

Hence, following equation (16e) the stress field decreases proportional to $1/r^2$ in contrast to the stress field of a dislocation which is proportional to $1/r$, see equation (20):

$$\sigma_{rr} = \sigma_{\varphi\varphi} = -\frac{\mu\, b}{2\,\pi\,(1-\nu)}\,\frac{\sin\theta}{r} \quad \text{(edge dislocation)}\,. \tag{20}$$

In Fig. 6 the stress distributions $\sigma_{\varphi\varphi}(r)$ and $\sigma_{rr}(r)$ are plotted for a fission spike in UO_2 at $t = t_0$, $9\,t_0$, and $49\,t_0$ as curves (a) and (b), respectively. Equation (20) is shown as curve (D) with $\sin\theta = 1$ (maximum of stress field of edge dislocation).

For these calculations $\alpha = 1.1 \times 10^{-5}$ and $2\,\mu = 1.47 \times 10^4$ kp/mm^2 $= 1.44 \times \times 10^{12}$ dyn/cm^2 have been used. For α it is not justified to use values of thermal expansion measured at high temperatures under static conditions since the lattice does not contain the equilibrium vacancy concentration.

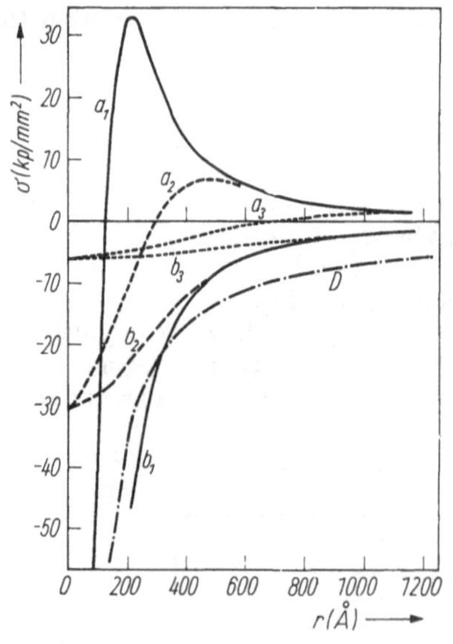

Fig. 6. Plot of stress components σ_{rr} (curves ($b_{1,2}$) and $\sigma_{\varphi\varphi}$ (curves ($a_{1,2}$)) for UO_2 as function of r at three different times $t = t_0$, $9\,t_0$, and $49\,t_0$ corresponding to 2.6×10^{-11}, 2.3×10^{-10}, and 1.3×10^{-9} s. The curve with label D corresponds to the stress field $\sigma_{\varphi\varphi} = \sigma_{rr}$ of an edge dislocation in UO_2

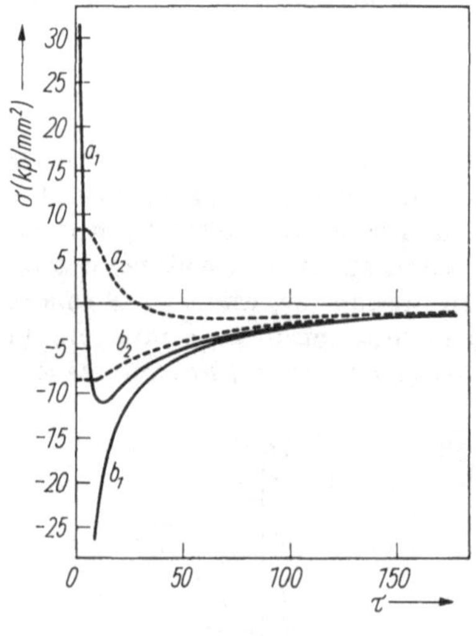

Fig. 7. Plot of σ_{rr} (curves ($b_{1,2}$)) and $\sigma_{\varphi\varphi}$ (curves ($a_{1,2}$)) as function of time at two different radii $r_1 = 200$ Å and $r_2 = 500$ Å in UO_2. The time scale is $\tau' = (t + t_0)/t_0$ with $t_0 = 2.6 \times 10^{-11}$ s

Obviously the strength of the stress fields of fission spike and dislocation are quite comparable up to $r \approx 500$ Å and times $10\, t_0$ for the fission spike. Even up to $r = 1000$ Å and $t \approx 10^{-9}$ s the stress field of the fission spike gives stresses of the order 1 to 2 kp/mm² which are of the order of engineering stresses acting as driving force for in-pile creep of nuclear fuels.

Examples of the time history of the stresses $\sigma_{\varphi\varphi}$ and σ_{rr} at two fixed radii are shown in Fig. 7.

The scale factor for all stresses according to equations (17) and (16d) is

$$\sigma_0 = \frac{2\,\mu\,\alpha'\,Q}{4\,\pi\,k t_0}. \tag{21}$$

For an estimate of the difference in σ_0 between UO_2 and UC we get with the values of Tables 1 and 2 and by equation (12b) and the results of Sections 4.2 and 4.3

$$\sigma_0 \approx \frac{610}{n^2}\,\frac{\text{kp}}{\text{mm}^2} \quad \text{with} \quad n \approx 1 \quad \text{for} \quad UO_2\,, \tag{21a}$$

$$\sigma_0 \approx \frac{650}{n^2}\,\frac{\text{kp}}{\text{mm}^2} \quad \text{with} \quad n > 10 \quad \text{for} \quad UC\,. \tag{21b}$$

With the values of n indicated it is obvious that the thermoelastic stress field in UC (and UN) is nearly negligible in comparison with that in UO_2.

6. Conclusions

The thermoelastic stress pulse produced in UO_2 by a fission spike lasts about 10^{-9} s with stress values between 150 and 1 kp/mm² covering a volume with radius of the order of 1000 Å and a length of nearly 15 µm. It should, however, be remembered, that the peak values of the stress diminish along the path of the fragment, see Fig. 1. The stress values mentioned here correspond to $(\mathrm{d}E/\mathrm{d}R)_{\mathrm{eff}} = 1000$ eV/Å. From Fig. 7 we see that at a distance of 200 Å from the core of the track still stress values of about 50 kp/mm² exist. Hence, the production of fission fragment tracks close to a crystal surface which are visible by TEM, can be understood by the combined action of this stress field and the elastic waves emitted from the core of the track at times $t' < t_0$ in the case of UO_2.

A similar effect will be produced if a fission fragment passes closely by a fission gas bubble, a pore, or a void in oxide fuel material.

The effects produced by thermoelastic stresses of fission spikes must, therefore, be taken into account, if one wants to understand the re-solution of fission gas by the destruction of fission gas bubbles in oxide fuels, also in-pile sintering and creep of oxide fuel should be affected by these stress pulses.

In UO_2 a cation vacancy is mostly accompanied by at least one oxygen vacancy. This results in a highly anisotropic point defect configuration. The thermoelastic stress pulses produced by fission spikes are oriented at random and must provoke a stress-induced change of orientation of these defects proportional to the fission rate. This should have consequences for the point defect entropy and act against their biased diffusion.

Another important implication of the stress field of fission spikes will be their interaction with dislocations.

References

[1] N. BOHR, Mat.-Fys. Medd. Dan. Vid. Selsk. **18**, No. 8 (1948).
[2] J. LINDHARD, M. SCHARFF, and H. E. SCHIØTT, Mat.-Fys. Medd. Dans. Vid. Selsk. **33**, No. 14 (1963).
[3] P. V. THOMSON and J. LINDHARD, private communication, 1971.
[4] A. D. WHAPHAM and M. J. MAKIN, Phil. Mag. **7**, 1441 (1962).
[5] A. D. WHAPHAM and B. E. SHELDON, J. nuclear Mater. (Amsterdam) **10**, 157 (1963).
[6] H. BLANK and S. AMELINCKX, J. appl. Phys. **34**, 2200 (1963).
[7] H. BLANK and S. AMELINCKX, Z. Metallk. **55**, 460 (1964).
[8] J. OZEROFF, AEC Research Report AECD-2973 (1949).
[9] L. T. CHADDERTON and I. MCTORRENS, Fission Damage in Crystals, Methuen, London 1969.
[10] B. A. BOLEY and F. H. WEINER, Theory of Thermal Stresses, Wiley, 1962.
[11] W. FULKERSON, T. G. KOLLIE, S. C. WEAVER, J. P. MOORE, and R. K. WILLIAMS, Nuclear Metallurgy (part I) **17**, 374 (1970).
[12] A. L. DAVIES, Nuclear Metallurgy (part I) **17**, 209 (1970).
[13] J. GRUNZWEIG-GENOSSAR, M. KUZNIETZ, and F. FRIEDMAN, Phys. Rev. **173**, 562 (1968).
[14] J. J. QUINN, Phys. Rev. **126**, 1453 (1962).
[15] W. F. KROLIKOWSKI and W. E. SPICER, Phys. Rev. **B 1**, 478 (1970).

(Received December 20, 1971)

phys. stat. sol. (a) **10**, 479 (1972)

Subject classification: 14.3; 11; 22.2.1

Physics Department, University of Athens

Carrier Lifetimes in Low-Resistivity GaAs upon Electron Bombardment and Annealing[1])

By

D. I. KLADIS and P. C. EUTHYMIOU

Single crystals of GaAs, n- and p-type, were bombarded with 1 MeV electrons at 78 °K. By photoconductivity and photomagnetoelectric measurements, a decrease of majority and minority carrier lifetime was detected in all cases. The bombardment dose was small and hence the changes of carrier concentration and mobility were insignificant. The decrease of lifetimes is hence due solely to the increase of the number of recombination and trapping centres. Isochronal annealing experiments showed a complete, more or less, recovery for n-type specimens at about 300 °K, while the revocery of p-type specimens is expected to occur at higher temperatures. The recovery takes place through three normal annealing stages for both types of specimens, in accordance with the Fletcher and Brown theory. We found also discrete reverse annealing stages at temperatures usually below each normal annealing stage. This is attributed to the increase of the activity of the recombination and trapping centres due to a rearrangement of the associated defects to more stable states.

Des cristaux de GaAs du type n et p ont été bombardés par des rayons β d'une énergie de 1 MeV à la température de 78 °K. Une diminution de la durée de vie des porteurs, mesurée par la méthode de photoconductivité et de phenoméne photomagnétoélectrique, était aperçue dans tous les cas. La dose du bombardment était assez petite pour avoir des variations sur la concentration et la mobilité des porteurs. Par consequence la diminution de la durée de vie des porteurs doit être attribuée seulement à l'accroissement du nombre des centres de récombinaison et des trappes. Des expériments de recuission isochrone montrent que le recouvrement des cristaux du type n est réalisé plus ou moins complètement à la température de 300 °K tandis que le recouvrement des cristaux du type p est attendu a été réalisé à une témperature plus haute. Le recouvrement suivit trois degrés normals pour tous les deux types des cristaux en accord avec la théorie de Fletcher et Brown. Nous avons trouvé aussi pendant la recuission isochrone des degrés renversés très distingués situés habituellement au-dessous de chaque degrés normale. Le dernier effet a été attribué à un accroissement de l'activité des centres de récombinaison et des trappes due à un déplacement des défauts aux états plus stables.

1. Introduction

The effect of electron bombardment and isochronal annealing on several properties (carrier concentration and mobility, thermal conductivity, crystal length etc.) of GaAs has been extensively studied [1 to 8]. Data on the effect of electron bombardment on the carrier lifetimes of GaAs at 300 °K are given only by Vitovskii et al. [9] and especially for the majority carrier lifetime.

In the present work we investigated the effect of electron bombardment on the minority and majority carrier lifetimes of GaAs at 78 °K. The recovery of

[1]) This research was supported in part by the University of Athens and by the National Hellenic Research Foundation.

the defects was also studied by isochronal annealing experiments up to 300 °K. The values of the carrier lifetimes τ_n and τ_p are deduced from photoconductivity (PC) and photomagnetoelectric (PME) effect measurements, and are related to the PC and PME lifetimes τ_{PC} and τ_{PME} according to the following expressions [10, 11]:

$$\tau_{PC} = \frac{\tau_n + \frac{\tau_p}{b}}{1 + \frac{1}{b}}, \tag{1}$$

$$\tau_{PME} = \frac{\tau_n + c\,\tau_p}{1 + c}. \tag{2}$$

In the general case when both recombination and trapping centres exist, the following expressions give the PC and PME currents per unit sample width i_{PC} and i_{PME}:

$$i_{PC} = \left(1 + \frac{1}{b}\right) \frac{e\,I\,\mu_n\,E\,\tau_{PC}}{1 + \dfrac{\tau_{PME}\,s}{\left(\dfrac{kT}{e}\mu_n\,\tau_{PME}\right)^{1/2}} \left(\dfrac{1 + b\,c}{1 + c}\right)^{1/2}}, \tag{3}$$

$$i_{PME} = \left(1 + \frac{1}{b}\right) \frac{e\,I\,\mu_n\,B\left(\dfrac{kT}{e}\mu_n\,\tau_{PME}\right)^{1/2} (1 + c)^{1/2}}{\left[1 + \mu_n^2\,B^2 + b\,c\left(1 + \dfrac{\mu_n^2\,B^2}{b^2}\right)\right]^{1/2}} \times$$

$$\times \frac{1}{1 + \dfrac{\tau_{PME}\,s}{\left(\dfrac{kT}{e}\mu_n\,\tau_{PME}\right)^{1/2} (1 + c)^{1/2}} \left[1 + \mu_n^2\,B^2 + b\,c\left(1 + \dfrac{\mu_n^2\,B^2}{b^2}\right)\right]^{1/2}}, \tag{4}$$

where I is the photon flux density, s is the surface recombination velocity, b is the electron to hole mobility ratio, and c is the electron to hole concentration ratio.

The value of b in our experiments has been found to be about 20, so that

$$1 + \frac{1}{b} \approx 1\,.$$

On the other hand the larger values of Hall mobilities, which we found, were such that

$$1 + \mu_H^2\,B^2 \approx 1\,.$$

Under the further assumption of a low surface recombination velocity for our specimens, the application of the expressions (1) to (4) for n-type specimens, where $c = \infty$, and for p-type specimens, where $c = 0$, gives the following:

a) n-type specimens:

$$\tau_{PC} = \tau_n , \tag{1a}$$

$$\tau_{PME} = \tau_p , \tag{2a}$$

$$i_{PC} = e\, I\, \mu_n\, E\, \tau_{PC} , \tag{3a}$$

$$i_{PME} = e\, I\, \mu_n\, B \left(\frac{kT}{e} \mu_p\, \tau_{PME}\right)^{1/2} . \tag{4a}$$

b) p-type specimens:

$$\tau_{PC} = \frac{\tau_p}{b} , \tag{1b}$$

$$\tau_{PME} = \tau_n , \tag{2b}$$

$$i_{PC} = e\, I\, \mu_n\, E\, \tau_{PC} , \tag{3b}$$

$$i_{PME} = e\, I\, \mu_n\, B \left(\frac{kT}{e} \mu_n\, \tau_{PME}\right)^{1/2} . \tag{4b}$$

The majority carrier mobilities are deduced from Hall and conductivity measurements. The minority carrier mobilities are calculated from the majority carrier ones, with the further assumption of a constant mobility ratio $b = 20$.

2. Experimental Part

Measurements were made on three specimens of undoped n-type GaAs single crystals (resistivity $\varrho = 0.1\ \Omega$cm and carrier concentration $n = 3 \times 10^{16}\ \text{cm}^{-3}$ at 300 °K), and on three specimens of Cd-doped p-type GaAs single crystals (resistivity $\varrho = 0.3\ \Omega$cm and carrier concentration $p = 2 \times 10^{17}\ \text{cm}^{-3}$ at 300 °K). The dimensions of the samples were $10 \times 3 \times 1\ \text{mm}^3$.

The surfaces of the samples were chemically treated in an alkali etchant (3 parts NaOH and 1 part H_2O_2), in order for a low surface recombination velocity to be achieved. Ohmic contacts were prepared by fusing In at about 300 °C for a few minutes. All measurements took place at liquid nitrogen temperature.

The GaAs samples were bombarded with β-particles from a Sr^{90}–Y^{90} source, having a nominal activity of 180 mCi. The average energy of the particles was 1.0 MeV, while the irradiation dose was estimated to be about 10^{13} electrons per cm^2 for a continuous exposure of 36 h. The duration of each annealing step was selected to be 30 min.

An incandescent 8 V/50 W lamp was used as the illumination source. A system of mirrors and a filter assured a uniform illumination of the specimens with a mean wavelength of 7300 Å.

All dc measurements were made with standard potentiometric techniques, while PC and PME measurements were performed with a synchronous amplifier at a frequency of about 20 Hz. The magnetic field induction was 0.87 Wb/m^2.

3. Results and Discussion

3.1 Electron bombardment

As we can see from Table 1, both carrier lifetimes τ_n and τ_p are decreased due to electron bombardment, the decrease of the minority carrier lifetime being always larger than that of the majority carrier one.

The above values of τ_n and τ_p prior to bombardment are all in agreement with those found by other investigators [12 to 14]. The fact that the values of the majority carrier lifetimes are generally found to be much larger than those of the minority carrier ones indicates a large number of minority carrier trapping centres [12].

Table 1

n-type specimens				
specimen No.	prior to bombardment		after bombardment	
	τ_n (10^{-7} s)	τ_p (10^{-8} s)	τ_n (10^{-7} s)	τ_p (10^{-8} s)
1	13.7 ± 0.6	1.09 ± 0.10	10.5 ± 0.4	0.50 ± 0.05
2	1.34 ±0.05	0.43 ± 0.04	1.11 ± 0.04	0.31 ± 0.03
3	8.6 ± 0.3	1.39 ± 0.13	5.5 ± 0.2	0.70 ± 0.06
p-type specimens				
specimen No.	prior to bombardment		after bombardment	
	τ_n (10^{-9} s)	τ_p (10^{-6} s)	τ_n (10^{-9} s)	τ_p (10^{-6} s)
4	1.23 ± 0.11	3.07 ± 0.12	0.51 ± 0.05	1.90 ± 0.08
5	19.9 ± 1.8	6.17 ± 0.30	10.1 ± 0.9	4.54 ± 0.20
6	68.0 ± 6	4.19 ± 0.16	40.0 ± 4	2.98 ± 0.12

Electron bombardment increases the number of both recombination and trapping centres. An increase of the number of the recombination centres causes a decrease of the lifetime of either types of carriers, while an increase of the number of the minority carrier trapping centres causes a decrease of the minority carrier lifetime and an increase of the majority carrier one. If we assume that the increment of the number of the recombination centres is larger than that of the trapping centres, the observed results are easily explained. This assumption is quite reasonable, because the number of the initially existing recombination centres is always small for low-resistivity GaAs [15].

We observed no change of the carrier concentration and mobility due to bombardment, although this was found by the others investigators. It is clear that the low irradiation dose, which we used, produced only a relatively small number of defects, which did not affect the less sensitive properties of the solid, such as the carrier concentration and mobility. Therefore the change of the carrier lifetimes is solely due to the change of the number of recombination and trapping centres.

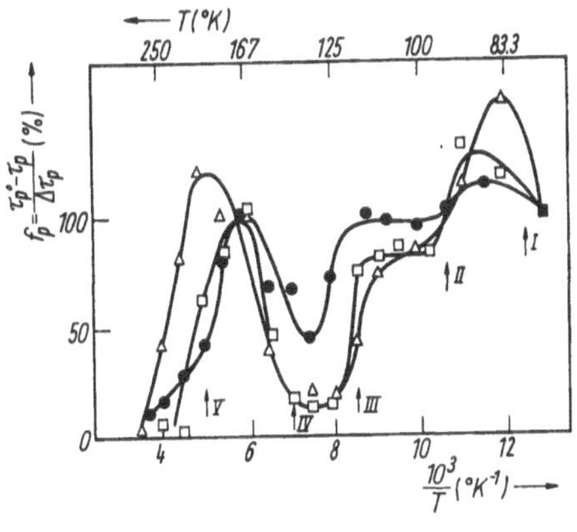

Fig. 1. Isochronal annealing of minority carrier lifetime for n-type GaAs crystals. △ sample No. 1; □ No. 2; ● No. 3

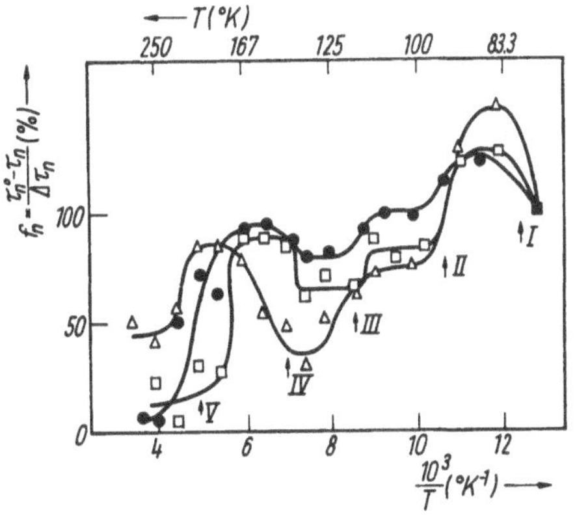

Fig. 2. Isochronal annealing of majority carrier lifetime for n-type GaAs crystals. △ sample No. 1; □ No. 2; ● No. 3

3.2 Isochronal annealing

The results of the isochronal annealing are shown in Fig. 1 to 4. The quantities f_n and f_p, defined as follows,

$$f_n = \frac{\tau_n^0 - \tau_n}{\Delta\tau_n}, \qquad f_p = \frac{\tau_p^0 - \tau_p}{\Delta\tau_p},$$

represent the fraction of the remaining damage after each annealing step; τ_n^0 and τ_p^0 are the carrier lifetime values prior to bombardment, $\Delta\tau_n$ and $\Delta\tau_p$ are the changes of carrier lifetimes due to bombardment, and τ_n and τ_p are the carrier lifetime values after each annealing step.

From Fig. 1 and 2 we can distinguish five annealing stages for n-type material at 81 °K (stage I), 94 °K (stage II), 117 °K (stage III), 143 °K (stage IV), and 200 °K (stage V). All specimens seem to have entirely recovered at room temperature. Stage II, III, and V are normal recovery stages, and the existence of them is in accordance with the theory of Fletcher and Brown [16], who have suggested that the recovery of the defects, produced by electron bombardment, occurs in three stages. The first stage corresponds to recombination between "close pairs". At the second stage the "correlated recombination" between interstitials and vacancies at distances larger than the capture radius takes place, and at the third stage the "uncorrelated recombination" between the defects, which do not recombine through the correlated mechanism, occurs.

From Fig. 3 and 4 we can also distinguish five annealing stages for p-type material at 82 °K (stage I), 91 °K (stage II), 108 °K (stage III), 164 °K (stage IV), and 250 °K (stage V). The stages II and IV are the only normal ones, but, since a fraction of defects still remained after annealing to room temperature, we can assume that the complete recovery would have occurred at a third normal annealing stage above 300 °K, in accordance with the theory of Fletcher and Brown.

The reverse annealing stages I, IV for n-type material and III, V for p-type material may be attributed to the increase of the capture cross section of both the recombination and trapping centres, resulting from some kind of rearrange-

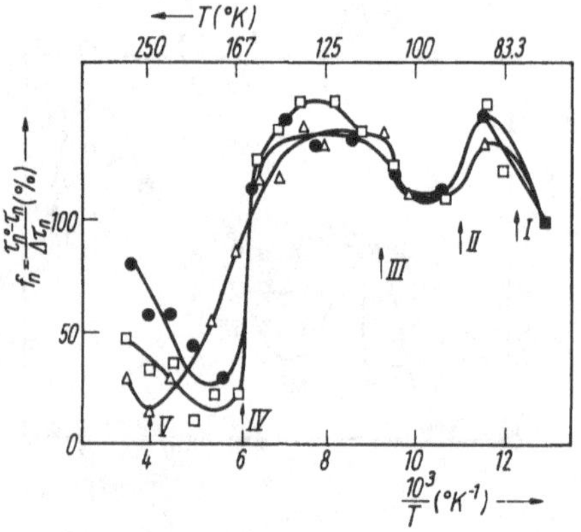

Fig. 3. Isochronal annealing of minority carrier lifetime for p-type GaAs crystals. △ sample No. 4; □ No. 5; ● No. 6

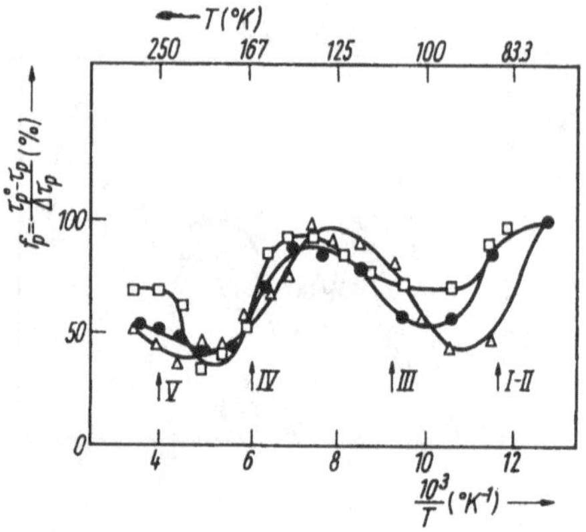

Fig. 4. Isochronal annealing of majority carrier lifetime for p-type GaAs crystals. △ sample No. 4; □ No. 5; ● No. 6

ment of the associated defects to more stable states. In order to reconcile the difference between annealing stages I of Fig. 3 and 4, we must assume a rearrangement of the defects for p-type material, by which only the capture cross section of the minority carrier trapping centres is increased. This would result indeed in a decrease of the minority carrier lifetime and an increase of the majority carrier one.

The present experiments show that the annealing behaviour of carrier lifetimes does not agree with that of other properties of GaAs [1 to 4, 6]. Although the previous annealing experiments in GaAs do not always agree among themselves, in general they show no reverse annealing. The only known exception occurred in experiments on crystal length carried out by Vook [3]. Reverse annealing seems to be an effect associated mainly with the activity of some kind of centres and not with the number of them. Changes purely due to the activity of centres are best observed with low bombardment doses as then the carrier concentration and mobility are practically unaffected by the bombardment and the whole effect of annealing is investigated through carrier lifetimes study only. It might be that reverse annealing stages occur in general before normal stages.

Acknowledgements

We are greatly indebted to Prof. K. D. Alexopoulos for many valuable discussions. We wish also to thank P. Bekris and N. Tsoupas for their help in measurements.

References

[1] J. A. Grimshaw and P. C. Banbury, Proc. Phys. Soc. **84**, 151 (1961).
[2] J. A. Grimshaw, Conf. Rad. Dam. Semicond., Academic Press, Paris 1964 (p. 376).
[3] F. L. Vook, J. Phys. Soc. Japan, Suppl. II, **18**, 190 (1963).
[4] F. L. Vook, Phys. Rev. **135**, A1742 (1964).
[5] V. S. Vavilov, A. F. Plotnikov, M. A. Selezneva, and A. A. Sokolova, Soviet Phys. — Solid State **8**, 2710 (1967).

[6] K. Thommen, Radiat. Eff. **2**, 201 (1970).
[7] E. Yu. Brailovskii and I. D. Konozenko, Fiz. Tekh. Poluprov. **5**, 641 (1971).
[8] E. Yu. Brailovskii and V. N. Brundyi, Fiz. Tekh. Poluprov. **5**, 1248 (1971).
[9] N. A. Vitovskii, T. V. Mashovets, S. M. Ryvkin, and R. I. Khansevarov, Soviet Phys. — Solid State **5**, 2575 (1964).
[10] S. W. Kurnick and R. N. Zitter, J. appl. Phys. **27**, 278 (1956).
[11] R. N. Zitter, Phys. Rev. **112**, 852 (1958).
[12] C. Hilsum and B. Holeman, Proc. Internat. Conf. Semicond., Academic Press, Prague 1960 (p. 962).
[13] N. M. Kolchanova and D. N. Nasledov, Soviet Phys. — Solid State **8**, 876 (1966).
[14] A. V. Dudenkova and V. V. Nikitin, Soviet Phys. — Solid State **8**, 2432 (1967).
[15] C. Hilsum, Progr. Semicond. **9**, 137 (1965).
[16] R. C. Fletcher and W. L. Brown, Phys. Rev. **92**, 585 (1953).

(Received December 17, 1971)

phys. stat. sol. (a) **10**, 487 (1972)

Subject classification: 20.1; 6; 22.6

Institute of Physics, Czechoslovak Academy of Sciences, Prague

The Transmittivity of Cu_2O Powder Samples in the Infrared Region

By

J. PASTRŇÁK and B. HEJDA

The transmittivity of Cu_2O powder samples has been measured in the infrared spectral region in the vicinity of the optically active vibrational modes $\omega_{TO\,1} = 610\ cm^{-1}$ and $\omega_{TO\,2} = 145\ cm^{-1}$ for various surrounding media. The theory of optical phonons in finite ionic crystals by Englman and Ruppin has been used for the calculation of the transmittivity. The comparison between theory and experiment shows, that this theory may be used in spite of the great portion of covalent binding in Cu_2O.

Die Transmission von Cu_2O-Pulverproben wurde im infraroten Spektralbereich in der Nähe der optisch aktiven Schwingungsmoden $\omega_{TO\,1} = 610\ cm^{-1}$ and $\omega_{TO\,2} = 145\ cm^{-1}$ bei verschiedenen Umgebungsmedien gemessen. Für die Berechnung der Transmission wurde die Theorie der optischen Phononen in endlichen Ionenkristallen von Englman und Ruppin benutzt. Der Vergleich zwischen Theorie und Experiment zeigt, daß diese Theorie trotz des großen kovalenten Bindungsanteils von Cu_2O benutzt werden kann.

1. Introduction

The method of pressed pellets and emulsions has been used for many years for measurements of infrared spectra on powder materials in infrared spectroscopy. Nevertheless, only recently the physicists became aware of the problem to what extent the spectra of powder samples correspond to those of single crystals. Now the number of papers devoted to this question is increasing rapidly [1, 2]. Englman and Ruppin [3 to 5] worked out the detailed theory of long-wavelength phonons in finite crystals for spherical and cylindrical geometry of particles neglecting all short-range effects caused by crystal boundaries and taking into consideration only the effect of long-range Coulomb forces. This theory was successfully used for the interpretation of the experimental curves with various materials [2, 6 to 10]; in separate cases the disagreement was recorded [11].

The theory of Englman and Ruppin shows that the absorption curve of powder material with grains much smaller than the wavelength of the absorbed optically active ir radiation will have a single maximum at the frequency ω_S in the range between ω_{TO} and ω_{LO}. This frequency will depend on the dielectric constant ε_M of the surrounding medium. With the increasing average dimensions of the particles, in the case of statistical distribution of the radii of the particles around some average radius R_M, another maximum will appear on the absorption curve at frequencies smaller than ω_{TO}. Its position will shift in the direction of longer wavelengths with increasing average radius of the particles and in the same time its intensity will increase [7].

In our paper we use the theory of Englman and Ruppin for the interpretation of the measurements on powder samples of Cu_2O. This material, unlike other materials the measurements on which has been described up to now, is characterised by a large degree of covalent binding which follows from the small difference between the optical ($\varepsilon_\infty = 6.46$) and static ($\varepsilon_0 = 7.54$) dielectric constant [12]. Just these constants, together with the dielectric constant of the surrounding medium ε_M, are the decisive parameters in the theory of Englman and Ruppin [3 to 5] for the surface mode frequencies. One of the authors measured the frequency of the optically active modes of Cu_2O in a powder sample pressed in a KBr pellet some years ago [13]. These measurements have been confirmed since [12, 14, 16], nevertheless the question remains, how the finite size of a crystal manifests itself in the accuracy of the results.

The vibrational spectrum of Cu_2O contains two optically active frequencies $\omega_{TO\,1} = 610 \pm 4\ cm^{-1}$ and $\omega_{TO\,2} = 145 \pm 2\ cm^{-1}$ found in ir absorption and reflection [12 to 17]; the errors are estimated from the data of different authors. These frequencies belong to the two optically active modes at the centre of Brillouin zone [18, 19].

In the vicinity of these frequencies a number of powder samples of Cu_2O pressed in various materials with different dielectric constants ε_M have been measured to clarify the influence of the size effect on the frequency of optical phonons and the transmission curve.

2. Experiment

The measurements have been performed on Cu_2O powder samples with grains of irregular shape, approximately spherical. The size distribution has its sharp maximum below a radius of $\approx 1\ \mu m$ and falls rapidly up to 12 to 13 μm where it is cut off. For the measurements in the region of 450 to 900 cm^{-1} the pellets of various alkali halides, in the region of 100 to 200 cm^{-1} air and polyethylen have been used as surrounding media. The effective thickness of the material was 3 and 10 μm. In the region 450 to 900 cm^{-1} the double-ray spectrometer Perkin-Elmer 421 has been used, in the long-wavelength region 100 to 200 cm^{-1} we used the Fourier spectrometer FS-620.

The results of the measurements are presented in Fig. 1 and 2. The absorption maxima are at the frequencies $620 \pm 4\ cm^{-1}$ and $146 \pm 2\ cm^{-1}$. For the higher frequency region pellets with KBr ($\varepsilon_M = 2.33$) and TlCl ($\varepsilon_M = 4.3$) have been chosen as examples. The value of the dielectric constant of TlCl was estimated by extrapolation from the values of the refraction index in the visible region [20].

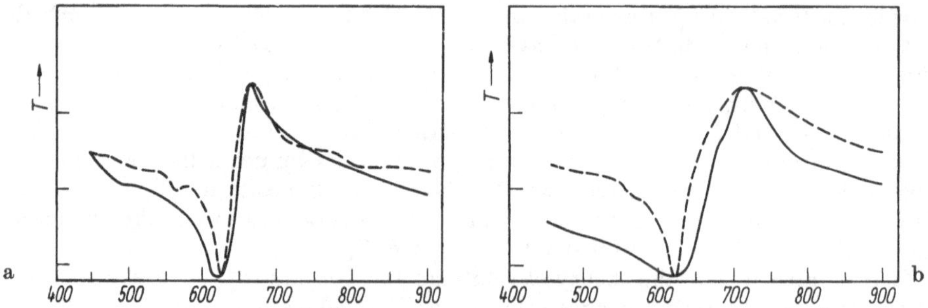

Fig. 1. a) The transmittivity (in arbitrary units) of a powder sample of Cu_2O pressed in a KBr pellet (solid curve); theoretical calculation with $\varepsilon_M = 2.33$ (dashed curve). b) The transmittivity (in arbitrary units) of a powder sample of Cu_2O pressed in a TlCl pellet (solid curve); theoretical calculation with $\varepsilon_M = 4.3$ (dashed curve)

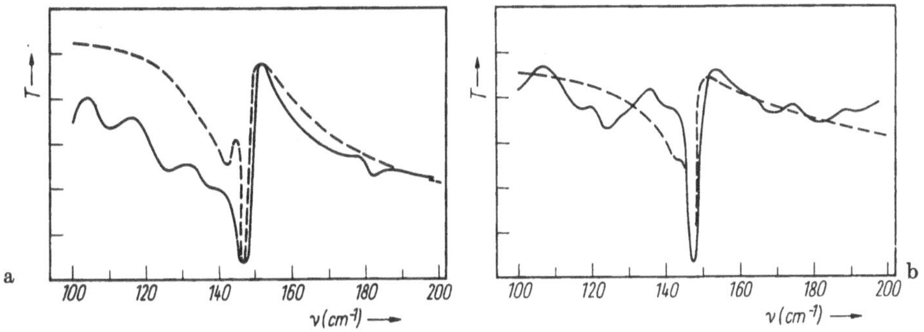

Fig. 2. a) The transmittivity (in arbitrary units) of a powder sample of Cu_2O sputtered on poylethylen (solid curve); theoretical calculation with $\varepsilon_M = 1$ (dashed curve). b) The transmittivity (in arbitrary units) of a powder sample of Cu_2O pressed in polyethylen (solid curve); theoretical calculation with $\varepsilon_M = 2$ (dashed curve)

The character of the curves was the same for effective thicknesses 3 and 10 μm and in the short-wavelength region also for other materials as matrix, therefore they are not presented.

3. The Discussion of the Results

The transmittivity curve of Cu_2O powder samples has the typical shape of a dispersion or reflection curve. Already Phillippi [21] drew attention to this similarity apparent with some materials; he interpreted the spectra of powder samples in such cases as reflection spectra. The agreement between experimental and theoretical data has been achieved by properly chosen parameters of absorbing classical oscillators.

We compared our experimental results with the theory of Englman and Ruppin [4]. The total effective cross section (absorption plus scattering cross section) averaged across the set of spherical particles, with the radii distributed around the mean radius R_M, has been calculated. The distribution has been estimated experimentally. The same method as in paper [7] has been used in the calculation. The values of the dielectric constants of Cu_2O are $\varepsilon_0 = 7.54$ and $\varepsilon_\infty = 6.46$ [12]. $\omega_{TO\,1} = 610\ cm^{-1}$ and $\omega_{TO\,2} = 145\ cm^{-1}$ have been used as most probable for optically active frequencies [12 to 17]. We took $S_1 = 0.75$, $S_2 = 0.33$, $\gamma_1 = 21$, $\gamma_2 = 1.3$ [12] in the expression of the dielectric constant

$$\varepsilon(\omega) = \varepsilon_\infty + \sum_{j=1}^{2} \frac{S_j}{1 - \left(\frac{\omega}{\omega_{TO_j}}\right)^2 - i \frac{\gamma_j}{\omega_{TO_j}} \frac{\omega}{\omega_{TO_j}}} .$$

The best agreement with experiment has been achieved in the case of KBr ($\varepsilon_M = 2.33$) as matrix.

In the case of TlCl the value of $\varepsilon_M = 5.1$ from [22] gave too large shift of the transmittivity maximum. This value corresponds to the squared refractive index of the visible region [20] and is obviously too high. The value $\varepsilon_M = 4.77$ obtained by extrapolation of refractive index values by the help of the Cauchy relation [23] led to a shape of the curve close to the experimental one. The complete agreement (of the position of the maximum) required the value $\varepsilon_M =$

$= 4.3$; it was obtained by fitting the calculation to the experiment by changing ε_M. This lowering of ε_M corresponds to the experimental fact, that the real value of ε_M in the pellet could be lowered by the presence of a great concentration of defects, as it is more difficult to press TlCl than KBr.

We used the theory for spherical particles in all calculations of the theoretical curves. In agreement with Martin [11], we believe that the deviations from spherical shape do not substantially express themselves in the shape of the absorption curves after averaging, as far as the particles are finite in all three dimensions.

The comparison of the results of theory and experiment demonstrates that even in the case of Cu_2O with a low portion of ionic binding and two optically active frequencies, the theory of Englman and Ruppin may be successfully used for the interpretation of measurements on powder samples. The absorption maximum corresponding to ω_S is not very much expressed in this case and its shift with the change of the dielectric constant of the matrix is negligible. It corresponds to the small difference of ε_0 and ε_∞. The minimum of extinction at the short-wavelength side of that maximum manifests itself much more expressively. Its shift in the direction of shorter wavelengths with the increase of the dielectric constant of the medium from $\varepsilon_M = 1$ to $\varepsilon_M = 4.3$ is $\approx 60\ cm^{-1}$.

Similarly, as it was shown in paper [7] for frequencies $\omega < \omega_{TO}$, the extinction increases with increasing mean radius R_M of the grains. At the same time this extinction is stronger in a matrix with smaller dielectric constant under otherwise identical conditions. As in the case of NaCl [7] the experimental curves have more spreaded and smoothed structure and greater halfwidths of the absorption maxima than the theoretical curves. This may be caused by spreading of the frequencies of individual modes over a wider range, deviations from the spherical shape, or a higher density of lattice defects with small grains.

Also, the theoretical curves are smoother when we use a greater number of values of R to find the average value of the cross section; in our calculation this number (thirteen) was limited by the computation time. So we do not consider the "fine structure" of calculated curves significant. On the other hand the small absorption maxima at $110 \pm 2\ cm^{-1}$, $126 \pm 2\ cm^{-1}$, and $168 \pm 2\ cm^{-1}$ visible on experimental curves (Fig. 2a, b) have been found in most measurements. They were observed in [17] too and they correspond to the optically inactive modes activated by the presence of impurities [16, 17].

In conclusion we can say that our measurements and calculations show, that even in the case when the shift of the extinction maximum is small and the result of the change of the dielectric constant of the medium is the shift of the extinction minimum it is possible to interpret the experimental data obtained with powder samples by the theory of Englman and Ruppin. For this purpose we must know the optical and static dielectric constant of the material in question and the optical dielectric constant of the matrix. Then, from the results, we can infer also about the mean size and the distribution of the particles of our sample.

Acknowledgements

The authors are indebted to L. Novák and J. Petzelt of the Institute of Physics of the Czechoslovak Academy of Sciences, Prague, for the help in measurements.

References

[1] A. A. MARADUDIN, Solid State Phys. **18**, 273 (1966); **19**, 1 (1966).
[2] R. RUPPIN, Rep. Progr. Phys. **33**, 149 (1970).
[3] R. ENGLMAN and R. RUPPIN, J. Phys. C (Solid State Phys.), ser. 2, **1**, 614 (1968) (part I).
[4] R. RUPPIN and R. ENGLMAN, J. Phys. C (Solid State Phys.), ser. 2, **1**, 630 (1968) (part II).
[5] R. ENGLMAN and R. RUPPIN, J. Phys. C (Solid State Phys.), ser. 2, **1**, 1515 (1968) (part III).
[6] J. PASTRŇÁK and B. HEJDA, phys. stat. sol. **35**, 941 (1969).
[7] J. PASTRŇÁK and K. VEDAM, phys. stat. sol. (a) **3**, 647 (1970).
[8] J. PASTRŇÁK, phys. stat. sol. (a) **3**, 657 (1970).
[9] J. T. LUXON, D. J. MONTGOMERY, and R. SUMMITT, Phys. Rev. **188**, 1345 (1969).
[10] V. V. BRYKSIN, YU. M. GERLSTEIN, and D. N. MIRLIN, Solid State Commun. **9**, 669 (1971).
[11] T. P. MARTIN, Solid State Commun. **9**, 623 (1971).
[12] E. C. HELLTEMES, Phys. Rev. **141**, 803 (1966).
[13] J. PASTRŇÁK, Optika i Spektroskopiya **6**, 107 (1959).
[14] M. O'KEEFFE, J. chem. Phys. **39**, 1789 (1963).
[15] NEIL T. MCDEVITT and A. D. DAVIDSON, J. Opt. Soc. Amer. **55**, 209 (1965).
[16] C. CARABATOS, A. DIFFINE, and M. SIESKIND, J. Physique (France) **29**, 529 (1968).
[17] A. I. STEKHANOV and J. F. MARKOV, Fiz. tverd. Tela **11**, 3330 (1969).
[18] K. HUANG, Z. Phys. **171**, 213 (1963).
[19] C. CARABATOS and B. PREVOT, phys. stat. sol. (b) **44**, 701 (1971).
[20] LANDOLT-BÖRNSTEIN, 8 B, Springer-Verlag, 1962 (p. 220).
[21] C. M. PHILLIPPI, Tech. Rep. AFML-TR-69-31, Air Force Materials Laboratory, Wright-Patterson Air Force Base, Ohio 1969.
[22] M. BORN and K. HUANG, Dynamical Theory of Crystal Lattices, Clarendon Press, Oxford 1956.
[23] See, e.g., G. S. LANDSBERG, Optika, GITTL, Moskva 1954 (p. 457).

(Received December 22, 1971)

phys. stat. sol. (a) **10**, 493 (1972)

Subject classification: 18.2; 5; 21.1.1

Institut für Metallphysik und Metallkunde, Universität des Saarlandes, Saarbrücken

Spin Orientation Determination by Correlating X-Ray Textures with Mössbauer Line Intensities in Fe_3C

By

U. GONSER, M. RON[1]), H. RUPPERSBERG, W. KEUNE, and A. TRAUTWEIN

In ferromagnetically ordered cementite, Fe_3C, the spin orientation was investigated by correlating the texture measured with X-rays and the relative line intensities of a Mössbauer hyperfine pattern. Formulae are derived to make this correlation method generally applicable. The spin in Fe_3C was found to be oriented within a solid angle of 20° along the crystallographic c-axis.

In ferromagnetisch geordnetem Zementit, Fe_3C, werden die Spinorientierungen durch Verknüpfung der gemessenen Textur mit Röntgenstrahlen und den relativen Intensitäten des Mössbauer-Hyperfeindiagramms untersucht. Formeln werden abgeleitet, die diese Korrelationsmethode allgemein anwendbar machen. Es wurde gefunden, daß der Spin in Fe_3C in einem Raumwinkel von 20° zur kristallographischen c-Achse orientiert ist.

1. Introduction

The angular dependence of a Mössbauer hyperfine pattern reflects the orientation of the principal axes of the electric field gradient (EFG) and the spin. Based on the latter, a number of methods have been developed to obtain information regarding the direction of the principal axes of the EFG and of the spin in various materials [1 to 5]. Usually single crystals are required.

A method for determining the spin orientation in a polycrystalline magnetic material is described which is based on the correlation between the relative Mössbauer line intensities and the preferred orientation of its crystallites (texture) determined by X-rays. The method has general applicability, in cases where the following conditions are fulfilled.

1. A preferred orientation of the material exists or can be obtained.
2. There exists a unique spin orientation in regard to a crystallographic direction (direction of easy magnetization).

2. Cementite Structure

Cementite, Fe_3C, which is ferromagnetically ordered below 210 °C, was studied in this investigation. To our knowledge the spin orientation is unknown and single crystals of reasonable size have not been prepared. Fe_3C has orthorhombic structure with space group Pnma [6, 7]. The carbon atoms are located in interstitial positions in a structure of iron atoms which can be described in terms of a pleated-layer hexagonal arrangement. In Fe_3C two crystallographically different iron sites can be distinguished: G (general) and S (special). A six-line Mössbauer pattern has been reported for Fe_3C [8]. Recently, the pattern

---
[1]) Guest scientist on leave from Technion, Haifa, Israel.

of Fe_3C has been further analysed in terms of two superimposed components corresponding to the two different iron sites [9]. However, the deduced values of the internal magnetic field on both lattice sites were found to be very similar ($H_G = 205$ and $H_S = 207$ kOe); thus, in the present investigation the spectrum of Fe_3C at room temperature is assumed to be consisting of a single six-line hyperfine pattern, i.e. the same internal magnetic field is assumed to exist at both lattice sites.

3. Sample Preparation

The Fe_3C used in this experiment was prepared by melting pure iron in a graphite crucible in an inductive furnace with an inert gas atmosphere. The melt was cooled at a rate of approximately 50deg/h in the critical range between 900 and 500 °C. The Fe_3C was chemically extracted from the iron–carbon alloy according to a procedure described elsewhere [10]. The particles obtained have the shape of flakes with diameters of 50 to 100 μm and a ratio of diameter to thickness of about 20:1.

The flake-like Fe_3C material was placed in a sample holder and slightly pressed. The X-ray and Mössbauer experiments were performed on the same sample. The X-ray studies were carried out with filtered CoK α and MoK α radiation in combination with a scintillation counter. In accordance with ASTM chart No. 6-0688 no lines other than those of Fe_3C were detected. The preferred orientation of the sample was determined by a standard procedure.

4. X-Ray Results

A strong texture exists. The crystallites of the sample are aligned in such a way that their c-axes, [001], are preferentially oriented perpendicular to the planes or the surface of the sample. Actually the orientation was determined from the (004) reflection. The (004) and (023) planes have nearly the same spatial d-value and therefore the same Bragg angle. From the angular distribution of the (103), (112), and (021) reflections a (023) texture could be ruled out. The X-ray diffraction intensities invariant to a rotation of the sample around the normal, i.e. the texture has cylindrical symmetry.

The angle between the c-axes of the crystals and the normal to the surface of the sample is symbolized by ϑ (see Fig. 1), and the orientation distribution of the Fe_3C crystals as a function of ϑ is plotted in Fig. 2a. $D_E(\vartheta)\, d\Omega$ represents the relative volume of the crystallites with c-axes oriented within the solid angle element $d\Omega$. For example, D_E is a constant for a randomly oriented assembly. For the correlation with Mössbauer data another representation of the tex-

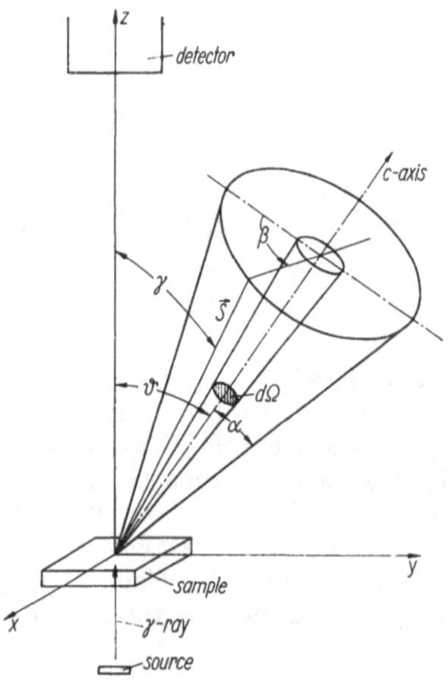

Fig. 1. Correlation between c-axis and spin, S, in the polycrystalline sample in conjunction with the experimental arrangement. The various angles are defined in the text

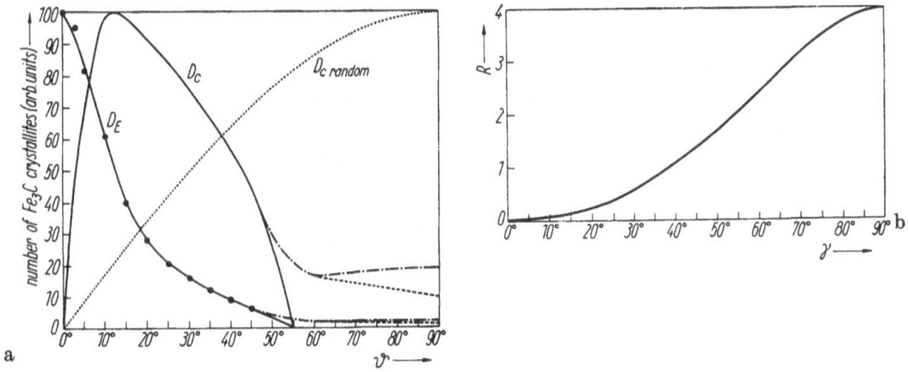

Fig. 2. a) Orientation distribution of the Fe_3C crystallites in regard to the angle, ϑ, formed by their c-axis and the normal to the sample. D_E represents the orientation within the solid angle element $d\Omega$ and D_c the orientation within the solid angle element between ϑ and $\vartheta + d\vartheta$ (cone shell). b) Ratio of the relative line intensities $R = I_{2,5}/I_{3,4}$ vs. angle, γ, between the γ-arrow and the spin $\boldsymbol{S}$

ture, $D_c(\vartheta)$, is preferred. $D_c(\vartheta)\, d\vartheta$ is defined as the relative volume of crystallites with c-axes in a particular cone shell angle element between ϑ and $\vartheta + d\vartheta$. Since in the usual notation with spherical coordinates (polar angle ϑ, azimuthal angle φ) $d\Omega$ is given by $\sin\vartheta\, d\vartheta\, d\varphi$, the relation between $D_E(\vartheta)$ and $D_c(\vartheta)$ can easily be obtained by integration of $D_E\, d\Omega$ over φ (rotational symmetry), and is given by $D_c(\vartheta) = D_E(\vartheta) \sin\vartheta$. For the measured D_E values D_c is plotted in Fig. 2a. For comparison, the distribution for an assembly of randomly oriented crystallites is also shown in this figure as $D_{c,\,random}$.

Due to the decreasing signal-to-noise ratio with increasing ϑ the D_E values could be measured only up to 45° (dots in Fig. 2a). In the range between 45 and 90° three different possibilities for D_E were assumed: i) linear extrapolation up to 55° (solid line), ii) $D_E =$ const between 60 and 90° (dashed-dotted line), iii) physically reasonable extrapolation with small non-vanishing contribution up to 90° (dashed line); the latter case corresponds to a background of randomly oriented crystallites at higher angles. The corresponding D_c curves for these cases are also shown in Fig. 2a.

For the analysis it is convenient to have $D_E(\vartheta)$ in analytical form. A polynomial approximation of the form

$$D_E(\vartheta) = \sum_{i=0}^{n} a_i\, \vartheta^i \tag{1}$$

was chosen, the coefficients a_i being determined by a least-squares fit computer program. All the curves drawn in Fig. 2a are analytical functions obtained in this way.

5. Mössbauer Results

The Mössbauer spectrum of the same Fe_3C at room temperature obtained with a Co^{57}–Pd-source is shown in Fig. 3. The arrangement of the experiment can be seen schematically in Fig. 1. The γ-ray propagation direction, γ-arrow, was perpendicular to the plane of the sample (absorber).

The relative line intensities in the Zeeman hyperfine pattern are a function of the angle, γ, between the γ-arrow and the direction of the magnetic field at the

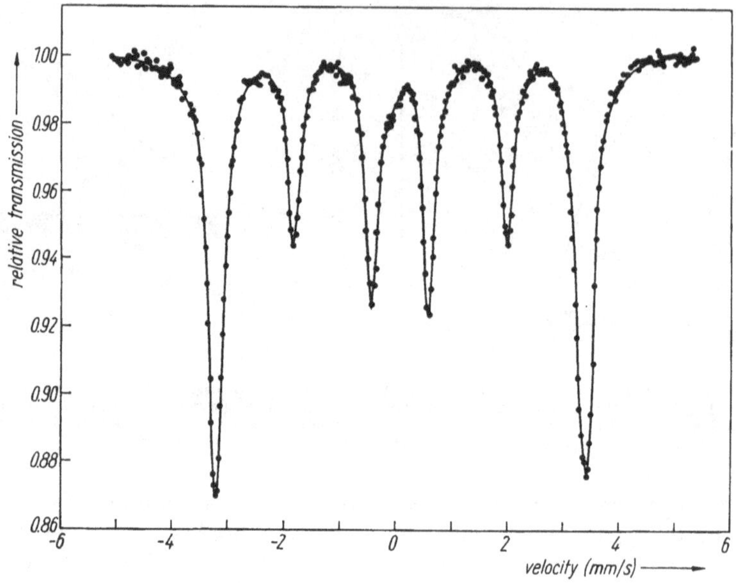

Fig. 3. Mössbauer spectrum of a Fe_3C absorber and a Co^{57}–Pd-source both at room temperature

Fe^{57} nucleus, which is known to be parallel (or antiparallel) to the atomic spin. Disregarding any lattice vibration anisotropy (Goldanskii-Karyagin effect), and assuming the thin absorber approximation, one knows, that the intensities of the $\Delta m = 0$ lines ($I_{2,5}$) are proportional to $\sin^2 \gamma$ and the inner $\Delta m = \pm 1$ lines ($I_{3,4}$) are proportional to $(1 + \cos^2 \gamma)/4$. The intensities of the outer lines ($I_{1,6}$) are proportional to the latter; however, they are not considered here because their greater sensitivity to saturation effects makes the thin absorber approximation less valid. The ratio

$$R = \frac{I_{2,5}}{I_{3,4}} = \frac{\sin^2 \gamma}{\frac{1}{4}(1 + \cos^2 \gamma)} \tag{2}$$

is plotted in Fig. 2b. Note that R ranges from zero for $\gamma = 0°$ to 4 for $\gamma = 90°$ and has a value of 2 at $\gamma \approx 55°$. For an assembly of randomly oriented spins one expects a six-line hyperfine pattern with the well-known relative line intensities of 3:2:1:1:2:3, i.e. $R_A = 2$. The subindex A refers to an assembly of spins.

The $\Delta m = 0$ lines in the hyperfine pattern of the Fe_3C absorber have noticeable lower intensities, and a value of the ratio of areas $R_A = 0.65$ was obtained. A small saturation effect was taken into account. The low value of R_A indicates that the c-axes of the Fe_3C crystals measured by X-rays as well as the spins in the absorber are directed toward the γ-arrow. The question arises: can the analysis of the combined X-ray and Mössbauer results determine the spin orientation in Fe_3C quantitatively?

6. Correlation between Mössbauer Line Inentsities R_A and Texture D_E

In the following an exact analysis of the Mössbauer intensity ratio R_A is given. In the case of colinearity of spin and c-axis the distribution of the spins coincides with the distribution of the c-axes, $D_c(\vartheta)$. Therefore $D_c(\vartheta)$ can be used as a weight function for calculating the intensities:

$$R_A = \frac{I_{2,5\,A}}{I_{3,4\,A}} = \frac{\int\limits_0^{\pi/2} (1 - \cos^2\vartheta)\, D_c(\vartheta)\, d\vartheta}{\frac{1}{4}\int\limits_0^{\pi/2} (1 + \cos^2\vartheta)\, D_c(\vartheta)\, d\vartheta}. \tag{3}$$

The integration over ϑ can be limited up to ϑ_{max} for cases where crystal orientations do not occur at higher angles. R_A might be obtained graphically by dividing the area under D_c (Fig. 2a) with the appropriate R values of Fig. 2b into one part constituting line $I_{2,5}$ and another part constituting line $I_{3,4}$.

Consider the more general case where colinearity does not exist. α is defined as the angle between a spin $\mathbf{S}$ and its corresponding c-axis (see Fig. 1). The spins are assumed to be randomly distributed with respect to the angle β. From Fig. 1 one can see that $\mathbf{S}$ makes the angle γ with the γ-ray. The correlation between α, β, γ, and ϑ is given by

$$\cos\gamma = \cos\alpha\cos\vartheta + \sin\alpha\sin\vartheta\cos\beta\,. \tag{4}$$

Deriving a formula for $R_A(\alpha)$ for the non-colinear case, the argument ϑ in $\sin^2\vartheta$ and $(1 + \cos^2\vartheta)/4$ has to be replaced by γ, and the additional integration over the angle β has to be taken into account. Integration over the azimuthal angle φ is not necessary because of rotational symmetry around the z-axis. One finds

$$R_A(\alpha) = \frac{\int\limits_0^{2\pi}\int\limits_0^{\vartheta_{max}} (1 - \cos^2\gamma)\, D_c(\vartheta)\, d\vartheta\, d\beta}{\frac{1}{4}\int\limits_0^{2\pi}\int\limits_0^{\vartheta_{max}} (1 + \cos^2\gamma)\, D_c(\vartheta)\, d\vartheta\, d\beta}. \tag{5}$$

Using equation (4) for substituting $\cos\gamma$, and carrying through the integration over β one obtains

$$R_A(\alpha) = \frac{\int\limits_0^{\vartheta_{max}} [2\sin^2\alpha + (2 - 3\sin^2\alpha)\sin^2\vartheta]\, D_c(\vartheta)\, d\vartheta}{\frac{1}{4}\int\limits_0^{\vartheta_{max}} [4 - 2\sin^2\alpha - (2 - 3\sin^2\alpha)\sin^2\vartheta]\, D_c(\vartheta)\, d\vartheta}. \tag{6}$$

$D_c(\vartheta)$ can be replaced by $D_E(\vartheta)\sin\vartheta$.

Two special cases are useful for checking equation (6):

1. Polycrystalline material with random distribution of c-axes (D_E = const). With the appropriate integrations over ϑ between 0 and $\vartheta_{max} = \pi/2$ R_A is found to be $R_{A,\,random} = 2$.

2. Colinearity between spin and c-axis ($\alpha = 0$): Introducing $\alpha = 0$ into equation (6) leads to the equation (3) formerly derived.

Combining equation (1) and (6), one obtains

$$R_A(\alpha) = 4\,\frac{2\sin^2\alpha \sum\limits_{i=0}^{n} a_i \int\limits_0^{\vartheta_{\max}} \vartheta^i \sin\vartheta\, d\vartheta + (2 - 3\sin^2\alpha) \sum\limits_{i=0}^{n} a_i \int\limits_0^{\vartheta_{\max}} \vartheta^i \sin^3\vartheta\, d\vartheta}{(4 - 2\sin^2\alpha) \sum\limits_{i=0}^{n} a_i \int\limits_0^{\vartheta_{\max}} \vartheta^i \sin\vartheta\, d\vartheta - (2 - 3\sin^2\alpha) \sum\limits_{i=0}^{n} a_i \int\limits_0^{\vartheta_{\max}} \vartheta^i \sin^3\vartheta\, d\vartheta}\,. \tag{7}$$

The integrals $\int\limits_0^{\vartheta_{\max}} \vartheta^i \sin\vartheta\, d\vartheta$ and $\int\limits_0^{\vartheta_{\max}} \vartheta^i \sin^3\vartheta\, d\vartheta$ were evaluated by computer.

7. Discussion and Conclusion

By inserting the orientation distribution of the crystallites, D_E or D_c, into equation (6) the angle between the spin and the c-axis can be calculated. Unfortunately, the X-ray determination of the texture is limited in accuracy and any contribution in the range $\vartheta > 45°$ was not any more distinguishable experimentally. This point is critical in view of the sensitivity of the crystals with such orientation to the R_A value as can be seen readily from Fig. 2. The three extrapolated orientation distributions D_E and D_c in Fig. 2 have been used to evaluate R_A and the obtained values are plotted in Fig. 4 as a function of α, the angle between the c-axis and the spin.

Case i) (solid line): No crystal orientation is allowed in the range between 55 and 90°, and also any possible contribution from spin deviation at the surface of the crystallites is neglected. These stringent assumptions are somewhat physically unrealistic, however, this orientation distribution sets a lower limit for R_A (plotted as lower curve in Fig. 4) and gives an upper limit of the angle, α, between spin and c-axis which is still consistent with our experiment: $R_A = 0.65$, thus $\alpha \approx 21°$.

Case ii) (dashed-dotted line): The upper curve in Fig. 4 represents the R_A value one expects with a small contribution of randomly oriented crystallites at $\vartheta > 60°$. Even under the condition of colinearity of $\mathbf{S}$ and c-axis ($\alpha = 0$), minimizing R_A, the theory for this orientation distribution predicts a value of $R_A = 0.75$ which is well outside the experimental result of $R_A = 0.65$.

Case iii) (dashed line): A few misoriented crystals at high angle undetected in the background of the X-ray measurements give an orientation distribution indicated by the dashed line in Fig. 2a. We believe that this curve is a good

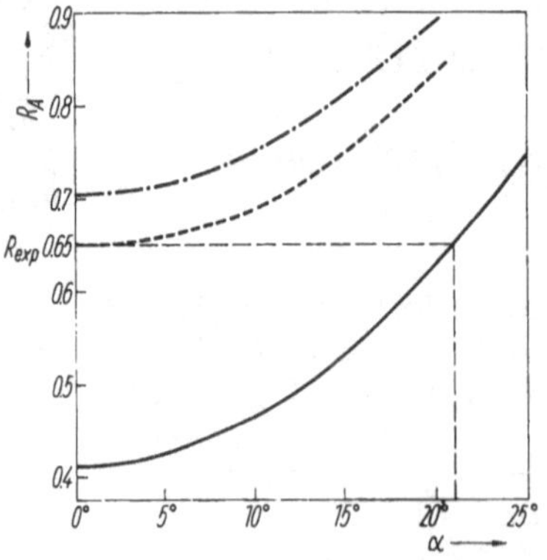

Fig. 4. Relative Mössbauer line intensities, R_A, vs. angle between spin and c-axis, α, for the three orientation distribution

representation of the actual condition of our experiment. With this distribution the experimental Mössbauer value of $R_A = 0.65$ can only be reconciled if colinearity of spin and c-axis exists.

From the correlation of the preferred orientation determined by X-rays with the Mössbauer line intensities we conclude that the spin in Fe_3C is colinear with the c-axis, however, the accuracy of our experiments does not exclude the possibility that spin and c-axis form an angle up to about 20°.

Acknowledgements

The financial contribution by the ,,Deutsche Forschungsgemeinschaft'' and "IRSID" are appreciated. We thank Mr. H.-D. Pfannes and Dr. D. Williamson for helpful discussions and Dr. Z. Mathalone and Mr. J. Pipman for assistance in preparing the sample.

References

[1] S. S. Hanna, J. Heberle, C. Littlejohn, G. J. Perlow, R. S. Preston, and D. H. Vincent, Phys. Rev. Letters **4**, 177 (1960).
[2] G. J. Perlow, S. S. Hanna, M. Hamermesh, C. Littlejohn, D. H. Vincent, R. S. Preston, and J. Heberle, Phys. Rev. Letters **4**, 74 (1960).
[3] H. Frauenfelder, D. E. Nagle, R. D. Taylor, D. R. F. Cochran, and W. M. Visscher, Phys. Rev. **126**, 1065 (1962).
[4] C. E. Johnson, W. Marshall, and G. J. Perlow, Phys. Rev. **126**, 1503 (1962).
[5] U. Gonser, R. W. Grant, H. Wiedersich, and S. Geller, Appl. Phys. Letters **9**, 18 (1966).
[6] E. J. Fasiska and G. A. Jeffrey, Acta cryst. **19**, 463 (1965).
[7] M. J. Duggin, Trans. MS AIME **242**, 1091 (1968).
[8] H. Bernas, I. A. Campbell, and R. Fruchart, J. Phys. Chem. Solids **28**, 17 (1967).
[9] M. Ron and Z. Mathalone, Phys. Rev. **B4**, 774 (1971).
[10] Z. Mathalone, M. Ron, J. Pipman, and S. Niedzwiedz, J. appl. Phys. **42**, 687 (1971).

(Received February 7, 1972)

phys. stat. sol. (a) **10**, 501 (1972)

Subject classification: 20.3; 11; 21; 21.1; 21.1.1; 21.3; 21.6; 22.6

Sektion Physik der Humboldt-Universität zu Berlin, Bereich Atomstoßprozesse der Festkörperphysik

Photon Emission during Bombardment of Solids with Alkali Ions in the Energy Range between 2 and 10 keV

By

H. KERKOW

Experimental results are reported regarding the emission of photons by the bombardment of solids with 2 to 10 keV alkali ions. The results about spatial and spectral distribution are interpreted with respect to a sputtering mechanism. In this process sputtered particles are excited and emitted by atomic collisions with the projectile ion. Outside of the solid in front of the surface the excited particles interact electronically with the surface and recombine radiatively into deeper electronic states. So the higher photon yields for insulators and semiconductors in regard to metals can be explained. An attempt was made to interpret the spatial distribution of photon emission.

Es werden experimentelle Ergebnisse zur Photonenemission von Festkörpern beim Ionenbeschuß mitgeteilt. Die Ergebnisse werden mit dem Zerstäubungsmechanismus erklärt. Bei diesem Prozeß werden die zerstäubten Teilchen durch Atomstöße mit den Projektilionen angeregt und emittiert. Außerhalb des Festkörpers nahe der Oberfläche treten die angeregten Teilchen in elektronische Wechselwirkung mit der Oberfläche und rekombinieren strahlend in tieferliegende Elektronenzustände. So kann man die höhere Photonenausbeute für Halbleiter und Isolatoren gegenüber Metallen verstehen. Es wird versucht, die räumliche Verteilung der Photonenemission zu erklären.

1. Introduction

The mechanism of photon emission at the bombardment of solids is discussed in several papers [1 to 13], but there is no consistent model about this phenomenon. The photon emission was interpreted by radiative recombination of the excited atoms outside [7 to 13] or by plasma oscillations of the electrons inside the solids [5, 6]. Because the authors obtained their results under different experimental conditions it is difficult to compare the results with those of others and to develop a consistent model.

In the present work photons emitted from polycrystalline or amorphous targets of metals, semiconductors, and insulators under bombardment by 2 to 10 keV alkali ions were studied. The spectral and spatial distribution of photon emission are explained within a simple model.

2. Experimental

The apparatus used for the measurements is shown schematically in Fig. 1. In the centre of a cylindrical vacuum chamber the target was localized. Thermally emitted potassium or cesium ions were directed on the target with current densities of about 2 to 3 mA/cm^2, accelerated to energies between 2 and 10 keV and having angles of incidence between $0° \leq \alpha \leq 90°$. Further details of the ion source are described elsewhere [14]. The vacuum chamber was evac-

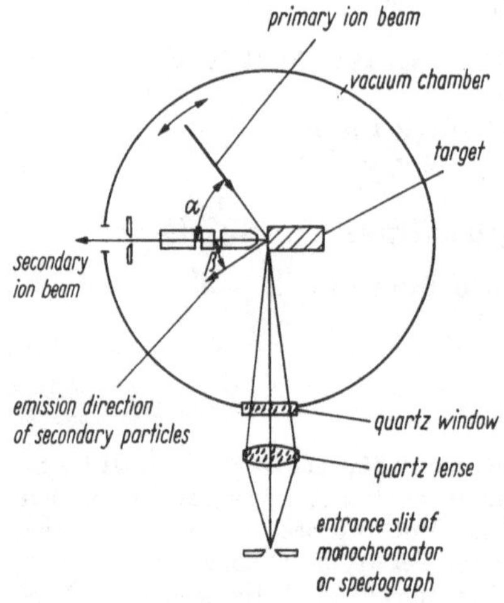

Fig. 1. Experimental arrangement

uated down to 5×10^{-6} Torr. Polycrystalline or amorphous targets of a number of materials (Pt, Au, Ag, Rh, Ta, W, Mo, Zn, Cd, Mg, Ni, Fe, Cr, Al, CuO, WO_3, CdO, BaO, CrO_3, MgO, Al_2O_3) were bombarded. The spectral distribution of photon emission was investigated parallel to the solid surface in the spectral region between 250 and 550 nm, either with a prism-spectrograph (aperture $f = 15$) and photoplate, or with a prism-monochromator (aperture $f = 6$) and multiplier. For determination of quantum yield the absolute spectral sensitivity of the monochromator was determined. In the place of the target a tungsten ribbon was mounted and, therefore, a correction for absorption and reflection was not necessary. The spectral radiation density of the tungsten ribbon was calculated from the well known radiation density of the black body [15] and the emissivity of tungsten [16]. The spatial distribution of photon emission was photographed both through a side window (side view, Fig. 1) and perpendicular through a quartz window at the top of the vacuum chamber (top view).

In some cases secondary ions were extracted perpendicular to the solid surface and analyzed with a mass spectrometer.

3. Results

For all target materials several atomic lines of the metallic components were observed. Exclusively the resonance and the arc lines at low excitation energies of the metal atoms show high intensities. The quantum yield for atomic lines of a metal target is about 10^{-5} quanta/ion. It is about one order of magnitude higher for the same line if the target is a semiconducting or an insulating chemical compound of the same atom. This is demonstrated in Table 1 for the case of aluminium and aluminium oxide.

Table 1

Absolute quantum yields for Al- and Al_2O_3-targets at bombardment with 8 keV K-ions in (10^{-6} quanta/ion).

wavelength (nm)	Al	Al_2O_3
256.80	0.32	5.60
257.51	0.55	10.80
266.04		
265.25	0.32	4.76
305.47	1.09	2.45
308.02	4.50	78.3
309.27	9.70	157
394.40	10.4	57.0
396.15	19.6	108

A side-view photograph of photon emission (Fig. 2) shows a symmetrical distribution with respect to the target normal. In this direction the extension of the emission is 1 to 2 mm, as can be seen from the monochromatically recorded distribution (Fig. 3).

Fig. 2. Side view photograph of the photon emission. On the right side of the picture is seen the target holder, on the left side the extraction optics for the secondary ions. The primary ions are coming from the left behind the plane under an angle of 45° to the target surface. *Parameters*: 8 keV Cs-ions, 100 μA target current, polycrystalline chromium target. Exposure 4 min with an aperture of 5.6 and 18/10° DIN film

Top-view photographs show asymmetric distributions (Fig. 4). The spatial distribution will have a preferred direction opposite to the incident ion beam. This preferred direction depends neither on the mass ratio between projectile and target atoms nor on the angle of incidence.

The spatial distributions described are only characteristic of solids. In Fig. 5 the distributions of a solid and a gas are compared. In the case of a gas the intensity is high if the product of ion current density and gas density is high. The emission extends preferentially in the direction of the incident ion beam and emergent vapour jet. A preference in the direction of reflection, as it was observed for solid targets, cannot be found.

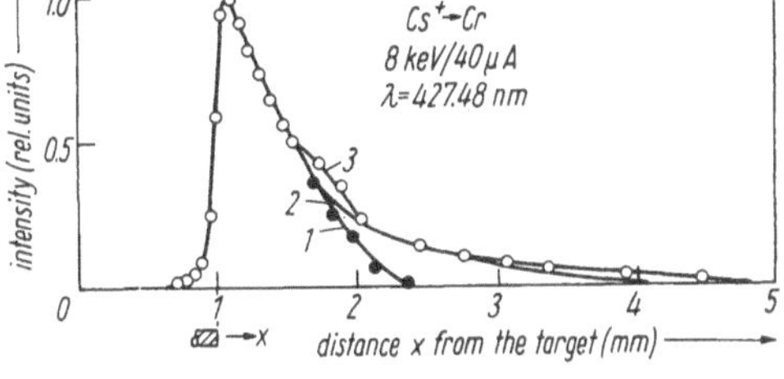

Fig. 3. Monochromatically recorded intensity distribution of photon emission along the target normal for different angles of incidence of primary ions. (1) : 0°; (2) : 30°; (3) : 70°

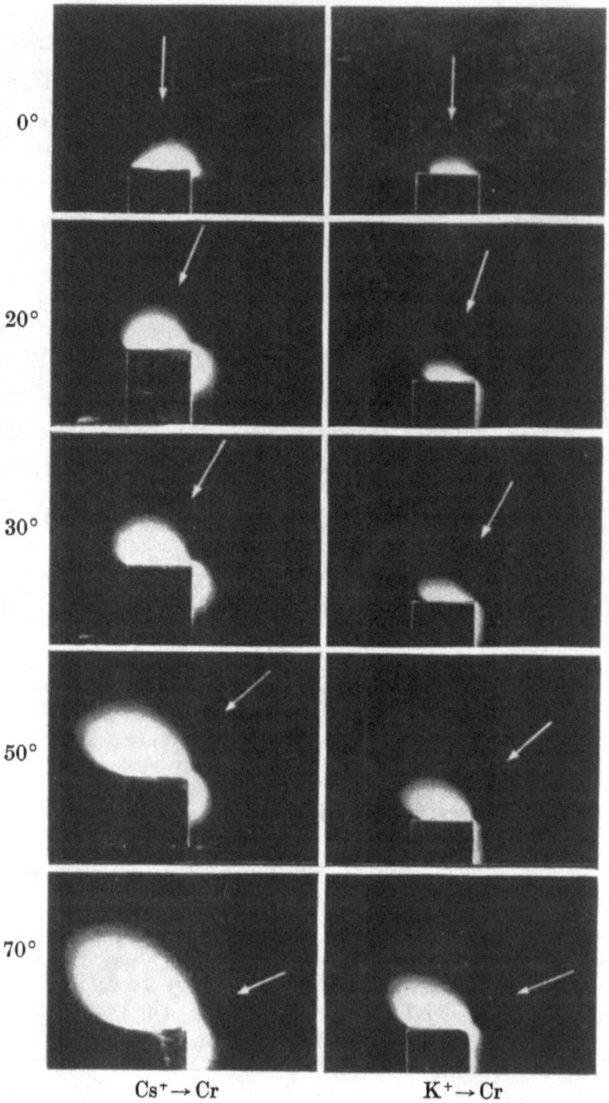

Fig. 4. Top view photographs of the photon emission. Bombardment of a polycrystalline chromium target with Cs-, respectively K-ions at different angles of incidence

4. Interpretation of the Results

4.1 Model

For the interpretation of our results we propose an emission model being based on essentially three assumptions:

1. The target atoms are excited by collisions with the projective ions in the first atomic layer. The estimated inelastic energy loss according to Firsov's formula [17] points out, that for each required deflection the transferred energy is much higher than the excitation energy of the target atoms. This estimation

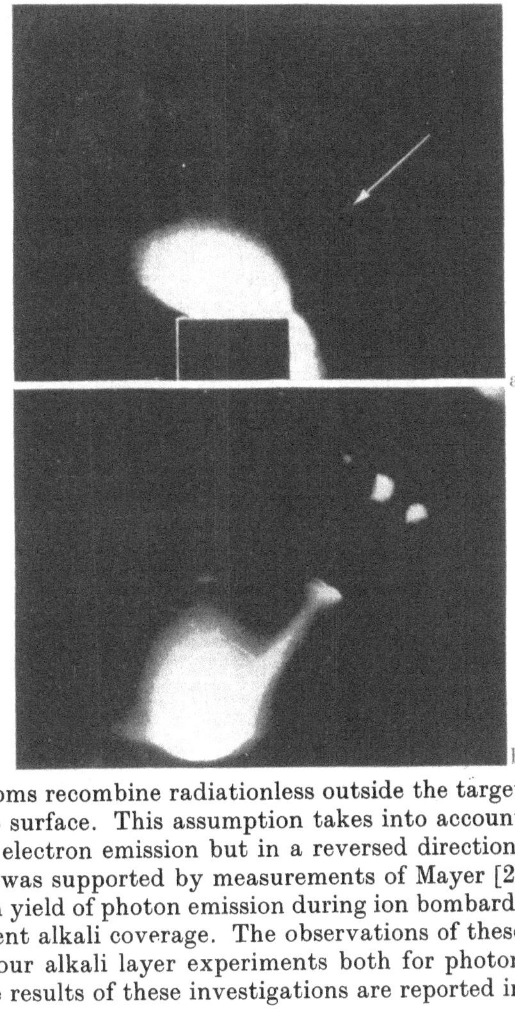

Fig. 5. Photon emission at bombardment of a) a solid Cr-target and b) a vaporious Cd-target

is supported by the results of Morgan and Everhart [18] about the inelastic energy losses by atomic collisions and by the measurements of Moe [19] about the threshold energy for excitation by atomic collisions (excitation energy lower than 100 eV).

2. Some excited atoms possess impulse components along the target normal. The reflected projectile ions, deflected by single collisions of the target surface, are observed in our mass spectrum of secondary ions. This observation together with results about ion reflection, reported in the literature [20] confirms our second assumption.

3. The excited and emitted atoms recombine radiationless outside the target under the influence of the target surface. This assumption takes into account the ideas of potential secondary electron emission but in a reversed direction. Experimentally this assumption was supported by measurements of Mayer [2] and Sporn [3] about the quantum yield of photon emission during ion bombardment of solids which have different alkali coverage. The observations of these authors could be confirmed by our alkali layer experiments both for photon and secondary ion emission. The results of these investigations are reported in another paper [21].

4.2 Calculation of the angular distribution

The characteristic spatial distribution of photon emission can be easily calculated within the outlined emission model.

By use of the hard-sphere approximation, a cosinusoidal distribution around the direction of primary ion beam results for the target atoms subjected to collision. The fraction dN_0^+ of N target atoms subjected to collision, which reach the solid angle $d\Omega$ with the angle of deflection θ, is given by the following formula:

$$dN_0^+ = N \cos\theta \, d\Omega \quad \text{with} \quad 0 \leqq \theta \leqq 90^\circ . \tag{1}$$

All atoms which reach a deflection angle θ within the plane of incidence of more than $90° - \alpha$, get into the vacuum, as it is evident from Fig. 6. A part of

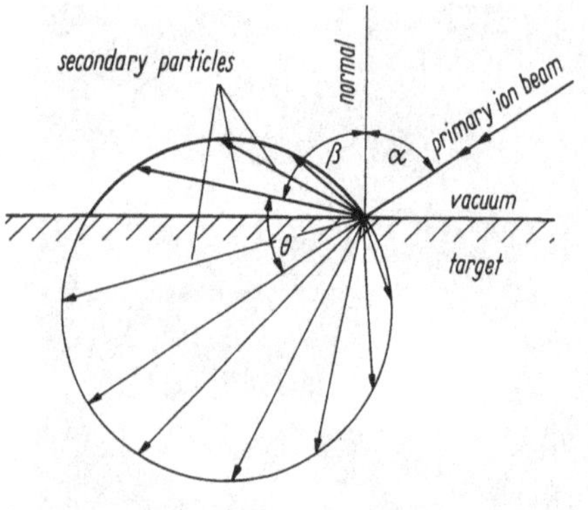

Fig. 6. Schematic representation of the intensity distribution of stroken target atoms after the first collision

these excited atoms may recombine to the ground state without radiation by recombination processes at the surface:

$$\frac{\mathrm{d}}{\mathrm{d}t}(\mathrm{d}N^{+}) = -R(x)\,\mathrm{d}N^{+}\,. \tag{2}$$

According to Hagstrum [22] for the recombination coefficient $R(x)$ the following statement is made:

$$R(x) = C \exp\,(-c\,x)\,, \tag{3}$$

with x being the distance of the excited atom from the target surface and C, c are constants. Substituting $v_{\perp}t$ for x ($v_{\perp}$ velocity normal to the target surface), integrating equation (2) over the time from $t = 0$ to $t = \infty$, and using equation (3) it follows:

$$\mathrm{d}N = \mathrm{d}N_0^{+} \exp\left(\frac{-C}{c\,v_{\perp}}\right), \tag{4}$$

where $v_{\perp}$ depends on the direction of emission of the excited atoms with respect to the target surface, and on the energy transferred by a single collision. The latter may easily be calculated from the laws of conservation of energy and momentum. For the velocity $v_{\perp}$ the following formula results:

$$v_{\perp} = 2\,\frac{(2\,m_1\,E_0)^{1/2}}{m_1 + m_2}\cos\theta\cos\beta\,, \tag{5}$$

where E_0 is the energy of primary ions, and m_1 and m_2 are the masses of the primary ions and target atoms, respectively, and β the angle of emission (see Fig. 1). Substituting equations (1) and (5) into (4) one obtains

$$\mathrm{d}N = N\cos\theta\exp\left(\frac{-C}{c\,b\cos\theta\cos\beta}\right) \tag{6}$$

with the abbreviated notation $b = (2\,m_1\,E_0)^{1/2}/(m_1 + m_2)$.

4.3 Discussion

The spatial distributions of photon emission calculated using equation (6) are shown in Fig. 7. An agreement between these calculated distributions and the photographs of Fig. 4 can be seen in following points:

1. The spatial distribution is asymmetric with respect to the target normal.
2. The extension of photon emission grows with increasing angle of incidence (see also Fig. 3).

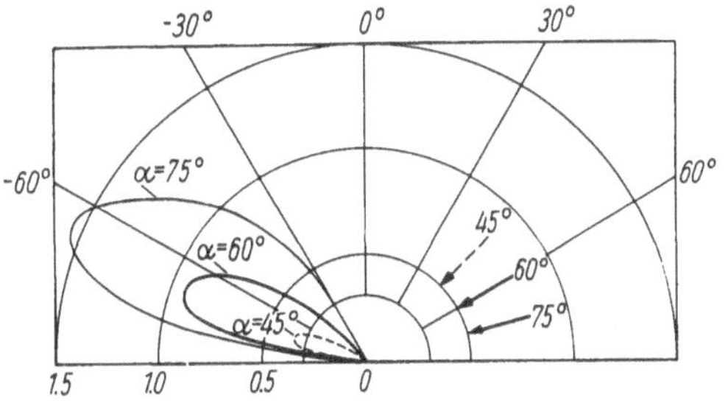

Fig. 7. Angular distribution of excited atoms, calculated according to equation (6)

3. The distribution is relatively independent of the projectile energy and mass.

4. The preferred direction of photon emission is hardly dependent on angle of incidence.

Similar spatial distributions investigated Gritsina et al. [11] by bombardment of tantalum with protons. In contrast to our results the photon spectrum consisted of spectral lines of hydrogen atoms, which are the atomic lines of the incident particles not the lines of the target material. Because the hydrogen partial pressure in their vacuum chamber was high, it cannot be excluded that adsorbed hydrogen atoms are excited and sputtered away like target atoms.

As mentioned earlier, we observed for all materials, also for all metals, arc lines of the metallic atoms. Tsong [13] detected photon emission only for some metals as Mg, Mn, Sc, Ta, Ti, V, and W, but not for Ag, Al, Cd, Cu, Fe, Mo, Nb, Ni, and Zn. The absence of a photon emission in the latter case was explained by the author with a resonance ionization of the excited atoms on the surface. This conclusion is unallowed, as Tsong's experimental conditions are insufficient because of a coverage of the windows during the measurement and on insensibility of his detection system. We conclude that electronic interaction on the surface causes different yields between metals, semiconductors, and insulators, but up to day too less is known about differences in quantum yield for metallic targets in order to proof the model.

The velocity of the excited particles could be estimated from the measured lateral distribution as seen in Fig. 3. Without knowledge of the exact angular distribution and lifetimes of the excited particles their energy amounts to about 100 eV. Similar high energies of excited particles calculated van der Weg and Biermann [9] and Gritsina et al. [11] from the Doppler broadening of spectral lines. All these results confirm the above assumption on an excitation and emission of target atoms by two-body collisions with projectile ions.

While Tsong [13] postulated that photon emission is produced by radiative recombination of excited, slowly sputtered particles, we conclude that fast sputtered particles contribute to photon emission. But with respect to the above mentioned model the emission of excited particles should disappear at perpendicular bombardment, in contrast to the observations shown in Fig. 4; in this case photon emission was also observed. From this it may be concluded that also after the collision with the projectile ion the excited atoms are knocked into the direction of the target normal by collisions with neighbouring atoms. The existence of such processes has been confirmed by van der Weg and Biermann [23]. Because these processes at oblique ion bombardment also exist, a few, not too fast paɹticles contribute in every case to photon emission.

Acknowledgements

Thanks are due to Prof. Dr. Bernhard for the suggestion of this work and his active interest and to Dr. S. Schwabe for valuable discussions.

References

[1] J. STARK and G. WENDT, Ann. Phys. (Germany) (4. Folge) **38**, 669 (1912).
[2] H. MAYER, Phil. Mag. (Ser. 7) **16**, 549 (1933).
[3] H. SPORN, Z. Phys. **112**, 278 (1938).
[4] V. I. VEKSLER, Fiz. tverd. Tela **5**, 2737 (1963).
[5] R. M. CHAUDRI and M. J. KHAN, Proc. V. Internat. Conf. Ion Phen. Gases, Part 2, Munich 1961 (p. 1195).
[6] R. M. CHAUDRI, M. M. CHAUDRI, S. M. KALIER, and M. RAFIQUE, Proc. VII. Internat. Conf. Ion. Phen. Gases, Part 1, Belgrad 1965 (p. 186).
[7] J. M. FLUIT, L. FRIEDMAN, J. VAN ECK, C. SNOEK, and J. KISTEMAKER, Proc. V. Internat. Conf. Ion. Phen. Gases, Part 1, Munich 1961 (p. 131).
[8] C. SNOEK and J. KISTEMAKER, Adv. Electronics and Electron Phys. **21**, 67 (1965).
[9] W. F. VAN DER WEG and D. J. BIERMAN, Physica (Utrecht) **44**, 206 (1969).
[10] J. TERZIĆ and B. PEROVIĆ, Atomic Collision Phenomena in Solids, North-Holland Publ. Co., Amsterdam 1970 (p. 536).
[11] V. V. GRITSINA, T. S. KIYAIN, A. G. KOVAL, and JA. M. FOGEL, Phys. Letters (Netherlands) A **27**, 292 (1968).
[12] W. F. VAN DER WEG and P. K. ROL, Nuclear Instrum. and Methods **38**, 274 (1965).
[13] I. S. T. TSONG, phys. stat. sol. (a) **7**, 451 (1971).
[14] F. BERNHARD and H. KERKOW, to be published.
[15] H. STAUDE, Physikalisch-Chemisches Taschenbuch, Bd. 2, Akademische Verlagsgesellschaft Geest & Portig K.-G., Leipzig 1949.
[16] R. D. LARRABEÉ, J. Opt. Soc. Amer. **49**, 619 (1959).
[17] O. B. FIRSOV, Zh. eksper. teor. Fiz. **36**, 1517 (1959).
[18] G. H. MORGAN and E. EVERHART, Phys. Rev. **128**, 667 (1962).
[19] D. E. MOE, Phys. Rev. **104**, 694 (1956).
[20] U. A. ARIFOV, Vzaimodeistvie atomnikh chastits s poverkhnostyu tverdovo tela, Izd. Nauka, Moskva 1968 (p. 295).
[21] H. KERKOW, to be published.
[22] H. D. HAGSTRUM, Phys. Rev. **96**, 336 (1954).
[23] W. F. VAN DER WEG and D. J. BIERMAN, Physica (Utrecht) **38**, 406 (1968).

(Received December 30, 1971)

phys. stat. sol. (a) **10**, 509 (1972)

Subject classification: 15; 22.9

Technische Hochschule Karl-Marx-Stadt, Sektion Physik, Elektronische Bauelemente

Measurements of the Thermoelectric Power of Copper Phthalocyanine Single Crystals

By

C. HAMANN

Measurements are made of the thermoelectric power of β-copper phthalocyanine single crystals. Sample preparation and measuring apparatus are described and the observed temperature dependence of the thermo-e.m.f. and of the measured absolute values are discussed using a model which considers trap-free solids and some trap distributions.

Es wurden Messungen der Thermokraft von Kupferphthalocyanin-Einkristallen der β-Modifikation durchgeführt. Die Probenherstellung und die Meßapparatur werden beschrieben und die gefundene Temperaturabhängigkeit des Seebeckkoeffizienten sowie die auftretenden Absolutwerte an Hand eines Modells diskutiert, das den haftstellenfreien Festkörper und bestimmte Haftstellenverteilungen berücksichtigt.

1. Introduction

For many years we have been engaged in the measurement of the thermoelectric power of organic semiconductors, especially of phthalocyanine (Pc) and its metal complexes [1 to 11]. Studying different metal complexes (H_2Pc, CuPc, NiPc, ZnPc) we have found that in any case the β-modification has a higher positive thermo-e.m.f. than the α-modification [9]. Investigations of eight variously prepared H_2Pc batches proved that the value of the thermo-e.m.f. sensitively depends on the method of preparation. Polycrystalline samples of α-modification showed a temperature-independent thermo-e.m.f. [2]. At investigations of metal complexes of phthalocyanine with a group of atoms in the centre of the molecule where these groups have different extensions (Cl–Al–Cl–Pc · 2 H_2O, OH–Al–Cl–Pc, HSO_4–Al–Cl–Pc, Cl–Al–Pc, OH–Al–Pc–H_2O, Pc–Al–O–Al–Pc, Cl–In–Pc, Cl–In–Cl–Pc, Cl_2–Sn–Pc, VO–Pc), we found relations between the structure, the electron density distribution of the molecule, and the occurrence of positive and negative values of the thermo-e.m.f. [11]. These investigations were continued on an increased scale with chlorinated CuPc (Cu–Cl_n–Pc; $n = 1$ to 16). There we pointed out a regular dependence of the value of the thermo-e.m.f. (continuous decrease of the thermo-e.m.f. from high positive to high negative values with a zero transition) on the chlorine content [5, 7, 8]. Investigations of single crystals and of polycristalline samples of β-CuPc in air and under purest nitrogen yielded always positive thermopowers, which have a strong temperature dependence [6].

Fielding and Gutmann [12] found a temperature-independent differential thermoelectric power of +50 μV/deg for β-CuPc single crystals in the temperature region from 70 to 130 °C. Kargin [13] gives for CuPc without determination of the crystal modification a value of +1100 μV/deg. In the work of Airapetyants and Rozenshtein [14] only the note exists that CuPc has a positive thermo-e.m.f. Delacote and Schott [15] gave for CuPc +700 μV/deg. Levina et al. [16] found that samples which consist of a metal film (Ni, Cu) covered with the corresponding phthalocyanine complex, have always negative thermopowers for CuPc and positive or negative thermopowers for NiPc. Interesting investigations were published by Vaisnys and Kirk [17]. They carried out measurements of the thermo-e.m.f. S of polycrystalline β-CuPc at pressures of $p = 0$ to 60 kbar. At atmospheric pressure they found −920 μV/deg. For phase I (1 to 30 kbar) and phase II (> 30 kbar) they found $S = +1100$ μV/deg, $dS/dp = -6$ μV/kbar deg and $S = -940$ μV/deg, $dS/dp = +6$ μV/kbar deg, respectively. Their results are compatible with the band model.

2. Preparation of Samples

CuPc single crystals were prepared by sublimation under inert gas [18]. They are very brittle and ravel out at applied mechanical stress along the needle axis. This process can be understood if one considers the strong binding forces in molecule stacks compared with the weaker binding forces between the stacks. At cleaving of the crystals along their b-axis all cleavage planes parallel to the stacking axis b are equivalent. There is no cleavage face which gives a surface comparable in quality with surfaces produced by crystal growth. Therefore, one has to study naturally grown surfaces. Cleavage of crystals in planes not parallel to the b-axis is impossible and results in a destruction of the crystal.

In order to produce suitable samples we applied the method developed in [5]. The CuPc single crystals are embedded in phthalopal. This embedding material has a melting point of 150 °C and does not attack phthalocyanine. Phthalopal wets phthalocyanine. It is transparent in solid state and has the same hardness as CuPc. Such prepared crystals can be sawed in each orientation with a fretsaw. After sawing the sectional area in the embedding material is grinded and polished. Finally, the treated crystal is dissolved out of the embedding material and is cleaned in organic solvents.

The sectional orientations chosen by us were

a) 2×(010), contacts on (010);
b) 2×(010) + 2×⊥ (001), contacts on ⊥ (001);
c) 2×(010) + 2×⊥ (001), contacts on (001).

3. Measuring Apparatus and Measuring Method

From the first it was clear that based on the geometric dimensions of the single crystal samples (≈ 2×2×2 mm³) being available and as a result of the low conductivity at room temperature and lower temperatures, only higher measuring temperatures could be applied. The upper temperature limit follows from the vapour pressure of

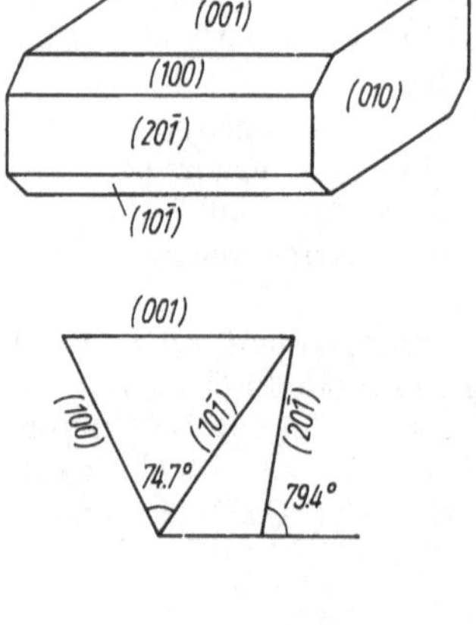

Fig. 1. Habit of the β-CuPc single crystal. (001) surfaces do not occur as a smooth surface

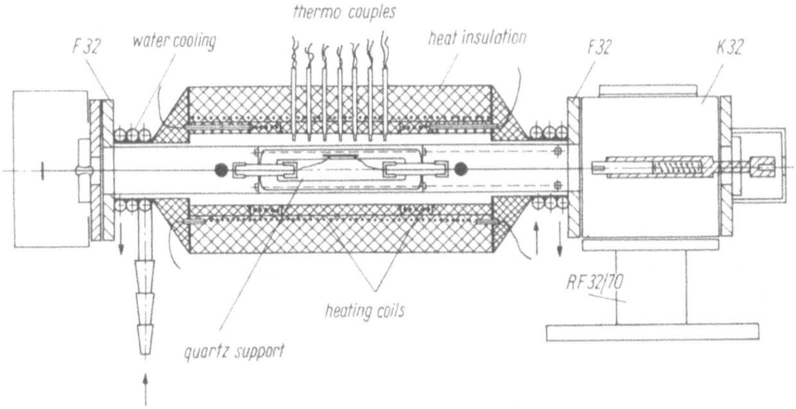

Fig. 2. Thermo-e.m.f. measuring cell for high-resistivity samples

CuPc, that is, 350 °C cannot be exceeded essentially in high vacuum without vaporizing the sample [19]. The lower limit is dependent on the input resistance of the measuring instrument, on the insulation of the measuring cell, and on the signal-to-noise ratio. We had carefully to avoid charging and drift phenomena.

Fig. 2 shows the used measuring cell, Fig. 3 the sample support. We renounced a measurement of the temperature inside the cell in order to avoid an influence on the temperature gradient and scattering voltages. The sample is on the sample support. The distance between sample and inner wall of the stainless steel tube (inner diameter 30 mm) is lower than 1 mm. The sample lies exactly under these seven measuring junctions, which are placed at a distance of 5 mm from the inner tube wall. The sample support consists of quartz, and it carries two silver contacts. In order to avoid heat transport to the sample it was put on a lattice made of thin quartz fibres. The electric connection near the sample was made of silver-plated quartz fibres. These silver contacts had no influence on the trap distribution on the surface.

The seven temperature measuring points which have a distance of 10 mm between each other are occupied by a differential thermocouple and five copper–constantan measuring junctions. One of these measuring junctions was used for a time control equipment. A measuring cycle (70 – 320 – 70 °C) continues 8 h and was carried out by the time control equipment with nearly linear temperature increase and decrease. By additional windings positive and negative temperature gradients could be produced. We used 5 to 10 deg/cm.

The temperature difference along the sample was calculated. This procedure was applicable because close to the sample a linear temperature gradient occurred over distances which are long related to the sample dimensions. The measurements were carried out under a vacuum of 10^{-5} Torr. The measuring conditions were very well reproducible. All contacts and conductors which are placed in the measuring cell were made of silver in order to avoid additional

Fig. 3. Sample support. The sample is held on a lattice of quartz fibres

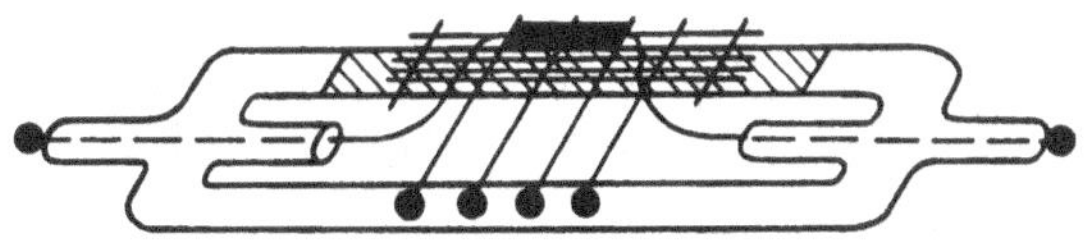

thermoelectric powers. The thermopower measured by a vibrating-reed electrometer was fed together with the voltages of the thermocouples to a galvanometer amplifier and then recorded. The thermopower measuring cell allowed to investigate samples with dimensions $\geqq 1$ mm and with a sample resistance $\leqq 10^{13}\ \Omega$. By the use of a sensitive galvanometer amplifier with 25 μV full scale, small temperature differences with an absolute error of ± 0.05 deg could be detected.

4. Results of Measurements

In Fig. 4 and 5 some of the experimental temperature dependences of differential thermoelectric power are represented. Generally it is valid that at the measuring method used, samples which are not annealed at high temperatures in a vacuum of 10^{-5} Torr show strong deviations of the thermoelectric power relative to a final value after heat treatment. This concerns above all samples which carry on the surfaces along the orientation of the temperature gradient (that means in the direction of charge carrier transport) surface defects. Crystals cut corresponding to section a) have the lowest defect density on surfaces (Fig. 4). The defect density increases for samples with section b) because there are two surfaces disturbed by cutting along the direction of carrier drift. The highest defect density is reached for samples with section c). There are only disturbed surfaces. The increase of this influence is expressed by the measuring results (Fig. 5). The clearest detection of the influence of surface defects could be obtained if the crystal was sand-blasted. An irreversible defect of the surface and of layers near the surface was obtained by mechanical treatment. If one measures samples cut by section a) but cleaved along the *b*-axis, the same results could be obtained as at sections b) and c).

Electric measurements on single crystals treated in a chlorine ambient for a long time showed a strong increase of conductivity. Due to chlorine addition the crystals have on the surface a green gleam. The activation energy analysis shows that the added chlorine atoms form a level 0.73 eV distant from the conducting band. By heat treatment in high vacuum chlorine begins at 120 °C to be released and the crystals show the typical values characteristic of β-CuPc without remaining alterations.

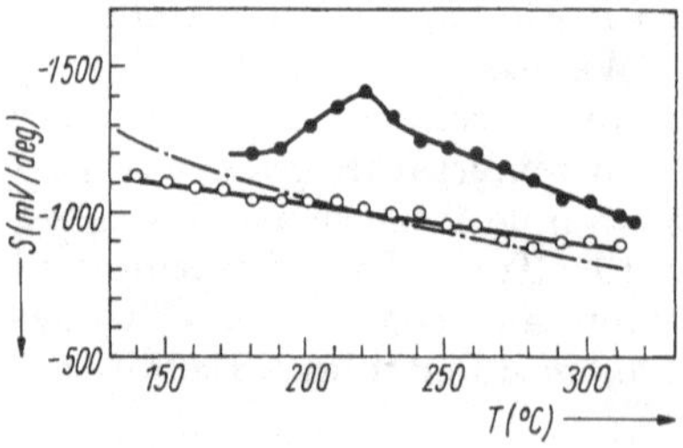

Fig. 4. Temperature dependence of the differential thermo-e.m.f. of β-CuPc single crystals. Sample E IV 25: temperature gradient || *b*-axis, dashed-dotted line: theoretical curve according to (8), $(b-1)/(b+1) = 0.58$; ● ● ● temperature rise, ○ ○ ○ temperature decrease

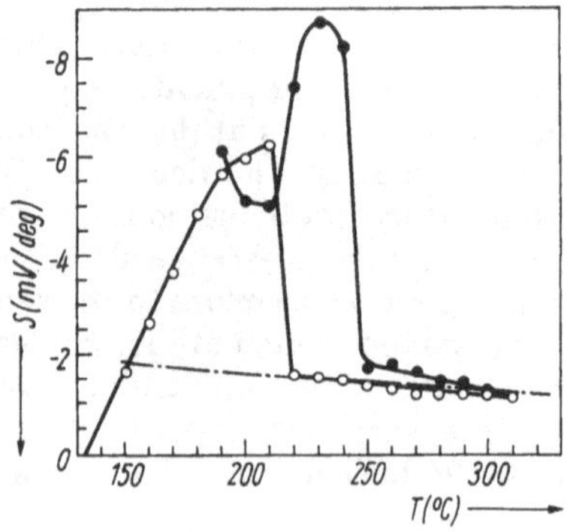

Fig. 5. Temperature dependence of the differential thermo-e.m.f. of β-CuPc single crystals. Sample E IV 119: ⊥ *b*-axis, ⊥ (001) plane, dashed-dotted line: theoretical curve according to (8), $(b-1)/(b+1) = 0.87$; ● ● ● temperature rise, ○ ○ ○ temperature decrease

At temperatures below 150 °C high positive differential thermoelectric powers were obtained, but they were recorded with high disturbances because the sample resistances were too high. At heat treatment of about 300 °C an annealing of surface defects produced mechanically was obtained and the oxygen was removed from the surface and from layers near the surface. Oxygen is again trapped below 150 °C and the samples got positive thermopowers. These observations support our model conception that oxygen forms an acceptor level 1.6 eV above the valence band and that discrete hole traps with a trap depth of (0.85 ± 0.05)eV are produced [5, 6].

5. Discussion of Measuring Results

We started with the equation of the differential thermo-e.m.f. valid for intrinsic semiconductors [20]:

$$S = -\frac{k}{e}\frac{1}{n\,\mu_n + p\,\mu_p}\left[n\,\mu_n\left(2 + \frac{E_F}{kT}\right) - p\,\mu_p\left(2 + \frac{E_c - E_F}{kT}\right)\right], \tag{1}$$

where $\mu_{n,p}$ is the mobility of electrons and holes, respectively, S the differential thermo-e.m.f., n, p the concentration of electrons and holes, respectively, E_F the Fermi level, and E_c the band gap. It follows in the first approximation with $E_F \approx E_c/2$, where $E_c = E_{c0} - a\,T$,

$$S = -\frac{k}{e}\frac{n\,\mu_n - p\,\mu_p}{n\,\mu_n + p\,\mu_p}\left(\frac{E_{c0}}{2\,kT} + 2 - \frac{a}{2\,k}\right). \tag{2}$$

(Here E_{c0} means the band gap at $T = 0$ °K.)

Using the temperature dependence of μ_n and μ_p found by Fillard and Schott [21, 22] and with the assumption that $n = p$ one obtains

$$S = -\frac{k}{e}\left[\frac{(\mu_{n0}/\mu_{p0}) - (T/T_0)}{(\mu_{n0}/\mu_{p0}) + (T/T_0)}\right]\left(\frac{E_{c0}}{2\,kT} + 2 - \frac{a}{2\,k}\right) \tag{3}$$

(T_0 reference temperature, μ_{n0}, $\mu_{p0} \mathrel{\hat{=}} \mu_{n,p}(T_0)$). For $\mu_{n0} \ll \mu_{p0}$ or $\mu_{n0} \gg \mu_{p0}$ equation (3) becomes

$$S \approx \pm\frac{k}{e}\left(\frac{E_{c0}}{2\,kT} + 2 - \frac{a}{2\,k}\right). \tag{4}$$

In the case of $\mu_{n0} \approx \mu_{p0}$ the second factor in (3) takes values between -1 and $+1$. Considering the trapping of charge carriers the following transformations can be applied:

$$(n - n_H)\,\mu_n \to n'\,\mu_{n,\text{eff}}\,, \tag{5}$$

$$(p - p_H)\,\mu_p \to p'\,\mu_{p,\text{eff}} \tag{6}$$

($n' = p$, $p' = n$, at the existence of one kind of traps, respectively; n_H, p_H trapped electron and hole concentration, respectively).

At dominating influence of electron traps that was found here for polycrystalline samples and thin films [23, 24], it holds

$$S = -\frac{k}{e}\left(\frac{E_{c0}}{2\,kT} + 2 - \frac{a}{2\,k}\right)\left[\frac{(\mu_{n,\text{eff}}/\mu_p) - 1}{(\mu_{n,\text{eff}}/\mu_p) + 1}\right]. \tag{7}$$

At dominating influence of hole traps, which was obtained for oxygen-free bulk conductivity of CuPc single crystals and for chlorinated samples (that means, chlorine added by chemical substitution), in equation (7) we have to transform $(\mu_{n,\mathrm{eff}}/\mu_p) \to (\mu_n/\mu_{p,\mathrm{eff}})$. For an exponential trap distribution of electrons,

$$h(E) = \frac{H}{kT_c} \exp\left(-\frac{E_c - E}{kT_c}\right),$$

equation (7) becomes

$$S = S_0(T)\left(\frac{b-1}{b+1}\right), \tag{8}$$

$$S_0(T) = -\frac{k}{e}\left(\frac{E_{c0}}{2\,kT} + 2 - \frac{a}{2\,k}\right), \tag{9}$$

$$b^{(1)} = \frac{\mu_{n,\mathrm{eff}}}{\mu_p} = \frac{\mu_n}{\mu_p}\left\{1 - \frac{H}{N_c}\exp\left[\frac{E_{c0} - a\,T}{2\,kT}\left(1 - \frac{1}{l}\right)\right]\right\} \tag{10}$$

(H integral trap concentration, N_c effective density of states in the conduction band, $l = T_c/T$, T_c trap distribution parameter). It is not justified to conclude from low values of μ_{eff} to the invalidity of the band model, as long as other measuring methods (e.g., direct drift measurements) on trap-free samples give mobilities which are compatible with the band model. If only discrete hole traps with a depth of E_H are present, (10) becomes

$$b^{(2)} = \frac{\mu_n}{\mu_p}\left[1 - \frac{N_H}{N_c}\exp\left(\frac{E_H}{kT}\right)\right]^{-1} \tag{11}$$

(N_H trap concentration, E_H energy distance of the trap level from the top of the valence band). When both kinds of traps are present, it follows

$$b^{(3)} = b^{(1)}\,b^{(2)}\,\frac{\mu_p}{\mu_n} \tag{12}$$

($\mu_{p,n}$ are the corresponding mobilities for the trap-free solid). If acceptor levels determine the conductivity, the differential thermo-e.m.f. is [20]:

$$S = \frac{k}{e}\left(2 + \frac{1}{2}\ln\frac{N_v}{N_A} + \frac{E_A}{2\,kT}\right) \tag{13}$$

(E_A activation energy of the acceptor level, N_v effective density of states in the valence band, N_A concentration of acceptor levels). Expression (13) is only weakly temperature-dependent, but the term N_v/N_A may cause a strong temperature dependence of $S(T)$ by adsorption or desorption of oxygen.

The equations (3) to (13) are limiting cases which are realized only under special conditions. They allow to explain positive and negative differential thermoelectric powers and cover the zero passage of $S(T)$. In (10) to (12) we have neglected the possible dependence of $\mu_{n,p}$ on temperature in form of a potential function because other terms show a much stronger exponential dependence on temperature.

These are the following temperature dependencies (values given after the second equality sign are calculated for β-CuPc): with (4) we get

$$\frac{dS}{d(1/T)} = \pm \frac{E_{c0}}{2e} = \pm 1 \text{ V},$$

$$S(500 \text{ °K}) = \pm 1700 \ \mu\text{V/deg},$$

with (3)

$$\frac{dS}{d(1/T)} = -\left(\frac{b-1}{b+1}\right)\frac{E_{c0}}{2e} + S_0(T)\frac{2\,b^{(0)}\,T_0}{[b^{(0)}(T_0/T)+1]^2},$$

$$\frac{dS}{d(1/T)}(500 \text{ °K}) = -1.3 \text{ V},$$

$$S(500 \text{ °K}) = +1500 \ \mu\text{V/deg},$$

$$(b^{(0)} \approx 0.1),$$

with (10)

$$\frac{dS}{d(1/T)} = -\left(\frac{b^{(1)}-1}{b^{(1)}+1}\right)\frac{E_{c0}}{2e} + S_0(T)\left(b^{(1)} - \frac{\mu_n}{\mu_p}\right)\left(\frac{E_{c0}}{2k} - \frac{a\,T}{2k}\,\frac{1}{l}\right)\frac{2}{(b^{(1)}+1)^2},$$

$$\frac{dS}{d(1/T)}(500 \text{ °K}) = -31 \text{ V},$$

$$S(500 \text{ °K}) = +1700 \ \mu\text{V/deg},$$

$$(\mu_n \approx \mu_p),$$

with (11)

$$\frac{dS}{d(1/T)} = -\left(\frac{b^{(2)}-1}{b^{(2)}+1}\right)\frac{E_{c0}}{2e} + S_0(T)\frac{2}{(b^{(2)}+1)^2}\,\frac{E_H}{k}\left(b^{(2)}\frac{\mu_p}{\mu_n} - 1\right),$$

$$\frac{dS}{d(1/T)}(500 \text{ °K}) = -1 \text{ V},$$

$$(N_H = 10^{14} \text{ cm}^{-3} \text{ [5, 6]},\ E_H = 0.85 \text{ eV}),$$

$$S(500 \text{ °K}) = -1700 \ \mu\text{V/deg},$$

and with (13)

$$\frac{dS}{d(1/T)} = \frac{E_A}{2e} = +0.8 \text{ V},$$

$$S(500 \text{ °K}) = +2200 \ \mu\text{V/deg}.$$

In Fig. 4 the theoretical curve is plotted with a dash-dotted line. It was found at use of (8) and has given the best approximation to the experimental values for $(b-1)/(b+1) = 0.58$. It is commonly known that for anorganic semiconductors the agreement between theory and experiment of thermoelectric measurements is unsatisfactory. We can qualitatively explain the observed differential thermopowers between +2200 and −1700 μV/deg and their temperature dependence.

Very high differential thermoelectric powers mostly observed at low temperatures could be caused by charging effects. Otherwise, they could give an explanation of the influence of defect levels (lattice defects, chlorine addition, oxygen). An extreme value would be reached if each CuPc molecule would carry a defect level. Then you find for chlorine addition using equation (13) ($E_D \triangleq E_A$) with $E_D = 0.73$ eV and $N_D = 1.7 \times 10^{21}$ cm^{-3}: $S = -700$ µV/deg. If we equate the concentration of charge carriers generated from the valence band with that generated from donor levels, we find the donor concentration $N_{D\,min}$, which represents a donor contribution of 50% to the conductivity. Then it becomes, taking into account (13),

$$S(500\ °K) = -2380\ \mu V/deg$$

($N_{D\,min} = 10^5$ cm^{-3}). By this it can be shown that the static experimental value can be explained, but no statement is possible about higher absolute values like $S = -40$ mV/deg with the model used here.

The anisotropy of the thermo-e.m.f. is low and its value is overlapped by the described influences of surface defects. A statement is possible if from the same single crystal some samples of different orientations can be made and these samples have nearly the same surface conditions. Our investigations at 200 to 300 °C resulted in

$$S^{(a)} : S^{(b)} = 1 : 1.6 \text{ to } 2.3$$

(a and b are crystal axes). Stronger anisotropies can be explained by orientation-dependent mobilities. It was shown theoretically by Kubarev and Ponomarev [25] that a stronger orientation dependence can take place. Our results show that the band width of β-CuPc calculated by Kubarev and Ponomarev is not exactly, because then there must be a positive thermo-e.m.f. along the crystallographic b-axis.

6. Some Remarks about the Model Used

The results of measurements of the thermo-e.m.f. on CuPc single crystals can be explained by the band model. The samples can be considered to be nearly field-free. Under such conditions carrier injection cannot take place. That means, the sample is free of stored charge. If one takes into account an exponential trap distribution of electron traps on surfaces and of discrete traps in the bulk of CuPc single crystals, it is possible to use some relations for the thermo-e.m.f. developed on the basis of the band model. These considerations were supported by electric, photoelectric, and dielectric measurements on the same samples, which are compatible with this model. Theoretical calculations of the band structure of CuPc showed a good agreement with the experimental results [1].

Acknowledgements

The author should like to express his gratitude to Dr. M. Starke, Mr. J. Heim, and Mrs. C. Tantzscher.

Reference

[1] C. Hamann, Thesis B, TH Karl-Marx-Stadt, 1970.
[2] C. Hamann and M. Starke, phys. stat. sol. **4**, 509 (1964).
[3] C. Hamann, phys. stat. sol. **12**, 483 (1965).
[4] C. Hamann and H. Schmidt, Plaste und Kautschuk **16**, 85 (1969).
[5] C. Hamann, Thesis A, TU Dresden, 1966.
[6] C. Hamann, phys. stat. sol. **20**, 481 (1967).

[7] M. Starke, Thesis A, TU Dresden, 1968.
[8] M. Starke and C. Hamann, Z. phys. Chem. **243**, 166 (1970).
[9] C. Hamann and I. Storbeck, Naturwissenschaften **50**, 327 (1963).
[10] H. Schmidt and C. Hamann, Z. Elektrochem., Ber. Bunsenges. phys. Chem. **69**, 391 (1965).
[11] M. Starke and C. Hamann, Z. anorg. allg. Chem. **354**, 1 (1967).
[12] P. E. Fielding and F. Gutmann, J. chem. Phys. **26**, 411 (1957).
[13] W. A. Kargin, Organische Halbleiter, Izd. Nauka, Moscow 1968.
[14] A. V. Airapetyants and L. D. Rozenshtein, Organicheskie poluprovodniki (Ser. Poluprovodniki), Vol. IV, Leningrad 1963.
[15] G. Delacote and M. Schott, phys. stat. sol. **2**, 1460 (1962).
[16] S. D. Levina, I. I. Astakhov, I. P. Lobanova, and Z. A. Rotenberg, Elektrokhimiya **2**, 1343 (1966).
[17] I. Vaisnys and R. Kirk, Phys. Rev. **141**, 641 (1966).
[18] C. Hamann, Kristall und Technik, **6**, 491 (1971).
[19] C. Hamann, Z. Chem. **9**, 195 (1969).
[20] A. I. Anselm, Einführung in die Halbleitertheorie, Akademie-Verlag, Berlin 1964.
[21] F. Fillard, Thesis, Université de Montpellier, 1967.
[22] J. P. Fillard and M. Schott, C. R. hebd. Séances Acad. Sci. **262**, 1287 (1966).
[23] C. Hamann, Wissensch. Z. TH Karl-Marx-Stadt **10**, 265 (1968).
[24] C. Hamann, phys. stat. sol. **26**, 311 (1968).
[25] S. I. Kubarev and O. A. Ponomarev, Teor. eksper. Khim. **4**, 498 (1968).

(Received February 21, 1971)

phys. stat. sol. (a) 10, 519 (1972)

Subject classification: 10.1 and 11; 21.3

C.N.E.A. Metallurgy Department, Buenos Aires (a), and Materials Science Division, Argonne National Laboratory, Argonne, Illinois (b)

Mechanical Properties of Low Temperature Irradiated Magnesium

By

C. POMAR (a), T. H. BLEWITT (b), and E. A. BISOGNI (a)

Magnesium crystals were irradiated at 77 °K in the RA-1 reactor to neutron doses up to 4×10^{18} ncm^{-2}, and the deformation properties were studied at this temperature without warm-up. The yield stress increased with dose to a value of (465 ± 45) p/mm^2 at 1×10^{17} ncm^{-2}. For neutron doses greater than this value, it was found that the deformation mechanism changed with an upper yield point being formed with an accompanying yield drop. The upper yield point and yield drop increased with dose and became very substantial at 4×10^{18} ncm^{-2} with the resolved yield stress increasing to 1000 p/mm^2 and the yield drop to the order of 50%. The lower yield stress was independent of dose in the higher dose region. Annealing experiments showed an annealing peak centered at 240 °K. For large doses, recovery first occurred in the upper yield stress with the lower yield stress being unaffected until complete recovery of the upper yield stress occurred. Strain markings were studied and found inhomogeneous in character but different from a typical Lüders band.

Magnesiumeinkristalle wurden bei 77 °K im RA-1-Reaktor mit Neutronendosen bis zu 4×10^{18} ncm^{-2} bestrahlt und die Deformationseigenschaften bei dieser Temperatur ohne Aufwärmung untersucht. Die Streckgrenze steigt mit der Dosis auf einen Wert von (465 ± 45) p/mm^2 bei 1×10^{17} ncm^{-2} an. Für Neutronendosen, die größer als dieser Wert sind, wurde gefunden, daß sich der Deformationsmechanismus mit einer oberen Fließgrenze, die mit einem begleitenden Flußabfall gebildet wird, ändert. Die obere Fließgrenze und der Flußabfall steigen mit der Dosis an und werden bei 4×10^{18} ncm^{-2} sehr wesentlich, wobei die aufgelöste Streckgrenze auf 1000 p/mm^2 ansteigt und der Flußabfall auf die Größenordnung 50%. Die untere Streckgrenze war unabhängig von der Dosis im Bereich der höheren Dosen. Temperungsexperimente zeigten ein Temperungsmaximum bei 240 °K. Für große Dosen tritt Erholung zuerst in der oberen Streckgrenze auf, wobei die untere Fließgrenze unbeeinflußt bleibt bis vollständige Erholung der oberen Streckgrenze eingetreten ist. Deformationsmarkierungen wurden untersucht und es wurde gefunden, daß sie inhomogen im Charakter, jedoch von typischen Lüdersbanden verschieden waren.

1. Introduction

The effects of neutron irradiation on the mechanical properties of the face-centered cubic metals have received widespread attention in the past twenty years [1, 2]. The effects are quite pronounced in these metals with the yield stress increasing by a decade or more for moderate irradiations. By comparison the hexagonal metals have been neglected, as the only work reported has been that of Kunz and Holden [3]. They studied the effect of room temperature irradiations on the stress/strain curve of zinc. This work, which was largely of the survey type, demonstrated a substantial increase in the yield stress of the irradiated zinc.

The present work was undertaken to attempt to fill at least in part this information gap in the field of irradiation effects. Magnesium was chosen for this study because it has low level radioactivity after reactor bombardment, it can be obtained in high purity, and it is ductile at 77 °K.

2. Experimental Procedure

Single magnesium crystals were grown from Johnson Mathey 99.99% magnesium, using the Bridgman technique, in the shape of square tensile specimens of 9 mm² cross-section and 3 cm long with spherical gripping heads 6.35 mm diameter in split molds made from reactor grade graphite. Their crystallographic orientation was determined by the back reflection Laue-technique. Although there were a few exceptions, most samples were selected with orientations given by $X = 58°$, $\lambda = 33°$, where X is the angle between the basal plane (0001) and the specimens axis, and λ is the angle between the $\langle 1120 \rangle$ direction (slip direction) and the crystal axis. Their yield stress was (40 ± 5) p/mm².

The irradiation was performed in the cryogenic facility located in the core of the RA-1 reactor of C.N.E.A. In this cryostat, which was designed and built at Grenoble [4], the samples are bombarded in a bath of liquid nitrogen. Although a complete description of this device is available in the literature [4], it is worth noting that refrigeration is provided by liquid nitrogen utilizing a reflux cycle. The samples were loaded into the cryostat in an aluminum can that was completely immersed in liquid nitrogen in order to permit withdrawal of the samples without warmup. The liquid nitrogen in this can was capable of keeping the samples cold for a period of time, which greatly exceeded the removal time.

The neutron flux in this location was determined by the yield stress of copper single crystals. The yield stress, measured as a function of the effective bombardment time, was then compared with yield stress dose curves previously published

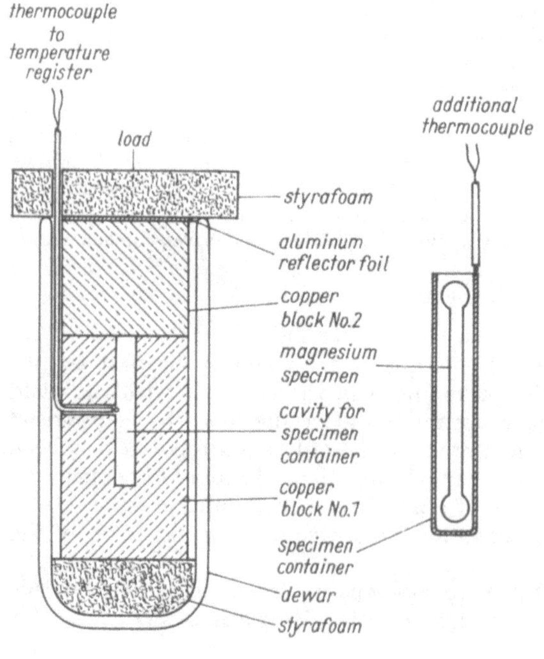

Fig. 1. Schematic diagram of the annealing cryostat. Both copper blocks were brought to the desired temperature. The sample, which is the specimen container and is stored in liquid nitrogen, is placed in the cavity of copper block No. 1. The heat capacity of the copper is so great compared to the sample that the temperature of the system was virtually unaffected by the insertion of the samples. A good square pulse can be obtained from this cryostat

to determine the flux. The dose values reported in this paper are then directly comparable to those reported in the Oak Ridge graphite reactor [1]. This technique yielded a flux of 3.3×10^{12} reactor neutrons/cm^2 s. Since it is estimated that only about 1×10^{12} n/cm^{2}s of these are above 0.5 MeV, the dose values reported here are about a factor of three greater than the fission neutron dose.

After removal from the cryostat, the crystals were stored in liquid nitrogen for a week or so to allow the activity to decay. They were then mounted in an Instron tensile machine (Model TTM) and deformed with a crosshead speed of 1×10^{-3} cm/min. The mounting was performed with the sample immersed in liquid nitrogen to prevent warmup. On some samples, deformation markings were examined by photomicrography as the deformation proceeded. The samples were removed from the tensile machine and photographed while immersed in liquid nitrogen. Good images could be obtained through the liquid nitrogen if the liquid layer were less than 3 mm thick.

Recovery measurements were performed in a very simple manner. A high mass copper block with a sample cavity in it was thermally isolated at the desired annealing temperature (Fig. 1). The sample was removed from the Instron machine, immersed in liquid nitrogen, and inserted in the cavity for the desired annealing time. Since the heat capacity was very high in comparison to that of the specimen, insertion of the specimen did not appreciably change the block temperature. A typical annealing curve is shown in Fig. 2. Following the annealing pulse, usually of ten minutes duration, the sample was removed from the cavity and quenched in liquid nitrogen. It was remounted in the Instron machine while immersed in liquid nitrogen, and the effect of the pulse on the yield stress was determined at the reference temperature of 77 °K. Since the same sample was used throughout the anneal, the plastic strain required to determine the yield stress was kept as small as possible.

3. Results

Load elongation curves obtained on crystals made to various doses at 77 °K and tested at the same temperature without intermediate warmup are shown in Fig. 3. Clearly from these data a fundamental change in the yielding phenomenon occurs after a critical dose is reached.[1]) For low doses a "normal" yielding phenomenon occurs, whereas for high doses the yield stress is multivalued with an "upper" and "lower" yield stress. The critical neutron dose for this transition is of the order of 1×10^{17} reactor neutrons/cm^2.

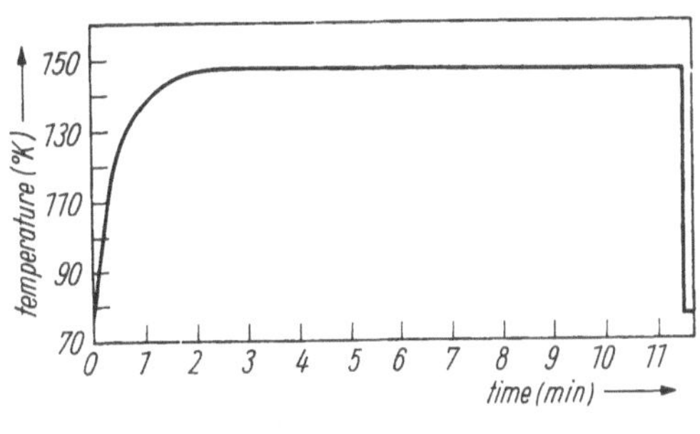

Fig. 2. Typical annealing pulse, $T = 148$ °K

[1]) The authors were guided by the results of our colleague, A. B. Gonzales, who made in situ measurements in the RA-1 reactor of the yield stress of magnesium as a function of dose. These results strongly suggested saturation effects and will be published soon.

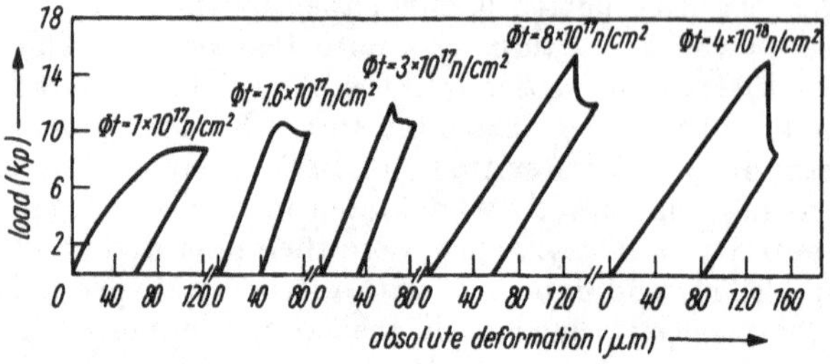

Fig. 3. Load–elongating curves at 77 °K for magnesium crystals bombarded to various doses. The samples have different orientations; it is obvious that the load drops increased as a function of dose

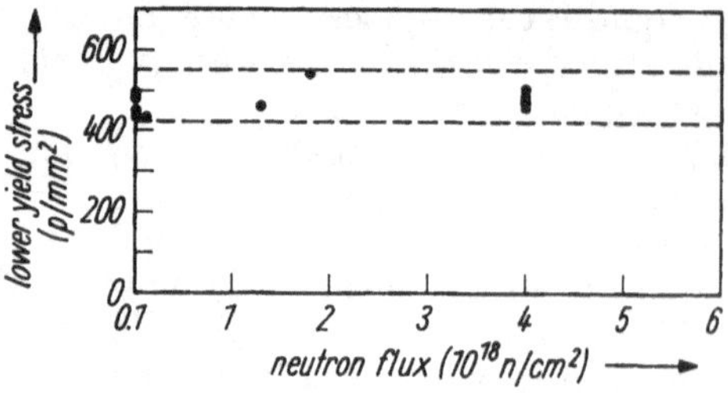

Fig. 4. The lower yield stress, measured on magnesium crystals bombarded and tested at 77 °K as a function of dose, shows that within experimental error, the lower yield stress is independent of dose

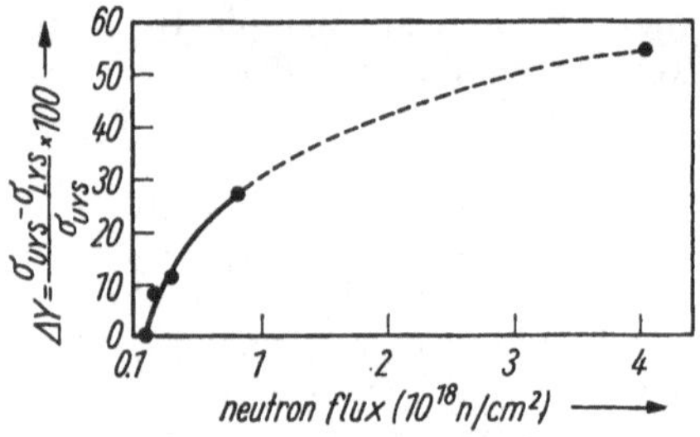

Fig. 5. The yield drop plotted as a function of neutron dose for magnesium crystals irradiated and tested at 77 °K

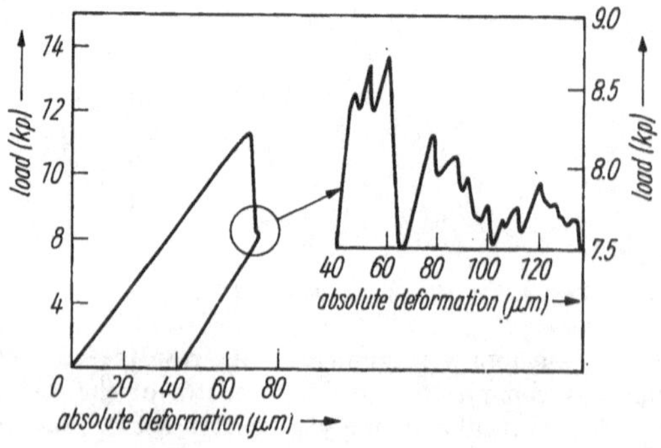

Fig. 6. The load drop for a sample bombarded to 4×10^{18} reactor neutrons/cm². The load–elongation curve for the lower yield stress was obtained on an expanded scale and is shown in the inset. The typical jerky flow makes it difficult to accurately define the lower yield stress

Consider the behavior of the crystals in the high-dose region, i.e. for doses greater than 1×10^{17} reactor neutrons/cm². As seen in Fig. 4, the lower yield stress is independent of the neutron dose for doses ranging from 1×10^{17} to 4×10^{18} reactor neutrons/cm² with a value of (485 ± 45) p/mm² compared to a value of (40 ± 5) p/mm² for the annealed crystal. However, the upper yield stress, and consequently the yield drop, increases with neutron dose. This is shown in Fig. 5. The yield drop, ΔY, is defined as

$$\Delta Y = \left(\frac{\sigma_{\mathrm{uys}} - \sigma_{\mathrm{lys}}}{\sigma_{\mathrm{uys}}}\right) \times 100,$$

where σ_{uys} is the upper yield stress and σ_{lys} is the lower yield stress. Details of the "yielding" are shown in Fig. 6. Here a sample bombarded to 4×10^{18} reactor neutrons/cm² is strained to the yield point, and the test stopped. The load cell of the tensile machine is then set on an expanded scale and the crystals tested again. Jerky flow follows, implying inhomogeneous straining and making lower yield stress difficult to define.

Annealing experiments seem to verify that the yield drop is a direct consequence of the defect concentration. That is, the removal of the defects by thermal migration reduces the yield drop without affecting the lower yield stress until the yield drop is completely recovered. The main annealing peak (for yield point recovery) occurs in the region above 200 °K with small, if any, recovery in the region from 77 to 200 °K. The annealing experiments were conducted by the study of the recovery of the yield drop and the recovery of the lower yield stress.

The yield drop experiments were rather limited in scope, as only two specimens were tested. Both were bombarded to 4×10^{18} reactor neutrons/cm² and tested at 77 °K. The first crystal deformed after a pulse anneal for 10 min at 170 °K with the yield stress, both upper and lower, being unaffected with experimental error (±10%).

The second crystal was deformed after pulse annealing to 273 °K for 30 min. In this case, practically complete recovery of the yield drop occurred without affecting the lower yield stress. The load elongation curve for this sample is shown in Fig. 7.

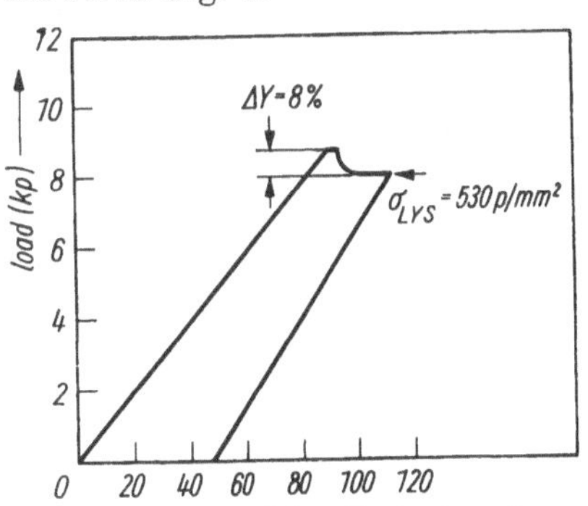

Fig. 7. The load-elongation curve for a sample bombarded to 4×10^{18} reactor neutrons/cm² at 77 °K after being annealed for 30 min at 273 °K

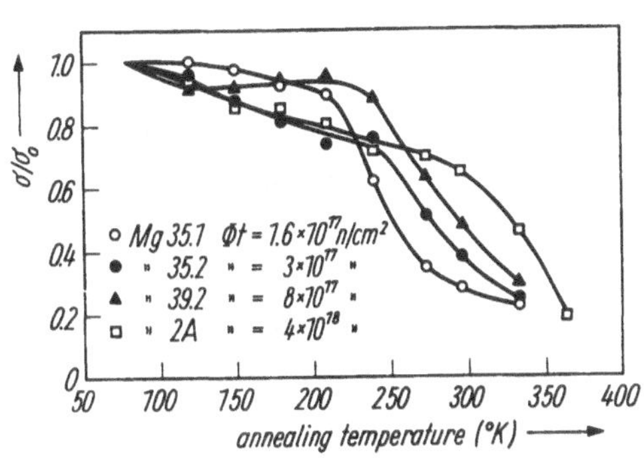

Fig. 8. Isochronal annealing curves of the lower yield stress in 77 °K bombarded magnesium. The samples were strained by about 0.2% at 77 °K prior to the recovery to remove the upper yield stress. Pulse annealing techniques were used. The temperature pulses failed to reactivate the upper yield stress

The effects of annealing on the lower yield stress were determined in a more detailed manner. Prior to annealing, all samples were prestressed to remove the upper yield stress with the aim of obtaining a reproducible lower yield stress. Pulse annealing techniques were used with 10 min pulses. Fig. 8 shows the yield stress at 77 °K as a function of the pulse temperature.

If the low-temperature annealing is ignored because of the lack of definition of the lower yield stress, apparently, a significant recovery of the yield stress occurs in the 200 to 350 °K temperature ranges. Furthermore, this annealing peak shifts to a higher temperature with increasing dose. The temperature shift of the annealing peak is in the wrong direction to be explained by the usual annealing kinetics, as either the peak should be unaffected by defect concentration (first order kinetics) or the peak should decrease with increasing concentration (second or higher order kinetics). This temperature shift can be explained by the concentration dependence of the lower yield stress and does, in fact, substantiate the annealing behavior of the upper yield stress.

In order to explain this shift, consider the lower yield stress to be unaffected by a defect concentration greater than a value C_c, which corresponds to that of a neutron dose of 1.5×10^{17} reactor neutrons/cm². Assuming that the defect concentration C is proportional to the neutron dose, a reasonable assumption for the doses considered, then a dose of 4×10^{18} reactor neutrons/cm² will have a defect concentration of about 25 C_c. Consequently, the effects of annealing will not be observed until the defect concentration has been reduced to 1/25th of its initial concentration, pushing the recovery of the yield stress into the high-temperature tail of the annealing peak, which accounts for a temperature shift. With these assumptions, it was possible to synthesize the isochronal annealing curve from the temperature shift as a function of dose (See Fig. 9.).

An attempt to determine the activation energy was made. The crosscut method was used with isothermal annealing curves of the lower yield stress being measured. Samples were bombarded to 1×10^{17} reactor neutrons/cm² so that the lower yield stress reflected the defect concentration. The isothermal

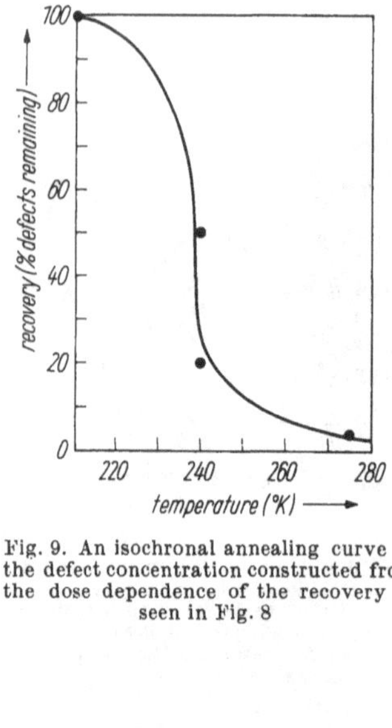

Fig. 9. An isochronal annealing curve of the defect concentration constructed from the dose dependence of the recovery as seen in Fig. 8

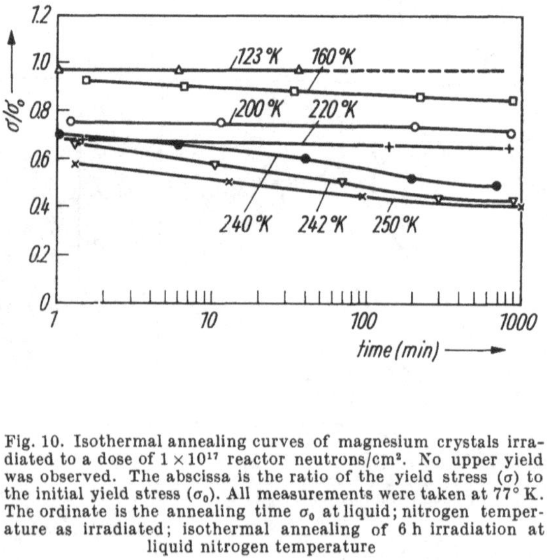

Fig. 10. Isothermal annealing curves of magnesium crystals irradiated to a dose of 1×10^{17} reactor neutrons/cm². No upper yield was observed. The abscissa is the ratio of the yield stress (σ) to the initial yield stress (σ_0). All measurements were taken at 77° K. The ordinate is the annealing time σ_0 at liquid; nitrogen temperature as irradiated; isothermal annealing of 6 h irradiation at liquid nitrogen temperature

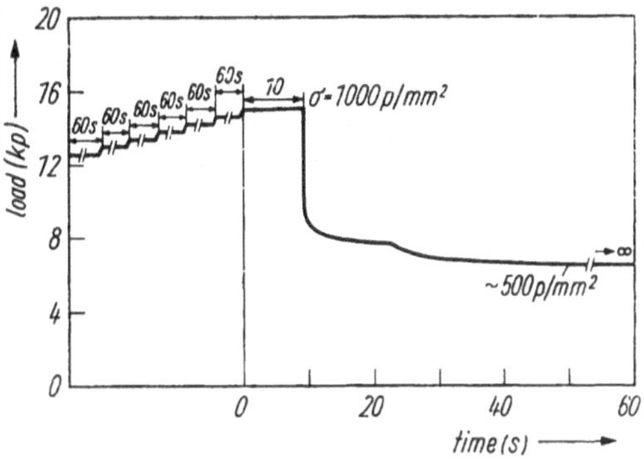

Fig. 11. Delayed yielding in magnesium crystals irradiated to 4×10^{18} reactor neutrons/cm² at 77 °K. The sample was incrementally loaded with delays of 60 s between load increments until yielding occurred. Two distinct events occurred — each with a delay time of about 10 s. Note that the lower yield stress is the same order of magnitude in the stated test as in the dynamic case. The entire test was carried out at 77 °K

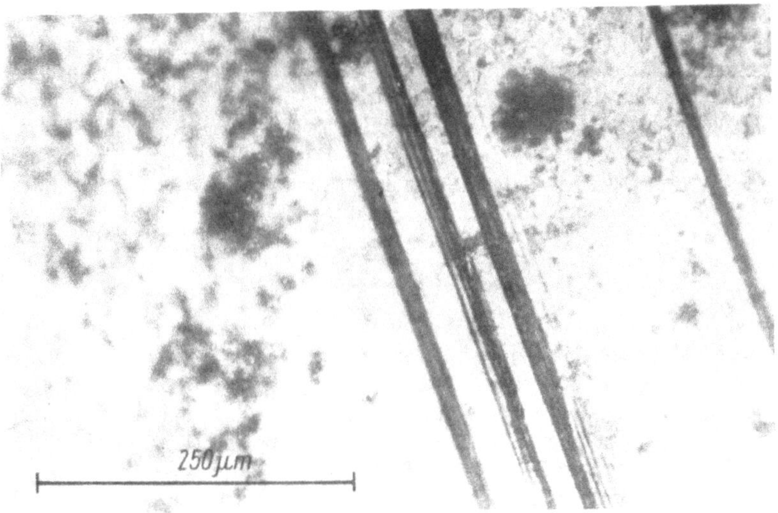

Fig. 12. The strain markings resulting from the yield drop shown in Fig. 11

Fig. 13. Slip lines in magnesium irradiated to 4×10^{18} reactor neutrons/cm² at 77°K as a function of the stran. The distance between successive markings is to scale. Deformation and photomicroscopy were performed at 77°iK a) $\varepsilon = 0.13\%$, $\Delta l = 40$ µm, b) $\varepsilon = 0.36\%$, $\Delta l = 108$ µm, c) $\varepsilon = 0.77\%$, $\Delta l = 231$ µm, d) $\varepsilon = 1.0\%$, $\Delta l = 300$ µm, e) $\varepsilon = 1.6\%$, $\Delta l = 580$ µm, f) $\varepsilon = 2.0\%$, $\Delta l =$ C00 µm

annealing curves are shown in Fig. 10. The shape of these curves makes it difficult to determine the activation energy, as it is hard to obtain overlap of the curves on a cut. It must be concluded that the activation is very high or the process is not uniquely activated. The latter conclusion is probably valid.

Some rough experiments were performed on the effects of crosshead speed on the yield stress. It was found that yield stress increased with the crosshead speed. This suggested that there might be a delay time in the yield point, and consequently, a search was made for it. The sample was loaded in increments with a waiting period of 60 s between each incremental increase in load. A delay time of the order of 10 s was observed. This experiment is illustrated in Fig. 11. Of particular interest was the second burst of slip lines that occurred some 10 s after the first with a stress of about one-half that of the yield stress.

Strain markings on the crystal were also studied. The photomicrographs of Fig. 12 were obtained after the time delay experiments of Fig. 11. It can be seen that four packets of slip lines were released during the two bursts, and most of the dislocations occurred during the first burst. Fig. 13 shows the progressive formation of the slip lines as a function of strain on a crystal bombarded to 4×10^{18} reactor neutrons/cm^2. Notice that the strain is inhomogeneous, yet it does not behave like a typical Lüders band, as new slip line packets form at a distance of some 500 μm with satellite lines forming on the old packets well behind the new slip lines.

4. Discussion

The effects of irradiation on the mechanical properties of magnesium are obviously greatly different from those in face-centered cubic metals. Development of large yield drop induced by irradiation sets magnesium apart from aluminum, gold, silver, and copper. This phenomenon has most of the important characteristics of yield phenomena observed in body-centered cubic [5, 6] and hexagonal [7, 8] metals resulting from interstitial impurities, i.e. large yield drops (for larger neutron doses of the order of 50% of yield stress); a significant delay time; and, inhomogeneous deformation. However, there are other aspects of the yield of irradiated magnesium that differ from the interstitial doped body-centered cubic metals. For example, in magnesium, the dislocation locking occurs without aging, and after deformation its behavior is atypical as aging did not restore the drop. Furthermore, deformation, while inhomogeneous, did not have the usual Lüders band characteristics. Nevertheless, it can probably be assumed that the basic yielding mechanism in irradiated magnesium is the same as that in interstitial impurity doped hexagonal and body-centered cubic metals.

Then if this is the case, the yield drop in irradiated magnesium would be expected to be the result of dislocation locking by an atmosphere of irradiation-inducted defects. Since vacant lattice sites and interstitial atoms are created in equal numbers during this bombardment, the atmosphere could be caused by either of these defects. It can probably be concluded that the atmosphere is composed of interstitial atoms. The argument is as follows. Since the upper yield point (and therefore the locking) occurred at 77 °K without aging, the atmosphere must arrive at the dislocation either as a result of thermal migration of the defects or as the result of displacement cascades intersecting the dislocation lines. Since all the dislocations must be locked, and the probability of this occurring is small for the latter case at the doses considered, it seems reasonably certain that the defects responsible for this locking must be mobile at 77 °K.

Examination of isochronal annealing curves for resistivity changes in neutron-irradiated magnesium [9, 10] showed annealing stages at 50, 75, and 140 °K. Of these three, the highest temperature peak was the largest and showed second or higher degree order, as it shifted to lower temperatures with increasing dose.

Apparently, either or both the 50 and 75 °K peaks are the result of long-range defect migration, because measurements of the elastic modulus show that the dislocation mobility decreased as the result of neutron bombardment at 77 °K [9]. On the basis of this experiment, it is feasible to assume that defects are moving to dislocations at 77 °K or below. Consideration of the migration energy of the defects and the number of jumps involved would indicate that migration defects are interstitials.

It is difficult to understand why a further increase in yield drop did not occur when the specimens were heated through the 140 °K peak. Measurements indicate that the modulus increased as the annealing pulse went through the 140 °K peak, whereas the load drop was either unaffected or decreased. Furthermore, recovery of the modulus occurred between 500 and 600 °K, whereas the load drop for equivalent doses occurred at least 100 °K lower in temperature. It must be concluded that there is little correlation between dislocation mobilities determining modulus with those determined by yield drops.

The increase in the yield stress in low dose regions and the constancy of the lower yield point in high dose regions is difficult to understand in light of yield point studies. The lower yield stress is frequently interpreted as stress required to move the unlocked dislocation through the lattice [11]. The results of Fig. 10 would seem to support this viewpoint for it can be seen that in the first yielding process there is a tail in the load-elongation curve where dislocations continue to move at the lower yield stress. If this is taken seriously, then it can be deduced that for low concentrations ($< 1 \times 10^{17}$ reactor neutrons/cm^2) where the yield drop is zero, the defects impede the motion of an unlocked dislocation, as the stress is raised from 40 ± 5 p/mm^2 to as much as 465 p/mm^2 at 1×10^{17} reactor neutrons/cm^2, whereas for doses above this value, defects lock the dislocation, as the upper yield increases. Apparently they do not further impede the motion of unlocked dislocations, as the lower yield stress remains at 465 p/mm^2. Therefore, it is difficult to avoid the conclusion that two different defect-dislocation interactions occur, one of which dominates at low defect concentrations, and the other at high defect concentrations.

Alternately, it can be assumed that the lower yield stress is a consequence of a stress magnification associated with the initial packet of the slip lines [12]. That is, the upper yield stress is required to move the locked dislocations, and the slip bands formed by this event act as a stress riser, enabling locked dislocating to subsequently move at a lower applied stress level. Since the back stress of dislocation pile-ups locks the dislocations generated in the slip lines, the only dislocations that are unlocked in any given instance are thus in motion. There is some evidence to support this model. For example, it was found in the pulse annealing experiments that the upper yield stress could not be restored by aging, implying that there were no unlocked dislocations available. Furthermore, the shift in annealing temperature of the lower yield stress with the dose offers collaborating evidence, as it suggests that defects were acting as locking atmosphere rather than as an impediment of their motion. While the uniform spacing of the new packets shown in Fig. 12 would tend to support a uniform lower yield stress, the large spacing between new slip packets makes it difficult to account for the magnitude of the stress magnification.[2])

All of this means that the sharp yielding phenomenon and yield drop are really not too well understood. It is hoped that irradiation effects in magnesium might ultimately prove to be a useful tool to study this phenomenon.

Acknowledgements

The authors would like to acknowledge the assistance of Mr. A. B. Gonzales and Mr. Rosenbaum in making the low temperature irradiations, and of Mr. J.E. Talia in growing the crystals. The cooperation of the Service de Basses Temperature of Centre Energie Nucleaire, Grenoble, France, in providing the low temperature facility through the International Atomic Energy Agency is gratefully acknowledged. Thanks are also due to Dr. J. Sabato for his interest and support of this work. The work was supported jointly by the U.S. Atomic Energy Commission and the Comision Nacional Energia Atomica, Argentina.

References

[1] T. H. Blewitt, R. R. Coltman, R. E. Jamison, and J. K. Redman, J. nuclear Mater. 2 (4), 277 (1960).

[2] J. Diehl, Vacancies and Interstitials in Metals, North-Holland Press, Amsterdam 1969.

[3] F. W. Kunz and A. N. Holden, Acta metall. 2, 816 (1954).

[4] L. Bochirol, J. Doulat, and L. Weil, Advances in Cryogenic Engineering 6, 130 (1961).

[5] H. Schwartzbort and J. R. Low, J. Metals 1, 637 (1949).

[6] A. T. Churchman and A. H. Cottrell, Nature (London) 167, 943 (1950).

[7] A. H. Cottrell and D. F. Gibbons, Nature (London) 160, 466 (1947).

[8] H. L. Wain and A. H. Cottrell, Proc. Phys. Soc. B63, 339 (1950).

[9] J. Delpalace, J. Hillairet, J. C. Nicoud, D. Schumacher, and G. Vogl, phys. stat. sol. 30, 119 (1968).

[10] T. H. Blewitt, R. R. Coltman, C. E. Klabune, D. K. Holmes, and J. K. Redman, ORNL-Report No. 2614, 1958.

[11] A. H. Cottrell, Dislocations and Plastic Flow in Crystals, Oxford University Press, London 1956.

[12] F. R. N. Nabarro, Theory of Crystal Dislocations, Oxford University Press, London 1967.

[13] P. Haasen and G. Leibfried, Nachr. Akad. Wiss. Göttingen, math.-phys. Kl. IIa, No. 2, 31 (1952).

(Received November 8, 1971)

[2]) Haasen and Leibfried [13] pointed out that a maximum stress magnification would occur at a distance from the slip line, which is about the same as the diameter of the dislocation pile-up. Something of this nature could account for these results. However, the separation between slip packets should then decrease with dose. Limited evidence suggests that this condition is not satisfied.

phys. stat. sol. (a) **10**, 529 (1972)

Subject classification: 1; 3; 22.3

Chemical Physics Division, Department of Physics, Eindhoven University of Technology

Structures and Compounds in the System $Ga_{1-x}Se_x$

By

J. C. J. M. TERHELL and R. M. A. LIETH

Structural investigations were made on the system $Ga_{1-x}Se_x$ for $0.48 < x < 0.60$ with powder and single crystal X-ray technique. It is checked, which possible structure types, known in the literature, could exist in the different regions of the T–x diagram of Ga–Se. Further is reported that a new compound Ga_3Se_2 was found.

Strukturuntersuchungen wurden am System $Ga_{1-x}Se_x$ mit $0{,}48 < x < 0{,}60$ mit Hilfe von Röntgendiffraktion an polykristallinem Material und Einkristallen durchgeführt. Geprüft wurde, welche möglichen Strukturtypen, die in der Literatur mitgeteilt werden, in den verschiedenen Bereichen des T–x Diagramms von Ga–Se existieren können. Weiter wird erwähnt, daß eine neue Verbindung Ga_3Se_2 gefunden wurde.

1. Introduction

In literature GaSe is reported to exist in three modifications namely the hexagonal β- and ε-structures and the rhombohedral γ-modification. The polytypes are built up from slabs containing Se and Ga in the sequence Se–Ga–Ga–Se [1, 2]. In the case of the ε-structure, one slab has a horizontal translation in the [11.0] direction with respect to the adjacent slab; the two layer repetition is of the type AββA, BγγB. The γ-structure has a three layer repetition of the type AββA, BγγB, CααC (see Fig. 1).

The third type of stacking, the so-called β-modification would be equal to the stacking found in gallium sulphide and would be the most stable of the three types, due to the fact that cations lie above anions in adjacent layers so that in this case a high stacking fault energy would be expected [3]. On the other hand this stacking fault energy is small in the case of the ε- and γ-modifications, because in these modifications not all the cations lie above anions in neighbouring slabs.

Evidence for the large stacking fault energy is found by electron microscopic investigations of GaS [3]. In literature there is a discrepancy about the predominance of the β-structure on the one hand and the ε-structure on the other hand. From the work of Mooser [4] it is evident that β is rarely found in GaSe single crystals, Jellinek and Hahn [2] on the other hand emphasized the predominance of the β-type.

In the course of our X-ray work on numerous GaSe single crystals we have found the ε-structure to be dominant, we have not been able to detect the β-modification in our GaSe single crystals, with the aid of rotation techniques.

In this paper we present our structure investigations on powders and single crystals carried out on the system $Ga_{1-x}Se_x$ with x having values between 0.48 and 0.60. It shows that it is impossible to explain the existence of the so-called β-structure for GaSe with the aid of powder pictures.

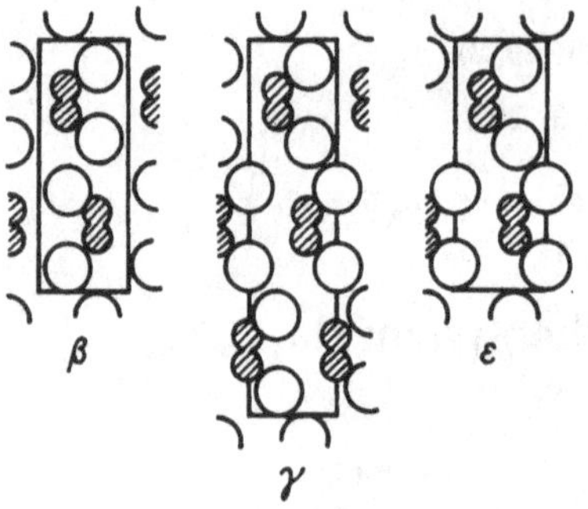

Fig. 1. Projection of the GaSe structure on the (11$\bar{2}$0) face [2]: β-type of stacking, a two layer repetition AββA, BααB, γ-type of stacking, a three layer repetition AββA, BγγB, CααC, ε-type of stacking, a two layer repetition AββA, BγγB

2. Preparation of the Samples

To obtain material having an exclusively hexagonal and an exclusively rhombohedral structure, the preparation and recrystallisation procedure of Schubert et al. [1] was closely followed. According to these authors non-stoichiometric material with a selenium content $50 < x < 52.5$ at% gives the γ-modification, while material with $48 < x \leq 50$ at% gives the ε-modification.

The presence of Ga_2Se_3 in Se-rich GaSe can be expected when one studies the phase diagram of the system Ga–Se as reported by Rustamov et al. [5] (Fig. 2). One should also expect from this diagram that the preparation and recrystallisation of Se-poor GaSe would give a mixture of Ga_2Se and GaSe. The diffractogram of GaSe ($50 < x < 52.5$) powdered and recrystallised for 24 h at 860 °C is presented in Fig. 3b. It shows clearly the sharp reflection of the γ-modification together with those of cubic Ga_2Se_3.[1]) The diffractogram of Se-poor GaSe, Fig. 3a, shows only reflections belonging to a hexagonal structure, which is in good agreement with the work of Schubert et al. No extra reflections indicating the presence of Ga_2Se have been found which

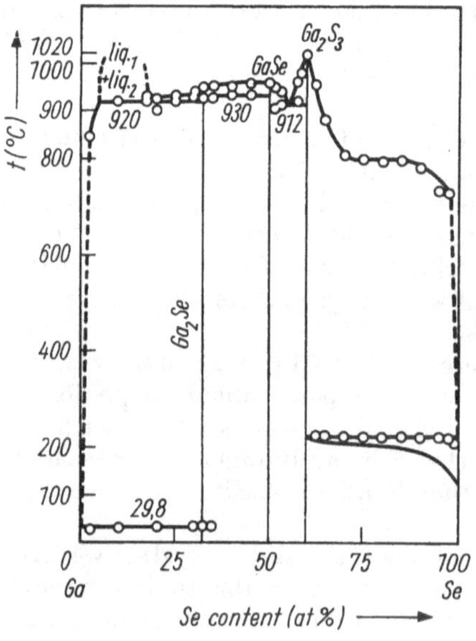

Fig. 2. The temperature-composition diagram of the system Ga–Se [5]. Newer data give the melting points [6]. Ga_2Se_3 937 ± 5 °C for GaSe and 1003 ± 5 °C for Ga_2Se_3. This is in excellent agreement with the work of Dieleman and Engelfriet [7]

[1]) Schubert et al. don't mention the cubic Ga_2Se_3 below 53 at% Se probably because the Ga_2Se_3 lines cannot be distinguished from the γ-lines in powder pictures.

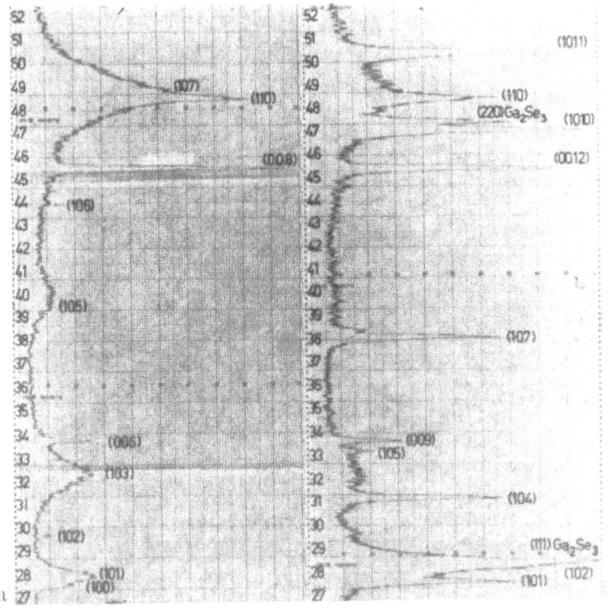

a b

Fig. 3. Diffractograms of recrystallised GaSe powders (24 h, 860 °C) which give the characteristic differences between an ordered an a disordered structure type. a) Se-poor GaSe, reflections of the hexagonal ε-modification are broadened, an indication that this stacking contains many faults. b) Se-rich GaSe, reflections of the rhombohedral γ-modification and of the cubic Ga_2Se_3 phase are sharp and close to each other

is not surprising as Klemm and von Vogel [8] already emphasized the fact that Ga_2Se cannot be obtained by direct synthesis from the elements. According to their paper the compound can only be obtained by sublimation, which means that it is formed by dissociation of the vapour. As a further result of the recrystallisation experiments with Se-poor GaSe in closed ampoules at 860 °C a few needles and ribbons were obtained. They had grown in a slightly cooler tip of the ampoule.

Among those sublimation products we could not detect Ga_2Se some of the ribbons however had an orthorhombic structure [9]. Chemical analysis of these ribbonlike crystals proved this compound to have the formula Ga_3Se_2. Just like in the case of Ga_2Se, we believe that Ga_3Se_2 cannot be obtained by direct synthesis, so that it cannot be found during thermal analysis of mixtures having the composition Ga_3Se_2. This also explains why no evidence is found for the existence of Ga_3Se_2 in the diffractogram of Se-poor GaSe, Fig. 3.

3. Disorder in GaSe-Polytypes

The characteristic differences between the sharp reflections belonging to the γ-structure and the broadened reflections of the ε-structure have already been mentioned (Fig. 3a and b). The broadening is the result of a faulty stacking and was observed by Schubert et al. [1]. A similar effect is seen when single crystals grown by way of the chemical transport process are powdered. It is however not only the powdering procedure which introduces the stacking faults; the Weissenberg picture of a GaSe single crystal plate demonstrates this in Fig. 4. This picture shows how smeared out the $(10.l)$ and the $(20.l)$ reflections are, while the $(00.l)$ reflections situated on the straight line are sharp.

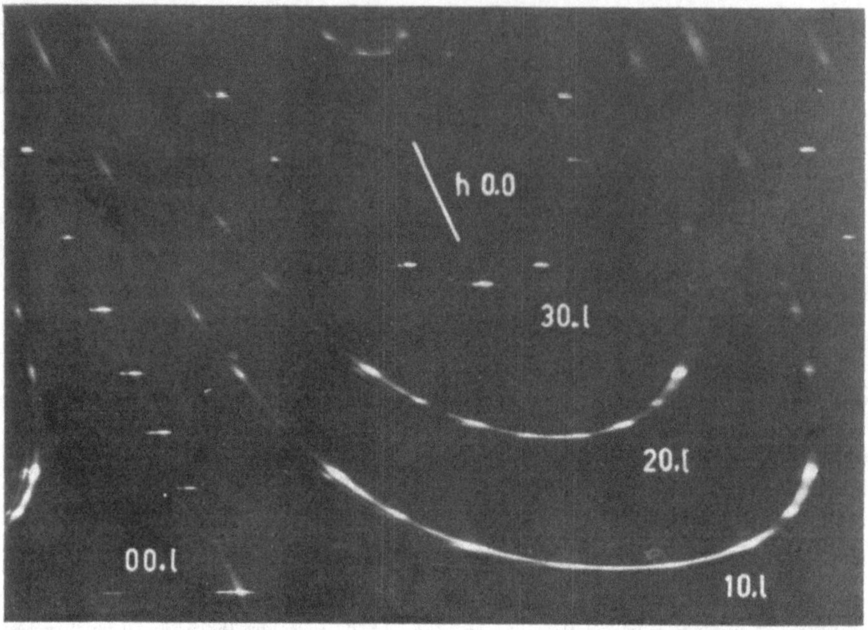

Fig. 4. An *a*-axis, zero layer Weissenberg picture of a GaSe single crystal grown by the iodine transport process, showing the high degree of disorder in the (10.*l*) and (20.*l*) and the sharp reflections on the (00.*l*) and the (30.*l*)

This broadening for reflections $h - k \neq 3\,n$ means that the stacking faults are made in the direction [11.0]. It indicates that this hexagonal GaSe single crystal has not an ordered stacking of slabs but is a disordered polytype. Powdering this crystal means simply that one introduces even more faults and that the reflections become more smeared out. This result is presented in the Debye-Scherrer powder picture, see Fig. 5a and b. In the case of the rhombohedral modification, where one would expect the same kind of broadening effect as observed for the ε-type, (on account of the similar stacking fault energy [3]) such a broadening is absent. The sharp γ-reflections obtained in powdered material (Fig. 3a) proof this γ-modification to be an ordered stable, polytype.

It is assumed that this ordered polytype ows its stability to the presence of cubic Ga_2Se_3. This cubic phase which can easily be detected in the powder diffractograms could not be detected in single crystals (rotation, Weissenberg pictures) [10].

4. Structure Type Determination

Schubert et al. [1] reported on the intensities of reflections of GaSe single crystal rotation pictures and concluded a hexagonal modification, the so-called ε-modification.

The GaSe powder pictures on which Hahn decided that GaSe had a β-structure were used by Jellinek [2] to compare the intensities with those obtained by Schubert et al. They showed a difference which led Jellinek to assume another structure for the powdered material of Hahn.

Taking a GaS-like β-structure for GaSe, Jellinek calculated the intensities. Especially the absence of the (10.2), (10.6), (20.6), and (21.6) reflections on the

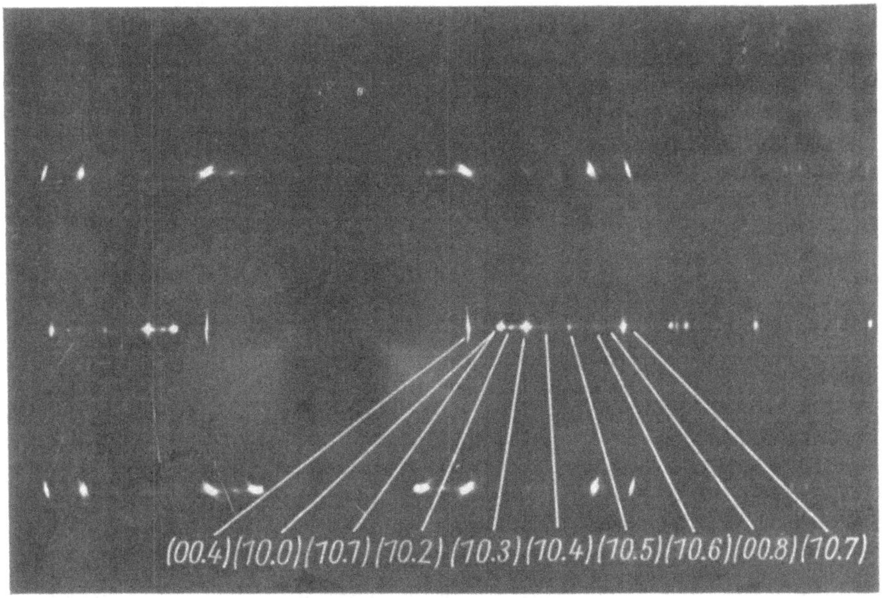

a

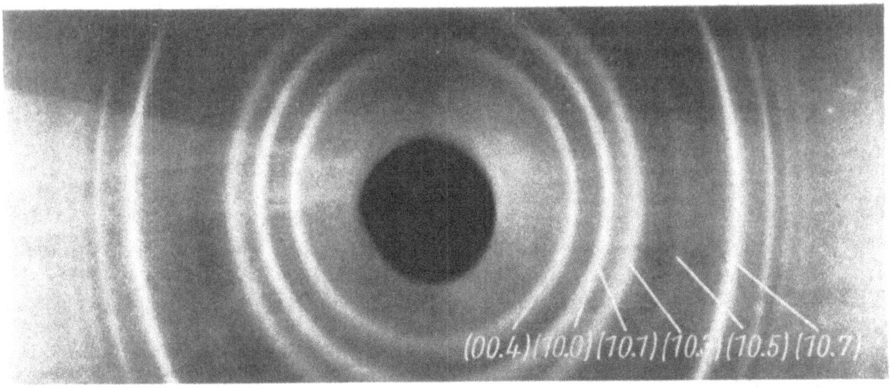

b

Fig. 5. a) Single crystal rotation picture of ε-GaSe presents the sharp reflections of (10.2) and (10.6). b) Debye-Scherrer picture of the same crystal, it shows the absence of the (10.2) and (10.6)

available powder pictures could be explained by the assumption of this β-structure. In the case of an ε-type stacking, these four reflections have to be present.

To make a justified decision we undertook the following experiment for which flakelike crystals, grown in the iodine transport process [11], were used.

From a plate crystal a small rim was cut parallel to the *a*- or *b*-axis and this rim was used for a rotation picture (Cu Kα radiation) and for Weissenberg pictures of the zero, the first, and the second layer lines. The intensities of the reflections obtained are in good agreement with the results published in reference [1] and indicate the ε-structure, space group $P\bar{6}$ (Fig. 5). The remainder of the single crystal from which the rim was taken, was powdered and used for a Debye-Scherrer picture; the lines on this powder picture showed clearly a difference in their intensities compared to those obtained in the rotation picture (Fig. 5a and b); the (10.2) and (10.6) reflections can clearly be seen on the rotation picture but they are completely absent on the powder foto. Obviously these intensities are affected by the large number of stacking faults introduced during powdering of the single crystal. In this procedure the relative low degree of order existing in the ε-single crystal is completely disturbed. This phenomenon has also been reported by Wildervanck [12] for MoS_2, evenstructural transitions can be introduced by powdering MoS_2 crystals.

A stable structure like the β-type should be, is not destroyed on grinding the flakes. This is demonstrated in our experiments on mixed crystals, reported earlier [13]. Crystals up to a composition of 75% GaSe had a β-structure and didn't lose their sharp reflections on powdering. Higher GaSe contents showed broadened reflections.

The possibility that in the course of the powdering procedure a β-structure is formed from the less ordered ε-structure can be rejected. The reflections in the Debye-Scherrer picture show the same broadening as observed in the diffractograms of recrystallised Se-poor GaSe, which is the ε-modification. Our results mentioned so far are in disagreement with the idea that a β-structure exists for GaSe or that the majority of the monocrystalline GaSe would adopt this structure [2]. We are convinced that the ε-type of stacking is predominant in platelet crystals; optical measurements [14] and measurements of exciton energies carried out on our single crystals [15] confirm this idea. However there is the possibility that in the stacking sequence of an ε-type crystal, some layers can form a β-region between the ε-layers, by simply rotating over 60°. Only by this assumption we are able to explain the results of the work of Aulich et al. [16]. They report the value of exciton energies measured in some GaSe crystals, which could be associated with a β-structure.

5. Conclusions

On employing Se-rich GaSe, recrystallisation produces always γ-GaSe powder mixed with Ga_2Se_3.

Diffractograms of Se-poor GaSe show only the hexagonal ε-structure.

The ε-structure is a polytype with a low degree of order, the γ-modification is an ordered polytype.

The new compound Ga_3Se_2 is formed during sublimation; in the same way as Ga_2Se probably is formed.

Powdering destroys the low degree of ordering in ε-type crystals; it does not produce the stable β-type.

Indicating the existence of a so-called β-type GaSe, with the aid of powder pictures is bound to introduce errors.

Acknowledgements

The authors thank Dr. J. A. Cras from the Dept. of Chemistry, Nijmegen University for the stimulating discussions and Mrs. J. C. M. van de Pol and C. A. Copraij for their help with the experimental work.

References

[1] K. Schubert, E. Dore, and J. M. Kluge, Z. Metall., **46**, 216 (1955).
[2] F. Jellinek and H. Hahn, Z. Naturf. **16b**, 713 (1961).
[3] L. S. Bassinski, D. B. Dove, and E. Mooser, Helv. phys. Acta **34**, 373 (1961).

[4] E. Mooser, private communications.
[5] P. G. Rustamov, B. K. Babaeva, and N. P. Luzhnaya, Izv. Akad. Nauk SSSR, Ser. neorg. Mater. **1**, 843 (1965).
[6] R. M. A. Lieth and C. W. M. v. d. Heyden, unpublished results.
[7] J. Dieleman and H. R. C. Engelfriet, J. less-common Metals **25**, 231 (1971).
[8] W. Klemm and H. H. von Vogel, Z. anorg. allg. Chem. **219**, 45 (1934).
[9] J. C. J. M. Terhell, J. H. Cras, and R. M. A. Lieth, publication in progress.
[10] J. C. J. M. Terhell and R. M. A. Lieth, to be published.
[11] R. M. A. Lieth, C. W. M. van der Heyden, and J. W. M. van Kessel, J. Crystal Growth **5**, 251 (1969).
[12] J. C. Wildervanck, Thesis, Groningen 1970.
[13] J. C. J. M. Terhell and R. M. A. Lieth, phys. stat. sol. (a) **5**, 719 (1971.)
[14] H. U. Boelsterli and E. Mooser, Helv. phys. Acta **35**, 538 (1962).
[15] I. Scholten and M. J. Gelten, T. H. E. Rep. Exciton Energies, Eindhoven 1970 (in Dutch).
[16] E. Aulich, J. L. Brebner, and E. Mooser, phys. stat. sol. **31**, 129 (1969).

(Received January 28, 1972)

phys. stat. sol. (a) **10**, 537 (1972)

Subject classification: 14; 22.9

Universität Karlsruhe, Institut für angewandte Physik, Sonderforschungsbereich 66 (a), and Institut für anorganische Chemie (b)

Conduction Mechanism in Square Planar Complexes of Platinum

By

P. Würfel (a), H. D. Hausen (b), K. Krogmann (b), and P. Stampfl (b)

In square planar complexes of platinum a strong dependence of the conductivity upon lattice distortions and humidity was detected. Besides the electronic conduction ionic conduction was found. It is probably related to the strong influence of humidity on the electronic conductivity.

In planaren Platinkomplexen wurde eine starke Abhängigkeit der Leitfähigkeit von Gitterfehlern und von der Luftfeuchtigkeit gefunden. Neben der elektronischen Leitfähigkeit wurde Ionenleitung nachgewiesen. Sie hängt wahrscheinlich zusammen mit dem starken Einfluß der Luftfeuchtigkeit auf die elektronische Leitfähigkeit.

1. Introduction

In some compounds of Pt II, e.g. $K_2(Pt(CN)_4) \cdot 3\,H_2O$, containing planar complexes, the central atoms which are surrounded by four ligands form linear chains perpendicular to the plane of the complex [1]. The weak bonding between the central atoms in the chains can be enhanced considerably in some cases by oxidation. This is proved by a decrease of the interionic distance within the chain. The Pt–Pt distance of the partly oxidized compounds (2.8 to 2.9 Å) is in the order of the distance of metallic Pt. After the oxidation hole levels are created and thus a one-dimensional metallic conductivity can be expected. In the oxidized compound electroneutrality is established either by a cation deficiency or — as in our case — by additional halide ions in interstitial lattice sites.

2. Electronic Conductivity

By measurement of the optical reflectivity in $K_2(Pt(CN)_4)Cl_{0.32} \cdot 2.6\,H_2O$ and $K_2(Pt(CN)_4)Br_{0.3} \cdot 2.3\,H_2O$ an electronic conductivity of $\approx 10^4\ (\Omega cm)^{-1}$ parallel to the Pt chains was determined for both compounds whereas perpendicular to the Pt chains both compounds appear to be insulators [2, 3].

Seebeck effect measurements show that the conductivity is caused by holes [4]. Thus the one-dimensional metallic hole conductivity in the Pt chains suggested by the crystal structure is established. The conductivity parallel to the Pt chains determined by electrical measurements, however, is smaller by at least one order of magnitude. The highest value determined by us at room temperature is $284\ (\Omega cm)^{-1}$. It also depends strongly on the quality of the crystal, contrary to the conductivity determined by optical measurements. Values of less than $1\ (\Omega cm)^{-1}$ have been observed.

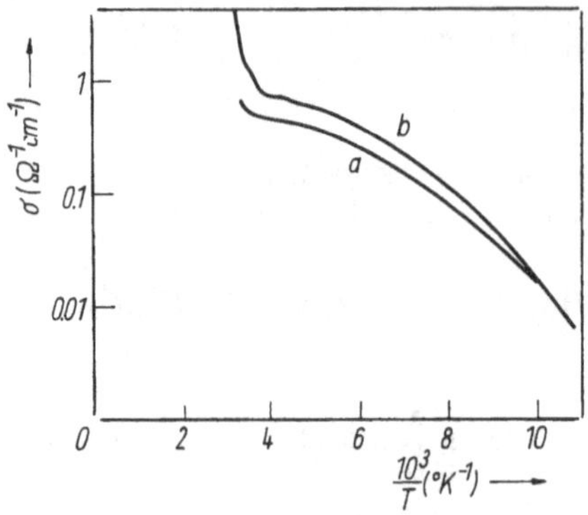

Fig. 1. Temperature dependence of the dc conductivity parallel to the Pt chains in $K_2(Pt(CN))_4Cl_{0.32}$, curve a and $K_2(Pt(CN)_4)Br_{0.3}$, curve b

The conductivity was measured by the four-probe technique. As contact material we used "aquadag" and gold, evaporated in vacuum onto the sample at ≈100 °K.

The temperature dependence of the conductivity for the compounds $K_2(Pt(CN)_4)Cl_{0.32} \cdot 2.6\,H_2O$ and $K_2(Pt(CN)_4)Br_{0.3} \cdot 2.3\,H_2O$ is shown in Fig. 1. The magnitude and the temperature dependence show that the dc conductivity is non-metallic, contrary to the conductivity determined by optical measurements.

3. Model for the Conduction Mechanism

The discrepancy between the conductivities determined by optical and electrical measurements can be explained by assuming that lattice distortions cause interruptions of the Pt chains. The corresponding potential barriers lead to an activation energy in the temperature dependence of the conductivity. The density of lattice distortions varies from one crystal to the other. This accounts for the wide range of observed conductivity values. If this model is correct the ac conductivity should increase at high frequencies where the interruptions of the chains are short circuited by their capacitance.

Berenblyum et al. [5] find a sharp increase in conductivity at temperatures below 100 °K at a frequency of 10^{10} Hz. Furthermore, these authors give values for the dielectric constant ranging from 10^3 to 10^4, measured at 100 °K and 10^{10} Hz.

From the reflectivity measurements [3] a dielectric constant of 2.25 is derived for the lattice disregarding the influence of free carriers. If this value is taken for the regions depleted from free carriers, i.e. for the interruptions in the chains, it follows from a dielectric constant of 10^3 to 10^4 that the total length of the interruptions is the fraction 10^{-4} to 10^{-3} of the entire crystal length.

If the low dc conductivity is caused by lattice distortions, the conductivity should change drastically upon creation of new lattice distortions.

A crystal of $K_2(Pt(CN)_4)Cl_{0.32} \cdot 2.6\,H_2O$ was bent slightly. Fig. 2 shows the resulting change in conductivity. Under the microscope no damages due to the bending could be detected. It is important to note the recovery of the conductivity in the bent crystal which shows the occurrence of a healing process. A sharp increase in conductivity to the initial value is found after blowing at the crystal with moist air of ≈30 °C.

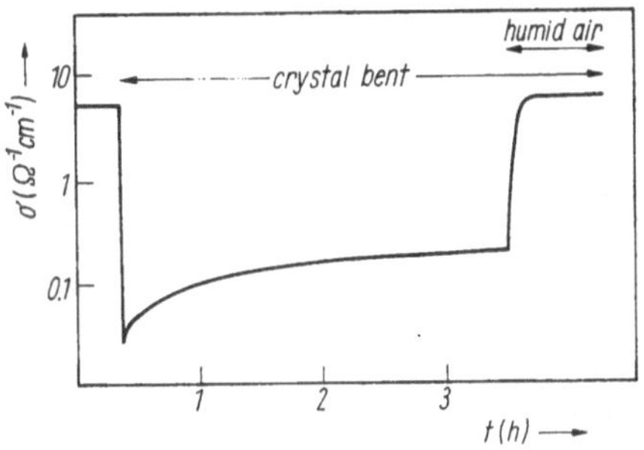

Fig. 2. Conductivity changes by generation of lattice defects and by change of humidity in $K_2(Pt(CN)_4)Cl_{0.32}$. The lattice defects occur by bending the crystal

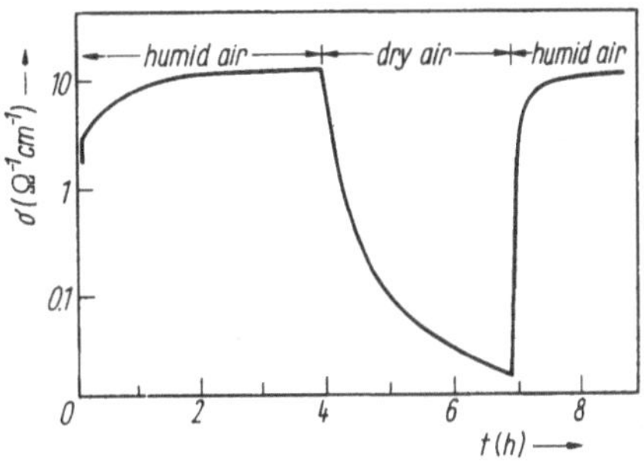

Fig. 3. Conductivity changes by change of humidity in $K_2(Pt(CN)_4)Cl_{0.32}$

4. Influence of Humidity

The dependence of the conductivity on the humidity is shown in Fig. 3. The conductivity of a crystal of $K_2(Pt(CN)_4)Cl_{0.32} \cdot 2.6\, H_2O$ increases in humid air (20 °C, $\approx$ 80%), and it decreases by several orders of magnitude in dry air (20 °C, $\approx$0%). This decrease is reversible as can be seen in Fig. 3.

The changes of the conductivity occur much more slowly in thick crystals than in thin crystals. The influence of humidity therefore seems to be a bulk effect.

During the variation of the conductivity the crystals retain their colour which is caused by the large electronic conductivity in the non-distorted regions of the platinum chains. Therefore humidity has an influence mainly on the number and/or size of the interruptions in the Pt chains and not on the free carrier concentration. From the fact that not even in very humid air a conductivity can be measured comparable to that obtained by optical methods we conclude that there are two kinds of lattice distortions:

1. Interruptions which are not influenced by humidity, e.g. lateral displacement of parts of Pt chains.
2. Interruptions which are influenced by humidity. From the crystal structure it seems possible that in regions with a water deficiency an otherwise water-coordinated Cl ion moves closer to a Pt chain where it forms a localized bond. Thereby a defect electron of the surrounding Pt ions is lost and not available for conduction any more. This mechanism yields insulating parts in the Pt chains.

5. Ionic Conduction

The relaxation of the lattice defects and the strong dependence of the conductivity on the humidity suggest an influence of mobile ions in the lattice.

As a matter of fact, ionic conductivity in $K_2(Pt(CN)_4)Cl_{0.32}$ and $K_2(Pt(CN)_4)Br_{0.3}$ can be shown to occur by using different contact materials.

Materials with a low work function give blocking contacts. Rectification would be expected when the conductivity is measured between a blocking contact and a non-blocking contact. In fact the crystal behaves like a battery. Open-circuit voltages from 0.8 to 1.2 V and short-circuit currents of several µA were measured. In this arrangement the contact with the lower work function is always negative. As non-blocking contact we used "Aquadag". As blocking contacts we used indium, evaporated onto the sample at 100 °K, liquid In–Ga alloy, or liquid and solid gallium. All blocking contact materials showed the same properties.

The open-circuit voltage is measured in the dark at constant temperature. Under these conditions electrons and holes in the crystal and in the metal contact would be in thermal equilibrium if they were the only mobile carriers. The Fermi energy then would be constant and no open-circuit voltage could be measured. The occurrence of an open-circuit voltage therefore points to ionic conductivity in addition to the electronic conductivity. By short circuiting the two contacts of a crystal, the Fermi energy is forced to be constant (Fig. 4a). Under this condition the gradient of the electrostatic potential at the blocking contact which is necessary for the equilibrium of the electrons and holes leads to an equal gradient of the electrochemical potential η_{ion} of the ions if their concentration and therefore their chemical potential μ_{ion} is assumed to be constant everywhere in the crystal. The short-circuit current j_{sc} caused by the gradient of η_{ion} is then given by

$$j_{sc} = \frac{\sigma_{ion}}{e} \frac{d\eta_{ion}}{dx} .$$

By disconnecting the blocking and non-blocking contact of a crystal an open-circuit voltage occurs. Under this condition the total current must disappear. Here we make the simplifying assumption that the electronic conductivity in the hole depletion layer at the blocking contact has become small compared to the ionic conductivity. Then the total current can disappear only when the ionic current disappears. So η_{ion} is constant and with it the electrostatic potential V. The hole concentration at the contact, however, is kept constant. This causes a difference between the Fermi level in the crystal and in the contact metal which is the open-circuit voltage. This consideration yields the correct sign for the open-circuit voltage. It follows that crystals doped with Br and Cl are p-conductors as was determined for Br-doped crystals by measurement of the Seebeck effect [4].

From the magnitude of the open-circuit voltage V_{oc} and the short-circuit current j_{sc} the order of magnitude of the ionic conductivity can be estimated. Assuming that in the space charge layer at the blocking contact the electronic

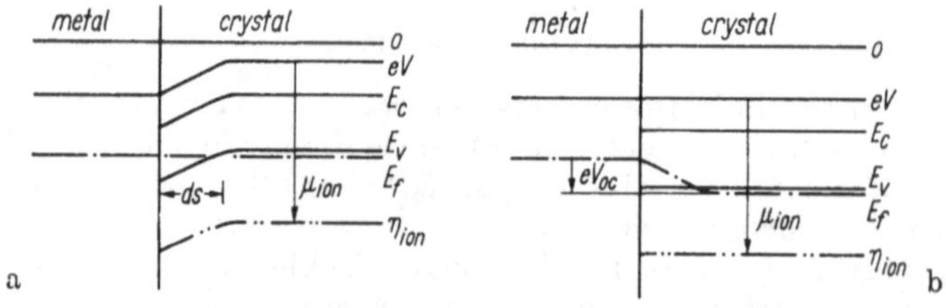

Fig. 4. $K_2(Pt(CN)_4)Cl_{0.32}$ in contact with a metal of small work function causing a hole depletion layer. a) Short-circuit condition. The gradient of the electro-chemical potential η_{ion} of a distinct sort of mobile ions gives rise to a short-circuit ionic current. b) Open-circuit condition. An open-circuit voltage V_{oc} is generated because the gradient of the electrostatic potential V, necessary for the equilibrium of the electrons and holes, is short circuited by ionic conduction

conductivity is small compared to the ionic conductivity, we find (cf. Fig. 4a and 4b)

$$j_{sc} = \frac{\sigma_{ion}}{e}\frac{d\eta_{ion}}{dx} \approx \frac{\sigma_{ion}}{e}\frac{\Delta\eta_{ion}}{d_s} = \sigma_{ion}\frac{V_{oc}}{d_s}.$$

d_s is the depth of the space charge layer at the blocking contact. Assuming d_s to be 10^{-6} to 10^{-5} cm, we find for the ionic conductivity $10^{-8}\ (\Omega cm)^{-1} < \sigma_{ion} < 10^{-7}\ (\Omega cm)^{-1}$.

The measurement of the short-circuit current has to be done quickly because of polarization effects: As the metal contact cannot supply the ions, the crystal will be destroyed at the contact when the number of transported charges becomes comparable to the total number of mobile ions in the space charge layer. The observation of [6] that the crystals were destroyed during the conductivity measurements need not be caused by overheating as was suggested in [2]. From a conductivity value of $10^{-3}\ (\Omega cm)^{-1}$ observed by the two-probe technique we conclude that Krogmann and Hausen [6] used blocking contacts causing ionic currents which lead to the destruction of the compound during the measurements.

The magnitude of the open-circuit voltage depends strongly on humidity just as the hole conductivity. In humid air it is slightly larger and it decreases sharply in dry air. This can be explained either by an increase of the hole conductivity or a decrease of the ionic conductivity in dry air. The observed bulk conductivity, however, decreases in dry air (Fig. 3). Therefore, the decrease of the ionic conductivity appears to be the correct explanation for the decrease of the open-circuit voltage.

The increase of the open-circuit voltage in humid air shows that the observed increase of the conductivity in humid air is not a surface effect, because a large surface conductivity would short circuit the open-circuit voltage.

Although the large bulk conductivity of the compounds under study is not caused by mobile ions, apparently it is influenced by them. Possibly the decrease in dry air of the concentration of a certain kind of mobile ions which must be related to the crystal water leads to interruptions of the electronic conductivity in the Pt chains which are removed when the concentration of this kind of ions is increased in humid air.

References

[1] K. Krogmann, Angew. Chem. **81**, 10 (1969).
[2] D. Kuse and H. R. Zeller, Phys. Rev. Letters **27**, 1060 (1971).
[3] H. P. Geserich, H. D. Hausen, K. Krogmann, and P. Stampfl, phys. stat. sol. (a) **9**, 187 (1972).
[4] M. J. Minot and J. H. Perlstein, Phys. Rev. Letters **26**, 371 (1971).
[5] A. S. Berenblyum, L. I. Burarov, M. D. Khidekel, I. F. Shchegolev, and E. B. Yakimov, Soviet Phys. — J. exper. theor. Phys., Letters **13**, 440 (1971).
[6] K. Krogmann and H. D. Hausen, Z. anorg. allg. Chem. **358**, 67 (1968).

(Received February 1, 1972)

phys. stat. sol. (a) **10**, 543 (1972)

Subject classification: 14.3.1 and 16; 13.4; 22.9

Groupe d'Interactions Hyperfines, Département de Recherche Fondamentale, Centre d'Etudes Nucléaires de Grenoble

Semiconductivity in Chromium(III) Trisacetylacetone

By

A. P. CHAPPE[1]) and J. I. VARGAS[2])

Thin films of chromium(III) trisacetylacetone complex are found to be p-type semiconductors. The analysis of the current–voltage characteristics in the dark and under illumination indicates that the results are consistent with the assumption of an exponential distribution of trapping states in the forbidden gap. Values of intrinsic and trap distribution parameters, and the position of the Fermi level as influenced by purity are reported.

Ce travail montre que les films minces de l'acetylacétonate de chrome III sont des semiconducteurs de type p. L'analyse de caractéristiques courant–tension en obscurité et sous illumination indique que l'hypothèse d'une distribution exponentielle de pièges dans la bande interdite rend bien compte des résultats, et l'on présente les valeurs des paramètres intrinsèques et de la distribution de pièges, ainsi que la position du niveau de Fermi en fonction de la pureté.

1. Introduction

Although considerable attention has been given to the study of transport phenomena in organic and organo-metallic compounds, so far very little work has been carried out on the electrical and photoelectrical properties of transition metal chelates [1]. This lack of information is surprising in view of the considerable potential of such compounds as model systems for the understanding of electronic processes in biologically interesting systems. In fact, the connection between energy balance and electron transport and transfer constitutes one of the central problems in bioenergetics [2]. Thus the metallic phtalocyanines, which display obvious similarities to haemoglobin, chlorophyll, and other vital materials, have indeed been examined in a series of pioneer works by Harrison, Heilmeier, and Warfield [3 to 5]. Unfortunately, the latter compounds are, however, too complex to afford at present a deeper understanding of the relationship existing between their electronic structure and the associated transport phenomena. It seemed therefore worthwhile to study the electrical and photoelectrical properties of the much simpler complexes of acetylacetone with transition metals: their relative structural simplicity, easiness of preparation, as well as their aromaticity, parallel many of the interesting features of the phtalocyanines. Among the acetylacetonates of the first transition series, the chromium(III) complex ($Cr(Ac)_3$), which has already been submitted to many optical [6] and EPR measurements [7], would be specially rewarding for such studies, on account of its extreme chemical and thermal stabilities.

Apart from the general interest bearing on the transport properties of the chelates, it should be remarked that quantitative data on these solids are

---

[1]) On partial fulfilment for doctor's degree.

[2]) On leave of absence from Universitade de Minas Gerais, Brazil.

needed to assist in the interpretation of phenomena associated to the optically and thermally induced reformations of chelate molecules broken down by thermal neutron capture in the coordinated metallic ion. In reality, most of these phenomena [8, 9, 10] have been interpreted in terms of an original suggestion given by Nath et al. [11]. According to this investigator, the annealing reactions of the thermal neutron capture generated recoil fragments would be piloted by the thermal release of electrons or holes from traps preexisting or created in the forbidden gap of these supposedly wide band semiconductors.

In this paper the dark and photoconduction in vacuum-deposited thin films of chromium(III) trisacetylacetone are examined.

2. Chromium (III) Trisacetylacetone

Chromium(III) trisacetylacetone crystallizes in a monoclinic lattice with alternate stacking of right- and left-handed molecules. The structure of the individual molecules, indicated on Fig. 1, has been worked out by three-dimensional X-ray diffraction analysis [12]. Three acetylacetonate ligands, CH_3–CO–CH–CO–CH_3, surround the chromium ion to form an octahedron of oxygen atoms with an overall D_3 symmetry. Therefore each chelate group forms a six-membered ring which includes the metal. In spite of the X-ray evidence for the octahedral arrangement of the oxygen atoms, EPR measurements revealed the presence of a strong axial field, which could be interpreted in terms of dipole moments associated with the CO groups, but which more likely would result from the effect of the π orbitals of the ligands on the d_ε orbitals of the metal [7]. The existence of π-bonding is correctly understood in terms of the electronic transitions [6, 13] confirmed by chemical evidence on the aromaticity in this system [14] and by Mössbauer studies on the iron(III) trisacetylacetone homologue [15]. $Cr(Ac)_3$ shows intense absorption bands in the near ultraviolet region due to $\pi_3 \rightarrow \pi_4$ and $d_\varepsilon \rightarrow \pi_4$ transitions occurring at 270 and 332 nm, and weak d–d bands in visible region due to the ${}^4A_2 \rightarrow {}^4T_2$ (560 nm) and ${}^4A_2 \rightarrow {}^4T_1$ ($\approx$ 400 nm) transitions [13]. Experiments conducted at 4 °K clearly show the vibronic components of the ${}^2E \leftrightarrow {}^4A_2$ transition [16].

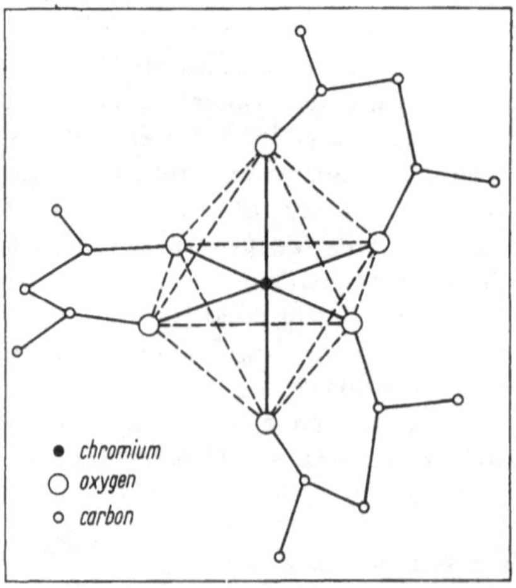

Fig. 1. Schematic representation of the molecular structure of $Cr(Ac)_3$

3. Experimental Techniques

The chromium (III)trisacetylacetone, supplied by FLUKA, was first purified by chromatography on a silica gel column with benzene–methanol mixture as eluent and recrystallized. Part of the material was pumped, washed in pure helium (> 99.99%), sealed in a 0.5 cm^2 × 40 cm glass tube, and then purified by zone-melting technique.

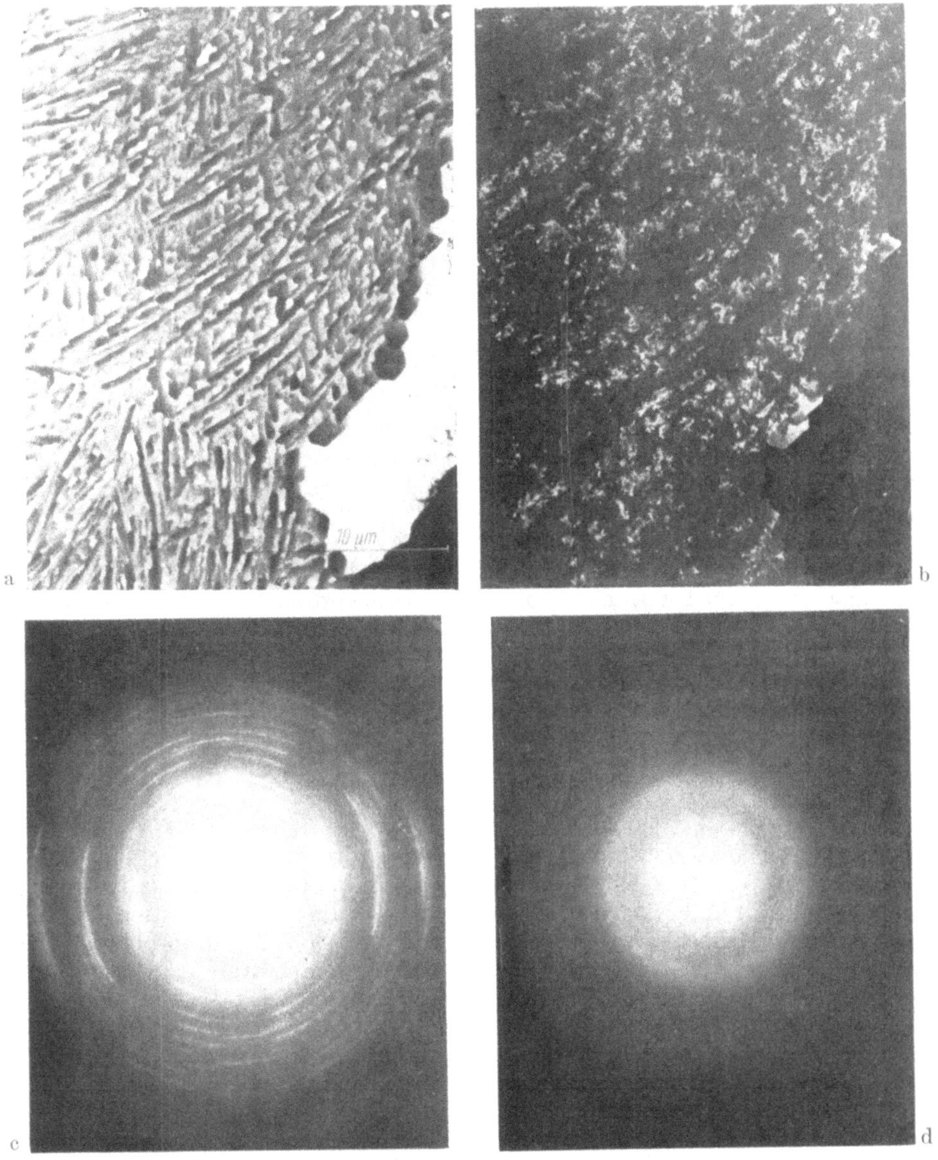

Fig. 2. Electron micrograph of a 0.1 µm thick evaporated film of $Cr(Ac)_3$ on a collodion substrate. a) Bright field and b) dark field picture of the same area; c) diffraction pattern of a polycrystalline domain, and d) diffraction pattern of an amorphous domain

For conductivity measurements, sandwich-type samples were used. The first gold electrode, the acetylacetonate film, and the second gold electrode were all deposited on a glass support by vacuum evaporation from electrically heated tantalum boats. To ensure against oxygen contamination, which may deeply influence the electrical characteristics of the material, any contact with the atmosphere was avoided. The thickness of each $Cr(Ac)_3$ film, ranging from 1 to 5 μm, was determined by dissolution in benzene and spectrophotometric measurements at 560 and 380 nm with an accuracy better than 5%.

The texture of the evaporated films was examined by electron microscopy. Fig. 2a shows an electron micrograph of a 0.1 μm thick $Cr(Ac)_3$ film evaporated on a collodion substrate. Fiber-like regions of two different preferential orientations can be observed. The diffraction pattern of Fig. 2c taken on polycrystalline domains of about 2 μm further confirms the fiber structure. These small domains of crystalline structure, shown on the corresponding dark field picture of Fig. 2b, are mixed with amorphous domains whose characteristic diffraction pattern is shown on Fig. 2d. Aligned microcrystals of 2 μm length can also be seen on the boundary of the film, showing the pseudo-hexagonal combination (001) (100) (110). Three of them are seen in Bragg position on Fig. 2b.

Electrical measurements were made in a controlled atmosphere cell, in which the substrate was maintained at constant temperature by clamping it to a perforated steel block connected to a circulating silicon oil thermostat. The temperature was maintained at the chosen value and was measured by two chromel–alumel thermocouples placed on the substrate with an accuracy better than 0.5 deg. Because of the sensitivity already mentioned of the conductivity to the presence of oxygen, the cell was carefully cleaned before each measurement with a flow of dry nitrogen gas (O_2, $H_2O < 10$ ppm) which was held during all measurements at a rate of 100 to 200 cm³/h. Prior to each measurement the samples were first annealed and kept at the highest working temperature (usually 110 °C) for 1/4 h. The measurements then proceeded as the temperature of the sample was lowered by stages to room temperature.

Currents were measured by a Keithley 417 high speed picoammeter. For the photoconductivity investigations the 4046 and 4358 Å filtered lines from a 250 W high pressure mercury vapour lamp were used.

4. Experimental Results and Discussion

An important consideration is the sign of the majority carrier. The mobilities in such organic materials are generally far too low to allow Hall effect measurements. However, the effect of exposure to well-known electronegative gases, such as O_2, NO, and NO_2, may often give some valuable information. In the case of $Cr(Ac)_3$ thin films, the introduction of oxygen in the atmosphere of the cell causes the conductivity to rise of 10^4 at room temperature indicating that, as in the case of copper phtalocyanine [19], this material is p-type in vacuo.

Temperature dependent measurements have been carried out from 20 to 120 °C. This range was limited at low temperature by the picoammeter sensitivity and at high temperature by the rate of sublimation of the sample [17].

The lg i–lg U characteristics for a 4.3 μm × 0.1 cm² sample at several temperatures are shown on Fig. 3. The obvious ohmic nature of the contacts may be explained by the presence of surface states that considerably lower the barrier height which would be expected from the comparison between the work func-

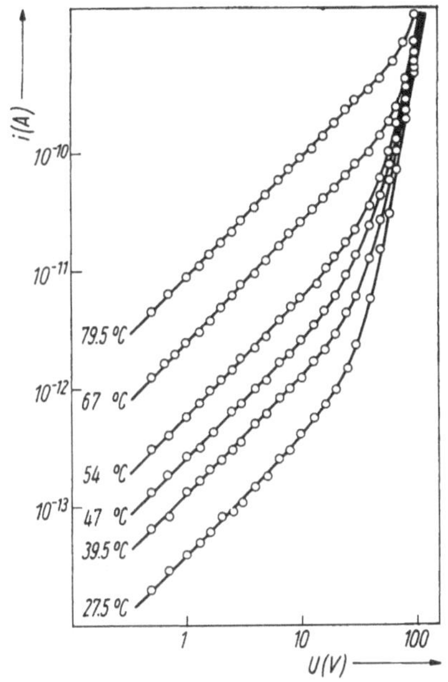

Fig. 3. Current–voltage characteristic for a 4.3 μm × × 0.1 cm² sample as a function of temperature

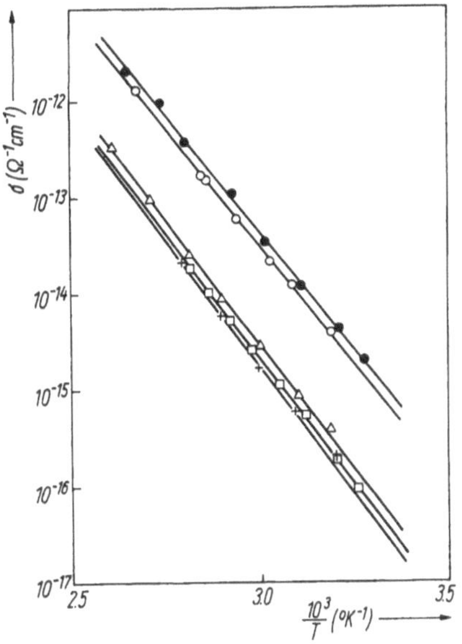

Fig. 4. $1/T$ dependence of the conductivity σ as function of purity. ●, ○ chromatographic purification; △, □, + chromatographic and zone-melting purification

tion of gold (4.8 eV) and the ionization potential of $Cr(Ac)_3$ (7.8 eV) [18]. The ohmic and space-charge-limited (SCL) behaviour are easily distinguished on Fig. 6. The linearity of the SCL lines has been tested over a range of 4 to 5 decades.

As usual, the ohmic current density is given by

$$j = e_0 \, \mu \, \frac{U}{d} \, N_v \exp \frac{F_0}{kT} , \tag{1}$$

where μ is the mobility in $cm^2 \, V^{-1} \, s^{-1}$, e_0 the elementary charge, U the voltage, d the film thickness in cm, N_v the density of states in the valence band, and F_0 the energy difference between the Fermi level and the valence band edge ($F_0 < 0$). Assuming N_v to be temperature independent and μ slowly varying with temperature, F_0 may be obtained from the slope of $\lg \sigma$ vs. $1/T$ plotted on Fig. 4 for various samples of different purities. The slopes have been calculated by a least squares fit leading to values of F_0 from 0.97 to 1.06 eV. The conductivity appears to be noticeably lowered by additional zone-melting purification. The correlation between the displacement of the Fermi level and the variation of the conductivity as function of purity at a given temperature (40 °C) is shown on Fig. 5. The experimental points are correctly fitted by the expected line of slope $1/kT$ whose extrapolation to $F_0 = 0$ gives the intrinsic conductivity

$$\sigma_0 = (17 \pm 8) \, (\Omega \mathrm{cm})^{-1}$$

leading to

$$N_v \, \mu = (1 \pm 0.5) \times 10^{20} \, (\mathrm{cm \, V \, s})^{-1} .$$

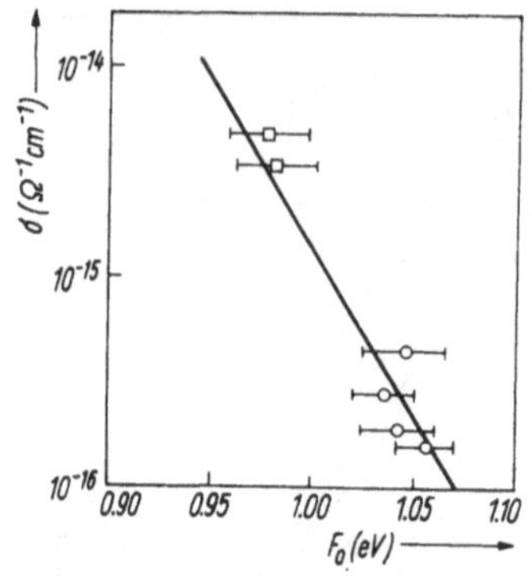

Fig. 5. Correlation between conductivity σ at 40 °C and F_0 for various samples of different purity. The slope is $1/kT$ as expected from equation (1). □ chromatographic purification, ○ chromatographic and zone-melting purification

For a more detailed discussion of the characteristics observed in the present investigation, it should be emphasized that evaporated films may noticeably deviate from crystallinity, and therefore this calls for a careful reexamination of the usual band models commonly employed in the description of the behaviour of well crystallized systems. In fact, in the former structures, the long and medium range orders are lacking and, while still keeping the band model, one may formally consider that lack of crystallinity produces merely a blurring of the edges of valence and conduction bands, giving rise to a distribution of more or less localized levels in the forbidden gap; the deeper the levels, the more localized becomes their character. Under these assumptions, the carrier propagation is then expected to be intermediate between impurity band and hopping conduction, the mobilities being, of course, expected to be typically low in both processes. To explore these ideas, it is essential to advance specific forms for the trap level distribution. Rose [19] has proposed the distribution to be exponential, which leads to a correct understanding of most of the one-carrier injection and photoconduction phenomena displayed by a large number of semiconductors. A particularly illuminating example of the high degree of self-consistency attained by this model is demonstrated by Sussman's work on the electrical properties of copper phtalocyanine [20]. In fact, assuming the distribution $P(E)$ as given by

$$P(E) = P_0 \exp \frac{E}{kT_c} \tag{2}$$

the complete current–voltage characteristic, first deduced by Lanyon [21], can be written in the form

$$j = \mu\, N_v\, e_0\, U \left(\frac{U + U_x}{U_m}\right)^l d^{-1}, \qquad l = \frac{T_c}{T}, \tag{3}$$

$$U_x = \frac{e_0\, d^2}{\varepsilon}\, kT_c\, P_0 \exp \frac{F_0}{kT_c}, \tag{4}$$

$$U_m = \frac{e_0\, d^2}{\varepsilon}\, kT_c\, P_0\,. \tag{5}$$

For $U \ll U_x$, (3) reduces to Ohm's law while for $U \gg U_x$ one obtains the well-known equation for the asymptotic behaviour of the j–U curve of slope $1 + l$:

$$j = \mu N_v e_0 \left(\frac{\varepsilon}{kT_c P_0 e_0}\right)^l U^{1+l} d^{-2l-1} . \tag{6}$$

T_c and U_x are first determined respectively from the slope of the SCLC lines and from their intercepts with the corresponding ohmic parts. Since U_m cannot usually be obtained by simple extrapolation of the convergent SCLC asymptotes, it is expedient to calculate it using U_x, T_c, and F_0 by

$$U_m = U_x \exp\left(-\frac{F_0}{kT_c}\right), \tag{7}$$

which is obtained from (4) and (5). This leads directly to

$$P_0 = \frac{\varepsilon U_m}{kT_c e_0 d^2}, \tag{8}$$

the density of trapping states per unit energy at the Fermi level is simply given by

$$P(F_0) = P_0 \exp\frac{F_0}{kT_c} = \frac{\varepsilon U_x}{kT_c e_0 d^2}, \tag{9}$$

which therefore depends only upon U_x, T_c, and d.

In the present case, the values of T_c were found to range from 800 to 1450 °K, and the computed values for the density of trapping states per unit energy at the Fermi level $P(F_0)$ to vary from 1.35 to 5.7×10^{15} cm^{-3} eV^{-1} for samples of

Table 1

Dark conductivity parameters for chromium(III) trisacetylacetone

	chromatographic purification	chromatographic and zone-melting purification
conductivity, σ, at 25 °C	$\approx 8 \times 10^{-16}$ $(\Omega cm)^{-1}$	$\approx 4 \times 10^{-17}$ $(\Omega cm)^{-1}$
Fermi level, F_0 (from the valence band)	$\approx$ 0.98 eV	$\approx$ 1.05 eV
intrinsic conductivity, σ_0	(17 ± 8) $(\Omega cm)^{-1}$	
density of states $\times$ mobility, $N_v \mu$	$(1.0 \pm 0.5) \times 10^{20}$ (cmVs)$^{-1}$	
characteristic temperature of trap distribution, T_c	800 to 1450 °K	
$kT_c P_0$	10^{18} to 10^{21} cm^{-3}	
density of traps per unit energy at the Fermi level $P(F_0)$	$(3.5 \pm 2) \times 10^{15}$ cm^{-3} eV^{-1}	

different thickness and purity. The two main characteristics of the results are the absence of correlation between T_c and the position of the Fermi level as influenced by purity, and the remarkable constancy of $P(F_0)$. It may be concluded that the trapping states responsible for SCLC conduction are those resulting from structural disorder rather than from impurity defects that would only alter the position of the Fermi level. The main parameters obtained in this investigation are summarized in Table 1.

To gain further insight into the nature of the localized levels as revealed by conductivity experiments in the dark, photoconductivity measurements are essential, because, while the former technique supplies informations only on the trapping character of the disorder centres, the latter may throw light on the occurrence of recombination processes.

A typical i–U characteristic under illumination at 25 °C is shown on Fig. 6 as compared to the corresponding dark current characteristic at the same temperature, and on Fig. 7 as a function of temperature. Three different regimes of conduction are easily distinguished. The ohmic region corresponds to a density of photogenerated free carriers resulting from the balance between optical excitation and recombination processes. The second and third region exhibiting U^2 and U^{1+l} dependence of j upon voltage corresponds to space charge conduction regimes involving relatively predominant and negligible photoexcitation of trapped carriers, respectively.

The density of photoexcited holes observed in the ohmic region, which obviously depends upon the effective number of recombination centres, has been obtained by Rose for the exponential trap distribution in terms of the existence

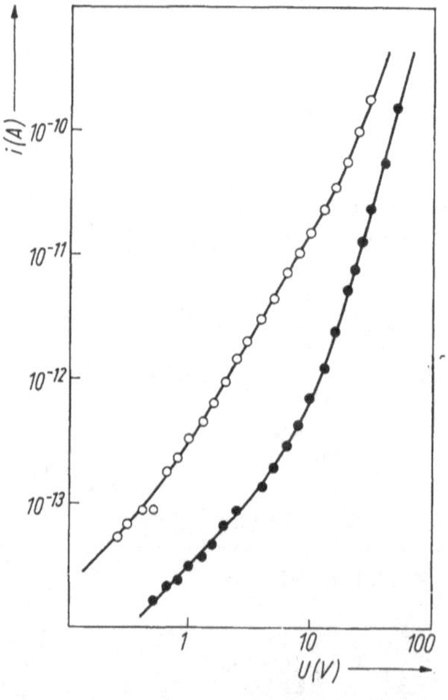

Fig. 6. Dark current–voltage characteristic at 25 °C (full points) and the corresponding characteristic under illumination (open points) at the same temperature

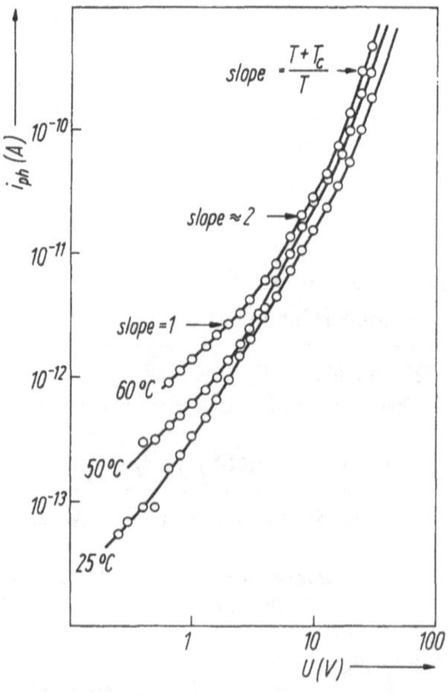

Fig. 7. Photocurrent–voltage characteristic as function of temperature

of demarcation levels for electrons and holes [19], and can be written in the form

$$p^* = p_0^{*\,T_c/(T+T_c)}\, N_v^{T/(T+T_c)}\,, \tag{10}$$

$$p_0^* = \frac{f}{kT_c\, P_0\, v\, s_p}\,, \tag{11}$$

where f is the rate of photogeneration of holes per unit volume, v the thermal velocity of holes, and s_p the capture cross section of recombination centres for holes. It is worth noting that the temperature dependence of p^* at constant light intensity has been misunderstood by Sussman [20], since (10) can be rewritten

$$p^* = p_0^* \left(\frac{N_v}{p_0^*}\right)^{T/(T+T_c)}\,, \tag{12}$$

where the temperature dependent term is clearly N_v/p_0^* to the power $T/(T + T_c)$, and not N_v to the power $T/(T + T_c)$. Fig. 8 shows the ohmic photocurrent at a given voltage plotted as a function of $T/(T + T_c)$ for a particular sample. The slope, determined by a least squares fit, is $N_v/p_0^* = 8 \times 10^{23}$. The extrapolation to $T = 0$ gives $i_0^* = p_0^*\, e_0\, \mu\, U S/d = 10^{-19}$ A.

Contrary to Sussman's assumption, neither μ nor N_v can be derived from these two quantities whose product

$$\frac{N_v}{p_0^*}\, i_0^* = \frac{e_0\, U\, S}{d}\, N_v\, \mu \tag{13}$$

only allows a new determination of $N_v\,\mu$. In our case, this product is $6 \times \times 10^{20}$ (V cm s)$^{-1}$, which is in good agreement with the previous determination given in Table 1. Additional computation of the other parameters is impossible at this stage because the quantum efficiency contained in p_0^* is unknown.

The validity of this model is not altered when assuming the light to be absorbed by the recombination centres themselves, as pointed out by Rose [19]. This type of photogeneration, however, plays a major role in the SCLC behaviour under illumination, for the absorption of energy by trapped carriers, which may take place directly by photon absorption or indirectly by means of exciton prop-

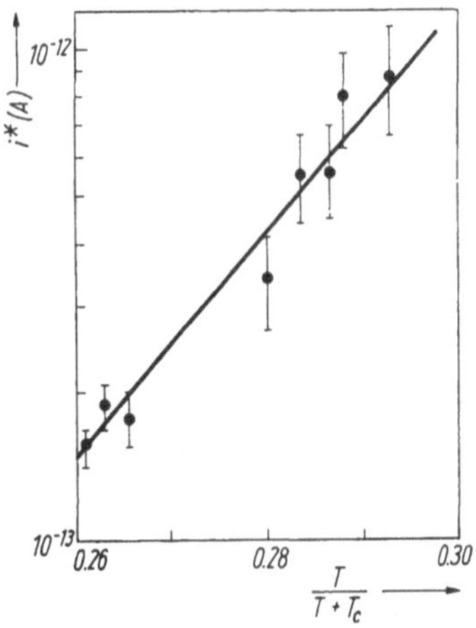

Fig. 8. Temperature dependence of the ohmic photocurrent. The slope is N_v/p_0^* according to equation (12)

agation, leads to a net increase of the release rate of trapped carriers and to a complete redistribution of the space charge among the trapping states. The SCLC model can be conveniently modified and then allows a correct understanding of the shape of the characteristics presented on Fig. 6 and 7. For this, the concept of demarcation level D is most helpful [22]. Assuming a constant photoionization cross section σ_{op} for all the states located in the vicinity of the Fermi level, the demarcation level is the one for which the probability of optical release of holes from traps is equal to the thermal release probability, and is thus defined by

$$\exp \frac{D}{kT} = \frac{\sigma_{op}\,\varphi}{N_v\, v\, s_p}, \tag{14}$$

where φ is the photon flux density. When the voltage is increased, the Fermi level moves from its ohmic value to the demarcation level, giving rise to the j–U relation first derived by Helfrisch and Mark [23]:

$$j = \frac{\mu\,\varepsilon}{kT_c\, P_0}\, N_v^{1/l} \left(\frac{\sigma_{op}\,\varphi}{v\, s_p}\right)^{1-(1/l)} \frac{U^2}{d^3}\,. \tag{15}$$

Equation (15) shows the typical quadratic dependence of j upon voltage which is observed on Fig. 6 and 7. A further increase of the voltage causes the Fermi level to surpass the demarcation level and the photocurrent to vary as U^{1+l}, as observed in the third region of the same figures.

The temperature dependence of the U^2 region has been investigated. The main limitation of the analysis arises from the narrowing of this region with increasing temperature and its disappearance at about 70 °C. This is in fact a simple consequence of the difference between the temperature dependence of the ohmic and U^2 photocurrents. The observed shift of the transition voltage between ohmic and U^2 regions can be easily interpreted using equations (10) and (15). Equation (15) may be rewritten as a linear relationship:

$$\log\left(\frac{j}{\mu\,\varepsilon}\,\frac{kT_c\, P_0}{N_v}\,\frac{d^3}{U^2}\right) = \left(1 - \frac{1}{l}\right) \log\left(\frac{\sigma_{op}\,\varphi}{N_v\, v\, s_p}\right), \tag{16}$$

which allows a precise determination of $\sigma_{op}\,\varphi / N_v\, v\, s_p$,which, under our experimental conditions of illumination, gives

$$\frac{\sigma_{op}\,\varphi}{N_v\, v\, s_p} = 10^{-15}$$

or alternatively, by use of (11),

$$D = 34.5\; kT\ (\text{eV})\,.$$

At 25 °C for example, $D = 0.886$ eV and $D - F_0 = 0.170$ eV. Although this value depends obviously upon light intensity, temperature, and purity of a given sample, it is interesting to compare it with that obtained by an independent calculation using the transition voltage between the U^2 and U^{1+l} region at a given temperature. At this voltage, the Fermi level coincides with the demarcation level, and the amplitude of the displacement of F from its original position to the new one is easily computed using the ratio of the corresponding thermal equilibrium and quasi-equilibrium free carrier densities. This calculation directly leads to $D - F_0 = 0.145$ eV which is in good agreement with the previous one, considering that $D - F_0$ was computed before by the difference of two nearly equal quantities. This agreement was observed in the whole range of temperatures studied, and seems to be one of the best arguments demonstrating the validity and the self-consistency of the present model.

5. Conclusion

The present investigation shows the solid chromium(III) trisacetylacetone complex to be a wide band semiconductor. The experimental results have been treated by a simple band model involving a trap distribution whose density exponentially decreases as the distance from the valence band increases. It is shown that the trap distribution is probably due to structural disorder rather than to impurity centres. The analogy between the behaviour of $Cr(Ac)_3$ and copper phtalocyanine towards oxygen suggests a p-type conduction. Photoconductivity experiments revealed the existence of intrinsic and extrinsic photogeneration processes whose analysis by means of ohmic and SCL conduction brought an additional confirmation to the validity of this model. Photoconductivity as function of light intensity and wavelength as well as dark conductivity of acetylacetonates of other transition and non-transition metals are presently under investigation. Further experimental work on conduction in single crystals and on neutron-irradiated samples will be soon reported.

Acknowledgements

The authors wish to thank Prof. L. Néel and Prof. A. Moussa for laboratory facilities, and Dr. D. Dautreppe for his interest in this investigation. Thanks are also due to Dr. A. Bourret who made the electron micrographs and to Dr. J. C. Pfister for many illuminating discussions.

References

[1] F. Gutman and L. E. Lyons, Organic Semiconductors, Wiley, 1967 (p. 458).
[2] R. J. P. Williams, in: Current Topics in Bioenergetics, Ed. R. Sanadi, Academic Press, 1969 (p. 79).
[3] G. H. Heilmeier and G. Warfield, J. chem. Phys. **38**, 163 (1963).
[4] G. H. Heilmeier, G. Warfield, and S. E. Harrison, J. appl. Phys. **34**, 2278 (1963).
[5] G. H. Heilmeier and S. E. Harrison, Phys. Rev. **132**, 2010 (1963).
[6] D. W. Barnum, J. inorg. nuclear Chem. **21**, 221 (1961).
[7] H. S. Jarrett, J. chem. Phys. **27**, 1298 (1957).
[8] R. M. Machado, Thesis, University of Grenoble, 1968.
[9] J. P. Mathieu, Thesis, University of Grenoble, 1969.
[10] A. Tissier, Thesis, University of Grenoble, 1970.
[11] A. Nath, S. Khorona, P. K. Mathur, and S. Sarup, Indian J. Chem. **51**, 4 (1966).
[12] B. Morosin, Acta cryst. **19**, 131 (1965).
[13] I. Hanazaki, F. Hanazaki, and S. Nagakura, J. chem. Phys. **50**, 276 (1969).
[14] J. P. Collman, in: Reactions of Coordinated Ligands, Amer. Chem. Soc., Washington (D.C.) 1963 (p. 78).
[15] A. Nath, M. P. Klein, W. Kündig, and D. Lichtenstein, Radiat. Eff. **2**, 211 (1970).
[16] P. X. Armendarez and L. S. Forster, J. chem. Phys. **40**, 273 (1964).
[17] T. P. Melia and R. Merrifield, J. inorg. nuclear Chem. **32**, 1489 (1970).
[18] S. M. Schildcrout, F. G. Pearson, and F. E. Stafford, J. Amer. Chem. Soc. **90**, 4006 (1968).

[19] A. ROSE, Concepts in Photoconductivity and Allied Problems, Interscience Publishers, 1963.
[20] A. SUSSMAN, J. appl. Phys. **38**, 2738 (1967).
[21] H. P. D. LANYON Phys. Rev. **130**, 134 (1963).
[22] M. A. LAMPERT and P. MARK, Current Injection in Solids, Academic Press, 1970 (p. 140).
[23] W. HELFRISCH and P. MARK, Z. Phys. **166**, 370 (1962).

(Received January 28, 1972)

phys. stat. sol. (a) **10**, 555 (1972)

Subject classification: 13.4 and 16; 14.3.3; 22.1.2

Départment de Recherche Fondamentale, Physique du Solide,
Centre d'Etudes Nucléaires de Grenoble,
and Université Scientifique et Médicale de Grenoble

Photoconductivity as a Tool for the Study of Deep Electronic Traps of Metal–Oxide–Silicon Sandwiches[1])

By

F. Barruel and J. C. Pfister

Photoconductivity measurements associated with capacitance–voltage curves have been used as a tool to study electronic traps in thermally grown SiO_2. The oxides are grown and metallized under very clean conditions and it is shown that the presence and amount of deep traps is directly related to the oxygen pressure during the last high temperature anneal after oxidation. When deep traps are present, the results are consistent with a trapped space charge model.

On a étudié, par photoconductivité et tracé des courbes capacité–tension, les pièges contenus dans la silice thermique d'une structure MOS. Les oxydes sont obtenus, puis métallisés, dans des conditions de propreté particulières et l'on montre que la présence et la concentration de pièges profonds dans le SiO_2 sont directement reliées à la pression d'oxygène du recuit qui suit immédiatement l'oxydation. Quand il y a des pièges profonds les résultats s'interprètent dans un modèle de charge d'espace piégée.

1. Introduction

The photoemission of electrons from both metal and silicon electrodes into the conduction band of SiO_2 has been used [1 to 3] to measure the thresholds associated with band structure relationships in MOS structures. We have associated photoconductivity measurements with capacitance–voltage curves to study electron traps in thermally grown SiO_2. The oxides were grown and metallized under very clean conditions, as described in [4], and we show that the presence and amount of deep traps is directly related to the oxygen pressure during the last high temperature anneal after oxidation.

2. Experimental Technique

2.1 Samples

The preparation of the samples has already been described [4, 5]. All treatments (ultra-high vacuum cleaning, oxidation, annealing, and electrode deposition) are made in the same ultra-high vacuum equipment and without any exposure to room air. The oxide films are grown on (111) oriented 10 Ωcm p-type wafers by heating at 1250 °C under 100 Torr of dry oxygen for 6 h. The resulting oxide layer is approximately 2500 Å thick and the final annealing treatment at 850 °C does not lead to any significant variation in thickness. The annealing pressures used in the present study were 10^{-9}, 10^{-4}, 1, 5, 10, 40,

[1]) The present work has been supported in part through DGRST contract No. 71 7 2604.

and 100 Torr. It has been shown [5] that this annealing is equivalent to a modification of the oxidation parameters. All other parameters in the preparation of the samples were kept constant. Semitransparent aluminium electrodes are evaporated after the last annealing and it has been verified that they effectively shield the interface against impurity diffusion from the ambient.

2.2 *Photoconductance measurements*

The sample is illuminated by the output of a Perkin-Elmer monochromator equipped with a high pressure xenon arc source giving an intense continuum in the ultraviolet range. The band width used is 0.05 eV for 0.5 mm slit width.

The current is measured by a vibrating reed electrometer (Cary 401) which allows measurements down to 10^{-17} A. The quantities that are systematically measured are:

dark current and equilibrium photocurrent versus bias with negative voltage applied on the metal electrode (leading to electron emission from aluminium being the dominant injection mechanism);

equilibrium photocurrent versus photon energy $h\nu$, for a constant bias.

These results are used to draw curves of the quantum yield versus photon energy, assuming that the photon absorption probability is independent of energy and the true quantum yield (photocurrent per absorbed photon) is proportional to the photocurrent per incident photon.

Transient currents are systematically registered but no attempt has been made to give a quantitative interpretation.

2.3 *Flat band voltage shift*

The shift in the flat band voltage is related to the amount and position of fixed charges in the oxide. Low-frequency capacitance–voltage curves are measured using the voltage ramp method with an electrometer. As no significant deformation of the C–U curves appears during bias illumination treatments, we consider that illumination does not change the surface state density and the position of the minimum of the C–U curve on the voltage axis is used to measure the shift in flat band voltage. The accuracy of this measurement is estimated to be ± 0.5 V or 3×10^{15} charges/cm^3 for charges uniformly distributed through the oxide layer.

2.4 *Release of trapped electrons*

Illumination using photons with energy below the threshold for photoemission from the electrode releases carriers from all traps shallower than the photon energy used. The current transients due to this effect can be integrated to yield the amount of trapped electrons versus trapping energy. If the density of trapped electrons is high enough, their release may also result in a shift of the flat band voltage. In all our samples the number of electrons released was below the threshold of the voltage shift measurements, so that only one measuring technique was available.

3. Results

3.1 *Main features*

The dark current is in the range of 10^{-15} to 10^{-14} A/cm^2 for fields between 10^5 and 10^6 V/cm. This low value is a result of the very clean and dry conditions prevailing during oxidation and metallization of the samples.

The photocurrents are in the range 10^{-13} to 10^{-9} A/cm^2 for the light intensities used. The quantum yield R corrected for interference effects [6] is shown in Fig. 1, and can be explained from the known band structure of MOS systems; the 3 eV threshold corresponds to electron injection from the aluminium electrodes whereas the increase for $h\nu > 4$ eV is due to both the onset of hole injection from Si and (at least for samples annealed under high oxygen pressures) bulk photoconductivity involving deep traps in SiO_2, as will be described below.

The photocurrents are always linear with light intensity for all operating points used in this work.

The samples can be separated into three groups, with qualitative differences:
samples annealed under low oxygen pressures: 10^{-9} and 10^{-4} Torr,
samples annealed under intermediate pressures: 1, 5, and 10 Torr,
samples annealed under high pressures: 40 and 100 Torr.

All three groups show the same behaviour as regards the release of trapped electrons: Some 10^{14} electrons/cm^3 are released for photon energies centered about 2 eV. No measurable shift in the capacitance–voltage curves can be detected for this very small amount of charge, so that no information is available regarding the distribution of trapped electrons within the oxide layer.

3.2 Samples annealed under low oxygen pressures

3.2.1 The current–field characteristics for steady-state photocurrent are linear on a log–log plot ($I \sim E^n$), with slopes slightly above two, except for a tendency towards saturation at very high fields.

3.2.2 No sizable current transients beyond the RC time constant of the measuring system are observed, for any direction of the change in illumination.

3.2.3 No shift of the capacitance–voltage curve is observed.

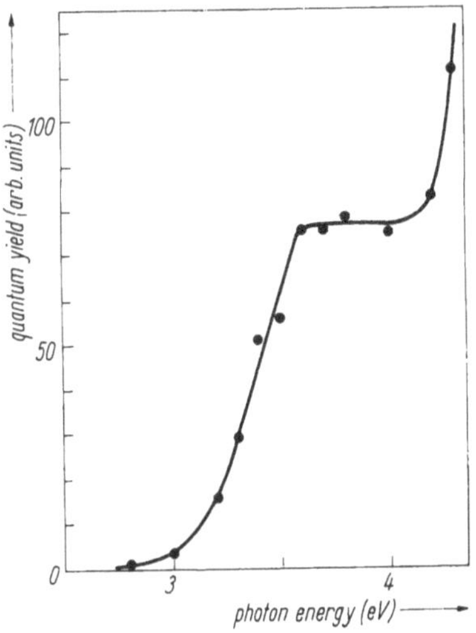

Fig. 1. Quantum yield versus photon energy

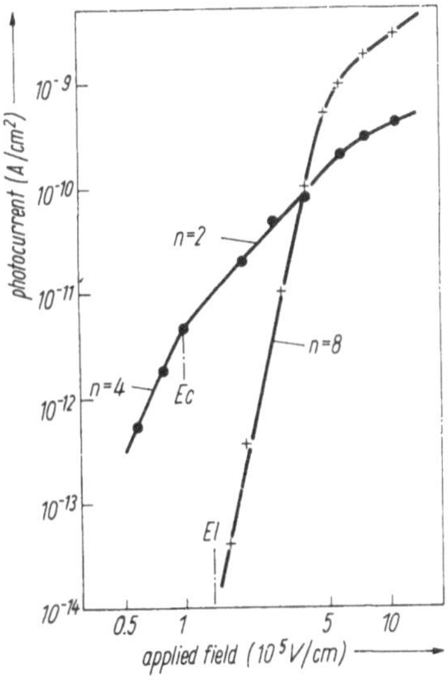

Fig. 2. Photocurrent versus applied field for various samples, annealed at oxygen pressure of 1 Torr (●) and 10 Torr (+)

3.3 Samples annealed under intermediate pressures

3.3.1 The I–E curves, shown in Fig. 2 on a lg–lg plot, show high slopes for low fields, up to a critical value E_c increasing with oxygen pressure. Above E_c the slope is about two and shows the same tendency towards saturation as seen in the preceding group. E_c is independent of photon energy. Below a threshold field E_l the photocurrent is very low and comparable with dark current.

3.3.2 Very long transients are observed after a change either in voltage or in photon energy. Typical values of stabilization time are a few days. The only exception is the very fast decay of the current when illumination is shut off. The transient current after an increase in voltage is higher than the steady-state value and decays slowly towards it. The transient current after a decrease in voltage is below the steady-state value and increases towards it.

3.3.3 The slope of the I–E curve below E_c decreases with increasing photon energy in the range 3 to 4 eV, as may be seen on Fig. 3.

3.3.4 The $C(U)$ curves measured after bias illumination treatments lead to the ΔU shift shown on Fig. 4 versus applied field. (Numbers refer to the order of successive bias–illumination treatments.) For long treatments ΔU reaches an asymptotic value which appears as a linear function of applied field E. Treatments 1, 2, 3 have not been applied for a time long enough and do not correspond to an equilibrium state. A distance ΔU_0 separates the initial state from the stable state. No shift in ΔU is observed after bias dark treatments (No. 11, 13).

3.3.5 The quantum yield in the range 3 to 4 eV shows hysteresis: The yield is higher after illumination with high-energy photons ($h\nu > 4$ eV) and decreases slowly during measurements, with an estimated time constant of a few weeks (see Fig. 5).

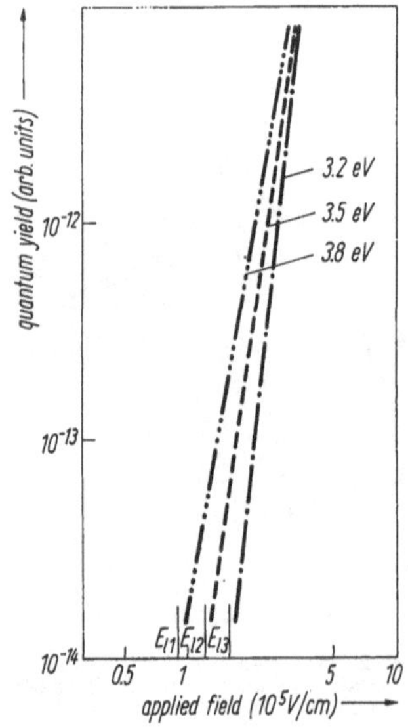

Fig. 3. Quantum yield verus applied field for three energy photons

Fig. 4. ΔU shift of the $C(U)$ curves versus applied field

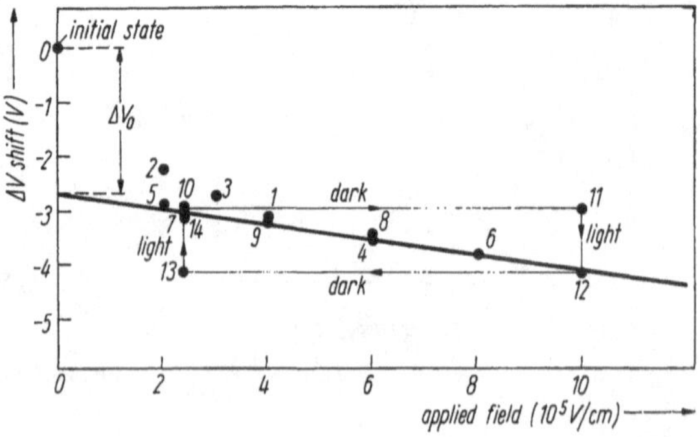

Fig. 5. Irreversibility of the quantum yield curve versus photon energy for an applied field lower than E_c

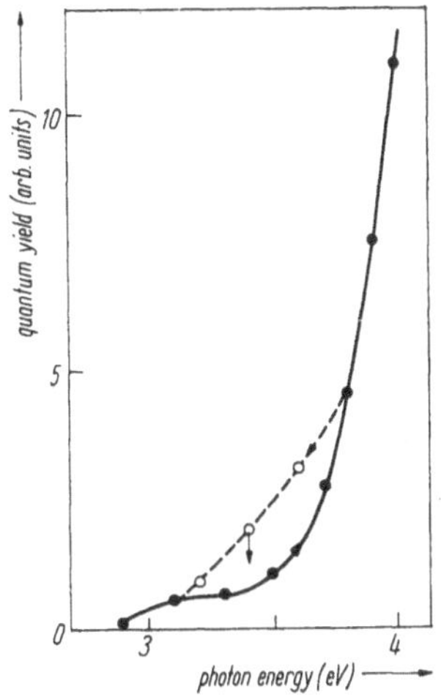

3.4 Samples annealed under high oxygen pressures

3.4.1 The I–E characteristics are also of the form $I \sim E^n$, with n between two and three and show the same tendency towards saturation for high field as in the preceding groups.

3.4.2 The current transients are shorter than in the preceding case (a few hours) but definitely observable and longer than the RC time constant of the measuring system.

3.4.3 The slopes of the ΔU–E curves are higher than in the preceding group and the time constants for equilibration of the ΔU shift and of the current are equal.

4. Discussion

4.1 Samples annealed under intermediate pressures

The long transients (3.3.2) as well as the variations of the ΔU shift (3.3.4) show clearly the existence of traps in the oxide and the direction of the voltage shift corresponds to electron trapping. If the trapped electrons are assumed to be uniformly distributed through the oxide, their densities are in the range 5×10^{15} to $2 \times 10^{16}/\text{cm}^3$, from ΔU_0 measurements.

Detrapping experiments show only very small amounts of charge in the "shallow" traps (depth below 2.8 eV), so that we may conclude to the existence of a high density of very deep traps near the midgap. Measurements of the I–E characteristics at nitrogen temperature confirm that no thermally activated process is involved.

The general behaviour of these samples is consistent with a trapped space charge model. The equilibrium between free and trapped electrons occurs only through photoexcitation and capture of carriers. In the steady state, injection from the metallic electrode is very small (the electric field being zero at the interface), and so serves to counterbalance the electrons excited out of the space charge volume. From a comparison between the trapped charge and the amount of charges traversing the oxide layer, the trapping probability of an njected electron is of the order of 1%. Once dynamic equilibrium is reached, one can write

$$n\,(N_t - n_t)\,C_e = f\,n_t\,C_{ph}$$

with n concentration of free electrons, n_t concentration of trapped electrons, N_t concentration of traps, C_e capture coefficient of electrons, C_{ph} photon absorption coefficient with excitation of a free carrier, and f light intensity (photons/cm^{2}s).

Thus, if $n_t \ll N_t$ the photocurrent is linear with light intensity for any space charge regime, which is consistent with experimental results.

The shape of the transient currents (3.3.2) also agrees qualitatively with this picture: After an increase in the applied field the trapped space charge is lower than the new steady-state value and the excess injected charge will be present as free carriers, leading to a high current which decays as more and more electrons are trapped into immobile space charge. On the other hand, after a decrease in applied field the trapped charge is higher than the new steady-state value and injection at the electrode will be blocked until the excess space charge is excited out of the traps.

The steady-state situation is the same as for a space-charge-limited current (SCLC) [7] with photoexcitation replacing thermal release of trapped carriers.

For an applied field $E < E_l$ and a given photon energy $h\nu < 4$ eV, the corresponding amount of trapped charge fills the deepest traps which cannot be photoexcited: dark current is measured (3.3.1) and E_l increases with decreasing photon energy (Fig. 3). The analogue in the space-charge-limited current model is the "deep trap" situation.

The threshold field E_c corresponds to a complete filling of the traps, which is the so-called trap-filled limit voltage in the corresponding SCLC model. E_c is independent of photon energy (3.3.1) and as E_l is not, it is clear that the slope of the I–E curves increases when photon energy decreases (3.3.3).

When photon energy reaches 4 eV, half the energy gap of SiO_2, all traps can communicate with the bands and bulk photoconductivity is observed.

The tendency towards saturation of the current at high fields is not readily explained in the framework of our model and may be due to a specific high field effect in amorphous SiO_2 or to finite injection rate at the electrode. The latter explanation is obviously consistent with the observed linear dependence of current on light intensity.

Using the assumption of space-charge-limited current with zero electric field at the metallic electrode along with the applied values of U, we can estimate the density of charged traps. In a simplified model assuming constant space charge density ϱ_0 for $0 \leqq x \leqq x_0$ and no charge for $x_0 < x \leqq L$ (L being the thickness of the oxide corresponding to capacitance C_0) elementary electrostatics yields

$$U = \frac{\varrho_0\,x_0}{C_0}\,\frac{(1 - x_0)}{2\,L} \qquad (U \text{ applied bias})\,.$$

This expression is obtained assuming that the field resulting from the applied voltage and the charge distribution described above is zero for $x = 0$. This is equivalent to assuming a perfectly ohmic metal electrode. The image charge of the same distribution leads to a ΔU shift

$$\Delta U = \frac{\varrho_0\,x_0}{C_0}\,\frac{x_0}{2\,L}\,.$$

The observed slope of ΔU vs. U yields a value of x_0 which is typically 250 Å associated with a space charge density $\varrho_0 = 10^{18}/\text{cm}^3$. We made no attempt at determining a more precise profile of this space charge density and the above values are to be considered as averages.

Irreversibility of the quantum yield curves (3.3.5) is also consistent with our model. During illumination with photons above 4 eV, the injected holes from the valence band of silicon neutralize the negative trapped space charge. If the next treatment is applied with photons below 4 eV (emission of holes is blocked), the negative space charge must be created again and photocurrents higher than the equilibrium values are momentarily measured. After long transients the stable ascending photocurrent curve is reached (Fig. 5).

4.2 Samples annealed under high oxygen pressures

From ΔU–E curves (3.4.3), the trapping of electrons is also effective in this group of samples, but no E_1 threshold or trap-filled limit voltage E_c are observed on the I–E curves. Transients for equilibration between free and trapped electrons are shorter than in the preceding group (3.4.2). One interpretation of these facts could be that the deepest traps have disappeared, leaving only traps very near to 3 eV. Exploration of these traps is difficult because the release from traps is mixed with the threshold for injection from the aluminium. Another explanation is more consistent with other data, in particular the increase in surface state density with oxygen pressure [8]. For high trap density, the Coulomb interaction between neighbouring traps is no longer negligible and will lead to a variation in the depth of each individual trap according to whether its neighbours are full or empty. For 10^{18} traps/cm^3, these fluctuations in electrostatic potential due to the neighbours may reach 0.5 eV. Accordingly, no trap can be considered to be inaccessible to photon excitation.

4.3 Samples annealed under low oxygen pressures

The very short transients and the absence of any measurable fixed space charge (3.2.2, 3.2.3) lead us to believe that these samples contain no traps analogous to those described above.

Thus the traps are probably associated with stoichiometric defects related to oxygen pressure, e.g. oxygen interstitial or silicon vacancy. This conclusion is in agreement with measurements of surface state density as a function of oxygen pressure in the same samples [5].

5. Conclusion

Photoconduction studies in MOS capacitors reveal the existence of deep electron traps in samples annealed in dry oxygen. The fact that other authors working on MOS capacitors have not reported any comparable trapping effect is probably due to their using samples with minimum oxide charge, which are obtained by annealing under reducing and/or wet conditions.

Our measurements show a correlation between the trap density and the oxide charge which vary in the same way with oxygen pressure.

Experiments on samples annealed under various pressures of water vapour, such as have been made to reduce surface state densities [9], are necessary to clarify the relation between electronic traps, oxide charge, and surface states. These experiments are now being prepared.

Acknowledgements

The authors are indebted to Mr. L. Revoil for preparing the samples and to Mr. M. Nescroi for helping in experiments.

References

[1] A. GOODMAN and J. O'NEILL, J. appl. Phys. **37**, 3580 (1966).
[2] R. WILLIAMS, Phys. Rev. **140**, A569 (1965).
[3] R. WILLIAMS, J. appl. Phys. **37**, 1491 (1966).
[4] K. SAMINADAYAR, J. L. PAUTRAT, and J. C. PFISTER, J. Phys. Chem. Solids **29**, 1709 (1968).
[5] J. L. PAUTRAT, Thesis, University of Grenoble, 1970.
[6] R. J. POWELL, J. appl. Phys. **40**, 5093 (1969).
[7] M. A. LAMPERT, Rep. Progr. Phys. **27**, 329 (1964).
[8] J. L. PAUTRAT and J. C. PFISTER, to be published.
[9] E. KOOI, Philips Res. Rep. **20**, 578 (1965).

(Received February 7, 1972)

phys. stat. sol. (a) **10**, 563 (1972)

Subject classification: 12.1; 21

Physics Department, Sri Venkateswara University, Tirupati, India

Third-Order Elastic Constants of Aluminium

By

V. P. N. Sarma and P. J. Reddy

The six independent third-order elastic constants of aluminium single-crystal were determined at 298 °K by measuring the uniaxial stress derivatives of the effective second-order elastic constants of the crystal in ⟨110⟩ direction. A pulse interference technique at 10 MHz was used, capable of measuring natural velocities to a few parts in 10^6. The specimen was irradiated with a neutron beam to pin down dislocation network under uniaxial stresses.

Die sechs unabhängigen elastischen Konstanten dritter Ordnung des Aluminium-Einkristalls wurden bei 298 °K durch Messung der Abhängigkeit der Konstanten zweiter Ordnung von einer in ⟨110⟩-Richtung angelegten Zugspannung bestimmt. Es wurde eine Impuls-Interferenztechnik bei 10 MHz und einer Präzision von einigen ppm angewandt. Die Probe wurde mit Neutronen bestrahlt, um das Versetzungsnetzwerk bei Anlegen der Spannung festzuhalten.

1. Introduction

In recent years studies on the anharmonicity of crystals are becoming extensive. A knowledge of the third-order elastic constants of a crystal helps to calculate many of its anharmonic properties. To determine these constants, earlier workers studied the second-order elastic constants of the crystal as a function of hydrostatic stress. These experiments did not give all the independent third-order elastic constants but a few combinations of them. A study of the sound velocity changes, however, as a function of uniaxial stress in appropriate directions gives a complete set of third-order elastic constants. Or, this data when combined with the data from hydrostatic pressure experiments give all the independent third-order elastic constants.

In this paper, third-order elastic constants of aluminium are reported. Aluminium is a cubic crystal and is characterised by six third-order elastic constants, C_{111}, C_{112}, C_{123}, C_{144}, C_{166}, and C_{456}. The natural sound wave velocities are determined along the ⟨110⟩ direction of the crystal with stress applied normal to the direction of propagation. From the velocity-stress derivatives, the six third-order elastic constants are evaluated. The notation by Thurston and Brugger [1] of the third-order elastic constants has been followed.

2. Experimental

2.1 Specimen and transducers

The aluminium sample used in the present investigations was obtained from Poland. It was of high purity (99.999%) and oriented with (001), (110), and ($1\bar{1}0$) faces. Acoustic surfaces were prepared by hand lapping. The orientation of the faces was then checked by Laue back-reflection and found to correspond to the crystalline planes within 1°.

X-cut and Y-cut quartz transducers of resonance frequency 10 MHz were employed for longitudinal and shear excitations, respectively. Phenyl salycilate (Salol) was used as the bonding agent because of low differential compressibility between quartz transducer and the crystal specimen. The aluminium sample was annealed and irradiated with a neutron beam at the APSARA reactor of the Bhabha Atomic Research Centre, Bombay, for one week with a total neutron flux of 5×10^{17} neutrons/cm^2 and of energy of about 200 keV. This was done in order to arrest the movement of the dislocation network when the specimen was under stress, and to raise the critical shear stress of the specimen. The maximum stress applied was about 35 kp cm^{-2} in all experimental runs.

2.2 *Velocity measurement*

Adiabatic sound velocity changes were measured using the pulse interferometer technique developed by us based on the principle of Williams and Lamb [2]. The technique is capable of measuring the changes in velocity to an accuracy of a few parts of 10^6. A block diagram of the electronic system for this technique is shown in Fig. 1. A stable rf oscillator is used to generate the carrier frequency of the input pulse and this frequency can be continuously varied in the range 9 to 11 MHz. The gate signal generator provides a pair of flat-topped af pulses each about 2 μs length and about 90 V height. The output of the rf oscillator is gated out to produce two phase coherent rf pulses. The amplitude and the time delay of the second pulse is variable relative to the first. The rf signal at the receiving transducer is amplified by a band pass amplifier and displayed on a Tektronix Oscilloscope (Type 545A) which has a flat rise calibrated preamplifier plug-in unit (Type L). The carrier frequencies are measured using a Beckmann Digital Frequency Counter (Type 6155).

The principle and method of the velocity measurement has been adequately described by Williams and Lamb [2]. This pulse interference method is capable of very high accuracy because the basic measurement involves the frequency value rather than time delay. Also, it makes possible to evaluate quantitatively the effect of phase shift associated with the reflection of the sound wave at the interface of the sample and the transducer.

2.3 *Uniaxial press system*

The uniaxial stresses are applied by a lever mechanism through a piston type arrangement. The piston arrangement is formed by a pair of stainless steel tubes, the inner tube being movable inside the outer fixed tube without any friction. The fixed tube supports the specimen, which is placed in between two flat

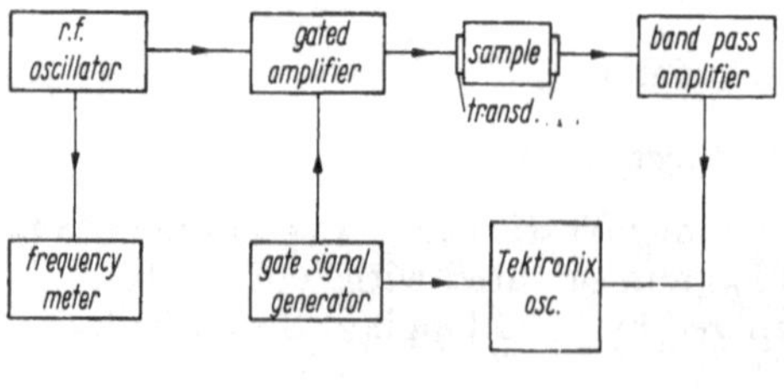

Fig. 1. Block diagram of the electronic system for the pulse interference technique

monel-alloy blocks. A ball bearing is used between the upper plate and the bottom of the movable stainless steel tube to keep the stress axis well aligned. The force load is furnished by a set of calibrated weights which are placed in a pan suspended from the lever arm made of carbon steel. The system can generate a pressure of 50 kp cm^{-2} on a half-inch cube.

3. Results and Discussion

The natural propagation velocities of longitudinal and shear waves were measured in ⟨110⟩ direction as a function of uniaxial stress applied normal to the direction of propagation. The details of the direction of propagation and the directions of the applied stress are given in Table 1. The measurements were

Table 1

Propagation modes and stress directions used in the present study

expt. no.	stress direction	propagation direction	polarisation direction	effective second-order constant w
1	$(1\bar{1}0)$	(110)	(110)	$\frac{C_{11} + C_{12} + 2\,C_{44}}{2}$
2	$(1\bar{1}0)$	(110)	$(1\bar{1}0)$	$\frac{C_{11} - C_{12}}{2}$
3	$(1\bar{1}0)$	(110)	(001)	C_{44}
4	(001)	(110)	(110)	$\frac{C_{11} + C_{12} + 2\,C_{44}}{2}$
5	(001)	(110)	$(1\bar{1}0)$	$\frac{C_{11} - C_{12}}{2}$
6	(001)	(110)	(001)	C_{44}

made with both increasing and decreasing stresses. In each run, constant temperature was maintained by checking with copper–constantan thermocouples. The data were normalized to 298 °K using the temperature derivatives of the velocities. The quantity, $\varrho_0 W^2$ calculated from the measured velocities are potted against stress in Fig. 2. Here ϱ_0 is the density of the unstressed specimen, W the natural velocity defined as $2\,L_0/t$ where L_0 is the length of the sound path in the unstressed medium and t, the round-trip transit time. It is seen that

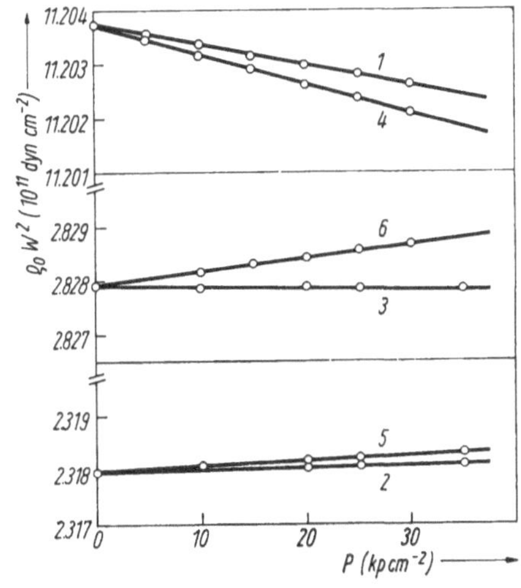

Fig. 2. Variation of the effective elastic constants with uniaxial stress for different modes of vibration and stress directions. The numbers indicated refer to the experiment numbers given in Table 1. – ● – increasing stress, – ○ – decreasing stress

the changes in $\varrho_0 W^2$ are linear up to the highest stress applied. This is an indication to the effect that there might not be any dislocation line movement in the crystal under stress. From the graph, the slopes $(\partial/\partial P)(\varrho_0 W^2)$ at $P = 0$ are determined and are presented in Table 2 with associated errors. The explicit

Table 2

Thurston-Brugger relations along with the measured slopes, $[(\partial/\partial P)(\varrho_0 W^2)]_{P=0}$ in the six experiments shown in Table 1

expt. no.	$[(\partial/\partial P)(\varrho_0 W^2)]_{P=0}$ effective values	measured value
1	$w(a-b+2c)+\frac{1}{4}(a-b)C_{111}+\frac{1}{4}(5a-3b)C_{112}+$ $+(a-b+4c)C_{166}+aC_{144}+\frac{1}{2}aC_{123}$	-3.77 ± 0.10
2	$w(a-b-2c)+\frac{1}{4}(a-b)C_{111}+\frac{1}{4}(a+b)C_{112}-\frac{1}{2}aC_{123}$	$+0.40 \pm 0.03$
3	$2wa+\frac{1}{2}(a-b)C_{144}+\frac{1}{2}(3a-b)C_{166}+2cC_{456}$	-0.13 ± 0.13
4	$2wa+\frac{1}{2}aC_{111}+\frac{1}{2}(3a-b)C_{112}+2aC_{166}-$ $-bC_{144}-\frac{1}{2}bC_{123}$	-5.75 ± 0.13
5	$2wa+\frac{1}{2}aC_{111}-\frac{1}{2}(a+b)C_{112}+\frac{1}{2}bC_{123}$	$+1.01 \pm 0.03$
6	$-2wb+aC_{144}+(a-b)C_{166}$	$+2.50 \pm 0.09$

$$a = \frac{C_{12}^{\mathrm{T}}}{3B^{\mathrm{T}}(C_{11}^{\mathrm{T}} - C_{12}^{\mathrm{T}})}, \quad b = \frac{C_{11}^{\mathrm{T}} + C_{12}^{\mathrm{T}}}{3B^{\mathrm{T}}(C_{11}^{\mathrm{T}} - C_{12}^{\mathrm{T}})}, \quad c = \frac{1}{4C_{44}},$$

$3B^{\mathrm{T}} = C_{11}^{\mathrm{T}} + 2C_{12}^{\mathrm{T}}$,

w is the effective second-order elastic constant of the particular mode.

relations developed by Thurston and Brugger [1] for these slopes in terms of the second- and third-order elastic constants are also given in Table 2. Using these relations, the six third-order elastic constants are solved by computer programming. The required second-order isothermal moduli are computed from the measured adiabatic values using the necessary heat capacity and thermal expansion data [3]. The third-order elastic constants obtained are presented in Table 3

Table 3

Third-order elastic constants of aluminium at 298 °K in units of 10^{12} dyn cm^{-2}

constant	present study	Thomas, Jr.
C_{111}	-12.24 ± 0.60	-10.76 ± 0.30
C_{112}	$-\ 3.73 \pm 0.38$	$-\ 3.15 \pm 0.10$
C_{123}	$+\ 0.25 \pm 0.25$	$+\ 0.36 \pm 0.15$
C_{144}	$-\ 0.64 \pm 0.08$	$-\ 0.23 \pm 0.05$
C_{166}	$-\ 3.68 \pm 0.13$	$-\ 3.40 \pm 0.10$
C_{456}	$-\ 0.27 \pm 0.07$	$-\ 0.30 \pm 0.30$

along with estimated errors. The values obtained earlier by Thomas, Jr. [4] for aluminium are also given for comparison. In his measurements, Thomas, Jr. has first measured $(1/W)(\partial W/\partial T)_P$ for a particular sound mode and then $(\partial T/\partial P)_W$ for the various stress derivatives of that mode. Combining these two measurements, he has evaluated the natural velocity–stress derivatives. Considering the fact that the experimental approach made in the investigations of Thomas and in the present study are different in evaluating the third-order elastic constants, the agreement between the present set of the constants which are obtained using uniaxial stresses alone and the values of Thomas, Jr. may be taken as good.

The three combinations of the constants, viz. $(C_{111} + 2\,C_{112})$, $(1/2)\,(C_{111} - C_{123})$, and $(C_{144} + 2\,C_{166})$ are evaluated from the hydrostatic stress measurements of Lazarus [5], Schmunk and Smith [6] and Ho and Ruoff [7] and are presented

Table 4

Comparison of the combinations $(C_{111} + 2\,C_{112})$, $\frac{1}{2}\,(C_{111} - C_{123})$, and $(C_{144} + 2\,C_{166})$ (in units of 10^{12} dyn cm^{-2}) as determined by different workers

type of measurement	$-(C_{111} + 2\,C_{112})$	$-\frac{1}{2}(C_{111} - C_{123})$	$-(C_{144} + 2\,C_{166})$	author
hydrostatic	22.5	—	7.4	Lazarus [5]
hydrostatic	20.14	6.20	7.81	Schmunk and Smith [6]
hydrostatic + uniaxial	17.06	5.56	7.03	Thomas, Jr. [4]
hydrostatic	20.15	6.26	8.0	Ho and Ruoff [7]
uniaxial	19.70	6.24	8.0	present study

in Table 4. These combinations are calculated from the independent third-order elastic constants determined by Thomas, Jr. and those determined in the present study and shown in the same table for comparison. The agreement seen is yet another indication that there was no dislocation movement in the specimen affecting the measurements in the present study.

Acknowledgement

The authors wish to thank Prof. J. Bhimasenachar, Professor of Physics for his interest in the present work and for helpful discussions. They are also thankful to Council of Scientific and Industrial Research for awarding research grants to carry out this project.

References

[1] R. N. Thurston and K. Brugger, Phys. Rev. **133**, A1604 (1964).
[2] J. Williams and J. Lamb, J. Acoust. Soc. Am. **30**, 308 (1958).
[3] American Institute of Physics Hand-book, McGraw-Hill Co., Inc., New York 1957.
[4] J. F. Thomas, Jr., Phys. Rev. **175**, 955 (1968).
[5] D. Lazarus, Phys. Rev. **76**, 545 (1949).
[6] R. F. Schmunk and C. S. Smith, J. Phys. Chem. Solids **9**, 100 (1959).
[7] P. S. Ho and A. L. Ruoff, J. appl. Phys. **40**, 3151 (1969).

(Received January 24, 1972)

phys. stat. sol. (a) **10**, 569 (1972)

Subject classification: 14.3.1; 1.5; 22.1.1

Electronics Research Laboratory, Norwegian Institute of Technology, Trondheim

Germanium Films on Sapphire and Germanium Substrates

By

J. S. JOHANNESSEN

Germanium films deposited simultaneously on sapphire and single crystal germanium substrates appear to have the same electrical transport properties. These results are in contrast to the observed crystalline structure, which show coherent crystalline film growth on germanium crystals while films on sapphire are polycrystalline with random crystallite orientation. Conduction in germanium films on germanium substrates is satisfactorily described by a composite conduction model in which the film is separated from the substrate by an interface barrier layer.

Auf Substraten von Saphir oder einkristallinem Germanium gleichzeitig aufgedampfte Germaniumschichten scheinen die gleichen elektrischen Transporteigenschaften zu haben. Diese Ergebnisse stehen im Widerspruch zur beobachteten Kristallstruktur der Schichten, die auf Germaniumkristallen ein kohärentes, kristallines Schichtwachstum aufweisen, während Schichten auf Saphir polykristallin mit beliebigen Kristallrichtungen sind. Die Leitung in Germaniumschichten auf Germaniumsubstraten wird durch ein zusammengesetztes Leitungsmodell, in dem die Schicht von dem Substrat durch eine Grenzflächenbarriere getrennt ist, in befriedigender Weise beschrieben.

1. Introduction

Considerable effort has been expended in optimizing the growth conditions in order to improve the structure and electrical properties of thin semiconducting films. The general observations are that comparatively high deposition (substrate) temperature and low deposition rate yield films of good crystalline quality with electrical properties comparable to that of the bulk material [1, 2]. Attention has been paid also to the influence of the substrate on the film properties [3], particularly the relation between the crystalline structure of the substrate and that of the deposit. Generally, observations indicate that the best crystalline structures are obtained for small lattice mismatch.

The present paper is concerned with an experimental investigation of the electrical transport properties of germanium deposited simultaneously under identical conditions on sapphire and on p-type single crystal germanium substrates. Germanium substrates provide a perfect lattice match while the sapphire substrates do not. However, the match in thermal expansion between germanium and sapphire is excellent over a wide range of temperature. X-ray diffraction patterns indicate that the crystalline structure of the deposits to a certain extent depends on the crystalline stucture of the substrate. The electrical transport properties of the deposits as deduced from measurement of resistance and Hall-effect are less sensitive to the crystalline properties of the substrate. The type of conduction of all deposited films is p-type. This is due to a high density of structural defects rather than autodoping from the substrate [4]. The experimental results are discussed on the basis of a model suggested for conduction in composite structures. The composite conduction model may be used to obtain information about the properties of the barrier which forms between the germanium film and the germanium substrate. The results of the present investigation show that this barrier cannot be regarded as a simple semiconductor junction.

2. Composite Conduction in a System of Germanium Film on Germanium Substrate

When the electrical transport properties of films deposited on insulating and semiconducting substrates are compared, care must be taken to account for the conduction in the semiconducting substrate.

Conduction along a thin semiconducting film deposited on a semiconducting substrate may be regarded as a composite structure represented by a film of resistance R_f and a substrate of resistance R_s separated by an interface barrier of resistance R_{sf}. The composite structure may be described by a set of self consistent differential equations relating currents and voltages as function of position along the interface. When solved subject to proper boundary conditions and in the limit of zero applied voltage the observable resistance of the composite structure is given by (see Appendix)

$$R = R_f \frac{R_s + R_f \chi(\eta)}{R_s + R_f}, \tag{1}$$

where

$$\chi(\eta) = \frac{\tanh \eta}{\eta} \tag{2}$$

is an interface barrier function related to the barrier resistance through

$$\eta = \left(\frac{R_s + R_f}{R_{sf}}\right)^{1/2}. \tag{3}$$

The composite Hall voltage can be obtained in terms of the Hall voltages V_H^f and V_H^s of film and substrate, respectively. The result is (see Appendix)

$$V_H = V_H^f + R_f \frac{V_H^s - V_H^f}{R_s + R_f} \Gamma(\eta), \tag{4}$$

where

$$\Gamma(\eta) = 1 - 2 \frac{\coth \eta - \operatorname{cosech} \eta}{\eta} \tag{5}$$

is a function of the interface barrier resistance through η.

The Hall voltages in the two regions of the composite structure are given by

$$V_H^f = V_{H0}^f \frac{R_s + R_f \chi(\eta)}{R_s + R_f} \tag{6}$$

and

$$V_H^s = V_{H0}^s \frac{R_f(1 - \chi(\eta))}{R_s + R_f}, \tag{7}$$

where V_{H0}^f and V_{H0}^s are the Hall voltages corresponding to the total current flowing in each of the separate regions.

For large values of the interface barrier resistance, such that $\eta < 1$, the barrier functions may be approximated by their asymptotic values

$$\chi(\eta) = 1 , \qquad \Gamma(\eta) = 0 . \tag{8}$$

For these conditions the composite resistance given by equation (1) and the composite Hall voltage given by equation (4) become

$$R = R_f , \qquad V = V^f_{H0} . \tag{9}$$

Thus, for sufficiently high interface barrier resistances the observed properties of the composite structure reduce to that of the deposited film.

On the other hand, small interface barrier resistances, that is $\eta \gg 1$, yield

$$\chi(\eta) \approx 0 , \qquad \Gamma(\eta) \approx 1 , \tag{10}$$

which corresponds to a resistance and Hall voltage of the two regions coupled in parallel

$$R = \frac{R_s\, R_f}{R_s + R_f} = R_{11} , \qquad V_H = \left(\frac{V^f_H}{R_f} + \frac{V^s_H}{R_s}\right) R_{11} . \tag{11}$$

For intermediate values of η, intermediate values of the observable quantities are obtained.

3. Experimental Methods

The germanium films are prepared by electron beam evaporation of intrinsic germanium onto single crystals of sapphire and p-type germanium.

The substrates are p-type germanium crystals, of resistivity of the order of 10 Ωcm, cut perpendicular to the [111] direction, and sapphire cut at an angle of 60° to the [0001] direction. All the substrates are given similar surface treatment. They are lapped and polished to an optical finish, degreased in a Teepol-water solution, rinsed in deionized water, and dehydrized in alcohol vapour. The germanium substrates are etched in CP4. Immediately before deposition the substrates are kept at 800 °C for several minutes in order to degas the substrates and to remove oxide contamination from the germanium surfaces. Above some 600 °C germanium oxide decomposes and an oxide free surface is obtained [5]. During all depositions the residual pressure has been less than 10^{-7} Torr, and the deposition rate kept approximately constant at a value of 10^3 Å/min. The substrates have been kept at fixed temperatures ranging from 700 to 800 °C. After deposition of the germanium films the substrate temperature is lowered towards room temperature and rhodium contacts are deposited before the samples are exposed to the ambient atmosphere.

The structure of the deposits is examined by X-ray powder technique and the surface topography observed by microscopy.

The electrical properties of the deposited films are deduced from measurement of resistance and Hall-effect in the temperature range between 80 and 300 °K. The Hall-effect is observed at a magnetic field of 10 kG.

4. Experimental Results and Discussion

4.1 Crystalline structure

For all depositions carried out in the present investigation, the observed X-ray diffraction patterns of germanium films deposited simultaneously on sapphire and germanium substrates are different. Typical diffraction patterns

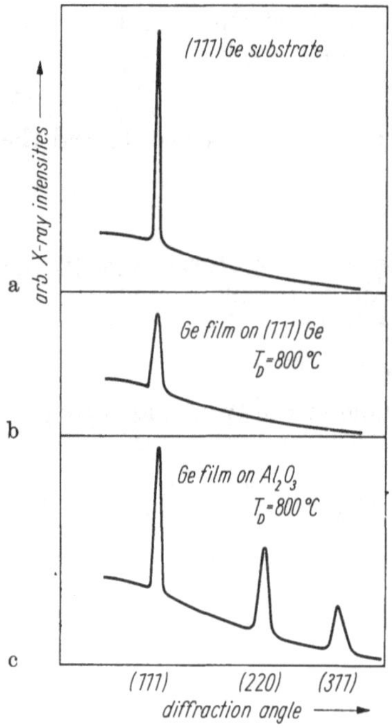

Fig. 1. Diffraction profiles for a) (111) oriented germanium substrate, b) germanium films deposited simultaneously on (111) germanium substrate, and c) sapphire substrate

are shown in Fig. 1. The films deposited on (111) germanium substrates show the (111) diffraction peak only (Fig. 1b). This result indicates coherent single crystal growth. However, the diffraction peak is broader than that of the substrate alone (Fig. 1a). Microscopic inspection of the surface of the films reveals a multicrystalline structure. Thus, the films deposited under the present conditions seem to be polycrystalline with crystallites of the same orientation. The broadened X-ray diffraction peak may be due to slight misorientation or structural defects in the film.

The diffraction pattern of films deposited simultaneously on sapphire is shown in Fig. 1c. This diffraction profile is similar to that of a germanium powder. In this case the crystallites nucleate and grow non-coherently with respect to the substrate, and there is a random mutual crystallite orientation. These results are in general agreement with previous studies of germanium on a number of substrate materials [6 to 8]. It is apparent that good lattice match between deposit and substrate promote coherent nucleation and growth of the deposit, while random nucleation and growth occur if the lattice match is poor.

4.2 Resistance

The temperature dependence of the resistance of pairs of simultaneously deposited films on sapphire and germanium substrates is shown in Fig. 2 for deposition temperatures of 750 and 800 °C. The open symbols represent deposition on sapphire and filled symbols represent deposition on germanium.

The resistance is measured in the limit of zero applied voltage. This is done because the resistance of films deposited on germanium substrates is voltage dependent, decreasing with increasing voltage, particularly at low temperatures. In contrast, the current–voltage characteristics of germanium films on sapphire and of the reference substrate are linear and symmetric at all temperatures. The nonlinear characteristics of germanium films on germanium substrates are, therefore, not due to contact phenomena or grain boundaries, but rather due to a voltage dependent interface resistance.

The resistance of films on sapphire is observed to be nearly temperature independent, because a decrease in hole density with decreasing temperature is almost compensated by a corresponding increase in hole mobility. These results are in general agreement with previous observations [9].

The observed resistance of the films deposited on germanium substrates shows, apart from the structure near room temperature, a characteristic increase at lower temperatures. In the low temperature region the resistance approaches values close to those of the films deposited simultaneously on sapphire. This is

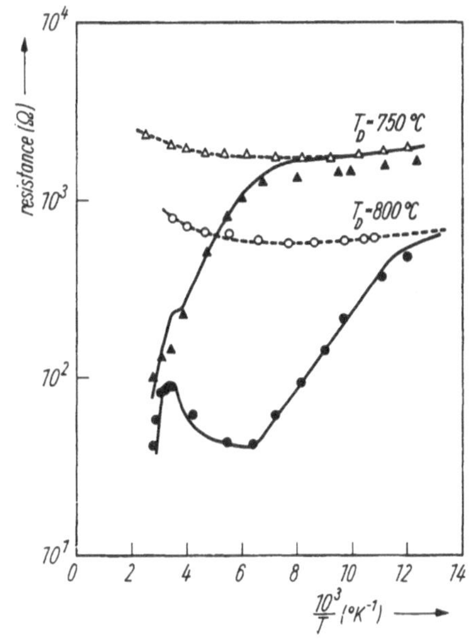

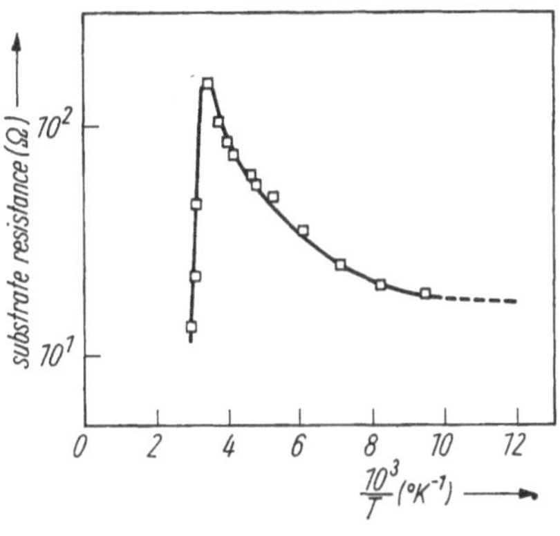

Fig. 3. Temperature dependence of resistance of the reference germanium substrate

Fig. 2. Temperature dependence of resistance of germanium deposited on sapphire (△, ○) and germanium substrates (▲, ●). The full lines are calculated composite resistance

particularly pronounced for the film deposited at 750 °C. The structure in the temperature dependence of the resistance that appears near room temperature is due to conduction in the substrate. For comparison and in order to illustrate this, the temperature dependence of the resistance of a reference substrate is shown in Fig. 3. The electrode arrangement is identical to that used on the thin films. The rapid decrease in substrate resistance above room temperature is due to intrinsic conduction. Below room temperature the conduction is extrinsic with approximately constant carrier density. The gradually decreasing resistance is therefore due to increasing hole mobility with decreasing temperature.

The observed temperature dependence of the composite resistance shown in Fig. 2 indicates three characteristic temperature regions:

i. A low temperature region of high interface resistance in which conduction takes place in the film only.

ii. A high temperature region of low interface resistance in which conduction takes place mainly in the substrate.

iii. An intermediate temperature region in which the interface resistance is of importance for the observed resistance.

The full lines of Fig. 2 are plots of the composite resistance as given by equation (1). In order to evaluate the composite resistance the following assumptions are made:

i. The resistance R_f of the composite structure is equal to the resistance of the film deposited simultaneously on sapphire at all temperatures.

ii. The substrate resistance R_s is equal to that of the reference substrate for all temperatures. The magnitude is adjusted by comparison of four point probe measurement of resistivity (at room temperature) of the substrate of the composite structure and the reference substrate.

iii. The temperature dependence of the interface resistance R_{sf} is obtained by fitting equation (1) to the observed composite resistance at temperatures where the transition occurs, and assumes that the temperature dependence is the same above and below this temperature region.

The first assumption is tied to the general observation that the electrical transport properties of a deposit is largely determined by the deposition temperature and deposition rate rather than the structure of the substrate as long as the thermal mismatch is small enough to prevent straining of the deposit upon cooling [4]. As shown in Fig. 2 the temperature dependence of the resistance of the films deposited on sapphire and germanium at 750 °C is practically the same at low temperatures. In fact, the difference in magnitude is less than 20% in the region below 125 °K. Since the electrical properties are defect controlled [4], it is reasonable to assume a similar temperature dependence of resistance of films deposited under identical conditions on substrates of different structure.

The second assumption is violated only if the electrical properties of that part of the substrate which is covered by the film is seriously affected by the deposited layer. That the substrate is comparatively unaffected by the deposited layer is inferred from the observed composite resistance of the deposition made at 800 °C (Fig. 2). The composite resistance at temperatures above 150 °K has a temperature dependence identical to that of the reference substrate. Thus the second assumption is justified.

There is no way of measuring the interface resistance and its temperature dependence directly. It is, however, reasonable to assume that the temperature dependence of the interface resistance deduced from composite resistance in a narrow region of temperature is not drastically different in adjacent temperature regions.

The temperature dependence of the interface resistance obtained in this manner may be written in the form

$$R_{sf} = R_0 \exp \frac{q\Phi}{kT}. \qquad (12)$$

In Table 1 are given the values of R_0 and Φ which are used in the computation of the composite resistances shown in Fig. 2. As is evident from Fig. 2 there is fair agreement between experimental values of the observed composite resistance and the values obtained by the composite model (equation (1)).

The present experimental results indicate that the conductivity of films deposited simultaneously under identical conditions on germanium and sapphire substrates is similar. The fair agreement obtained between the composite conduction model and experimental results for all temperatures indicate that the assumptions on which the computation rests are sound.

4.3 Hall effect

The Hall-effect observed in the films discussed in the previous section are shown in Fig. 4. The Hall voltage was positive in the region of temperatures between 80 and 300 °K for all films deposited on sapphire and germanium substrates. The open symbols represent experimental values for films on sapphire, while filled symbols represent experimental values for films on germanium. The Hall mobility corresponding to the observed Hall-effect is also shown in Fig. 4.

Fig. 4. Temperature dependence of Hall voltage (left) and corresponding Hall mobility (right) of germanium films deposited on sapphire (△, ○) and germanium substrates (▲, ●). The full lines are calculated composite Hall voltage. The upper dashed line indicates the temperature dependence of the Hall voltage of the reference substrate

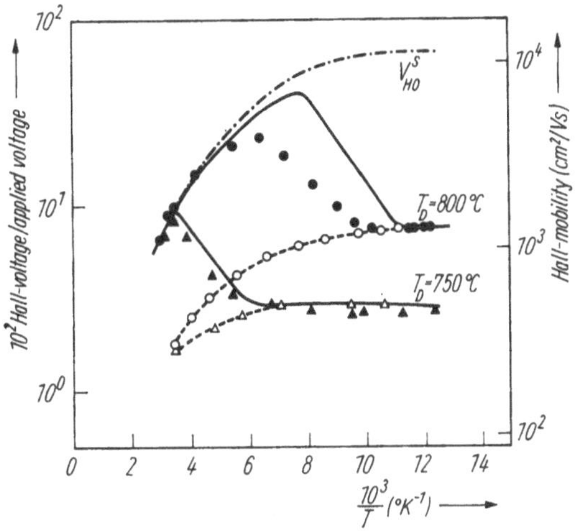

The full lines show the Hall voltage evaluated from the composite model equation (4), where it is assumed that the Hall-effect in films on germanium substrates is the same as that in films deposited simultaneously on sapphire. The temperature dependence of the Hall voltage of the substrates is deduced from the temperature dependence of the resistance of the reference substrate, under the assumption that the changes in resistance is mainly due to changes in mobility at temperatures below room temperature. The resistance data used in this computation are the same as those used in evaluating the composite resistance of Fig. 2.

The agreement between experimental results and computed values is fair. At temperatures near room temperature the apparent Hall-effect is mainly due to conduction in the substrates. The corresponding Hall mobilities are close to 1850 cm²/Vs, which is the phonon limited hole mobility in bulk p-type germanium [10]. In the low temperature region the Hall voltage of the composite structure approaches values close to the Hall voltage observed in films deposited simultaneously on sapphire substrates.

The low temperature Hall mobilities and carrier densities are $\mu_H = 1300$ cm²/Vs and $P_H = 1.4 \times 10^{16}$ cm^{-3} for deposition at 800 °C, and $\mu_H = 500$ cm²/Vs and $P_H = 5.2 \times 10^{16}$ cm^{-3} for deposition at 750 °C. These results show the critical dependence of the electrical transport properties on deposition temperature. The acceptors present in these films are structural defects such as dislocations and vacancy types of defects [4]. The denisty of these types of defects is strongly dependent on deposition temperature.

In the intermediate temperature region, where the transition is observed, there is a discrepancy between model and observations, particularly for the deposition at 800 °C. This discrepancy may be due to several sources. Firstly, the Hall effect is measured at a finite applied voltage, at which the interface resistance is voltage dependent. Thus the validity of the composite conduction model is violated. This will of course be particularly pronounced at temperatures where the interface resistance is of importance. Secondly, an overestimation of the Hall-effect in the substrate below room temperature may partly explain the discrepancy between model and observations. It is, however, felt that the agreement obtained between model and experimental results near room temperature and at low temperatures supports the conclusion arrived at in the discussion of the observed resistance in the same structures.

4.4 *Interface resistance*

The nonlinearity in current–voltage characteristic of the films deposited on germanium substrates is observed neither in films on sapphire nor in the germanium reference substrate. The observed nonlinearity may therefore be assigned to the interface resistance of the composite structure.

The dependence of the interface resistence on applied voltage may be used to gain further knowledge about the interface barrier between film and substrate. The observed voltage dependence of the composite resistance does provide such information only in an indirect manner, as is evident from the expression for the composite resistance equation (1). However, if the experimental conditions are such that the following conditions are satisfied:

$$\eta > 1 , \qquad R_f \gg R_s , \tag{13}$$

then the expression for the composite resistance as given by equation (1) may be rearranged and simplified to yield the interface resistance

$$R_{sf} \approx \frac{(R - R_g)^2}{R_f} . \tag{14}$$

The films deposited at 800 °C (Fig. 2) satisfy these conditions for all temperatures within the experimental range.

In Fig. 5 is plotted the interface resistance R_{sf} as function of the square root of applied voltage V_A for the film deposited at 800 °C. The curves correspond to the nonlinear interface resistance as obtained through equation (14) for several lattice temperatures. The straight line character of these curves indicate an exponential $-V_A^{1/2}$ dependence. It is observed also that the slope of these curves is proportional to $1/T$.

When the temperature dependence of the interface resistance at zero applied voltage is taken into account, the following approximate expression yields the interface resistance

$$R_{sf} = R_0 \exp \frac{q\,\Phi - \beta\, V_A^{1/2}}{kT} , \tag{15}$$

where β is a constant.

Attempts have been made to measure the interface resistance by alloying a contact to the back of the substrates. The characteristics of this configuration confirmed the exponential $-V_A^{1/2}$ dependence, but also showed polarity effects, with higher resistance when the film was biased negatively with

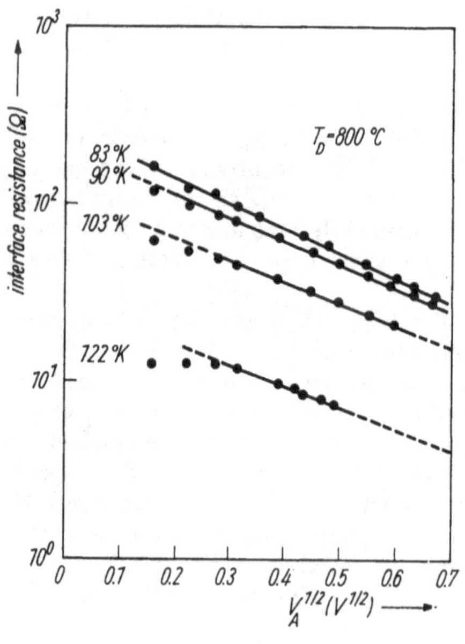

Fig. 5. Plot of evaluated interface resistance vs. square root of applied voltage for germanium deposited on germanium at 800 °C. The parameter is lattice temperature

respect to the substrate. The expression for the interface resistance equation (15) does not correspond to a simple semiconductor junction as far as the voltage dependence is concerned [11]. There is no simple relation between the applied voltage V_A of equation (15) and the voltage V of equation (15) and (16). However, for experimental conditions which satisfy the restrictions of equation (13), the voltage across the barrier may be approximated by V_A.

Conduction by electrode limited Schottky emission across a barrier [12]

$$i = i_0 \exp \frac{\beta_S V^{1/2} - q \Phi_S}{kT} \tag{16}$$

is of a similar form as that observed, and this is also the case for Poole-Frenkel conduction in the bulk of an interface layer [13]

$$\sigma = \sigma_0 \exp \frac{\beta_{PF} V^{1/2} - q \Phi_{PF}}{kT}, \tag{17}$$

where

$$\beta_{PF} = \left[\frac{q^3}{\pi \varepsilon \varepsilon_0 d}\right]^{1/2} \tag{18}$$

and

$$\beta_{PF} = 2 \beta_S . \tag{19}$$

In these expressions the electronic charge is q, the dielectric permittivity is $\varepsilon \varepsilon_0$, and the thickness of the barrier is d.

The Poole-Frenkel conduction mechanism immediately lends itself to an interface resistance as given by equation (15). This conduction mechanism is characteristic for conduction in amorphous semiconductors [14], but it is independent of polarity. The potential barrier Φ_{PF} is the ionization potential of localized levels in the bandgap of the amorphous semiconductor. Values reported for the conductivity activation energy of amorphous germanium are ranging from 0.2 eV near room temperature down to 0.05 eV at 50 °K [15]. Note that the values of Φ given in Table 1 lie within this range of activation energies.

Table 1

Zero voltage interface resistance parameter R_0 and potential barrier height Φ for germanium deposited on germanium substrates at different deposition temperatures

T_D (°C)	R_0 (Ω)	Φ (V)
750	4.6×10^{-2}	0.145
800	3.2×10^{-4}	0.110

Previous investigations of germanium deposited on single crystal germanium substrates reveal a heavily distorted interface layer [16]. By etching techniques the thickness of these layers is estimated to be between 0.5 and 3 μm, strongly dependent on the surface treatment of the substrate. Careful treatment reduced the thickness of the distorted layers. These observations show that the film–substrate interface is not a simple semiconductor junction, but rather a region of material with structural properties different from the bulk of the deposit and the substrate itself.

A similar etching technique was used in the present investigation, but an interface region was not observed. It is, however, possible to evaluate the barrier thickness d of equation (18) from the slopes of the curves in Fig. 5. The values of d arrived at is of the order of 350 Å, which is considerably less than those observed by etching technique [15]. This may also explain why an interface region was not resolved by chemical etching.

A distorted interface layer supports a Poole-Frenkel type of conduction mechanism across the interface. Nonlinear conduction in amorphous semiconductors is observed at electric fields of the order of 10^4 to 10^5 V/cm [16]. In the present investigation, nonlinear conduction occurs at applied voltages of 0.1 V, which for barrier thicknesses of 350 Å corresponds to electric fields between 10^4 to 10^5 V/cm at the interface.

For electric field strengths such that

$$V \geqq \left(\frac{kT}{\beta_{\mathrm{PF}}}\right)^2 \tag{20}$$

also the Schottky mechanism equation (16) may be reformulated to yield an interface resistance similar to that of equation (15). The electrode limited Schottky emission is polarity dependent and the potential Φ_{S} may be regarded as the difference in work function across the interface. For two semiconductors with the same electron affinity the difference in work function is equal to the difference in intrinsic Fermi level, which is related to the ratio of carrier densities in the semiconductor. The potential barrier evaluated from carrier densities in film and substrate is 0.135 eV for deposition at 750 °C and 0.117 eV for deposition at 800 °C. These values correspond closely to the values quoted in Table 1 for the respective deposition temperatures.

Both conduction mechanisms do allow for the temperature dependence of the interface resistance, but only the Schottky mechanism accounts for the observed polarity phenomena. However, this does not completely exclude the Poole-Frenkel type of conduction in the interface.

It is felt that both conduction mechanisms may account for the interface resistance in the composite structure of the present investigation. In this case the germanium film–substrate interface may be regarded as a thin region of distorted material separated from film and substrate by Schottky barriers, the height of which are determined by differences in work function.

5. Conclusion

The similarity of electrical transport properties of pairs of films deposited under identical conditions on sapphire and germanium, are in contrast to the observed dissimilarity in crystalline structure. These results are, however, in agreement with previous investigations of germanium deposited simultaneously on several insulating substrate materials [4].

The electrical properties of sputtered germanium films indicate a distorted region at the film–substrate interface with high carrier density and low carrier mobility [18]. These results are in agreement with the present observations.

The electrical properties of germanium deposited on germanium substrates are not directly observed, but deduced from a proposed model for conduction in composite structures. Evaluation of data from the model and its comparison with experimental results rest on the assumptions that the substrate is unaffected by the deposited layer and that the interface consists of a barrier described by Schottky emission, Poole-Frenkel effect, or a combination of the two mechanisms. The model suggested for conduction in composite semiconductor structures in its present form does not give a complete description of conduction in germanium films on germanium substrates. However, the agreement between experimental results and model does to a large extent justify the assumptions on which the comparison rests, and it supports the general conclusion arrived at in the previous discussion.

Acknowledgements

The author is indebted to K. Bløtekjær for helpful discussions and advice, and the Royal Norwegian Council for Scientific and Industrial Research is acknowledged for the financial support.

Appendix

The composite resistance of equation (1) is obtained from a similar expression for conduction in inversion layers in semiconductor surfaces [17].

The composite conduction model may be extended to include Hall effect in similar structures.

Consider the composite structure in Fig. 6 which consists of a thin film (R_f) on a substrate (R_s) separated by a junction or barrier (R_{sf}). The equivalent circuit of Fig. 6 may be regarded as the distributed circuit shown in Fig. 7. The differential film resistance and substrate resistance are $R_f \frac{\Delta x}{l}$ and $R_s \frac{\Delta x}{l}$, respectively. The differential resistance of the interface barrier is $R_{sf} \frac{l}{\Delta x}$. The differential Hall voltages of film and substrate are $V_H^f \frac{\Delta x}{l}$ and $V_H^s \frac{\Delta x}{l}$. Applying Kirchhoff's law to successive loops in the equivalent circuit, a set of self-consistent differential equations is arrived at (in the limit of $\Delta x \to 0$), these are

$$\frac{d^2}{dx^2} I(x) = (\eta/l)^2 I(x) - \frac{V_H^s - V_H^f}{R_{sf} l^2}, \tag{A1}$$

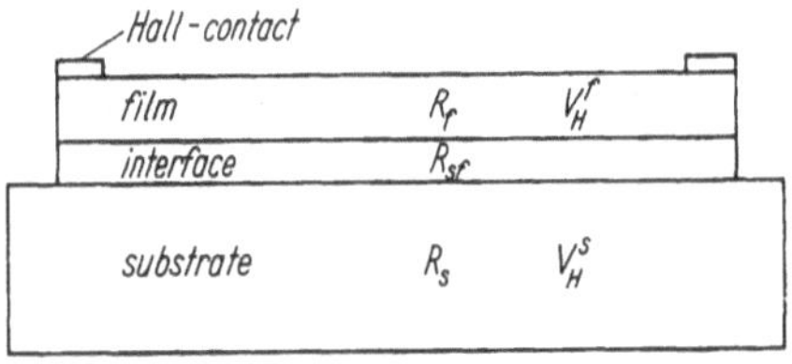

Fig. 6. Schematic cross section of composite structure of germanium film and substrate separated by an interface region

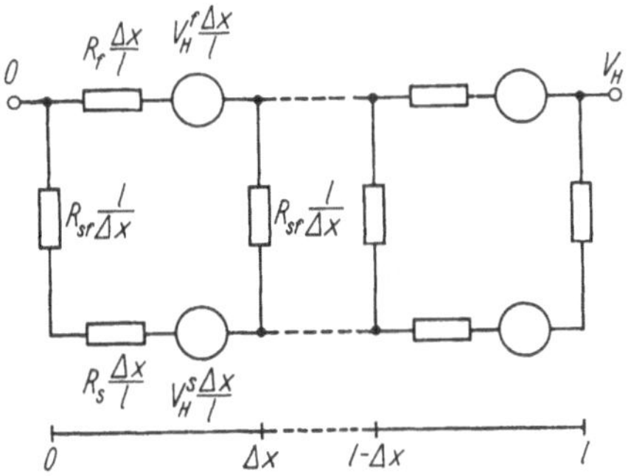

Fig. 7. Distributed equivalent circuit of the composite structure

$$\frac{\mathrm{d}}{\mathrm{d}x} V_f(x) = \frac{R_f}{l} I(x) + \frac{V_H^f}{l}, \tag{A2}$$

$$\frac{\mathrm{d}}{\mathrm{d}x} V_s(x) = -\frac{R_s}{l} I(x) + \frac{V_H^s}{l}, \tag{A3}$$

$$V_f(x) - V_s(x) = R_{sf}\, l \frac{\mathrm{d}}{\mathrm{d}x} I(x)\,. \tag{A4}$$

The observable Hall voltage of the composite structure is given by

$$V_H = \int_0^l \frac{\mathrm{d}}{\mathrm{d}x} V_f(x)\,\mathrm{d}x = V_H^f + R_f \frac{1}{l} \int_0^l I(x)\,\mathrm{d}x\,. \tag{A5}$$

The solution of (A1) subject to the boundary conditions $I(0) = I(l) = 0$, yields the loop current

$$I(x) = \left(\frac{V_H^f - V_H^s}{R_s + R_f}\right)\left[\frac{(1 - \mathrm{e}^{-\eta x/l}) \sinh \eta\, x/l}{\sinh \eta} + \mathrm{e}^{\eta x/l} - 1\right]. \tag{A6}$$

Upon substitution of (A6) into (A5) and integrating one gets for the composite voltage equation (4) where the barrier function $\Gamma(\eta)$ is small when $\eta > 1$, and approaches unity for large values of η.

References

[1] B. W. SLOOPE and C. O. TILLER, J. appl. Phys. **36**, 3174 (1965).
[2] E. KRIKORIAN and R. J. SNEED, J. appl. Phys. **37**, 3665 (1965).
[3] R. F. ADAMSKY, J. appl. Phys. **40**, 4301 (1969).
[4] J. S. JOHANNESSEN, to be published.
[5] J. E. DAVEY, J. appl. Phys. **33**, 1015 (1962).
[6] J. E. DAVEY, R. J. TIERNAN, T. PANKEY, and M. D. MONTGOMERY, Solid State Electronics **6**, 205 (1963).
[7] J. D. WILLIAMS and L. E. TERRY, J. Electrochem. Soc. (Solid State Sci.) **114**, 158 (1967).
[8] R. L. SCHALLA et al., Single Crystal Films, Ed. M. H. FRANCOMBE, MacMillan Co., New York 1964 (p. 301).
[9] B. W. SLOOPE and C. O. TILLER, J. appl. Phys. **38**, 140 (1967).
[10] D. M. BROWN and R. BUY, Phys. Rev. **127**, 1593 (1962).
[11] S. M. SZE, Physics of Semiconductor Devices, Wiley/Interscience Publishers, 1969.
[12] N. CROITORU and L. VESCAN, Thin Solid Films **3**, 269 (1969).
[13] J. FRENKEL, Phys. Rev. **54**, 647 (1938).
[14] A. K. JONSCHER, Vacuum Sci. Technol. **8**, 135 (1971).
[15] P. A. WALLEY and A. K. JONSCHER, Thin Solid Films **1**, 367 (1968).
[16] T. C. COURVOISIER, H. HAIDINGER, P. J. W. JOCHEMS, and L. J. TUMMERS, Solid State Electronics **6**, 265 (1963).
[17] A. MANY, Y. GOLDSTEIN, and N. B. GROVER, Semiconductor Surfaces, North-Holland Publ. Co., Amsterdam 1965 (p. 216).
[18] G. WALLIS, E. E. PITTELLI, and J. PANTAMO, Thin Solid Films **4**, 309 (1969).

(Received February 11, 1972)

phys. stat. sol. (a) **10**, 581 (1972)
Subject classification: 7 and 14.3.4; 6; 22.4.1

Physics Laboratory III, Technical University of Denmark, Lyngby

Influence of Elastic Anisotropy on Acoustoelectric Domain Propagation

By
O. KELLER

The effect of elastic anisotropy on acoustoelectric domain propagation in semiconducting CdS has been studied by Brillouin scattering. Two configurations were examined. In the longitudinal configuration ($\boldsymbol{E} \parallel \boldsymbol{c}$) V-shaped domains with the two branches containing different acoustic wave vector distributions were observed. As the domain propagates, the domain tilt angle (θ_D) changes and saturates at an angle corresponding to the angle of energy propagation (θ_E) for the off-axis modes of maximum gain. Spatial splitting of the localized flux into two domains propagating near the surfaces has been observed. In the 45° configuration (angle between $\boldsymbol{E}$ and $\boldsymbol{c}$ 45°) angular focusing of energy is obtained experimentally, and phonon packets of quasi-longitudinal modes have been discovered upon reflection of the domain from the anode.

Die Einwirkung der elastischen Anisotropie auf die Fortpflanzung akustoelektrischer Domänen in halbleitendem CdS wurde durch Brillouinstreuung in zwei Fällen untersucht. Bei der longitudinalen Konfiguration (elektrisches Feld $\boldsymbol{E}$ parallel zur $\boldsymbol{c}$-Achse) wurden V-förmige Domänen, deren Zweige verschiedene Verteilungen akustischer Wellenvektoren enthielten, gefunden. Während der Fortpflanzung der Domäne ändert sich der Neigungswinkel der Domäne (θ_D), und zwar nähert sich θ_D dem Wert θ_E, d. h. dem Winkel der Fortpflanzung der Energie in den außeraxialen Schwingungsformen maximaler Verstärkung. Es wurde beobachtet, daß sich die Domäne allmählich in zwei Domänen aufspaltet, die sich in der Nähe der Oberfläche ausbreiten. Bei der 45°-Konfiguration (Winkel zwischen $\boldsymbol{E}$ und $\boldsymbol{c}$ 45°) ergeben die Experimente eine Winkelfokussierung der Energie. Bei Domänenreflexion an der Anode wurden aus quasi-longitudinalen Schwingungstypen bestehende Phononenpakete gefunden.

1. Introduction

The present paper deals with the effect of elastic anisotropy on acoustoelectric propagating domains. Deviations from the isotropic case are supposed to occur in the longitudinal configuration, where the domain consists of off-axis shear waves [1 to 3]. Several authors have studied off-axis modes by different methods [4 to 8]. So far, however, serious attempts to decide how much angular dispersion influences the domain characteristics have not been made.

By means of Brillouin scattering we have found strong experimental evidence for the existence of anisotropy effects in semiconducting CdS crystals.

In the longitudinal configuration ($\boldsymbol{E} \parallel \boldsymbol{c}$) we observe that the domain is formed as a V, and that the tilt angles of the two branches are mainly determined by the group velocity corresponding to the off-axis angle for the phases of maximum gain. The wave vector distribution is experimentally found to be different in the two branches. Propagation of the acoustic flux towards the surfaces of the sample often splits up the domain into two wave packets travelling near the surfaces. We also observe the formation of a second domain in the bulk of the crystal related to the splitting of the flux.

In a 45° configuration ($\sphericalangle \boldsymbol{E}, \boldsymbol{c} = 45°$) the acoustic energy is distributed perpendicular to the current direction independently of the applied voltage as a result of anisotropy. A theoretical analysis shows that a distribution of the wavevector cone symmetrically around the field direction can be obtained in ZnO and not in CdS.

In GaAs reflected domains of longitudinal type have been seen [9]. We report here on the observation of this kind of reflected flux also in CdS in a crystal direction, where the piezoelectric coupling vanishes. The longitudinal lattice damping at 600 MHz is obtained from these results.

2. General Description of Anisotropic Acoustoelectric Domains Studied by Brillouin Scattering

Amplification of thermal lattice vibrations in the gigacycle region will occur in piezoelectric semiconductors like CdS and ZnO, if the electron drift velocity due to an external electric field exceeds the phase velocity of the piezoelectric active shear wave. A travelling acoustic high field domain may be formed, when the amplified acoustic flux has reached such a level that non-linear acoustoelectric effects become important [10 to 13].

In the transverse configuration ($\boldsymbol{E} \perp \boldsymbol{c}$) the domain propagation takes place in the basal plane with the particle displacement parallel to the c-axis. In this special case, where the acoustic disturbance consists of pure transverse modes, the acoustic wave vectors form a narrow cone along the direction of the applied field [14]. This leads to a description of the domain, where all physical quantities of interest can be treated as scalars.

In the longitudinal configuration ($\boldsymbol{E} \parallel \boldsymbol{c}$) the analysis of the domain propagation becomes more involved. The domain now consists of off-axis quasi-transverse modes (T2), because these modes in contrast to the pure shear modes (T1) will give rise to a longitudinal piezoelectric field, which couples strongly to the conduction electrons. The quasi-longitudinal modes (L) will usually not be amplified because their phase velocity is more than twice the phase velocity of the shear modes.

In a previous paper [7] we have studied theoretically the propagation of anisotropic elastic waves in ZnO and CdS. We found the acoustic energy to propagate through space with a group velocity

$$\boldsymbol{V}_\text{g} = \nabla_{\boldsymbol{k}}\omega(\boldsymbol{k}) = \nabla_{\boldsymbol{k}}\left(k\, V_\text{p}(\boldsymbol{k})\right) \tag{1}$$

which can be approximated by

$$\boldsymbol{V}_\text{g} \approx V_\text{p}\frac{\boldsymbol{k}}{|\boldsymbol{k}|} + \hat{k}_{\theta_\text{p}}\frac{\text{d}V_\text{p}(\theta)}{\text{d}\theta}, \tag{2}$$

where we are left with the effect of angular dispersion only, since the phonon packet should be stable against velocity dispersion, as long as the electromechanical coupling constant is small [7, 15]. The symbols have their usual meaning: $\boldsymbol{k}$ is the acoustic wave vector, $V_\text{p}(\boldsymbol{k})$ is the phase velocity in the $\boldsymbol{k}$-direction, θ_p is the angle between the direction of phase propagation and the optical axis and $\hat{k}_{\theta_\text{p}}$ is a unit vector perpendicular to $\boldsymbol{V}_\text{p}$ in the plane defined by $\boldsymbol{V}_\text{p}$ and the c-axis. The angle between the c-axis and the direction of energy propagation for the off-axis mode θ_p will be called θ_E and the domain tilt angle θ_D. In order to determine experimentally the influence of elastic anisotropy on the domain characteristics we have chosen to study CdS, which according to our previous calculations should show the most pronounced angular dispersion. Two configurations have been examined. The longitudinal configuration, which gives rise to a strong angular spread in energy and the 45° configuration ($\sphericalangle \boldsymbol{E}, \boldsymbol{c} = 45°$), which causes a focusing of the energy for the off-axis modes in the plane defined by the field direction and the c-axis.

The high acoustic energy density connected with the wave packets can be studied in detail by Brillouin scattering technique [3, 7, 12, 14, 19, 20]. The scattering formula for optically anisotropic crystals [16] directly enables one to select a given plane wave component (ν, $\boldsymbol{k}$) of the acoustoelectric domain, ν being the frequency of the acoustic component in question. The intensity of the scattered signal will be proportional to the acoustic energy density in the plane wave components under consideration over a broad frequency range [17]. By using Brillouin scattering as a two-dimensional optical probe we are thus in a position to study the spatial distribution of energy as a function of time of the single plane wave component (ν, $\boldsymbol{k}$) in the domain.

It should be pointed out that a Brillouin scattering investigation of the different off-axis modes requires a knowledge of the angular dependence of the acoustic phase velocity. As long as the electromechanical coupling constant is small, the main contribution to the anisotropy of the phase velocity may be found by solving the eigenvalue problem for the pure elastic modes. In the case of a hexagonal crystal (6 mm symmetry) the phase velocity can be expressed in analytical form, which simplifies the calculation of the scattering geometry.

3. Anisotropy Effects in the Longitudinal Configuration

Until now only a few Brillouin scattering experiments have been dealing with travelling acoustoelectric domains in the longitudinal configuration [3, 7]. The previous investigations have been concerned with the determination of the time and space development of the angular spectrum and the frequency spectrum of the domain in an essentially one-dimensional model. No investigations of the distribution of the acoustic plane wave components perpendicular to the current direction in the longitudinal configuration have been reported so far.

Two other methods have been applied to observe the off-axis domain directly. Moore et al. [4] use a stroboscopic strain-birefringent technique. This method enables them to see the local concentration of the strain in some mean square sense. The results show a straight domain inclined to the c-axis. Furukawa et al. [5] use electrical probe measurements of the two-dimensional potential distribution on the surface of the crystal. The shape of the high field domain is here observed to be in the shape of the letter V. The last result seems to agree with our more detailed Brillouin scattering data on the domain shape, although it is not obvious that one might directly compare the electrical and acoustical aspects of the domain [18].

Our scattering investigations were performed with a He–Ne laser ($\lambda_0 = 0.6328$ μm). In all the experiments the experimental set-up was a conventional Brillouin scattering arrangement [19]. The scattering plane was horizontal. In order to detect the different off-axis modes we mounted the crystal so that it was rotable about a horizontal axis (h) perpendicular to the plane of optical probing. This enabled us to place a selected $\boldsymbol{k}$-vector in the horizontal scattering plane (see Fig. 1). Furthermore, the crystal was mounted so that we could perform translations of the sample both along and perpendicular to the c-axis

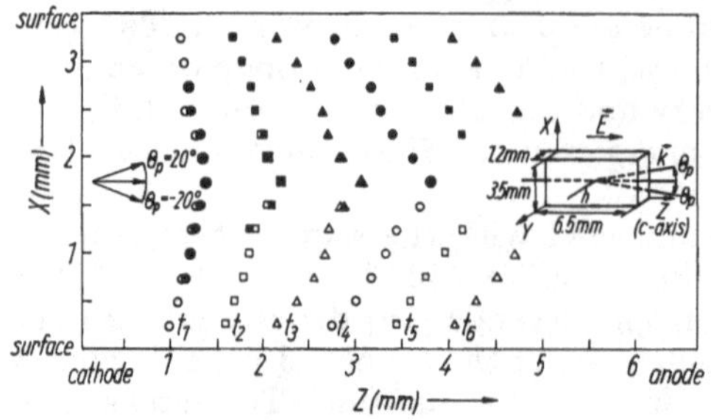

Fig. 1. Spatial distribution of peak acoustic intensity for the two plane wave components: $(\nu, \theta_p) = (1\ \text{GHz}, \pm 20°)$ in acoustoelectric domain in CdS at 300 K. The energy distribution is plotted at the following times after onset of the voltage pulse: t_1; t_2; ... t_6 = 0.4, 0.7, ... 1.9 μs. Crystal configuration: Longitudinal. Full signatures corresponds to $\theta_p = +20°$ the open signatures to $\theta_p = -20°$

in the plane of optical probing. In both directions the acoustic energy density was probed in points with a spacing of 100 to 200 μm. The resolution of the optical probe was of the same order of magnitude. Total internal reflection of light makes it impossible to detect several off-axis modes directly. In a previous paper [7] this problem was solved by surrounding the crystal with a highly refractive liquid. The present method of mounting the sample enables us to detect off-axis modes as easily as on-axis modes.

Our CdS crystals were cut to flat rods all having a length of about 6 to 7 mm in the direction of the c-axis. The cross sections had the following values: $1.2 \times 3.5\ \text{mm}^2$, $1.2 \times 6.0\ \text{mm}^2$, and $1.2 \times 8.0\ \text{mm}^2$. The resistivity of the samples was approximately 16 Ωcm.

In a recently published paper [7] we have shown experimentally that the off-axis angle of maximum gain for the phase can be varied as we change the applied voltage. In the present experiment the voltage was chosen so as to give an off-axis angle of maximum gain for phase propagation at $\theta_p = 20°$ corresponding to an angle of $\theta_E = 37°$ between the group velocity of the mode and the c-axis. This choice was made in order to obtain the maximum angular dispersion, which is convenient for studying anisotropy effects.

On Fig. 1 are shown the results obtained by two-dimensional optical probing in a plane containing the c-axis. Two symmetric plane wave components $(\nu, \theta_p) = (1\ \text{GHz}, \pm 20°)$ were examined. Notice that the off-axis angles $\theta_p = \pm 20°$ correspond to the directions of maximum gain for phase propagation. As may be seen from the figure, we have examined the spatial distribution of peak acoustic intensity at six different time instants (t_1, t_2, t_3, t_4, t_5, and t_6) measured relative to the onset of the voltage pulse.

It appears that the selected acoustic mode $\theta_p = 20°$ contributes only to the energy density of the travelling domain in part of the crystal. At the time t_1 the energy is spread over most of the cross section, and the mode considered thus contributes to the domain energy density in both of the two branches ($\pm\theta_p$). At t_3 the mode contributes only to the energy density in slightly more than half of the cross section. At a later time instant (t_4) the acoustic energy concentrates in one of the branches. The tilt angle of the spatial energy distribution of the mode at this time has the same sign as the off-axis angle for phase propagation (θ_p). Both angles are measured as positive to one side of the c-axis and negative to the other. By changing the scattering geometry so as to examine the symmetric wave vector ($-\theta_p$) we have found a contribution to the acoustic flux as shown on Fig. 1. It is is seen that this distribution is an almost symmetric image of the $+\theta_p$ wave vector distribution about a line parallel to the c-axis in the middle of the sample.

Fig. 2. Experimental values of domain tilt angle (θ_D) in CdS at 300 K as a function of time measured from the onset of the electric field. ○ Longitudinal configuration. Selected acoustic mode: (ν, θ_p) = (1 GHz, 20°). □ 45° configuration. Selected acoustic mode: (ν, θ_p) = (0.6 GHz, 33°)

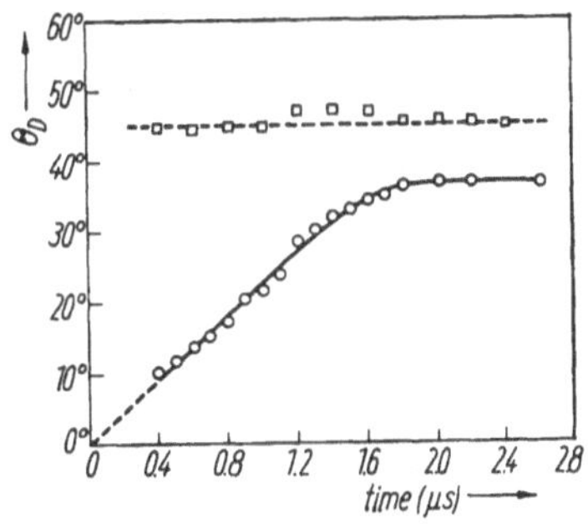

An investigation of different acoustic frequencies results in the same spatial energy distribution as discussed above. The results thus indicate that the acousto-electric domain is shaped as a V as long as we are not too near the cathode. Each of the two straight branches of the wave packet mainly consists of plane wave components with different frequencies having the same sign of the off-axis angle for phase propagation (θ_p) as the domain tilt angle (θ_D). The width of the two branches is 200 to 300 μm.

From Fig. 1 it is evident that the tilt angle of the energy distribution changes as the domain propagates along the sample. On Fig. 2 is shown the experimentally obtained results for the tilt angle (θ_D) as a function of time. The circles represent the longitudinal configuration. The time is measured from the onset of the electric field. In the first stages of propagation, the domain tilt angle changes linearly with time. After a period of 1.8 μs the domain tilt angle saturates at an angle of 37°, which equals the off-axis angle for the group velocity corresponding to a wave vector direction 20° from the c-axis. At this time the domain is 4.0 to 4.5 mm from the cathode. If we extrapolate the curve backwards in time it seems as if the acoustic energy density distribution in the domain is perpendicular to the c-axis near the cathode. The frequency examined on Fig. 2 was 1.0 GHz.

In the following a possible qualitatively explanation of the shift in θ_D as a function of time is proposed. Near the cathode the domain front is perpendicular to the c-axis indicating that the intensity of the $\boldsymbol{k}$-vector distribution must be symmetric around the rod axis. This is plausible because the amplification factor is symmetric around the c-axis and the "pathlengths" along which the flux has been amplified, are equal for symmetric $\boldsymbol{k}$-vectors. This holds as long as we are in a region where the distance from the cathode is small compared to the sample dimension perpendicular to c. The extension of this region depends on the direction of the group velocity of the domain. Outside this region the acoustic flux coming from the lower part of the crystal and moving in the direction given by the group velocity $\boldsymbol{V}_g$ will dominate in the upper part of the crystal, because of the difference in "pathlengths" for acoustic flux moving in the directions θ_E and $-\theta_E$. This gives rise to the tilt of the domain branches. This model is supported by the fact that in several crystals we have observed an overlap region in the middle of the crystal where both symmetric $\boldsymbol{k}$-vector distributions are present. The domain front in the middle of the crystal is in this case perpendicular to the current direction and the domain consequently looks like a "boomerang".

The domain tilt angle will thus be determined by the total strain occurring from a $\boldsymbol{k}$-vector cone around the direction of maximum gain. This might give

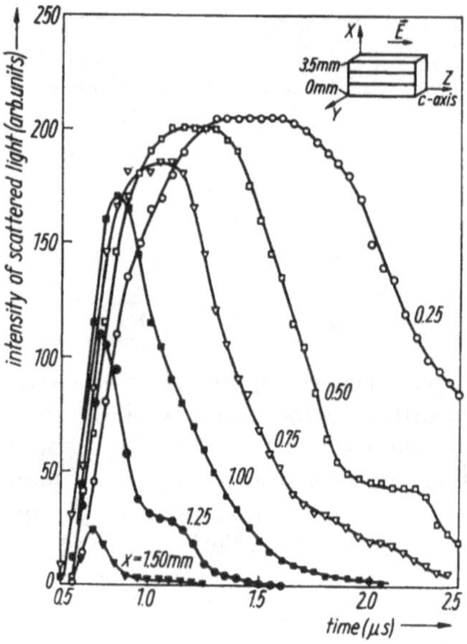

Fig. 3. Acoustic peak intensity as a function of time for the off-axis mode: $(\nu, \theta_p) = (1\ \text{GHz}, 20°)$ in acousto-electric domain in CdS at 300 °K. Configuration: Longitudinal. Probing was performed along the c-axis, and the parameter (x) gives the distance from the surface, perpendicular to the c-axis. Time is measured relative to the onset of the electric field

rise to a dispersion of the phonon packet in the first stages of the domain propagation. Later on wave vectors at the angle of maximum gain will dominate, if one can neglect non-collinear parametric effects [20]. This agrees well with the observed saturation of the tilt angle at $\theta_D = 37°$ corresponding to the direction of energy propagation for the off-axis modes of maximum gain $\theta_p = 20°$.

Another interesting question still remains open. How will the acoustic energy be distributed in the two branches of the domain as it moves towards the anode? In order to solve this problem Brillouin scattering experiments were performed along straight lines parallel to the c-axis at different distances from the crystal boundary. The results for one of the branches are shown on Fig. 3. We have not on this plot tried to follow the acoustic flux back into the linear regime, but have only been interested in the well established domain.

The measurements were performed at a fixed point in a coordinate system moving with the inclined V-formed domain. The parameter on the figure is the distance from the crystal surface in mm. The intensity of the scattered light corresponds to the acoustic peak intensity of the mode considered in the domain as a function of time.

From the figure it is evident that the acoustic energy distribution varies as a function of time. If we look at a particular parameter value we see that the acoustic peak intensity as a function of time has a maximum value, and as we move closer to the boundary, this maximum point is shifted to the right, indicating that the acoustic wave packet moves outwards towards the surface. As time goes on, the absolute maximum value of the energy increases and is shifted towards the crystal surface. Notice that after a certain time we obtain a situation where the acoustic density remains constant in two thin layers near the crystal surfaces over a relatively long time.

4. Spatial Splitting of the Acoustic Flux and Formation of a Second V Formed Domain

As the acoustic energy density tends to concentrate near the surfaces, we get an appreciable drop in the flux level in the centre of the crystal, and after a while the acoustic flux has disappeared completely in the middle of the sample, as shown on Fig. 3. At this time the V-formed domain has split up into two

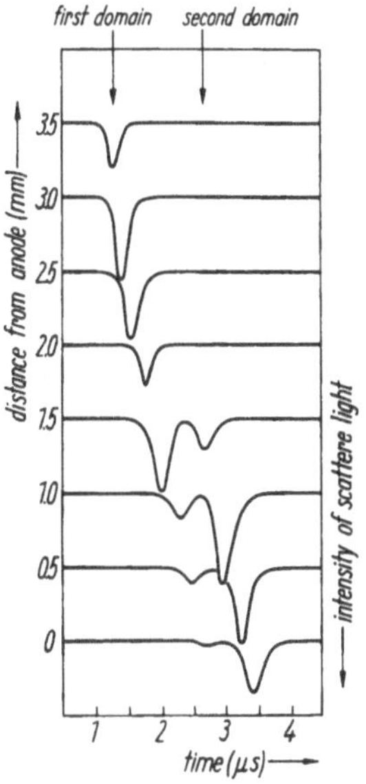

Fig. 4. Redrawn oscilloscope pictures showing the flux intensity as a function of time at different distances from the anode. The crystal and probing plane was the same as in Fig. 3. Selected plane mode: $(\nu, \theta_p) = (1\ \text{GHz}, 20°)$. Distance from the surface: 0.75 mm. The sensitivity on the oscilloscope is not the same for the different traces

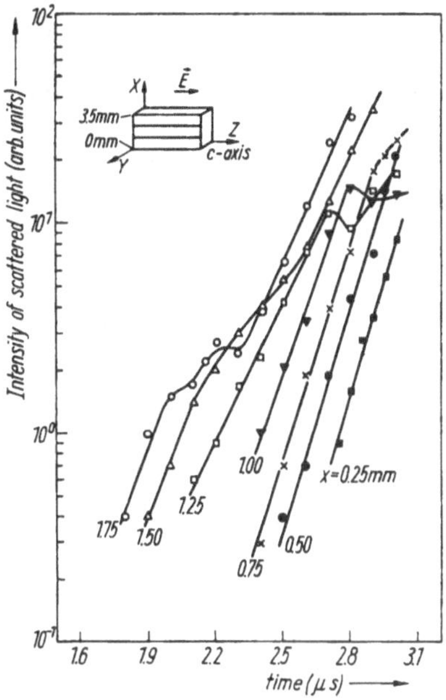

Fig. 5. Acoustic peak intensity for second domain as a function of time measured from onset of the voltage pulse. The crystal and the probing plane was the same as in Fig. 3. Parameter (x): Position from the surface perpendicular to the c-axis. Investigated acoustic mode: $(\nu, \theta_p) = (1.0\ \text{GHz}, 20°)$

single domains propagating near the surfaces of the crystal. We have found this behaviour to be quite general in all our broad CdS samples. If the crystal had a circular cross section, we would thus expect the domain to be annular.

We suggest that the disappearance of the excited flux in most of the cross section creates a field distribution in the centre, which enables the electrons to drift faster than the transverse velocity of sound and thereby amplify the background flux. This leads to the formation of a second domain as seen on the redrawn oscilloscope pictures of Fig. 4. The distance from the surface is 0.75 mm. The time difference between the appearance of the two domains on a given position is about 0.7 μs corresponding to a distance of around 1.5 mm between the domains. The second phonon packet is formed at a distance of several mm from the cathode, and exhibits the characteristic V-shape. The domain tilt angle is stable as a function of time and equals $\theta_D = 37°$.

On Fig. 5 is shown the experimentally obtained points for the flux level as a function of time for the 1 GHz plane wave component of the second domain with $\theta_p = 20°$. The time is measured from the onset of the high voltage pulse, and the scattered light intensity is a direct expression for the relative spatial acoustic energy density. The distance from the surface in mm is again a convenient parameter.

We have only been able to follow the second domain in its first stages of growth because of its appearance relatively close to the anode. The growth rate of the plane mode at the different distances from the surface is nearly the same, and the scattered signal first becomes visible in the centre of the sample. These facts demonstrate that the formation of the second acoustoelectric domain probably starts in the central regions of the cross section and not at the cathode contact. In the case of the second domain this seems to indicate that the cathode region does not necessarily act as a dominant source for the growing acoustic flux. We thus do not believe that propagating acoustoelectric domains can only originate from the shock produced at the cathode boundary by the pulse onset [21]. The reason for the deviations from straight lines on Fig. 5 is not clear at the moment. The "error" is, however, systematic in the sense that it seems to occur along a straight line on the figure.

The domain tilt as well as the formation of a second domain are reflected in the current waveform. We shall postpone a discussion of the current waveform, however, until we have treated the 45° configuration.

5. Anisotropy Effects in the 45° Configuration

Strong focusing of the elastic energy will occur in a 45° configuration [7]. Calculations for CdS have shown that for wave vector directions in the interval from 30 to 60° from the c-axis the corresponding energy cone will be $\theta_E \pm \Delta\theta_E = 45.3 \pm 0.9°$ and the spread in the group velocity $\Delta V_{g,45°}/V_{g,45°}$ will be less than 1%. If we look at the $\boldsymbol{k}$-vektor distribution in this configuration for a given applied electric field, it is found that in general the angular distribution of the off-axis modes will not be symmetric around the current direction.

Linear theory predicts a shift in the angle of maximum gain for phase propagation in the longitudinal configuration as we change the applied voltage. Our calculations on CdS in the 45° configuration based on the linear macroscopic model have shown a rapid shift with electric field in the angle of maximum gain from 67°, corresponding to a direction where the electron–phonon coupling vanishes, down to a saturated value near 30°, where the electromechanical coupling constant exhibits a maximum [10, 15]. In addition the wave vector cone changes its shape and becomes broader. When the cone has a maximum at 45° it is, however, very unsymmetric. In ZnO we get the same rapid shift in θ_p towards the angle of maximum electromechanical coupling as we change the applied electric field. In this case we can, however, obtain a situation, where the $\boldsymbol{k}$-vectors are symmetrically distributed around the current direction. This is mainly due to the fact that the electromechanical coupling constant exhibits its maximum closer to the 45°-direction and has a bigger halfwidth in ZnO than in CdS. The halfwidth of the phase cone in ZnO is $\Delta\theta_p = 16°$. The normalized applied voltage for this distribution is $V/V_{th} = 1.21$, where V_{th} is the threshold voltage.

It should be pointed out that we cannot expect linear theory to predict the angular distribution precisely, but it appears that the simple theory qualitatively predicts the shift in the angle of maximum gain, which is obtained experimentally when we change the applied voltage. When non-linearities become important the drift velocity tends to decay towards the transverse velocity of sound [10]. This is, however, not followed by a corresponding change in the angle of maximum gain [4, 5, 7]. The domain tilt angle is thus still determined by linear theory.

Based on this behaviour we conclude that in ZnO a focusing of elastic energy in the domain may occur in the 45° configuration, and one may choose the applied voltage, so that the distribution of the wave vectors becomes symmetric

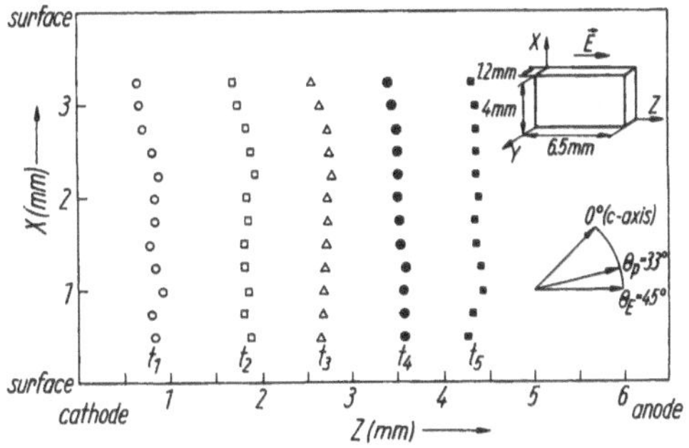

Fig. 6. Spatial distribution of peakacoustic energy in the domain as a function of time in 45° configuration. Investigated plane wave component in CdS at 300 K: $(\nu, \theta_p) = (0.6$ GHz, $33°)$. The energy distribution is plotted at the following times after onset of the electric field: $t_1, t_2, \ldots t_5 = 0.4, 0.8 \ldots 2.0$ µs

around the rod axis. In this case the average value of $\boldsymbol{k}$ will consequently be in the same direction as the group velocity. The phonon packet has thus obtained very simple properties. In CdS a focusing of energy may also be obtained, but in this case the $\boldsymbol{k}$-vectors cannot be distributed symmetrically around the current direction.

Experiments have been performed on semiconducting CdS crystals which have dimensions: $6.0 \times 4.0 \times 1.2$ mm³. The optic axis, which were placed in the 6.0×4.0 mm² plane, formed an angle of 45° with the longest rod axis. The high voltage was applied along the 6 mm side.

Fig. 6 shows our results for the spatial distribution of acoustic flux at different times for a 600 MHz component of the domain. The chosen off-axis angle for phase propagation (θ_p) equals 33°. The energy is distributed perpendicular to the current direction at the different times. The domain tilt angle equals $\theta_D = 45°$. This is expected because of the focusing effect. The squares on Fig. 2 represent the tilt angle as a function of time.

The intensity distribution perpendicular to the current direction at three different times is shown on Fig. 7. The domain was already well established at t_1. The intensity distribution across the sample is obviously not very constant. In one part of the cross section the intensity increases exponentially with position towards the middle of the sample and remains nearly unchanged in time. In the other part of the cross section the distribution of energy is more complicated and there is a slight increase in intensity as a function of time. In the complicated part of the distribution some of the flux propagates apparently

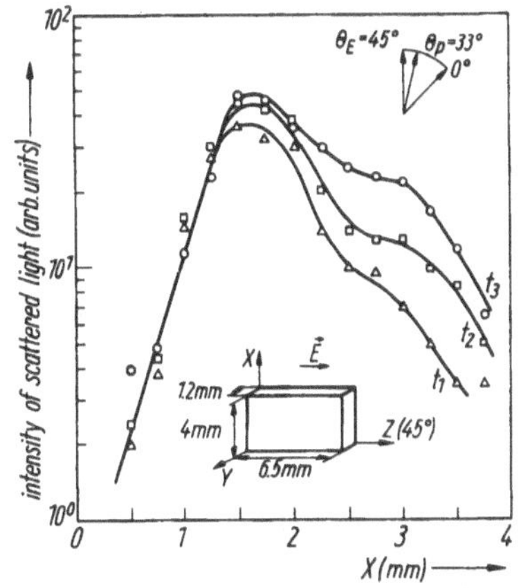

Fig. 7. Distribution of acoustic energy perpendicular to the current direction as a function of time in a 45° configuration (CdS, 300 K). t_1, t_2, and t_3 corresponds to 0.4, 0.6, and 0.8 µs after high voltage termination. Investigated plane wave: (ν, θ_p): (0.6 GHz, 33°)

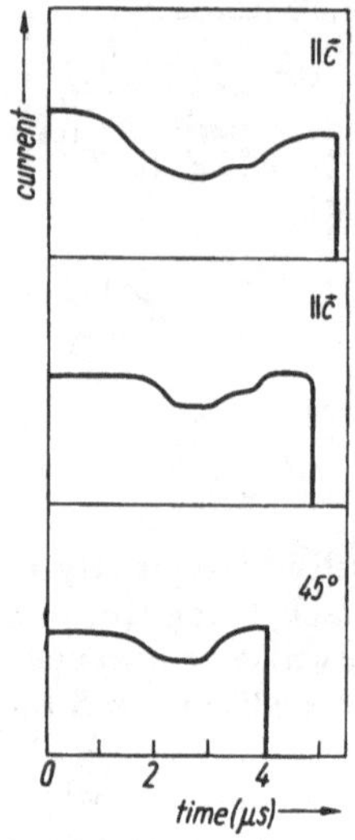

Fig. 8. Characteristic current wave forms. The distance from cathode to anode is the same for the three samples. The domain tilt angles are equal for the crystals in the two upper figures, but the top figure corresponds to a broader crystal than the figure in the middle

outwards. In the 45° configuration acoustic modes with phase propagation at angles less than 30° with respect to the c-axis will be present. The acoustic energy in these modes will not be focused in the current direction. This may give rise to the complicated spatial energy distribution on Fig. 7.

The differences in current wave forms between the longitudinal and the 45° configuration are illustrated in Fig. 8 which shows the wave forms occuring in three different crystals. In the 45° configuration the current returns very rapidly to its ohmic value at the end of the domain transit. This indicates that the domain collapses at the same time over the entire cross section. The rise time for the current is thus determined by the width of the domain. In the longitudinal configuration the current increases much more slowly towards the ohmic value, because of the inclined domain shape. For the broadest sample the rise time is longest, as we would expect. The formation of the second domain is reflected as a plateau in the current, when it has started to return to the ohmic value.

We have observed for a number of samples that the incubation time is about a factor of two smaller in the 45° configuration than in the longitudinal configuration. This may be due to the smaller acoustic losses at the surfaces in the 45° configuration, where the energy is focused along the rod axis.

The most essential results obtained in Section 3, 4, and 5 are summarized in schematic form in Fig. 9. We refer otherwise to the text below this figure.

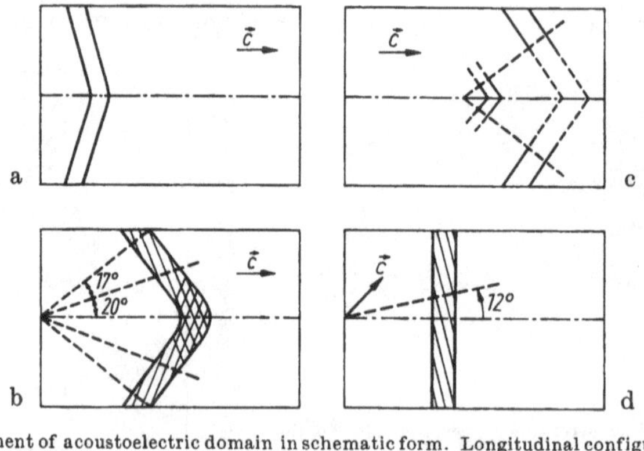

Fig. 9. Time development of acoustoelectric domain in schematic form. Longitudinal configuration: a) Formation of V-formed domain. Domain tilt angle changes as a function of time. b) Tilt angle saturates at an angle corresponding to the angle of energy propagation for the off-axis modes of maximum gain. Domain "wings" consists of different $\boldsymbol{k}$-vector distributions. c) Acoustic energy propagates towards the crystal surface, and this results in a splitting of the domain. Formation of a second domain is closely related to the splitting. 45° configuration: d) Angular focusing of the acoustic flux. The $\boldsymbol{k}$-vector distribution is not symmetric about the current direction

6. Longitudinal Reflected Domains

Acoustoelectric domains (T2) in ZnO and CdS can be converted into domains of non-coupling sound waves (T1) by reflection at an unpolished anode surface [22]. After reflection the pure transverse (T1), the quasi-transverse (T2) and the quasi-longitudinal (L) mode should be present, but the L and T2 mode is in general damped out rapidly due to their high electronic losses.

Until now only the inactive reflected domain has been seen in ZnO and CdS. In GaAs also a longitudinal reflected phonon packet has been observed [9].

In our 45°-configuration in CdS we have now been able to observe a longitudinal reflected domain propagating 4 to 5 mm towards the cathode in a direction 40° from the c-axis, which is close to the rod axis in this configuration. In the 40° direction the electromechanical coupling vanishes for the longitudinal mode, and we are left with only the lattice losses. The 45° configuration is also convenient when studying inactive reflected domains, because the energy is focused along the rod axis, so that surface losses are reduced. Besides the longitudinal flux we have observed the usual inactive domain.

On Fig. 10 are shown some redrawn oscilloscope pictures of the three different acoustic domains. We observe that the reflected inactive domain has a half-width, which is somewhat greater than that of the incoming domain. The measured domain velocities compared with the corresponding calculated values in brackets are: $V_{act} = 2.1 \times 10^5$ cm/s $(2.1 \times 10^5$ cm/s), $V_{inact.} = 1.7 \times 10^5$ cm/s $(1.8 \times 10^5$ cm/s), and $V_{long} = 4.0 \times 10^5$ cm/s $(4.2 \times 10^5$ cm/s). The agreement is seen to be quite good.

The observed longitudinal plane wave component had a frequency of 600 MHz, and the corresponding lattice loss constant was found to be $\alpha_{L,long} = 28$ dB/cm. If we extrapolate the results of Bateman and McFee [23], for longitudinal lattice losses perpendicular to the c-axis into the higher frequency region, we find, that our measured point lies on their extrapolated curve. The lattice losses for the longitudinal mode thus seem to be independent of direction in the crystal.

The possibility of observing longitudinal reflected domains can be used to measure lattice losses by optical probing. This seems to be possible in the region from 0.2 to 1.0 GHz for CdS and in a regime extended somewhat upwards in frequency for ZnO. Work on this line is in progress and will be published later.

7. Summary

Brillouin scattering investigations in semiconducting CdS have revealed that elastic anisotropy strongly influences the propagation of acoustoelectric domains in broad crystals.

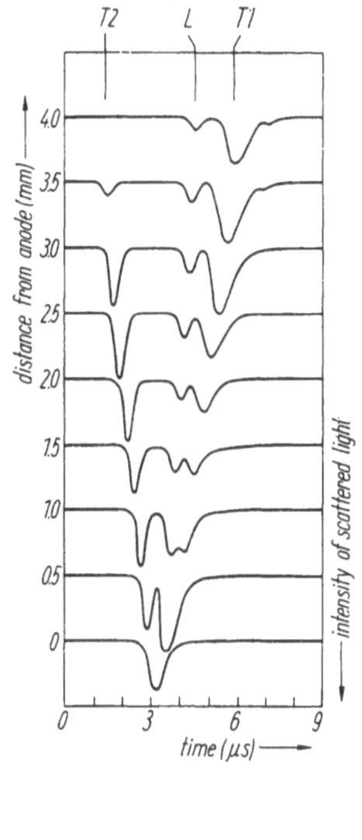

Fig. 10. Redrawn oscilloscope pictures of the three possible polarized domain types of a hexagonal crystal (CdS) in the 45° configuration: a) Acoustoelectric active quasi-transverse domain (T2). b) Reflected inactive pure transverse domain (T1). c) Reflected quasi-longitudinal domain (L). The reflected L-domain had a frequency of 600 MHz and the k-vector formed an angle of 40° with the c-axis. The sensitivity on the oscilloscope is not the same for the different traces

In the longitudinal configuration ($\boldsymbol{E} \parallel \boldsymbol{c}$) we obtain a V-formed domain. The branches consist mainly of plane wave components with off-axis angles for phase propagation (θ_p) having the same sign as the tilt angle of the spatial energy distribution in the domain. The domain tilt angle (θ_D) changes as a function of time and saturates at an angle corresponding to the angle of energy propagation (θ_E) for the off-axis modes of maximum gain. After a certain time the domain usually splits up into two domains, because the acoustic energy propagates away from the centre of the crystal. Formation of a second domain is closely related to this spatial spread of energy.

In the 45° configuration ($\sphericalangle \boldsymbol{E}, \boldsymbol{c} = 45°$) the energy is distributed perpendicular to the rod axis, and the entire acoustic energy propagates in the current direction, because of focusing of the energy for a broad angular spectrum of wave vectors.

Reflected domains consisting of longitudinal plane wave components have been observed for the first time, in a direction where the electromechanical coupling vanishes, and the longitudinal lattice loss for the 600 MHz acoustic mode in CdS is found to be 28 dB/cm.

Acknowledgements

I wish to thank Prof. N. I. Meyer for his interest in this work and his proposals regarding the form of presentation. Dr. E. Mosekilde is thanked for stimulating discussions.

References

[1] J. H. McFee, P. K. Tien, and H. L. Hodges, J. appl. Phys. **38**, 1721 (1967).
[2] A. R. Moore, Appl. Phys. Letters **13**, 126 (1968).
[3] W. Wettling and M. Bruun, Phys. Letters (Netherlands) **27A**, 123 (1968).
[4] A. R. Moore, R. W. Smith, and P. Worcester, IBM J. Res. Developm. **13**, 503 (1969).
[5] S. Furukawa, T. Hata, and T. Motosugi, Appl. Phys. Letters **16**, 455 (1970).
[6] J. C. Huber, Appl. Phys. Letters **16**, 458 (1970).
[7] O. Keller, phys. stat. sol. (a) **8**, 61 (1971).
[8] B. L. Timan, V. K. Komar, and V. D. Ryzhikov, Soviet Phys. — Solid State **13**, 723 (1971).
[9] E. D. Palik and R. Bray, Phys. Rev. **B3**, 3302 (1971).
[10] For extensive references see: N. I. Meyer and M. H. Jørgensen, Adv. Solid State Phys. **10**, 21 (1970).
[11] I. Yamashita, T. Ishiguro , and T. Tanaka, Japan J. appl. Phys. **4**, 470 (1965).
[12] W. Wettling and E. Pohlendt, Phys. Letters (Netherlands) **19**, 268 (1965).
[13] W. H. Haydl and C. F. Quate, Phys. Letters (Netherlands) **20**, 463 (1966).
[14] W. Wettling and M. Bruun, phys. stat. sol. **34**, 221 (1969).
[15] R. Klein, Phys. Letters (Netherlands) **28A**, 428 (1968).
[16] R. W. Dixon, IEEE J. Quantum Electronics **3**, 85 (1967).
[17] L. L. Hope, Phys. Rev. **166**, 883 (1968).
[18] Y. Mita, J. appl. Phys. **42**, 2886 (1971).
[19] J. Zucker and S. Zemon, Appl. Phys. Letters **9**, 398 (1966).
[20] S. Zemon and J. Zucker, IBM J. Res. Developm. **13**, 494 (1969).
[21] A. Many and V. Gelbart, Appl. Phys. Letters **19**, 192 (1971).
[22] F. Siebert, O. Keller, and W. Wettling, phys. stat. sol. (a) **4**, 67 (1971).
[23] T. B. Bateman and J. H. McFee, J. appl. Phys. **39**, 4471 (1968).

(Received December 28, 1971)

phys. stat. sol. (a) **10**, 593 (1972)

Subject classification: 19; 22.8.1

Institute of Chemical Physics, Academy of Sciences of the USSR, Moscow

The Calculation of ESR Spectrum of Mn^{2+} Ions in Polycrystalline Samples

By

N. N. TIKHOMIROVA, S. N. DOBRYAKOV, and I. V. NIKOLAEVA

A program is composed to simulate the ESR spectra of Mn^{2+} ions ($S = 5/2$, $J = 5/2$) in polycrystalline samples at X- and Q-band frequencies taking into account nonvanishing HFS eigenwidth and the 3-rd order terms of perturbation theory for cubic crystalline field with a small axial distortion ($D \ll g\,\beta\,H$). Only the lines of allowed transitions ($\Delta M = \pm 1$, $\Delta m = 0$) have been considered. The probability of various HFS transitions was assumed to be the same and independent of orientation. The resonance field value $H_m(\theta)$ was determined for every component. The form of the ESR spectrum of the Mn^{2+} ion in a polycrystalline sample was found to depend strongly on interrelations between ΔH_i, D, and H_0. The analysis of the calculated spectra permits to estimate the value of D. A comparison was made between the calculated and experimental spectra of Mn^{2+} ion in powdered calcite. The theoretical spectrum with the parameters $A = -93.5$ Oe, $D = 85.5$ Oe, and $\Delta H_i = 2.5$ Oe was found to show all the peculiarities of the experimental spectrum.

Es wird ein Programm zur Simulation der ESR-Spektren von Mn^{2+}-Ionen ($S = 5/2$, $J = 5/2$) in polykristallinen Proben für das X- und Q-Band zusammengestellt, das nichtverschwindende HFS-Linienbreiten und Terme 3. Ordnung der Störungstheorie für kubische Kristallfelder mit einer kleinen axialen Störung ($D \ll g\,\beta\,H$) berücksichtigt. Nur die Linien erlaubter Übergänge ($\Delta M = \pm 1$, $\Delta m = 0$) werden berücksichtigt. Die Wahrscheinlichkeit verschiedener HFS-Übergänge wurde als gleich und unabhängig von der Orientierung angenommen. Die Resonanzfeldstärke $H_m(\theta)$ wurde für jede Komponente bestimmt. Es wurde gefunden, daß die Form des ESR-Spektrums von Mn^{2+}-Ionen in polykristallinen Proben stark von den Beziehungen zwischen ΔH_i, D und H_0 abhängt. Die Analyse des berechneten Spektrums erlaubt den Wert von D zu bestimmen. Ein Vergleich wurde mit dem experimentellen Spektrum von Mn^{2+}-Ionen in Kalzitpulver durchgeführt. Das theoretische Spektrum mit den Parametern $A = -93{,}5$ Oe, $D = 85{,}5$ Oe und $\Delta H_i = 2{,}5$ Oe zeigt alle Eigenschaften des experimentellen Spektrums.

1. Introduction

Mn^{2+} ($S = 5/2$, $J = 5/2$) ion is often used as a spin-label to investigate the structure of elementary cells and the distribution of electric fields in polycrystalline solids. For this purpose the Mn^{2+} ions originally contained in the samples in the form of natural admixture may be used as well as those especially introduced in the course of preparation. At the same time the analysis of Mn^{2+} ESR spectra for obtaining information on the lattice structure in a polycrystalline sample is rather difficult. The HE interaction ($J = 5/2$), the zero field splitting ($D \neq 0$), the forbidden transitions ($\Delta m_J \neq 0$) determine the spectrum shape. All the factors mentioned when combined together make the spectrum extremely complicated. Thus, to analyse them the computed Mn^{2+} spectra in polycrystalline samples are necessary to get the values of the parameters responsible for the above-mentioned interactions.

Nearly isotropic g-factor ($g = 2$) and HF interaction constant (A), axial symmetry of the D tensor and a relatively small D constant ($D \ll g\,\beta\,H$, $E = 0$) are characteristic of Mn^{2+} spectra in polycrystalline samples [1].

Two theoretical papers have been published on the subject discussed. One is related to the infrequent case of $D, E \gg g\,\beta\,H$ [2], the other contains calculation for the case of the usual D values $D \ll g\,\beta\,H$ [3], but the linewidth ΔH_i of the individual HFS components are taken equal to zero. Such a calculation does not seem to be sufficient for certain cases since it fails to reproduce some important details of the spectrum.

The aim of the present work was to calculate the ESR spectra of Mn^{2+} in polycrystalline samples for the case of $\Delta H \neq 0$ and to analyse the variations in the spectrum shape when D is varied. Attempts to determine D from Mn^{2+} ion spectra in polycrystalline samples have been already made [4, 5]. However, they were limited to measurement of the line amplitudes without taking into account the effect of D on the shape of HFS components.

2. Calculation of the Spectrum Shape

We have repeated the calculation made in [3] for an infinitely small width of the HFS component, making use of the δ-function and of expression

$$A(H) = \sum_{m=-5/2}^{+5/2} \int_0^{\pi} \delta\left(H - H_m(\theta)\right) \sin\theta \, d\theta \,, \tag{1}$$

where $H_m(\theta)$ is the resonance field value.

The HFS component shape was found to depend essentially on D. The line has three peculiar points, of which two are of the δ-function type, while the central one represents a step function. This is in agreement with the published data [3, 6].

A calculation taking into account the non-vanishing HFS eigenwidth were performed for D values in the range from 0 to 400 Oe (to satisfy the condition $D \ll g\,\beta\,H$). The HF interaction constant was taken to be isotropic and equal to -96 Oe. In accordance with the theory the HF interaction constant was taken to be negative, the m_J values were assumed to be negative for the low-field spectrum region and positive for the high-field region [7, 8]. Only the HFS lines for allowed transitions ($\Delta m = 0$) have been considered. The probability of all HFS transitions was taken to be the same, the resonance field value $H_m(\theta)$ was determined for every component.

The Hamiltonian of Mn^{2+} ions for a crystalline field of axial symmetry is known [7]. As it will be shown later the most essential features of the ESR spectra of Mn^{2+} ions in powder for crystalline field of axial symmetry are satisfactorily described by the Hamiltonian in the following simple form:

$$\mathcal{H} = g\,\beta\,(\boldsymbol{H} \cdot \boldsymbol{S}) + D\left[S_Z^2 - \frac{1}{3}\,S\,(S+1)\right] + A\,(\boldsymbol{S} \cdot \boldsymbol{J})\,.$$

The calculation of energy levels and HFS line positions in the Mn^{2+} spectrum allowing for the fine structure (D) and HF(A) structures, on condition D and $A \ll g\,\beta\,H$, has been made earlier [3, 9 to 12] using perturbation theory methods and taking into account second and third order terms. An estimate of the third order term has shown that the most essential contribution to the displacement of energy levels is that from terms proportional to $D^2 A$ [3]. Thus, other third order terms have not been used in expression (2) for the resonance field:

$$\frac{H_0}{g_0\,\beta} = \frac{H}{g\,\beta} + \frac{D}{g\,\beta}\left(M_S - \frac{1}{2}\right)(3\cos^2\theta - 1) -$$
$$- \frac{D^2}{2\,g\,\beta\,H}\cos^2\theta\sin^2\theta\,[4\,S\,(S+1) - 24\,M_S\,(M_S - 1) - 9] +$$
$$+ \frac{D^2}{8\,g\,\beta\,H}\sin^4\theta\,[2\,S\,(S+1) - 6\,M_S\,(M_S-1) - 3] + \frac{A}{g\,\beta}\,m_J +$$
$$+ \frac{A^2}{2\,g\,\beta\,H}[J\,(J+1) - m_J^2 + (2\,M_S - 1)\,m_J] + \frac{D^2\,A}{2\,g\,\beta\,H^2}\sin^2\theta\cos^2\theta\,\times$$
$$\times \left\{\frac{[M_S^2 - S\,(S+1)]^2 - M_S^2}{M_S} - \frac{[(M_S-1)^2 - S\,(S+1)]^2 - (M_S-1)^2}{M_S - 1}\right\} m_J +$$
$$+ \frac{D^2\,A}{8\,g\,\beta\,H}\sin^4\theta\,\{M_S\,[2\,M_S^2 + 1 - 2\,S\,(S+1)] -$$
$$- (M_S - 1)\,[2\,(M_S-1)^2 + 1 - 2\,S\,(S+1)]\}\,m_J\,. \tag{2}$$

For powders with electric fields randomly oriented relative to H_0, all components of the fine structure except that with $M_S = 1/2 \leftrightarrow -1/2$ are broadened to such an extent that they can not be recorded. Thus, with D and $A \ll H_0$ only the central component will be recorded since perturbation of electronic levels for this transition are only of second (D^2/H, A^2/H) and third ($D^2\,A/H^2$) order perturbations. The spectrum intensity will then represent 9/35 of that of a single crystal. A weak angular dependence of the HFS component for the central electronic transition has been found for Mn^{2+} in $CaCO_3$ single crystal [13]. However, as will be shown later, even at a small increase of D of a cubic field with axial distortion, for instance, from 100 to 200 Oe, these weak perturbations may distort and broaden the ESR lines of a polycrystalline sample to such an extent that recording of the spectrum in the 3 cm region becomes impossible.

The spectrum shape has been calculated using

$$A'(H) = \sum_{m_J}\int\limits_0^{\pi/2} f'\left(\frac{H - H_m(\theta)}{\Delta H_i}\right)\sin\theta\,\mathrm{d}\theta\,, \tag{3}$$

where $H_m(\theta)$ is the field value determined from the resonance condition (2), ΔH_i is the eigenwidth of the Lorentzian line, $A'(H)$ is the first derivative of the absorption curve. Summation was made over all m_J. The integrals were calculated using a M-20 computer and the expressions given in [14]. The number of nodes was determined for every value of ΔH_i and D. For instance, this number was 500 for $\Delta H_i = 2.5$ Oe and $D = 250$ Oe. The count of one spectrum took from 10 to 15 min (Fig. 1a, b).

3. Results

The obtained spectra show that the shape of HFS components at a constant ΔH_i value is essentially dependent on the D and H_0 values. When D is 50 Oe at $\Delta H_i = 2.5$ Oe and $H_0 = 3300$ Oe the spectrum represents a sextet of non-

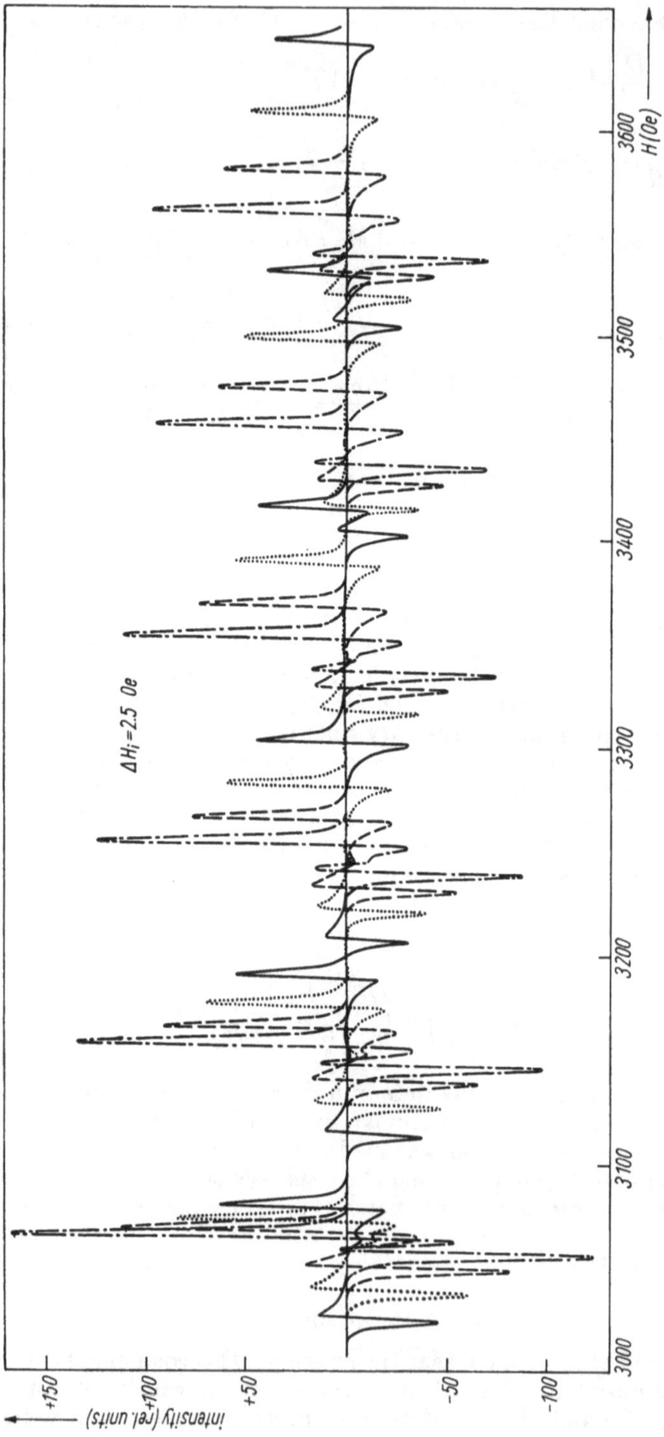

ΔH_l = 2.5 Oe
3000
3100
3200
3300
3400
3500
3600
H (Oe)
+150
+100
+50
-50
-100
intensity (rel. units)

Fig. 1a

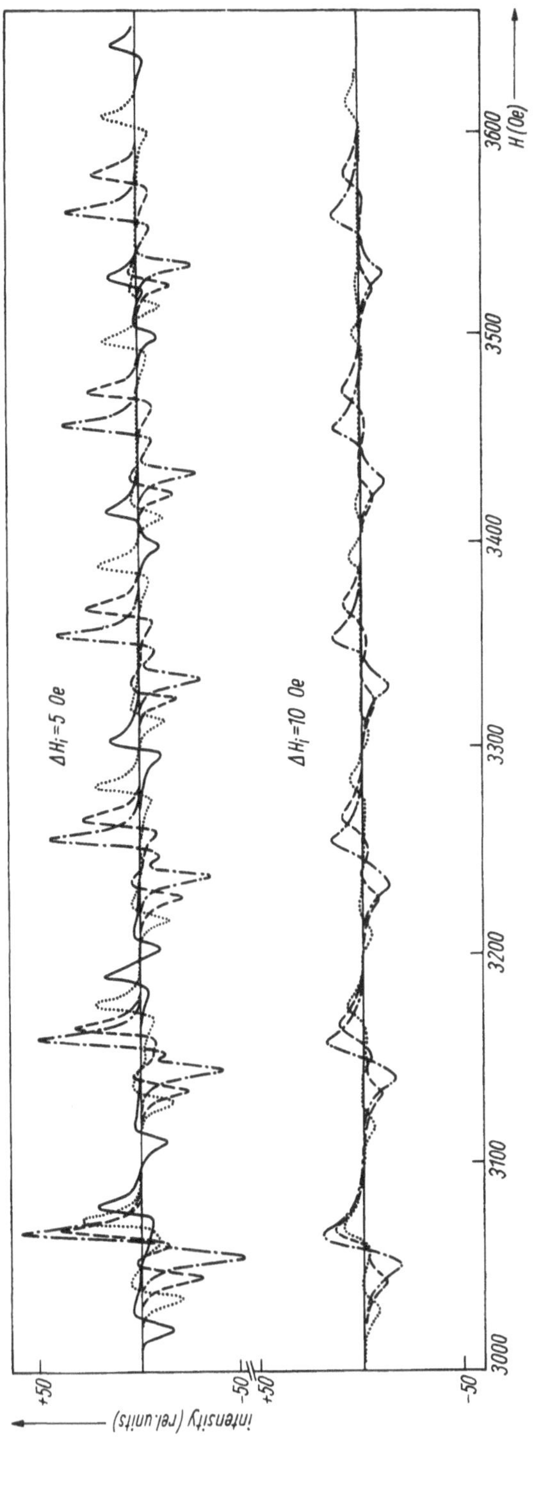
ΔH_i=5 Oe
ΔH_i=10 Oe
3000
3100
3200
3300
3400
3500
3600
H (Oe)
+50
-50
+50
-50
intensity (rel. units)

Fig. 1a

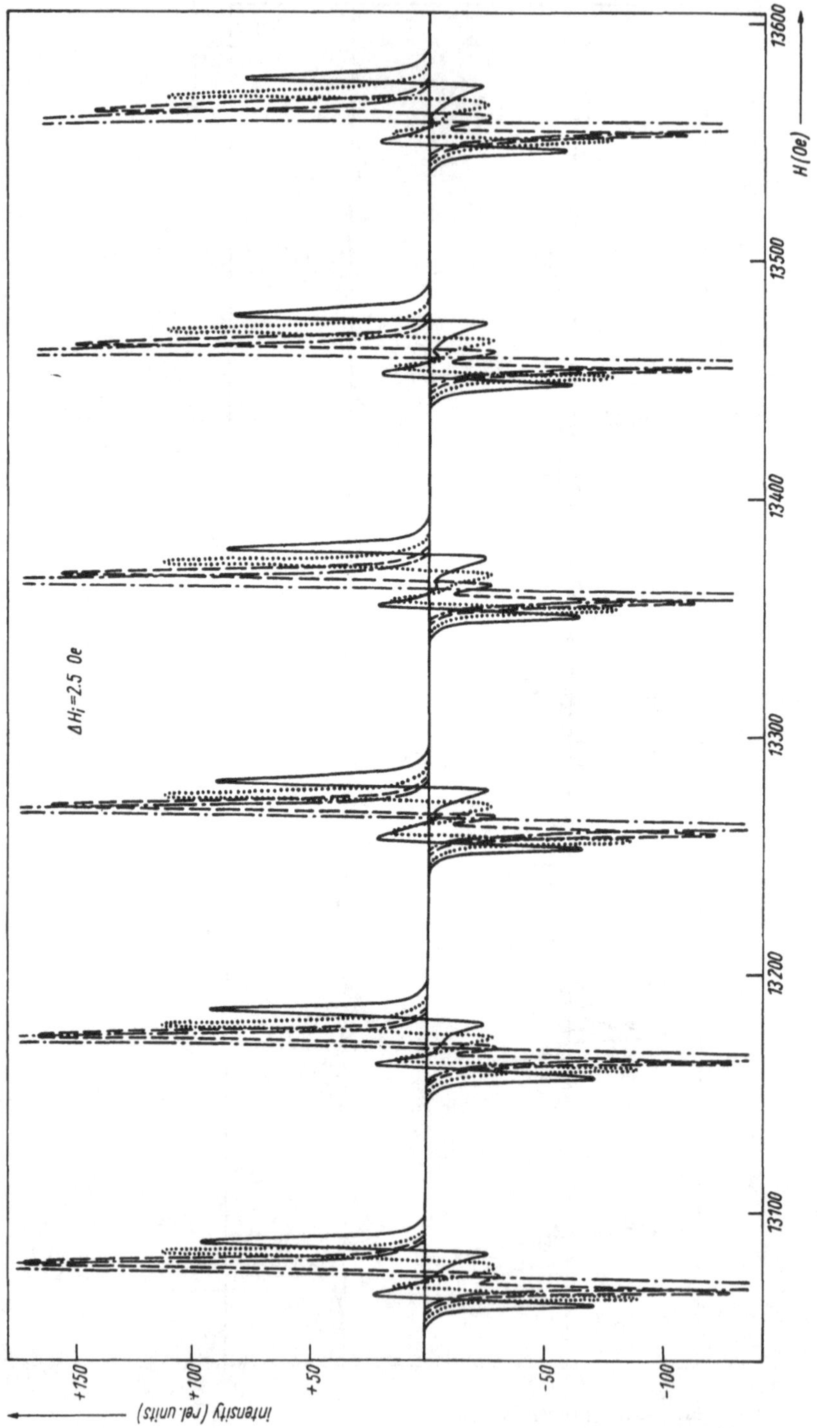

ΔH_i=2.5 Oe
13100
13200
13300
13400
13500
13600
H(Oe)
+150
+100
+50
-50
-100
intensity (rel. units)

Fig. 1 b

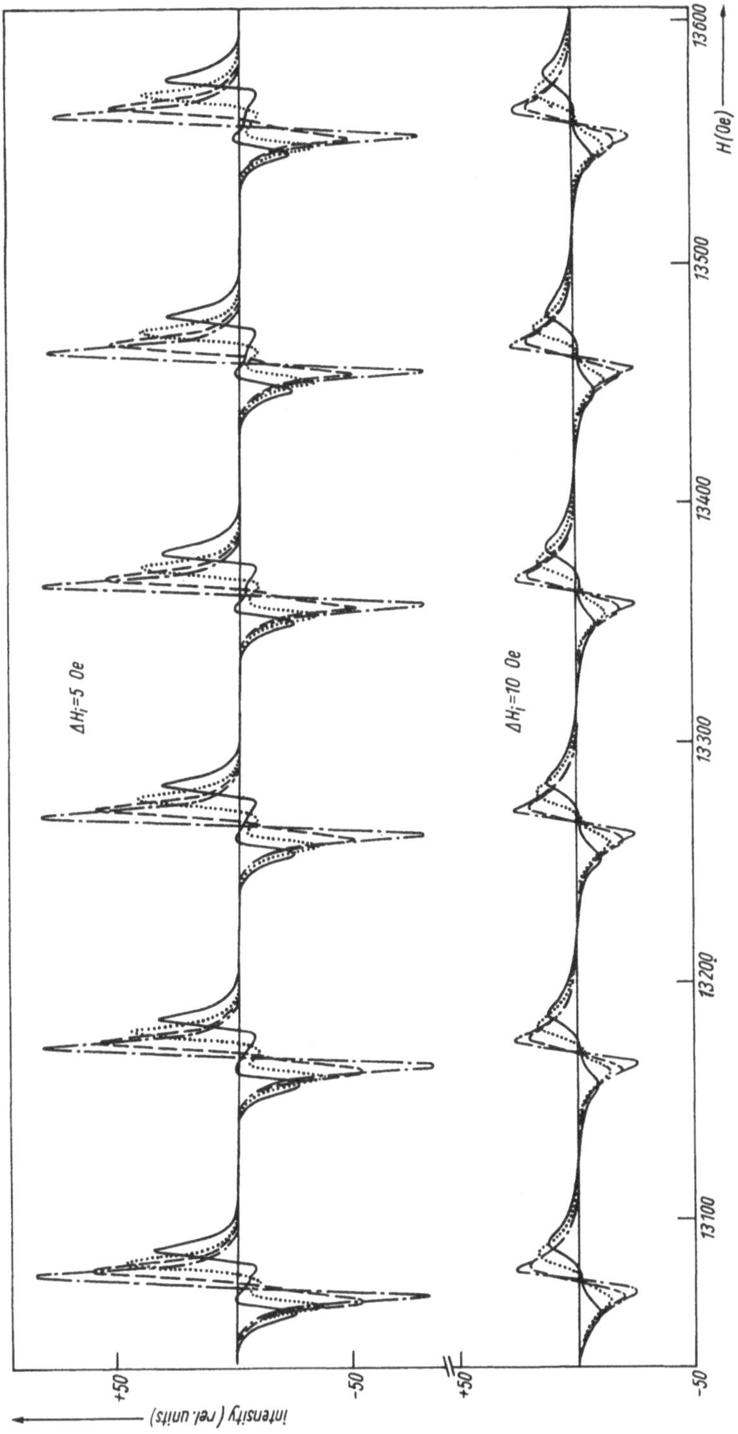

Fig. 1 b

Fig. 1. The calculated ESR spectra of Mn^{2+} ions in polycrystalline samples. a) $H_0 = 3300$ Oe and b) $H_0 = 13300$ Oe. $D = 100$ Oe (dash-dotted curves), $D = 150$ Oe (dashed curves), $D = 200$ Oe (dotted curves), $D = 250$ Oe (solid curves)

Table 1

$H_0 = 3300$ Oe

D	$\Delta H_i = 2.5$ Oe					$\Delta H_i = 5$ Oe					$\Delta H_i = 10$ Oe					$\Delta H_i = 15$ Oe					$\Delta H_i = 20$ Oe				
	$m_J = -5/2$		$m_J = +5/2$		$\Delta = \Delta H_e - \Delta H_1$	$m_J = -5/2$		$m_J = +5/2$		$\Delta = \Delta H_e - \Delta H_1$	$m_J = -5/2$		$m_J = +5/2$		$\Delta = \Delta H_e - \Delta H_1$	$m_J = -5/2$		$m_J = +5/2$		$\Delta = \Delta H_e - \Delta H_1$	$m_J = -5/2$		$m_J = +5/2$		$\Delta = \Delta H_e - \Delta H_1$
	J_1	H_1	J_e	ΔH_e		J_1	ΔH_1	J_e	ΔH_e		J_1	ΔH_1	J_e	ΔH_e		J_1	ΔH_1	J_e	ΔH_e		J_1	ΔH_1	J_e	ΔH_e	
50	608	4	361	7	3	183	6	141	8	2	48	10	43.6	12	2	21.8	15	20.9	15	0	12.2	20	12	20	0
100	287	11	167	24	13	110	11	61	26	15	39	14	24	28	14	20	17	13	30	13	11	21	8	34	13
150	192	21	96	53	32	71	23	39	54	31	25.6	25	14.3	55	30	14.4	28	8	60	32	9.2	30	5.2	60	30
200	142	37	80	91	54	49	37.5	28	92.5	55	18	40	9	95	55	10	45	4	98	53	7	45	2	100	55
250	107	57	64	140	83	38.3	60	23	142	82	14.2	60	9	144	84	7.7	65	5	150	85	4.8	65	3.3	150	85
300	59	82.5	50	200	117						9	85	6.7	205	120						3.1	102			

Table 2

$H_0 = 13300$ Oe

D	$\Delta H_i = 2.5$ Oe					$\Delta H_i = 5$ Oe					$\Delta H_i = 10$ Oe					$\Delta H_i = 15$ Oe					$\Delta H_i = 20$ Oe				
	$m_J = -5/2$		$m_J = +5/2$		$\Delta = \Delta H_e - \Delta H_1$	$m_J = -5/2$		$m_J = +5/2$		$\Delta = \Delta H_e - \Delta H_1$	$m_J = -5/2$		$m_J = +5/2$		$\Delta = \Delta H_e - \Delta H_1$	$m_J = -5/2$		$m_J = +5/2$		$\Delta = \Delta H_e - \Delta H_1$	$m_J = -5/2$		$m_J = +5/2$		$\Delta = \Delta H_e - \Delta H_1$
	J_1	ΔH_1	J_e	ΔH_e		J_1	ΔH_1	J_e	ΔH_e		J_1	ΔH_1	J_e	ΔH_e		J_1	ΔH_1	J_e	ΔH_e		J_1	ΔH_1	J_e	ΔH_e	
50	748	3	725	3	0	193	5	193	5	0	49.4	10	94.4	10	0	22	15	22	15	0	12.4	20	12.4	20	0
100	492	5	389	6	1	165	7	154	7	0	48	12.5	47	12.5	0	22	17.5	22	17.5	0	12	20	12	20	0
150	296	10	249	12.5	2.5	113	10	100	13	3	39.7	12.5	36.6	15	2.5	19.9	17.5	18.9	17.5	0	11.6	22.5	11.3	22.5	0
200	183	17.5	186	20	2.5	80	17	71	21	4	31	20	28	22.5	2.5	16.5	22.5	15.5	25	2.5	9	25	7.5	27.2	2.2
250	132	25	128	30	5	60	26	53	32	6	23.4	27.5	20	32.5	5	12	30	11	35	5	8.6	32.5	7.6	37.5	5
300	137	35	120	42.5	7.5	50	36	44	44	8	19	37.5	17	45	7.5	11	42.5	9	50	7.5	5.4	45	4.2	52.5	7.5
400	90	60	82	75	15	36	62	30	76	14	13	65	11	80	15	7.5	67.5	5	82	14.5	4.5	70			

equidistant components. At $D > 50$ Oe, and the same H_0 and ΔH_i values the singlet HFS components split to doublets. The width of the doublets increases with increasing D. For higher HFS eigenwidths the line becomes a doublet at D correspondingly higher. As the term $((D^2 A/H^2)\, m_J\, F(\theta))$ contributes to the HFS component width and m_J changes its sign along the spectrum, a monotonic increase in the HFS component width is observed for all spectra with a corresponding decrease in the HFS component amplitude. As all perturbations of the electronic level $M_s = 1/2$ changing the line shape are proportional to the second and third order terms, for higher frequencies they will occur with higher D.

Our additional calculation of the spectrum shape taking into consideration the parameter E ($E = 1/5\, D$) in the second order term of the perturbation theory manifests a change of relative intensity of doublet of the HF component. The monotonous alteration of the widths of HFS components typical for the ESR spectrum of Mn^{2+} ions in powder, is retained only if the calculation takes into account the third order terms in the parameter E.

Table 1 and 2 show the widths of the first and sixth HFS components at $H_0 = 3300$ Oe and $H_0 = 13300$ Oe for various D and ΔH_i. It will be seen from the tables and from Fig. 1a, b that the resolution of the Mn^{2+} spectrum in a polycrystalline sample, particularly in the high-field region, becomes less resolved with increasing D, or with a constant D and increasing HFS line widths within the 3 cm range. Taking into account that the HFS component amplitude decreases simultaneously it will be evident that observation of the spectrum at $H_0 = 3300$ Oe will become impossible at $D > 150$ and $\Delta H_i = 2.5$ Oe. In the 8 mm region ($H_0 = 13300$ Oe) the spectrum can be recorded over a considerably wider range of D and ΔH_i.

Another conclusion can be made from the data tabulated: for a given value of H_0 the difference in the widths of the first and sixth HFS components (Δ) is constant and is independent of ΔH_i. The dependence of Δ on D^2 is linear (Fig. 2). The data in Fig. 2 can be used for the estimate of D from experimental Mn^{2+} ESR spectra, if the difference in the width of the first and the sixth HFS components (Δ) is known. The accuracy of the estimate is dependent on that for the measurement of the widths of the first and the sixth HFS component. With an ESR-2M spectrometer it amounts to $\pm$ 10%.

4. Comparison with Experimental Data

It also follows from the results obtained that over a certain range of D and ΔH_i the HFS components show a double splitting and a twelve-component spectrum will thus be obtained. This spectrum can be erroneously explained as due to two non-equivalent ions in the system. Such an erroneous interpreta-

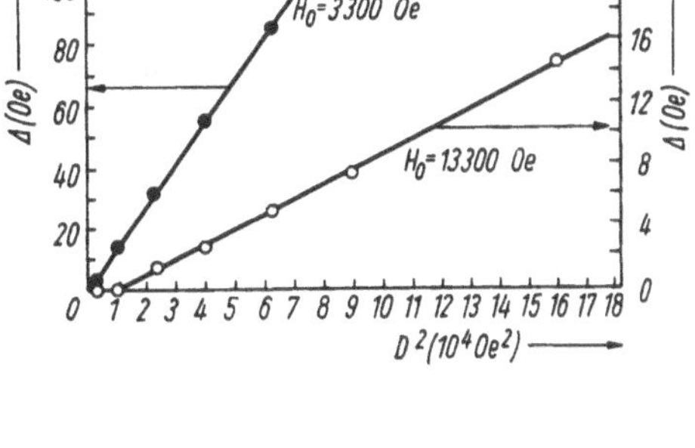

Fig. 2. The dependence of the difference in widths of the first and sixth HFS components Δ on D^2

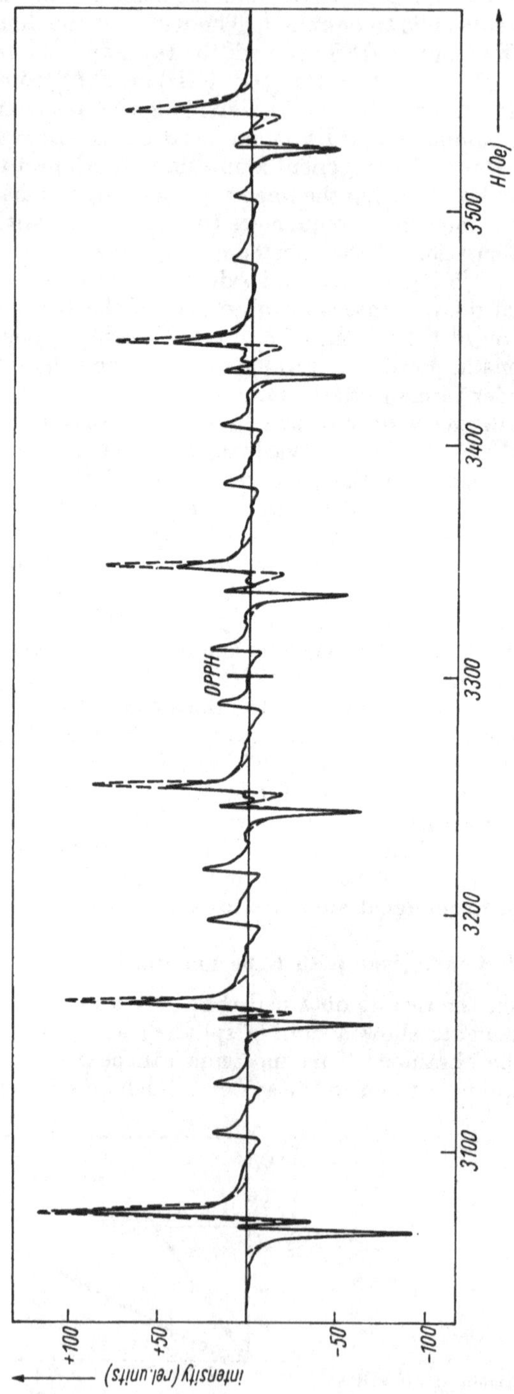

Fig. 3. The observed (solid curves) and simulated (dashed curves) spectrum of Mn^{2+} in powdered calcite. As was mentioned above in the theoretical spectrum only the HFS lines accounting for allowed transitions ($\Delta m = 0$) have been considered

tion actually took place for the spectrum of Mn^{2+} in calcite [15]. The observed spectrum is given in Fig. 3 in comparison with a simulated spectrum with parameters $A = -93.5$ Oe, $D = 85.5$ Oe, and $\Delta H_i = 2.5$ Oe. The virtually complete coincidence of these spectra shows that calcite contains only one type of paramagnetic Mn^{2+} ions. (As shown above, the forbidden transitions have not been taken into account in the calculated spectra and consequently they were omitted in the experimental spectrum of Mn^{2+} in $CaCO_3$.)

5. Conclusion

Thus the non-zero widths of the individual HFS components and zero field splitting may result in very complicated and unusual Mn^{2+} spectra in polycrystalline samples. This could lead to wrong interpretation. To make their interpretation more reliable one should record these spectra at several frequencies and take into account the dependence of the spectrum shape on frequency as noted above.

Acknowledgements

The authors wish to thank Prof. Y. S. Lebedev for his continuous interest in this work and Prof. L. A. Blumenfeld for helpful discussions.

References

[1] A. Carrington and A. D. McLachlan, Introduction to Magnetic Resonance with Applications to Chemistry and Chemical Physics, New York/London 1967.
[2] R. D. Dowsing and D. J. E. Ingram, J. Mag. Res. **1**, 517 (1969).
[3] H. W. de Wijn and F. R. van Balderen, J. chem. Phys. **46**, 1381 (1967).
[4] B. T. Allen, J. chem. Phys. **43**, 3820 (1965).
[5] R. D. Dowsing, J. Mag. Res. **2**, 42 (1970).
[6] B. Bleaney and R. S. Rubins, Proc. Phys. Soc. (London) **77**, 103 (1961).
[7] W. Low, Paramagnetic Resonance in Solids, Academic Press, New York/London 1960.
[8] Hyperfine Interactions, Academic Press, New York/London 1967.
[9] L. M. Matarrese, J. chem. Phys. **34**, 336 (1961).
[10] V. J. Folen, Phys. Rev. **125**, 1581 (1962).
[11] J. E. Drumheller and R. S. Rubins, Phys. Rev. **133**, A1099 (1964).
[12] J. Wolga and R. Tseng, Phys. Rev. **133**, A1563 (1964).
[13] F. K. Hurd, M. Sachs, and W. D. Hershberger, Phys. Rev. **93**, 373 (1954).
[14] R. W. Hamming, Numerical Methods for Scientists and Engineers, McGraw-Hill Publ. Co., New York 1962 (§ 21.5).
[15] C. Kikuchi, J. chem. Phys. **33**, 601 (1960).

(Received February 8, 1972)

phys. stat. sol. (a) **10**, 605 (1972)

Subject classification: 14.3.3; 22

Department of Electrical Engineering, University of Minnesota, Minneapolis (a) and Centre de recherches mathématiques, Université de Montréal (b)

General Transport Theory of Noise in p–n Junction-Like Devices

III. Junction Noise in p$^+$–n Diodes at High Injection[1])

By

H. S. MIN (a), K. M. VAN VLIET (b), and A. VAN DER ZIEL (a)

The noise in diodes is considered at high injection for an arbitrary geometry, and for the most general Shockley-Read type generation–recombination noise source. The Green's function technique of an earlier paper [3] is used. By algebraic manipulation of the basic-response-type hole density spectra, results are obtained which can be expressed in the network parameters Y_j (junction admittance), Z_j (junction impedance), and I_0 (steady state current) without calculation of these parameters. Only part of the noise, the so-called Shockley-Read excess, requires explicit computation. For a long diode the results are in agreement with previous one-dimensional results [4]. The total noise is expressible in network parameters in this case via a simple expression. For a short diode, as well as for other geometries, the Shockley-Read excess is modified, but the total noise is not drastically affected.

Es wird das Rauschen in Dioden bei hoher Injektion für beliebige Geometrie und für die allgemeinste Generation–Rekombinations-Rauschquelle vom Shockley-Read-Typ betrachtet. Dazu wird die Methode der Greenschen Funktion aus einer früheren Arbeit [3] benutzt. Durch algebraische Behandlung der Spektren der Löcherdichte vom Grund-response-Typ werden Ergebnisse erhalten, die durch die Netzwerkparameter Y_j (Sperrschicht-Admittanz), Z_j (Sperrschicht-Impedanz) und I_0 (stationärer Strom) ausgedrückt werden können, ohne diese Parameter zu berechnen. Nur ein Teil des Rauschens, der sogenannte Shockley-Read-Überschuß, erfordert explizite Berechnung. Für eine lange Diode sind die Ergebnisse mit früheren eindimensionalen Ergebnissen [4] in Übereinstimmung. Das Gesamtrauschen läßt sich mit den Netzwerkparametern ausdrücken, in diesem Falle mit einem einfachen Ausdruck. Sowohl für eine kurze Diode als auch für andere geometrische Anordnungen wird der Shockley-Read-Überschuß modifiziert, jedoch wird das Gesamtrauschen nicht wesentlich beeinflußt.

1. Introduction

The theory of noise in junctions and transistors for low-injection levels is well known [1] and has been substantiated by experiment [2]. At high-injection levels, i.e., when the density of injected minority carriers exceeds the density of the majority carriers of the equilibrium state, only recently a general approach was devised [3, 4] and experiments to explore the situation are still scarce [5].

At high injection a corpuscular approach is no longer possible since the carriers are generally not independent. Thus, only a collective approach, based on the transport equations for the fluctuating carrier densities, together with relatation-

---

[1]) Supported in part by the National Science Foundation.

ships for the currents and voltages in terms of the carrier densities, is feasible. To this purpose, a consistent Green's function approach for such problems was worked out in the first paper of this series [3] (referred to as I). From a device point of view, this is a generalization of van der Ziel's transmission line method of 1955 [6], but valid for an arbitrary three-dimensional geometry and general circuit conditions (open-circuited terminals, short-circuited terminals or hybrid arrangements). From a mathematical point of view, the theory is an extension of the methods employed by van Vliet and Fassett [7] to non-homogeneous systems.

1.1 Previous work

Part I dealt, besides with the principles of the method, with the detailed application to low-injection theory for diodes and transistors, and with the problem of high injection for structures in which the recombination is a band–band process or a "symmetric" Shockley-Read-Hall process (i.e., the capture constants of the levels for electrons and holes, τ_{n0} and τ_{p0}, are equal). For these cases (and for these cases *only*), the carrier fluctuations appear to be spatially uncorrelated, as follows from the solution of the "Λ-theorem" [8]. Curiously, it is not necessary to solve for the Green's function in these cases. It is sufficient to express both the noise and the network parameters (impedances, etc.) in the Green's functions and general, three-dimensional relationships follow. In the next paper [4] (referred to as II) the problem of high-injection noise in junction diodes with asymmetric Shockley-Read-Hall centers was investigated. It was initially believed that neither analytical nor three-dimensional results were possible for this case. Hence, the one-dimensional long diode was examined in detail and explicit results for the correlation function were established. Analytical results were obtained, at last, with rather simple high-frequency limits (as indicated in II). Moreover, if the long-diode network parameters are also computed, a unique relationship between the noise and these parameters is once more obtained (not indicated in II). Because of the explicit calculations the road to this result is simultaneously clear in objective and laborious in algebra (a situation also prevailing in van der Ziel's original transmission line method [6] for low injection).

1.2 Present work

The present article continues with the methods of the first paper. We will show that the problem of high-injection noise in junctions with general Shockley-Read-Hall recombination processes can be handled nearly completely for an arbitrary geometry. Again, we formulate the noise sources and the transport equations. By manipulation, rather than by explicit algebra, results can be found which are largely expressible in the network parameters. Only one distinct part, the Shockley-Read excess has to be considered separately for various geometries; even so, the complete result is expressible in network parameters for a simple geometry. A larger difficulty for high-injection noise stems from the fact that a fluctuating electric field is set up in the bulk regions. This gives rise to bulk modulation noise, which is quite complex. The contemplated study is therefore divided into two parts. The present Part III deals with the noise of the junction voltage and junction current proper, whereas the next Part IV completes this study by considering the noise of the terminal voltage and terminal current of the device.

1.3 Assumptions

As in the previous articles, the following basic assumptions are made. We consider asymmetric diodes, such as p$^+$–n (or n$^+$–p) structures. The hole current fraction is denoted by γ and assumed to be close to unity for a p$^+$–n device. The dependence on the current level $\gamma(I)$ is left in the results as an empirical expression. We consider the high-injection limit p $\approx$ n. Quasi-space-charge neutrality leading to ambipolar transport [9] is presumed, as in most high-injection theories. However, there are indications that this approach gives trouble for the bulk-modulation noise of long diodes, as we discuss in the next article in detail.

1.4 Relevance

A parallel exists between the development of the noise theory for high injection in bipolar devices and the noise theory of space-charge-limited electron tubes. The latter was developed by Spenke, North, and Rack some 20 years after Schottky's original theory of shot noise for saturated emission currents. The space-charge-limited theory was very complex, though the final results were expressible in simple tube parameters, e.g., as 0.64 times the thermal noise of the conductance. Likewise, the theory for high injection considered here is considerably more complex than the low-injection theory, which can be based on elementary corpuscular arguments. Nevertheless, the results are expressible in simple quasi-thermal and quasi-shot noise like expressions, with some parameters of well understood physical origin remaining in the results; compare in particular this article, equations (4.9) to (4.11).

2. Hole Fluctuation Spectra

We consider the ambipolar transport of electron and hole fluctuations, $\Delta n(\boldsymbol{r}, t) \approx \Delta p(\boldsymbol{r}, t)$, in the n-region of the p$^+$–n diode. The basic transport equation is, cf. I, equation (6.1),

$$\mathcal{L}[\Delta p(\boldsymbol{r},t)] \equiv \frac{\partial \Delta p}{\partial t} + \frac{\Delta p}{\tau_\mathrm{H}} - D_\mathrm{a} \nabla^2(\Delta p) = \gamma(\boldsymbol{r}, t) + \nabla \boldsymbol{\eta}_\mathrm{a}(\boldsymbol{r}, t) = \xi(\boldsymbol{r}, t), \quad (2.1)$$

where τ_H is the high injection lifetime for Shockley-Read recombination, D_a is the ambipolar diffusivity, γ is the generation–recombination source, and $\boldsymbol{\eta}_\mathrm{a}$ is the ambipolar diffusion source. The high-injection lifetime is $\tau_\mathrm{H} = \tau_{\mathrm{n}0} + \tau_{\mathrm{p}0}$, where $\tau_{\mathrm{n}0}$ and $\tau_{\mathrm{p}0}$ refer to the extreme values of the electron and hole capture lifetimes for the centers, respectively. The high-injection diffusivity is a constant at high injection, $D_\mathrm{a} = 2\, D_\mathrm{p}\, b/(b+1)$, where b is the mobility ratio and where we assume the validity of the Einstein relation for the electrons and holes, separately. We also write $\mathcal{L} = \Lambda + \partial/\partial t$, and for the Laplace transformed operator $L = \Lambda + s$.

The relations for the junction current and voltage were discussed in I, Section 6. For the steady state and fluctuating parts of the current we have, respectively,

$$I_0 = \frac{(b+1)\, e\, D_\mathrm{a}}{\gamma + \gamma\, b - 1} \int \mathrm{d}\boldsymbol{\sigma}_j\, \nabla_0\, p_\mathrm{s}(\boldsymbol{r}_0), \quad (2.2)$$

$$\Delta I(t) = \frac{b+1}{\gamma + \gamma\, b - 1} \int \mathrm{d}\boldsymbol{\sigma}_\mathrm{j} \{e\, D_\mathrm{a}\, \nabla_0[\Delta p(\boldsymbol{r}_0, t)] + e\, \boldsymbol{\eta}_\mathrm{a}(\boldsymbol{r}_0, t)\}, \quad (2.3)$$

where the integration is over the surface of injection $\boldsymbol{r}_0 \in S_j$ (assuming that the junction is abrupt), ∇_0 refers to the outer surface gradient, and $p = p_s + \Delta p$ is decomposed in steady state hole density and fluctuating hole density. For the density junction voltage relationship we assume Fletcher's results [10]; these relations were recently once more discussed and extended by Hauser [11]. The corresponding result for the fluctuations is simply

$$\Delta V(\boldsymbol{r}_0, t) = \frac{kT}{e\, p_s(\boldsymbol{r}_0)} \Delta p(\boldsymbol{r}_0, t) , \tag{2.4}$$

where k is the Boltzman constant, T the hole temperature, and where $p_s(\boldsymbol{r}_0) = p_{sj}$ does not depend on $\boldsymbol{r}_0$ if the junction surface is an equipotential surface (we so assume).

At the ohmic constant surface S_c ,and on the free surface S_f we set $\Delta p(\boldsymbol{r}_0, t) = 0$ (other conditions for surface recombination were discussed in I). Two distinct boundary conditions at S_j arise depending on whether we consider the short-circuited case, $\Delta V = 0$, or the open-ciruited case, $\Delta I = 0$. These conditions will be handled somewhat differently than in part I. We write, with $\boldsymbol{\nu}(\boldsymbol{r}_0)$ being the outward normal,

$$a(\boldsymbol{r}_0)\, \Delta p(\boldsymbol{r}_0, t) + b(\boldsymbol{r}_0)\, \boldsymbol{\nu}(\boldsymbol{r}_0) \cdot \nabla_0 \Delta p(\boldsymbol{r}_0, t) = \theta(\boldsymbol{r}_0, t) , \tag{2.5}$$

where $b(\boldsymbol{r}_0) = 0$, $a(\boldsymbol{r}_0) \neq 0$, $\theta \equiv 0$ on the surfaces S_c and S_f, while on S_j:

$$a(\boldsymbol{r}_0) = 1 , \quad b(\boldsymbol{r}_0) = 0 , \quad \theta \equiv 0 , \qquad \text{(short circuit)} , \tag{2.6a}$$

$$a(\boldsymbol{r}_0) = 0 , \quad b(\boldsymbol{r}_0) = 1 , \quad \theta = -\frac{\eta_a\, \nu}{D_a} , \qquad \text{(open circuit)} . \tag{2.6b}$$

The Green's function $G(\boldsymbol{r}, \boldsymbol{r}', s)$ for the Laplace-transformed operator L is introduced in the standard way [7], satisfying the homogeneous boundary conditions corresponding to (2.5); it will be superscripted by O or S for the open and short circuited case, respectively. The solution of the Laplace transformed equation (2.1), with $\Delta\hat{p}(\boldsymbol{r}, s)$ denoting the Laplace transform of $\Delta p(\boldsymbol{r}, t)$, and similar for $\hat{\xi}$ and $\hat{\theta}$, is then (compare I, Appendix I or [12, 13])

$$\begin{aligned}\Delta\hat{p}(\boldsymbol{r}, s) &= \int dv''\, G(\boldsymbol{r}, \boldsymbol{r}'', s)\, \hat{\xi}(\boldsymbol{r}'', s) - D_a \oint d\boldsymbol{\sigma}''\, [\Delta\hat{p}(\boldsymbol{r}_0'', s)\, \nabla_0''\, G(\boldsymbol{r}, \boldsymbol{r}_0'', s) - \\ &\quad - G(\boldsymbol{r}, \boldsymbol{r}_0'', s)\, \nabla_0''\, \Delta\hat{p}(\boldsymbol{r}_0'', s)] = \int dv''\, G(\boldsymbol{r}, \boldsymbol{r}'', s)\, \hat{\xi}(\boldsymbol{r}'', s) + \\ &\quad + \oint d\sigma''\, [Q(\boldsymbol{r}_0'')\, G(\boldsymbol{r}, \boldsymbol{r}_0'', s)]\, \theta(\boldsymbol{r}_0'', s) ,\end{aligned} \tag{2.7}$$

where the volume integral is over the entire n-region, the surface integral over all surfaces bounding this region, and where

$$Q(\boldsymbol{r}_0'') = -\frac{D_a}{a(\boldsymbol{r}_0'')}\, \boldsymbol{\nu}(\boldsymbol{r}_0'')\, \nabla_0'' \quad \text{if} \quad a(\boldsymbol{r}_0'') \neq 0 , \tag{2.8a}$$

$$Q(\boldsymbol{r}_0'') = \frac{D_a}{b(\boldsymbol{r}_0'')} \quad \text{if} \quad b(\boldsymbol{r}_0'') \neq 0 . \tag{2.8b}$$

In the terminology of Part I, G is the volume Green's function, and $Q\,G$ is the surface Green's function.

2.1 Response method

We can now find the spectra $S_{\Delta p}(\boldsymbol{r}, \boldsymbol{r}', \omega)$ from (2.7), by multiplying with $\Delta p(\boldsymbol{r}', s)$ and using the standard procedures (see I, Appendix I and [8], Appendix B). The result is the response expression, which is quadratic in the Green's function:

$$S_{\Delta p}(\boldsymbol{r}, \boldsymbol{r}', \omega) = 2 \int\int dv'' \, dv''' \, G(\boldsymbol{r}, \boldsymbol{r}'', j\omega) \, \Xi(\boldsymbol{r}'', \boldsymbol{r}''') \, G(\boldsymbol{r}', \boldsymbol{r}''', -j\omega) +$$
$$+ 2 \int \oint dv'' \, d\sigma \, G(\boldsymbol{r}, \boldsymbol{r}'', j\omega) \, \Omega(\boldsymbol{r}'', \boldsymbol{r}_0) \, Q(\boldsymbol{r}_0) \, G(\boldsymbol{r}', \boldsymbol{r}_0, -j\omega) +$$
$$+ 2 \oint \int d\sigma \, dv'' \, G(\boldsymbol{r}', \boldsymbol{r}'', -j\omega) \, \Omega(\boldsymbol{r}'', \boldsymbol{r}_0) \, Q(\boldsymbol{r}_0) \, G(\boldsymbol{r}, \boldsymbol{r}_0, j\omega) +$$
$$+ 2 \oint\oint d\sigma \, d\sigma' \, \Theta(\boldsymbol{r}_0, \boldsymbol{r}_0') \, Q(\boldsymbol{r}_0) \, G(\boldsymbol{r}, \boldsymbol{r}_0, j\omega) \, Q(\boldsymbol{r}_0') \, G(\boldsymbol{r}', \boldsymbol{r}_0', -j\omega) \,. \tag{2.9}$$

In this expression we introduced the correlation strengths (or white spectra) of the sources

$$\Xi(\boldsymbol{r}, \boldsymbol{r}') = \tfrac{1}{2} S_{\xi\xi}(\boldsymbol{r}, \boldsymbol{r}') \equiv \tfrac{1}{2} S_{\text{bulk}}(\boldsymbol{r}, \boldsymbol{r}') \,, \tag{2.10}$$

$$\Omega(\boldsymbol{r}, \boldsymbol{r}_0) = \tfrac{1}{2} S_{\xi\theta}(\boldsymbol{r}, \boldsymbol{r}_0) \equiv \tfrac{1}{2} S_{\text{cross}}(\boldsymbol{r}, \boldsymbol{r}_0) \,, \tag{2.11}$$

$$\Theta(\boldsymbol{r}, \boldsymbol{r}_0') = \tfrac{1}{2} S_{\theta\theta}(\boldsymbol{r}_0, \boldsymbol{r}_0') \equiv \tfrac{1}{2} S_{\text{surf}}(\boldsymbol{r}_0, \boldsymbol{r}_0') \,, \tag{2.12}$$

where the last r.h.s. indicates explicitly that these are the spectra from the bulk sources, surface sources, or their cross correlation. Clearly,

$$S_{\text{bulk}} = S_y + \nabla \nabla' : \mathbf{S}_{\eta a} \,, \quad S_{\text{cross}} = D_a^{-1} \boldsymbol{\nu} \nabla : \mathbf{S}_{\eta a} \,, \quad S_{\text{surf}} = D_a^{-2} \boldsymbol{\nu}(\boldsymbol{r}_0') \, \boldsymbol{\nu}(\boldsymbol{r}_0) : \mathbf{S}_{\eta a} .$$

The sources were derived before (see reference [3]), with result [everywhere $\nabla = \nabla_{\boldsymbol{r}}$, $\nabla' = \nabla_{\boldsymbol{r}'}$, etc.]

$$\Xi(\boldsymbol{r}, \boldsymbol{r}') = \frac{p_s(\boldsymbol{r})}{\tau_H} \, \frac{\tau_{n0}^2 + \tau_{p0}^2}{(\tau_{n0} + \tau_{p0})^2} \, \delta(\boldsymbol{r} - \boldsymbol{r}') + D_a \nabla \nabla' [p_s(\boldsymbol{r}) \, \delta(\boldsymbol{r} - \boldsymbol{r}')] =$$

$$= \frac{p_s(\boldsymbol{r})}{2\tau_H} \delta(\boldsymbol{r} - \boldsymbol{r}') + D_a \nabla \nabla' [p_s(\boldsymbol{r}) \delta(\boldsymbol{r} - \boldsymbol{r}')] + \left(\frac{\tau_{n0} - \tau_{p0}}{\tau_{n0} + \tau_{p0}}\right)^2 \frac{p_s(\boldsymbol{r})}{2\tau_H} \delta(\boldsymbol{r} - \boldsymbol{r}') \,, \tag{2.13}$$

$$\Omega(\boldsymbol{r}, \boldsymbol{r}_0) = -\boldsymbol{\nu}(\boldsymbol{r}_0) \cdot \nabla [p_s(\boldsymbol{r}) \, \delta(\boldsymbol{r} - \boldsymbol{r}_0)] \,, \qquad \boldsymbol{r}_0 \in S_j \text{ (open circuit only)}, \tag{2.14}$$

$$\Theta(\boldsymbol{r}_0, \boldsymbol{r}_0') = \frac{1}{D_a} p_s(\boldsymbol{r}_0) \, \delta(\boldsymbol{r}_0 - \boldsymbol{r}_0') \, \boldsymbol{\nu}(\boldsymbol{r}_0) \, \boldsymbol{\nu}(\boldsymbol{r}_0') , \qquad \boldsymbol{r}_0 \in S_j \text{ (open circuit only)}, \tag{2.15}$$

$$\Omega \equiv 0 \,, \quad \Theta \equiv 0 \qquad \text{(short circuited case)} \,. \tag{2.16}$$

The bulk sources (2.13) are more complete than in Part I, in that we have three contributions which are referred to as the quasi-bimolecular generation–recombination source, the ambipolar diffusion source, and the Shockley-Read (SR) excess, respectively. We set again $y = \tau_{n0}/\tau_{p0}$, as in Part II.

Equations (2.13) to (2.16) are now substituted into (2.9), which is rearranged using Green's and Gauß's theorems. The process is lengthy and is carried out in the Appendix. The final result, observing the boundary conditions (b.c.) $a\,G + b\,\boldsymbol{\nu} \nabla_0 G = 0$ on S_j, cf. equations (2.6), and $G = 0$ on the other surfaces, is then as follows. For the short-circuited case we find

$$S_{\Delta p,\,\text{short}}(\boldsymbol{r}, \boldsymbol{r}', \omega) = p_s(\boldsymbol{r}') \, G^S(\boldsymbol{r}, \boldsymbol{r}', j\omega) + p_s(\boldsymbol{r}) \, G^S(\boldsymbol{r}', \boldsymbol{r}, -j\omega) +$$
$$+ \left(\frac{1-y}{1+y}\right)^2 \int dv'' \, \frac{p_s(\boldsymbol{r}'')}{\tau_H} G^S(\boldsymbol{r}, \boldsymbol{r}'', j\omega) \, G^S(\boldsymbol{r}', \boldsymbol{r}'', -j\omega) \tag{2.17}$$

and for the open-circuited case

$$S_{\Delta p,\,\mathrm{open}}(\boldsymbol{r}, \boldsymbol{r}', \omega) = p_s(\boldsymbol{r}')\, G^O(\boldsymbol{r}, \boldsymbol{r}', j\,\omega) + p_s(\boldsymbol{r})\, G^O(\boldsymbol{r}', \boldsymbol{r}, -j\,\omega) - \\ - D_a \int d\boldsymbol{\sigma}_j \cdot G^O(\boldsymbol{r}, \boldsymbol{r}_0, j\,\omega)\, G^O(\boldsymbol{r}', \boldsymbol{r}_0, -j\,\omega)\, \nabla_0\, p_s(\boldsymbol{r}_0) + \\ + \left(\frac{1-y}{1+y}\right)^2 \int dv''\, \frac{p_s(\boldsymbol{r}'')}{\tau_H}\, G^O(\boldsymbol{r}, \boldsymbol{r}'', j\omega)\, G^O(\boldsymbol{r}', \boldsymbol{r}'', -j\omega)\,. \tag{2.18}$$

This result is entirely in accord with the results of I (equations (6.22) and (6.23)) except that another term is added stemming from the SR excess.

2.2 Pseudo-correlation method

The close connection with the results of I led us to examine whether the correlation function method would not yield the answer faster. Basically, the answer is no. We have seen in II that the correlation function $\Gamma = \langle \Delta p(\boldsymbol{r})\, \Delta p(\boldsymbol{r}')\rangle$, as determined by the sources via the Λ-theorem, is a very complex function for the one-dimensional case. The three-dimensional case has no closed form analytical solution to our knowledge. Yet, we can find the above results also by the following procedure. We split the source function $\Xi(\boldsymbol{r}, \boldsymbol{r}')$ into two parts, whereby $\Xi_1(\boldsymbol{r}, \boldsymbol{r}')$ denotes the first two terms of (2.13) and $\Xi_2(\boldsymbol{r}, \boldsymbol{r}')$ denotes the SR excess. Next, we *define* a function Γ_1, such that

$$(\Lambda_{\boldsymbol{r}} + \Lambda_{\boldsymbol{r}'})\, \Gamma_1(\boldsymbol{r}, \boldsymbol{r}') = \Xi_1(\boldsymbol{r}, \boldsymbol{r}')\,. \tag{2.19}$$

A particular solution for Γ_1 is the infinite domain solution of the correlation function for the symmetric recombination case, i.e., $\Gamma_1 = \frac{1}{2}\, p_s(\boldsymbol{r})\, \delta\,(\boldsymbol{r} - \boldsymbol{r}')$, cf. I, equation (6.21). The bulk term (first term of (2.9)) is split into a part involving Ξ_1 and a part involving Ξ_2. The Ξ_1-part is converted to the extended van Vliet-Fassett form which yields the parts linear in the Green's function, as is done in Part I; the Ξ_2-part is left in the response from which is quadratic in the Green's function. The total result is then again equations (2.17) and (2.18). We note that this is a purely *mathematical* procedure, reflecting the fact that the Λ-theorem is a purely mathematical theorem. The function Γ_1 for this case has no physical significance, nor has the partitioning of Ξ, since there are no (uncorrelated) noise sources ξ_1 and ξ_2 corresponding to Ξ_1 and Ξ_2. For more details we refer to [14].

3. Spectra of Short-Circuited Junction Current Fluctuations and Open-Circuited Junction Voltage Fluctuations

In order to find the junction fluctuation spectra, we proceed as in Part I. We stipulate either that the junction is short-circuited, in which case we have the b.c. $\Delta p = 0$ on the junction surface and we use the Green's function G^S, or we stipulate that the junction is open-circuited for each element $d\sigma_j$ of the junction, in which case we have the b.c. $D_a\, \nabla_0\, (\Delta p)\, \boldsymbol{\nu} = -\, \eta_a\, \boldsymbol{\nu}$ and we use the Green's function G^O.

3.1 Junction current

The spectrum of the junction current $\Delta I_j(t)$ is

$$S_{\Delta I_j}(\omega) = \iint d\boldsymbol{\sigma}_j\, d\boldsymbol{\sigma}_j' : \mathbf{S}_{\Delta \boldsymbol{J}}(\boldsymbol{r}_0, \boldsymbol{r}_0', \omega)\,, \tag{3.1}$$

where $\Delta \boldsymbol{J}$ is the fluctuating current density vector. We define a position-dependent current density $\Delta \hat{\boldsymbol{J}}$ by[2])

$$\Delta \hat{\boldsymbol{J}}(\boldsymbol{r}, t) = -\frac{b+1}{\gamma + \gamma b - 1} \left[e D_{\mathrm{a}} \nabla \left(\Delta p(\boldsymbol{r}, t)\right) + e\, \boldsymbol{\eta}_{\mathrm{a}}(\boldsymbol{r}, t)\right], \tag{3.2}$$

see also equation (2.3). Then

$$\mathbf{S}_{\Delta \boldsymbol{J}}(\boldsymbol{r}_0, \boldsymbol{r}_0', \omega) = \lim_{\substack{\boldsymbol{r} \to \boldsymbol{r}_0 \\ \boldsymbol{r}' \to \boldsymbol{r}_0'}} \mathbf{S}_{\Delta \hat{\boldsymbol{J}}}(\boldsymbol{r}, \boldsymbol{r}', \omega). \tag{3.3}$$

From equation (3.2) we obtain

$$\mathbf{S}_{\Delta \hat{\boldsymbol{J}}}(\boldsymbol{r}, \boldsymbol{r}', \omega) = \\ = \left(\frac{b+1}{\gamma + \gamma b - 1}\right)^2 [e^2 D_{\mathrm{a}}^2 \nabla_{\boldsymbol{r}} \nabla_{\boldsymbol{r}'} S_{\Delta p}(\boldsymbol{r}, \boldsymbol{r}', \omega) + e^2 D_{\mathrm{a}} \nabla_{\boldsymbol{r}} \boldsymbol{S}_{\Delta p, \eta_{\mathrm{a}}}(\boldsymbol{r}, \boldsymbol{r}', \omega) + \\ + e^2 D_{\mathrm{a}} \nabla_{\boldsymbol{r}'} \boldsymbol{S}_{\eta_{\mathrm{a}}, \Delta p}(\boldsymbol{r}, \boldsymbol{r}', \omega) + e^2 \mathbf{S}_{\eta_{\mathrm{a}}}(\boldsymbol{r}, \boldsymbol{r}', \omega)], \tag{3.4}$$

where we used the relations[3])

$$\mathbf{S}_{\nabla(\Delta p)}(\boldsymbol{r}, \boldsymbol{r}', \omega) = \nabla_{\boldsymbol{r}} \nabla_{\boldsymbol{r}'} S_{\Delta p}(\boldsymbol{r}, \boldsymbol{r}', \omega), \tag{3.5}$$

$$\mathbf{S}_{\nabla(\Delta p), \eta_{\mathrm{a}}}(\boldsymbol{r}, \boldsymbol{r}', \omega) = \nabla_{\boldsymbol{r}} \boldsymbol{S}_{\Delta p, \eta_{\mathrm{a}}}(\boldsymbol{r}, \boldsymbol{r}', \omega). \tag{3.6}$$

For $S_{\Delta p}$ we now use equation (2.17). The cross spectra are the same as in I, Section 7:

$$\boldsymbol{S}_{\Delta p, \eta_{\mathrm{a}}}(\boldsymbol{r}, \boldsymbol{r}', \omega) = -2 D_{\mathrm{a}} p_{\mathrm{s}}(\boldsymbol{r}') \nabla_{\boldsymbol{r}'} G^{\mathrm{S}}(\boldsymbol{r}, \boldsymbol{r}', j\omega), \tag{3.7}$$

$$\boldsymbol{S}_{\eta_{\mathrm{a}}, \Delta p}(\boldsymbol{r}, \boldsymbol{r}', \omega) = -2 D_{\mathrm{a}} p_{\mathrm{s}}(\boldsymbol{r}) \nabla_{\boldsymbol{r}} G^{\mathrm{S}}(\boldsymbol{r}', \boldsymbol{r}, -j\omega). \tag{3.8}$$

The spike term resulting from $\mathbf{S}_{\eta_{\mathrm{a}}}$ can be omitted if we maintain $\boldsymbol{r} \neq \boldsymbol{r}'$ in the limiting process indicated by equation (3.3); this is implemented by letting $\boldsymbol{r} \to \boldsymbol{r}_0^+$, where the super $+$ denotes a position vector just inside the surface of the junction. We then obtain for the spectrum of the current fluctuations from (2.17) and (3.1)[4])

$$S_{\Delta I_{\mathrm{j}}}(\omega) = \left(\frac{b+1}{\gamma + \gamma b - 1}\right)^2 e^2 D_{\mathrm{a}}^2 \Big\{ \iint \mathrm{d}\boldsymbol{\sigma}_{\mathrm{j}}' \, \mathrm{d}\boldsymbol{\sigma}_{\mathrm{j}}^+ : [p_{\mathrm{s}}(\boldsymbol{r}_0') \nabla_0^+ \nabla_0' G^{\mathrm{S}}(\boldsymbol{r}_0^+, \boldsymbol{r}_0', j\omega) + \\ + p_{\mathrm{s}}(\boldsymbol{r}_0^+) \nabla_0^+ \nabla_0' G^{\mathrm{S}}(\boldsymbol{r}_0', \boldsymbol{r}_0^+, -j\omega)] + \iint \mathrm{d}\boldsymbol{\sigma}_{\mathrm{j}}' \, \mathrm{d}\boldsymbol{\sigma}_{\mathrm{j}}^+ : [\nabla_0^+ G^{\mathrm{S}}(\boldsymbol{r}_0^+, \boldsymbol{r}_0', j\omega) \nabla' p_{\mathrm{s}}(\boldsymbol{r}_0') + \\ + \nabla' G^{\mathrm{S}}(\boldsymbol{r}_0', \boldsymbol{r}_0^+, -j\omega) \nabla_0^+ p_{\mathrm{s}}(\boldsymbol{r}_0^+)] + \left(\frac{1-y}{1+y}\right)^2 \iint \mathrm{d}\boldsymbol{\sigma}_{\mathrm{j}}' \, \mathrm{d}\boldsymbol{\sigma}_{\mathrm{j}}^+ : \\ : \int \mathrm{d}v'' \, \frac{p_{\mathrm{s}}(\boldsymbol{r}'')}{\tau_{\mathrm{H}}} \nabla_0^+ G^{\mathrm{S}}(\boldsymbol{r}_0^+, \boldsymbol{r}'', j\omega) \nabla' G^{\mathrm{S}}(\boldsymbol{r}_0', \boldsymbol{r}'', -j\omega) \Big\}. \tag{3.9}$$

---

[2]) Only near the junction surface the total current equals the diffusion current $\hat{\boldsymbol{J}}$.

[3]) $\mathbf{S}_{\Delta(\Delta p)}$ is the tensor spectrum of $\langle \nabla_{\boldsymbol{r}}[\Delta p(\boldsymbol{r})] \nabla_{\boldsymbol{r}'}[\Delta p(\boldsymbol{r}')] \rangle$ and $\mathbf{S}_{\nabla(\Delta p), \eta_{\mathrm{a}}}$ is the tensor spectrum of $\langle \nabla_{\boldsymbol{r}}[\Delta p(\boldsymbol{r})] \boldsymbol{\eta}_{\mathrm{a}}(\boldsymbol{r}') \rangle$.

[4]) For brevity we write $\nabla_0^+ \Phi = [\nabla_{\boldsymbol{r}} \Phi(\boldsymbol{r})]_{\boldsymbol{r} = \boldsymbol{r}_0^+}$; similar for $\nabla_0' \Phi$, etc.

Using the expression for the admittance (I, equation (1.31)), and the integral property for G^S (I, equation (7.4)) together with (2.2), we obtain now

$$S_{\Delta I_j}(\omega) = \frac{b+1}{\gamma + \gamma b - 1} [2\, kT \operatorname{Re} Y_j - e\, I_0] + S_{\Delta I_{SRE}}, \tag{3.10}$$

where SRE denotes the contribution resulting from the Shockley-Read excess, i.e., the last term of equation (3.9).

3.2 Junction voltage

The cross fluctuation spectrum $S_{\Delta V}(\boldsymbol{r}_0, \boldsymbol{r}_0', \omega)$ of the junction voltages in series with surface elements $d\sigma_j(\boldsymbol{r}_0)$ and $d\sigma_j(\boldsymbol{r}_0')$ is found from (2.4):

$$S_{\Delta V}(\boldsymbol{r}_0, \boldsymbol{r}_0', \omega) = \left[\frac{kT}{e\, p_{sj}}\right]^2 S_{\Delta p}(\boldsymbol{r}_0, \boldsymbol{r}_0', \omega). \tag{3.11}$$

As in I we introduce the current distribution factors for dc and ac currents over the junction surface, by the following equations, respectively:

$$d\gamma_0(\boldsymbol{r}_0) = \frac{dI_0(\boldsymbol{r}_0)}{I_0} = -\frac{\boldsymbol{J}_0(\boldsymbol{r}_0)\, d\boldsymbol{\sigma}_j(\boldsymbol{r}_0)}{I_0},$$

$$d\gamma_1(\boldsymbol{r}_0, \omega) = -\frac{\boldsymbol{J}_1(\boldsymbol{r}_0, \omega)\, d\boldsymbol{\sigma}_j(\boldsymbol{r}_0)}{I_1(\omega)} = \frac{dY_j(\boldsymbol{r}_0, \omega)}{Y_j(\omega)},$$

where $dY_j(\boldsymbol{r}_0, \omega)$ is the admittance in series with an element $d\sigma_j$ if each element is separately open-circuited and $Y_j(\omega)$ is the total admittance of the junction S_j. The total open-circuit equivalent voltage noise is then given by the Stieltjes integral

$$S_{\Delta V_j}(\omega) = \iint S_{\Delta V}(\boldsymbol{r}_0, \boldsymbol{r}_0', \omega)\, d\gamma_1(\boldsymbol{r}_0, \omega)\, d\gamma_1^*(\boldsymbol{r}_0', \omega). \tag{3.12}$$

We use (2.18) and obtain

$$S_{\Delta V_j}(\omega) = \left(\frac{kT}{e\, p_{sj}}\right)^2 \iint d\gamma_1(\boldsymbol{r}_0, \omega)\, d\gamma_1^*(\boldsymbol{r}_0', \omega) \Big\{[p_s(\boldsymbol{r}_0')\, G^O(\boldsymbol{r}_0, \boldsymbol{r}_0', j\,\omega) +$$

$$+ p_s(\boldsymbol{r}_0)\, G^O(\boldsymbol{r}_0', \boldsymbol{r}_0, -j\,\omega)] - D_a \int d\sigma_j''\, G^O(\boldsymbol{r}_0, \boldsymbol{r}_0'', j\,\omega)\, G^O(\boldsymbol{r}_0', \boldsymbol{r}_0'', -j\,\omega)\, \nabla_0''\, p_s(\boldsymbol{r}_0'') +$$

$$+ \left(\frac{1-y}{1+y}\right)^2 \int dv''\, \frac{p_s(\boldsymbol{r}'')}{\tau_H}\, G^O(\boldsymbol{r}_0, \boldsymbol{r}'', j\,\omega)\, G^O(\boldsymbol{r}_0', \boldsymbol{r}'', -j\,\omega)\Big\}. \tag{3.13}$$

Using the expression for the impedance (I, equation (1.33)), and equation (2.2), we finally find

$$S_{\Delta V_j}(\omega) = \frac{b+1}{\gamma + \gamma b - 1} [2\, kT \operatorname{Re} Z_j - e\, I_0\, Z_j\, Z_j^*] + S_{\Delta V_{SRE}}, \tag{3.14}$$

where the Shockley-Read term denotes the last term of (3.13).

We must finally prove the equivalence of the results (3.9) and (3.10) with (3.13) and (3.14). This proof is based on the identity

$$G^S(\boldsymbol{r}, \boldsymbol{r}', j\,\omega) - G^O(\boldsymbol{r}', \boldsymbol{r}, j\,\omega) = -D_a \int d\boldsymbol{\sigma}_j\, G^O(\boldsymbol{r}', \boldsymbol{r}_0, j\,\omega)\, \nabla_0 G^S(\boldsymbol{r}, \boldsymbol{r}_0, j\omega). \tag{3.15}$$

With this result and the expression for Z_j in terms of G^0, it is easily demonstrated that

$$S_{\Delta V_{SRE}} = S_{\Delta I_{SRE}} Z_j Z_j^* . \tag{3.16}$$

4. One Dimensional Evaluation and Summary

For the SRE term we restrict ourselves to a one-dimensional geometry, $0 \leqq x \leqq L$. In that event

$$G^S(x, x', j\omega) = \begin{cases} \dfrac{\sinh(h x') \sinh[h(L - x)]}{D_a h \sinh(h L)} & (x > x') \\ \dfrac{\sinh(h x) \sinh[h(L - x')]}{D_a h \sinh(h L)} & (x < x'), \end{cases} \tag{4.1}$$

where

$$h = \frac{(1 + j\omega\tau_H)^{1/2}}{l}, \quad l = (D_a \tau_H)^{1/2} . \tag{4.2}$$

The steady state solution is further

$$p_s(x) = \frac{p_{sj} \sinh[(L - x)/l] + p_0 \sinh(x/l)}{\sinh(L/l)}, \tag{4.3}$$

where $p_s(x = 0) = p_{sj}$ and $p(x = L) = p_0$.
Also, from (2.2),

$$I_0 = \frac{b + 1}{\gamma + \gamma b - 1} e D_a \frac{1}{l} \left[p_{sj} \coth\left(\frac{L}{l}\right) - p_0 \operatorname{csech}\left(\frac{L}{l}\right) \right]. \tag{4.4}$$

For our present purpose we can neglect the terms in p_0; actually, the high-injection limit cannot be reached close to the ohmic contact anyway, but one can ascertain that this region does not contribute much to the SR integral in (3.9). With

$$\left[\frac{\partial G^S(x, x'', j\omega)}{\partial x}\right]_{x=0} = \frac{\sinh[h(L - x'')]}{D_a \sinh(h L)}$$

we obtain, if τ_H does not depend on position,

$$S_{\Delta I_{SRE}} = \left(\frac{b + 1}{\gamma + \gamma b - 1}\right)^2 \left(\frac{1 - y}{1 + y}\right)^2 \frac{e^2 p_{sj}}{\tau_H} \times$$

$$\times \left[-\frac{2}{l} \frac{h h^*}{(h^2 + h^{*2} - 1/l^2)^2 - 4 h^2 h^{*2}} \frac{1}{\sinh(h L) \sinh(h^* L) \sinh(L/l)} + \right.$$

$$+ \frac{h + h^*}{2} \frac{\coth(h^* L) + \coth(h L)}{(h + h^*)^2 - 1/l^2} -$$

$$- \frac{1}{2l} \frac{\coth(L/l) \coth(h L) \coth(h^* L) + \coth(L/l)}{(h + h^*)^2 - 1/l^2} +$$

$$+ \frac{h - h^*}{2} \frac{\coth(h L) - \coth(h^* L)}{(h - h^*)^2 - 1/l^2} +$$

$$\left. + \frac{1}{2l} \frac{\coth(L/l) \coth(h L) \coth(h^* L) - \coth(L/l)}{(h - h^*)^2 - 1/l^2} \right]. \tag{4.5}$$

The admittance for this model is (from I, equations (1.31) and (4.1) or from [15])

$$Y_j = G_j(0) \frac{h\,L \coth(h\,L)}{(L/l) \coth(L/l)}, \qquad G_j(0) = \frac{e\,I_0}{kT}. \tag{4.6}$$

Only for the case $L/l \gg 1$, $|h\,L)| \gg 1$, the spectrum is simple

$$S_{\Delta I_{\mathrm{SRE}}} \approx \left(\frac{b+1}{\gamma+\gamma\,b-1}\right)^2 \left(\frac{1-y}{1+y}\right)^2 \frac{e^2\,p_{\mathrm{sj}}\,l}{\tau_{\mathrm{H}}} \frac{1}{[2+2\,(1+\omega^2\,\tau_{\mathrm{H}}^2)^{1/2}]^{1/2}+1} =$$

$$= \frac{b+1}{\gamma+\gamma\,b-1} \left(\frac{1-y}{1+y}\right)^2 \frac{G_j(0)}{2\,G_j(\omega)+G_j(0)}\,e\,I_0\,. \tag{4.7}$$

Clearly, the SR contribution disappears for sufficiently large ω.

For low frequencies h, $h^* \to 1/l$. Then (4.5) yields

$$S_{\Delta I_{\mathrm{SRE}}}(\omega \to 0) =$$

$$= \frac{b+1}{\gamma+\gamma\,b-1} \left(\frac{1-y}{1+y}\right)^2 e\,I_0 \left\{\frac{2}{3}\operatorname{sech}\lambda\operatorname{csech}^2\lambda - \frac{2}{3}\coth^2\lambda + 1\right\}, \tag{4.8}$$

where $\lambda = L/l$. The factor in { } is denoted as $\Psi(0, \lambda)$. For large λ its value is 1/3, in accord with (4.7), and it decreases slowly with decreasing λ. Some values appear in Table 1.

Table 1

Values of $\Psi(0, \lambda)$ for various $\lambda = L/l$

$\lambda \to$	1.0	2.0	3.0	5.0	10.0
$\Psi \to$	0.163	0.296	0.329	0.333	0.333

We summarize the results. The full junction current noise for arbitrary geometry is given by (3.10). For a *long, one-dimensional diode* this takes the explicit form

$$S_{\Delta I_j}(\omega)|_{\text{long diode}} = \frac{b+1}{\gamma+\gamma\,b-1}\left\{2\,kT\,G_j(\omega) - e\,I_0\left[1 - \left(\frac{1-y}{1+y}\right)^2 \frac{G_j(0)}{2\,G_j(\omega)+G_j(0)}\right]\right\}. \tag{4.9}$$

Using (4.6), we see that the high- and low- frequency limits are in complete harmony with the results of II. For the *short diode* the high-frequency limit is still determined by the term $2\,kT\,G_j(\omega) \propto \sqrt{\omega}$. The intermediate frequency behavior is complex, see equation (4.5), but the low-frequency limit is simple. We give an interpolation type formula

$$S_{\Delta I_j}(\omega)|_{\text{short diode}} = \frac{b+1}{\gamma+\gamma\,b-1}\left\{2\,kT\,G_j(\omega) - e\,I_0\left[1 - \left(\frac{1-y}{1+y}\right)^2 \Psi(\omega, \lambda)\right]\right\}, \tag{4.10}$$

where $\Psi(\omega, \lambda) \to 0$ if $\omega\,\tau_{\mathrm{H}} \gg 1$ and where $\Psi(0, \lambda)$ varies between 1/6 and 1/3 for $1 \leq \lambda < \infty$, with $\lambda = L/l$. For intermediate frequencies, numerical calculations indicate that the "Ansatz"

$$\Psi(\omega, \lambda) = \Psi(0, \lambda) \frac{3\, G_j(0)}{2 G_j(\omega) + G_j(0)} \tag{4.11}$$

provides a reasonable fit to the detailed behavior of (4.5) (to within 5%). We conjecture that the same is probably true for other arbitrary geometries, with λ depending somewhat on the current crowding in the active region of the conduction process.

Appendix

Pertaining to the Response Method

Substituting the sources Ξ into the bulk term of (2.9), this term reads

$$2 \iint dv''\, dv'''\, G(\boldsymbol{r}, \boldsymbol{r}'', j\,\omega)\, \Xi(\boldsymbol{r}'', \boldsymbol{r}''')\, G(\boldsymbol{r}', \boldsymbol{r}''', -j\,\omega) =$$

$$= 2 \iint dv''\, dv'''\, G(\boldsymbol{r}, \boldsymbol{r}'', j\,\omega)\, G(\boldsymbol{r}', \boldsymbol{r}''', -j\,\omega)\, D_a \nabla_{\boldsymbol{r}''} \nabla_{\boldsymbol{r}'''}[p_s(\boldsymbol{r}'')\, \delta(\boldsymbol{r}'' - \boldsymbol{r}''')] + \quad \text{(a)}$$

$$+ \iint dv''\, dv'''\, G(\boldsymbol{r}, \boldsymbol{r}'', j\,\omega)\, G(\boldsymbol{r}', \boldsymbol{r}''', -j\,\omega) \frac{p_s(\boldsymbol{r}'')}{\tau_H} \delta(\boldsymbol{r}'' - \boldsymbol{r}''') + \quad \text{(b)}$$

$$+ \iint dv''\, dv'''\, G(\boldsymbol{r}, \boldsymbol{r}'', j\,\omega)\, G(\boldsymbol{r}', \boldsymbol{r}''', -j\,\omega) \left(\frac{1-y}{1+y}\right)^2 \frac{p_s(\boldsymbol{r}'')}{\tau_H} \delta(\boldsymbol{r}'' - \boldsymbol{r}''')\,. \quad \text{(c)}$$

$$\text{(A1)}$$

Using the steady state continuity equation $(p_s/\tau_H) - D_a \nabla^2 p_s = 0$ the second term, (b), of (A1) is transformed into

$$\text{(b)} = \int dv''\, G(\boldsymbol{r}, \boldsymbol{r}'', j\,\omega)\, G(\boldsymbol{r}', \boldsymbol{r}'', -j\,\omega) \frac{2\, p_s(\boldsymbol{r}'')}{\tau_H} - D_a \nabla^2_{\boldsymbol{r}''} p_s(\boldsymbol{r}'')\,.$$

We apply Green's theorem for the ∇^2-operator in the second integral; we then obtain

$$\begin{aligned}
\text{(b)} = {} & 2 \int dv'' \frac{p_s(\boldsymbol{r}'')}{\tau_H} G(\boldsymbol{r}, \boldsymbol{r}'', j\,\omega)\, G(\boldsymbol{r}', \boldsymbol{r}'', -j\,\omega) - \\
& - D_a \int dv''\, p_s(\boldsymbol{r}'')\, [G(\boldsymbol{r}, \boldsymbol{r}'', j\,\omega)\, \nabla^2_{\boldsymbol{r}''} G(\boldsymbol{r}', \boldsymbol{r}'', -j\,\omega) + G(\boldsymbol{r}', \boldsymbol{r}'', -j\,\omega) \times \\
& \times \nabla^2_{\boldsymbol{r}''} G(\boldsymbol{r}, \boldsymbol{r}'', j\,\omega) + 2\, \nabla_{\boldsymbol{r}''} G(\boldsymbol{r}, \boldsymbol{r}'', j\,\omega)\, \nabla_{\boldsymbol{r}''} G(\boldsymbol{r}', \boldsymbol{r}'', -j\,\omega)] - \\
& - D_a \oint d\sigma\, G(\boldsymbol{r}, \boldsymbol{r}_0, j\,\omega)\, G(\boldsymbol{r}', \boldsymbol{r}_0, -j\,\omega)\, \nabla_0\, p_s(\boldsymbol{r}_0) + \\
& + D_a \oint d\sigma\, p_s(\boldsymbol{r}_0)\, [G(\boldsymbol{r}, \boldsymbol{r}_0, j\,\omega)\, \nabla_0 G(\boldsymbol{r}', \boldsymbol{r}_0, -j\,\omega) + G(\boldsymbol{r}', \boldsymbol{r}_0, j\,\omega) \times \\
& \times \nabla_0 G(\boldsymbol{r}, \boldsymbol{r}_0, j\,\omega)\,.
\end{aligned} \tag{A2}$$

For term (a) of (A1) we use the relation $f\, \nabla \cdot \boldsymbol{A} = \nabla \cdot (f\, \boldsymbol{A}) - \boldsymbol{A} \cdot \nabla f$; then,

$$\begin{aligned}
\text{(a)} = {} & 2\, D_a \int dv'''\, G(\boldsymbol{r}', \boldsymbol{r}''', -j\,\omega) \int dv'' \{\nabla_{\boldsymbol{r}''} [G(\boldsymbol{r}, \boldsymbol{r}'', j\,\omega)\, p_s(\boldsymbol{r}'') \times \\
& \times \nabla_{\boldsymbol{r}'''} \delta(\boldsymbol{r}'' - \boldsymbol{r}''')] - p_s(\boldsymbol{r}'')\, \nabla_{\boldsymbol{r}'''} \delta(\boldsymbol{r}'' - \boldsymbol{r}''')\, \nabla_{\boldsymbol{r}''} G(\boldsymbol{r}, \boldsymbol{r}'', j\,\omega)\}\,.
\end{aligned}$$

Next, we apply Gauß' theorem[5]) and obtain

$$(a) = 2\,D_a \int dv'''\, G(\boldsymbol{r}', \boldsymbol{r}''', -j\,\omega) \oint d\boldsymbol{\sigma}\, G(\boldsymbol{r}, \boldsymbol{r}_0, j\,\omega)\; p_s(\boldsymbol{r}_0)\, \nabla_{\boldsymbol{r}'''}\, \delta\,(\boldsymbol{r}''' - \boldsymbol{r}_0) -$$
$$- 2\,D_a \int dv''\, p_s(\boldsymbol{r}'') \int dv'''\, \nabla_{\boldsymbol{r}'''} [G(\boldsymbol{r}', \boldsymbol{r}''', -j\,\omega)\, \delta\,(\boldsymbol{r}'' - \boldsymbol{r}''') \times$$
$$\times \nabla_{\boldsymbol{r}''} G(\boldsymbol{r}, \boldsymbol{r}'', j\,\omega)] + 2\,D_a \int dv''\, \delta\,(\boldsymbol{r}'' - \boldsymbol{r}''')\, p_s(\boldsymbol{r}'') \times$$
$$\times \int dv'''\, \nabla_{\boldsymbol{r}''} G(\boldsymbol{r},\boldsymbol{r}'', j\,\omega)\, \nabla_{\boldsymbol{r}'''} G(\boldsymbol{r}', \boldsymbol{r}''', -j\,\omega)\,.$$

We apply Gauß' theorem once more for the second term and integrate the resulting surface delta function, noticing that $\int dv''\, f(\boldsymbol{r}'')\, \delta\,(\boldsymbol{r}'' - \boldsymbol{r}_0) = \frac{1}{2}\, f(\boldsymbol{r}_0)$ (I, Appendix II). We also carry out the delta function integration in the third term,

$$(a) = 2\,D_a \int dv'''\, G(\boldsymbol{r}', \boldsymbol{r}''', -j\,\omega) \oint d\boldsymbol{\sigma}\, G(\boldsymbol{r}, \boldsymbol{r}_0, j\,\omega)\, p_s(\boldsymbol{r}_0)\, \nabla_{\boldsymbol{r}'''}\, \delta\,(\boldsymbol{r}''' - \boldsymbol{r}_0) -$$
$$- D_a \oint d\boldsymbol{\sigma}'\, p_s(\boldsymbol{r}_0')\, G(\boldsymbol{r}', \boldsymbol{r}_0', j\,\omega)\, \nabla_0'\, G(\boldsymbol{r}, \boldsymbol{r}_0', j\,\omega) +$$
$$+ 2\,D_a \int dv''\, p_s(\boldsymbol{r}'')\, \nabla_{\boldsymbol{r}''} G(\boldsymbol{r}, \boldsymbol{r}'', j\,\omega)\, \nabla_{\boldsymbol{r}''} G(\boldsymbol{r}', \boldsymbol{r}'', -j\,\omega)\,.$$

Finally, the first term of this expression is changed by writing $\nabla_{\boldsymbol{r}'''}$ in front of the whole integral and by using Gauß' theorem again. This yields

$$(a) = 2\,D_a \oint\!\!\oint d\boldsymbol{\sigma}\, d\boldsymbol{\sigma}'\, G(\boldsymbol{r}, \boldsymbol{r}_0, j\,\omega)\, G(\boldsymbol{r}', \boldsymbol{r}_0', -j\,\omega)\, p_s(\boldsymbol{r}_0)\, \delta\,(\boldsymbol{r}_0 - \boldsymbol{r}_0') -$$
$$- D_a \oint d\boldsymbol{\sigma}\, p_s(\boldsymbol{r}_0)\, G(\boldsymbol{r}, \boldsymbol{r}_0, j\,\omega)\, \nabla_0\, G(\boldsymbol{r}', \boldsymbol{r}_0, -j\,\omega) -$$
$$- D_a \oint d\boldsymbol{\sigma}\, p_s(\boldsymbol{r}_0)\, G(\boldsymbol{r}', \boldsymbol{r}_0, -j\,\omega)\, \nabla_0\, G(\boldsymbol{r}, \boldsymbol{r}_0, j\,\omega) +$$
$$+ 2\,D_a \int dv''\, p_s(\boldsymbol{r}'')\, \nabla_{\boldsymbol{r}''} G(\boldsymbol{r}, \boldsymbol{r}'', j\,\omega)\, \nabla_{\boldsymbol{r}''} G(\boldsymbol{r}', \boldsymbol{r}'', -j\,\omega)\,. \tag{A3}$$

Term (c) remains unchanged, except for the delta integration:

$$(c) = \left(\frac{1-y}{1+y}\right)^2 \int dv''\, \frac{p_s(\boldsymbol{r}'')}{\tau_H}\, G(\boldsymbol{r}, \boldsymbol{r}'', j\,\omega)\, G(\boldsymbol{r}', \boldsymbol{r}'', -j\,\omega)\,. \tag{A4}$$

Adding up the terms of equations (A2), (A3), and (A4) we find after cancellation and rearrangement:

$$S_{\Delta p}(\boldsymbol{r}, \boldsymbol{r}', \omega) =$$
$$= \int dv''\, p_s(\boldsymbol{r}'')\, G(\boldsymbol{r}, \boldsymbol{r}'', j\,\omega) \left[\frac{G(\boldsymbol{r}', \boldsymbol{r}'', -j\,\omega)}{\tau_H} - D_a\, \nabla^2_{\boldsymbol{r}''} G(\boldsymbol{r}', \boldsymbol{r}'', -j\,\omega)\right] +$$
$$+ \int dv''\, p_s(\boldsymbol{r}'')\, G(\boldsymbol{r}', \boldsymbol{r}'', -j\,\omega) \left[\frac{G(\boldsymbol{r}, \boldsymbol{r}'', j\,\omega)}{\tau_H} - D_a\, \nabla^2_{\boldsymbol{r}''} G(\boldsymbol{r}, \boldsymbol{r}'', j\,\omega)\right] -$$
$$- D_a \oint d\boldsymbol{\sigma}\, G(\boldsymbol{r}, \boldsymbol{r}_0, j\,\omega)\, G(\boldsymbol{r}', \boldsymbol{r}_0, -j\,\omega)\, \nabla_0\, p_s(\boldsymbol{r}_0) +$$
$$+ 2\,D_a \oint\!\!\oint d\boldsymbol{\sigma}\, d\boldsymbol{\sigma}'\, G(\boldsymbol{r}, \boldsymbol{r}_0, j\,\omega)\, G(\boldsymbol{r}', \boldsymbol{r}_0', -j\,\omega)\, p_s(\boldsymbol{r}_0)\, \delta\,(\boldsymbol{r}_0 - \boldsymbol{r}_0') +$$
$$+ \left(\frac{1-y}{1+y}\right)^2 \int dv''\, \frac{p_s(\boldsymbol{r}'')}{\tau_H}\, G(\boldsymbol{r}, \boldsymbol{r}'', j\,\omega)\, G(\boldsymbol{r}', \boldsymbol{r}'', -j\,\omega) + \text{terms in } \Omega \text{ and } \Theta. \tag{A5}$$

---

[5]) We use two "Gauß theorems", for the gradient and the divergence:
a) $\int dv\, \nabla\, \boldsymbol{F}(\boldsymbol{r}) = \oint d\boldsymbol{\sigma}\, \boldsymbol{F}(\boldsymbol{r}_0)$; b) $\int dv\, \nabla\, \Phi(\boldsymbol{r}) = \oint d\boldsymbol{\sigma}\, \Phi(\boldsymbol{r}_0)$.

From the defining equation of the Green's function the factors in the first two terms are found to be

$$L(\pm j\,\omega)\,[G(\pm j\,\omega)] \mp j\,\omega\, G(\pm j\,\omega) = \delta\,(\boldsymbol{r} - \boldsymbol{r}') \mp j\,\omega\; G(\pm j\,\omega)\,. \quad \text{(A6)}$$

The first two terms thus result in

$$S_{\Delta p}(\boldsymbol{r}, \boldsymbol{r}', \omega) = p_s(\boldsymbol{r}')\, G(\boldsymbol{r}, \boldsymbol{r}', j\,\omega) + p_s(\boldsymbol{r})\, G(\boldsymbol{r}', \boldsymbol{r}, -j\,\omega)\,. \quad \text{(A7)}$$

For the short-circuited case $G = 0$ on the complete boundary and the surface integrals of (A5) vanish. The result (2.17) follows.

For the open-circuited case we must take the complete expression (2.9) and take along surface integrals, observing (2.14), (2.15), (2.8b), and (2.6b). Surface integrals over $d\boldsymbol{\sigma}_j$ remain. For the terms in Ω we bring again the ∇ in front of the integrand and apply Gauß' theorem. Whence,

$$\begin{aligned} &\text{terms in } \Omega \to -2\,D_a \int dv'' \int d\boldsymbol{\sigma}_j\, G(\boldsymbol{r}, \boldsymbol{r}'', j\,\omega)\, G(\boldsymbol{r}', \boldsymbol{r}_0', -j\,\omega) \times \\ &\times \nabla_{r''}\, [p_s(\boldsymbol{r}'')\, \delta(\boldsymbol{r}'' - \boldsymbol{r}_0')] + \text{similar term } (\boldsymbol{r} \to \boldsymbol{r}', \boldsymbol{r}' \to \boldsymbol{r}, j\,\omega \to -j\,\omega) = \\ &= -4\,D_a \int\int d\boldsymbol{\sigma}_j\, d\boldsymbol{\sigma}_j'\, G(\boldsymbol{r}, \boldsymbol{r}_0, j\,\omega)\, G(\boldsymbol{r}', \boldsymbol{r}_0', -j\,\omega)\, p_s(\boldsymbol{r}_0)\, \delta\,(\boldsymbol{r}_0 - \boldsymbol{r}_0') + \\ &+ D_a \int d\boldsymbol{\sigma}_j\, p_s(\boldsymbol{r}_0)\, G(\boldsymbol{r}', \boldsymbol{r}_0, -j\,\omega)\, \nabla_0\, G(\boldsymbol{r}, \boldsymbol{r}_0, j\,\omega) + \\ &+ D_a \int d\boldsymbol{\sigma}_j\, p_s(\boldsymbol{r}_0)\, G(\boldsymbol{r}, \boldsymbol{r}_0, j\,\omega)\, \nabla_0\, G(\boldsymbol{r}', \boldsymbol{r}_0, -j\,\omega)\,. \qquad \text{(A8)} \end{aligned}$$

Further,

$$\text{term in } \Theta = 2\,D_a \int\int d\boldsymbol{\sigma}_j\, d\boldsymbol{\sigma}_j'\, G(\boldsymbol{r}, \boldsymbol{r}_0, j\,\omega)\, G(\boldsymbol{r}', \boldsymbol{r}_0', -j\,\omega)\, p_s(\boldsymbol{r}_0)\, \delta\,(\boldsymbol{r}_0 - \boldsymbol{r}_0')\,. \quad \text{(A9)}$$

Adding (A5), (A8), and (A9) we notice that the surface double integrals (which are singular) cancel. The final result is

$$\begin{aligned} S_{\Delta p}(\boldsymbol{r}, \boldsymbol{r}', \omega) &= p_s(\boldsymbol{r}')\, G(\boldsymbol{r}, \boldsymbol{r}', j\,\omega) + p_s(\boldsymbol{r})\, G(\boldsymbol{r}', \boldsymbol{r}, -j\,\omega) - \\ &- D_a \int d\boldsymbol{\sigma}_j\, G(\boldsymbol{r}, \boldsymbol{r}_0, j\,\omega)\, G(\boldsymbol{r}', \boldsymbol{r}_0, -j\,\omega)\, \nabla_0\, p_s(\boldsymbol{r}_0) + \\ &+ D_a \int d\boldsymbol{\sigma}_j\, p_s(\boldsymbol{r}_0)\, \nabla_0\, [G(\boldsymbol{r}, \boldsymbol{r}_0, j\,\omega)\, G(\boldsymbol{r}', \boldsymbol{r}_0, -j\,\omega)] + \\ &+ \left(\frac{1-y}{1+y}\right)^2 \int dv''\, \frac{p_s(\boldsymbol{r}'')}{\tau_H}\, G(\boldsymbol{r}, \boldsymbol{r}'', j\,\omega)\, G(\boldsymbol{r}', \boldsymbol{r}'', -j\,\omega)\,. \qquad \text{(A10)} \end{aligned}$$

The third integral is zero since ∇G is zero on the boundary and (2.18) results.

References

[1] A. VAN DER ZIEL, Fluctuation Phenomena in Semiconductors, Butterworths, London 1959.

[2] E. R. CHENETTE and A. VAN DER ZIEL, IRE Trans. Electron Devices 9, 123 (1962).

[3] K. M. VAN VLIET, General Transport Theory of Noise in p–n Junction-Like Devices, Part I: Three-Dimensional Green's Function Formulation, Solid State Electronics, (in the press).

[4] M. L. TARNG and K. M. VAN VLIET, ibid., Part II: Carrier Correlations and Fluctuations for High Injection, Solid State Electronics (in the press).

[5] B. SCHNEIDER and M. I. O. STRUTT, Proc. IRE 48, 1731 (1960).

[6] A. VAN DER ZIEL, Proc. IRE 43, 1639 (1955); 45, 1011 (1957).

[7] K. M. VAN VLIET and J. R. FASSETT, in: Fluctuation Phenomena in Solids, Ed. R. E. BURGESS, Academic Press, 1965 (p. 267).

[8] K. M. VAN VLIET, J. Math. Phys. 12, 1981 (1971).

[9] W. VAN ROOSBROECK, Phys. Rev. 91, 282 (1953).

[10] N. H. FLETCHER, J. Electronics 2, 609 (1957).
[11] J. R. HAUSER, Solid State Electronics 14, 133 (1971).
[12] P. M. MORSE and H. FESHBACK, Methods of Theoretical Physics, Vol. I, Chap. 7, McGraw-Hill Publ. Co., 1953.
[13] M. D. GREENBERG, Applications of Green's Functions in Science and Engineering, Prentice-Hall, Inc., Englewood Cliffs (N.J.) 1971.
[14] H. S. MIN, Ph.D. Thesis, University of Minnesota, 1971.
[15] K. M. VAN VLIET, Solid State Electronics 9, 185 (1966).

(Received February 22, 1972)

phys. stat. sol. (a) **10**, 619 (1972)

Subject classification: 18.2 and 18.3

Institute of Physics, Czechoslovak Academy of Sciences, Prague

Radial Oscillations of Cylindrical Magnetic Domains — Bubbles

By

J. KACZÉR and I. TOMÁŠ

The equations of motion for radial oscillations of cylindrical domains (bubbles) are formulated for both a close packed hexagonal arrangement of bubbles and an isolated bubble. The non-linear differential equation is then linearized and solved in the small oscillation approximation. For the case of the isolated bubble the solution is presented together with graphs and an extensive discussion. Equations and graphs determining the resonance frequency and the wall damping as a function of bubble diameter and material parameters are presented.

Es werden die Bewegungsgleichungen für radiale Schwingungen von zylindrischen Elementarbereichen (Bubbles) sowohl für eine dichtgepackte hexagonale Anordnung als auch für eine isolierte Bubble formuliert. Die nichtlinearen Differentialgleichungen werden linearisiert und dann für den Fall einer isolierten Bubble in der Approximation kleiner Schwingungen gelöst. Die Lösung wird zusammen mit einer Reihe von Diagrammen und einer umfangreichen Diskussion angegeben. Gleichungen und Kurven, welche die Resonanzfrequenz und die Wanddämpfung als Funktion von Bubble-Durchmesser und Materialkonstanten bestimmen, sind beigefügt.

1. Introduction

Since the discovery of isolated cylindrical magnetic domains [1] and of close packed hexagonal arrays of these (honeycomb domain structure) [2] in thin platelets of uniaxial materials and especially after the establishment of the real possibility of their utilization in memory devices ([3, 4]) a great number of papers dealing with these domains have been published. Thiele [5] has dealt in detail with the theory of the isolated bubble, giving the energy balance and presenting a simple graphic method for the solution of the static equilibrium and stability. Extensive calculations of the statics of close packed arrays of bubbles are given in the paper of Druyvesteyn and Dorleijn [6] where other references to this subject will also be found.

The dynamics of bubbles was studied by Callen and Josephs [7] who neglected the kinetic energy of the moving domain wall and calculated an expression for the wall mobility. However, if we investigate the motion of domain walls undergoing large accelerations, e.g. under the action of high-frequency magnetic fields with the high frequency component parallel to the domain axis, the wall inertia cannot be neglected anymore and the whole problem becomes that of a damped oscillator with finite mass. A first attempt of such a calculation for given bubble dimensions was carried out by Henry [8], however, his basic equation of motion is incorrect (see below) and we cannot expect an agreement with his results.

For the increasing wall velocities used in memory devices the question of wall dynamics becomes more important and thus arises the necessity of an extensive analysis of the high-frequency behaviour of bubbles. Such an analysis leads to a better understanding of the properties of bubbles, it can serve as a guide for improving the domain mobility, and finally using the resulting equations permits to evaluate some of the material parameters by means of resonance or relaxation measurements. The aim of this paper is just such an analysis.

Independently of paper [8] we investigated the oscillational properties of bubble domains as a function of the material parameters and the geometry of the system. An intuitive view leads to the conclusion, that for given material constants the resonance frequency of the bubble will increase (assuming a slowly varying potential minimum) with decreasing diameter of the bubble since the mass of the domain wall is proportional to the diameter. However, at the same time the disappearance of the minimum in the potential energy at the collapse diameter means that for this equilibrium diameter the restoring force is zero leading to zero resonance frequency. We should, therefore, expect a maximum in the resonance frequency versus equilibrium diameter dependence. This view was actually confirmed in our calculations.

The calculations, which are at present carried out for the close packed hexagonal array of bubbles will be published in the nearest future. The present paper contains an extensive analysis of the isolated bubble which is the limiting case for large spacing of the hexagonal array. The comparatively simple expression for the potential energy of an isolated bubble given by Callen and Josephs [7] enables a lucid and clear-cut solution without the aid of a computer.

2. The Characteristic of the Problem

Consider an infinite thin plate of thickness h having zero coercivity and a uniaxial anisotropy with the easy axis in the z-direction which is normal to the plate. The plate is located in an external field H the direction of which corresponds to the positive z-axis. Let the conditions for the existence of stable cylindrical domains hold and let them be arranged in a close packed hexagonal array as shown in Fig. 1 [6]. Let the spontaneous magnetization M_s inside the bubbles point into the negative z-direction; outside in the opposite direction.

Let us further assume that the domain walls form cylinders of diameter $2r$ and height h and the wall thickness in comparison with the domain dimensions is negligible. Let D denote the smallest centre to centre distance. For $r = \text{const.}$ and D increasing to infinity we arrive at the case of the isolated bubble [5].

The basic equation of motion for zero mode radial oscillations of the domain walls can be written in the form

$$m_T \ddot{r} + \beta_T \dot{r} + \frac{\partial \epsilon_T}{\partial r} = 0\,, \tag{1}$$

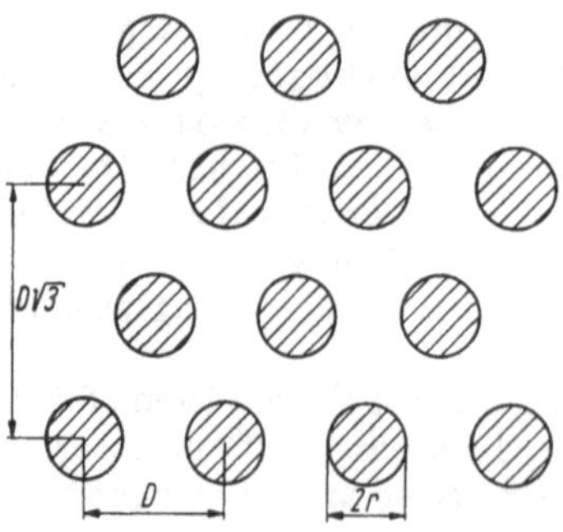

Fig. 1. Close packed hexagonal array of bubble domains

where m_T and β_T are the total mass and the total damping coefficient, respectively, and ϵ_T is the total free (potential) energy connected with the existence of bubble domains.

For $\ddot{r} = \dot{r} = 0$ equation (1) becomes the condition for static equilibrium

$$\frac{\partial \epsilon_T}{\partial r} = 0 , \tag{1a}$$

which permits to evaluate the equilibrium radius r_0 of the bubbles.

Let us now consider in detail the individual terms of equation (1).

2.1 Inertial mass of the walls

For a translational motion of the domain wall an additional term appears in the wall energy which is proportional to v^2. This is due to a demagnetizing field being set up inside the wall (because of the change of the angular momentum) which at least for small velocities is proportional to v. In analogy with the expression for the kinetic energy $\frac{1}{2} m v^2$ we can define an "inertial mass" m^* of the domain wall. We assume that the wall thickness δ is negligibly small compared to the radius of the bubble. We can then approximate the motion of a cylindrical wall by that of a 180° plane wall[1]) and according to [9, 10] we obtain an expression for the inertial mass per unit area of the wall:

$$m^* = \frac{1}{8 \pi \gamma^2} \frac{\sigma_W}{A} , \tag{2}$$

where $\gamma = \frac{1}{2} g e/(m_e c)$ is the gyromagnetic ratio and $\sigma_W = 4 (A K_u)^{1/2}$ is the rest energy of a 180° plane Bloch wall in a uniaxial material having an exchange constant A and a uniaxial anisotropy K_u. The total mass m_T is then obtained by multiplying m^* by the wall area of a bubble $2 \pi r h$ and the number N of bubbles present.

2.2 The wall damping

The question of the individual wall damping mechanisms is at present not completely solved. However, it is certain that there exists damping caused by the magnetic after-effect, relaxational damping analogous to the Landau-Lifshitz damping in ferromagnetic resonance and damping due to eddy currents. The present paper does not deal with the above problems, we only try a phenomenological approach. According to [11] we can express the damping constant per unit wall area β by means of the wall mobility μ as follows:

$$\beta = \frac{2 M_s}{\mu} . \tag{3}$$

Similarly to the wall mass m^* it can be shown that the damping constant β is a function of r. (The value of β increases for decreasing radius.) However, we

---

[1]) In the first approximation it can be shown that the energy of a curved domain wall (with div $\boldsymbol{M} = 0$) contains an additional term as compared to the plane wall. This is due to a contribution from the exchange energy. A rough calculation shows this correction to be

$$\sigma_W^{cyl} = \sigma_W^{plane} \left[1 + \frac{1}{4 \pi} \frac{\delta^2}{r^2} \right] .$$

shall neglect this dependence and put $\beta \neq \beta(r)$. We obtain again the total damping β_T by multiplying β by the wall area and the number of domains involved.

2.3 The restoring force

The total restoring force f_T acting on the walls displaced from the equilibrium position is given by the expression $f_T = -\partial\epsilon_T/\partial r$. The total potential energy connected with the existence of bubbles ϵ_T is given by the sum of three components:

$$\epsilon_T = \epsilon_W + \epsilon_D + \epsilon_H , \tag{4}$$

where ϵ_W is the total wall (rest) energy, ϵ_D is the total demagnetizing energy (the total energy of the stray fields produced by the bubbles), and ϵ_H is the total field energy of the bubbles in an external field H. If H is a dc field then equation (1) describes free oscillations of the bubbles, if, however, H contains an additional ac term then equation (1) describes forced oscillations.

3. Results and Their Discussions

3.1 Close packed bubble array

Using the expressions defined above for the total inertial mass m_T, the total wall damping β_T, and the total restoring force f_T we can rewrite (1) in the following normal form:

$$\ddot{r} + \frac{\beta}{m^*}\,\dot{r} + \frac{1}{2\pi N h m^*}\,\frac{1}{r}\,\frac{\partial\epsilon_T}{\partial r} = 0 . \tag{5}$$

From the expressions (9), (12), and (12a) for the derivative of the total energy given below it is obvious that the last term on the left-hand side of (5) is not a linear function of r. Equation (5) is therefore a non-linear differential equation of second order and it is not likely that an analytic solution can be found or even exists. For the purpose of analysis it is sufficient to limit the investigations to small oscillations of the walls about their equilibrium positions which involves a linearization of (5). We accomplish this by developing the last term on the left-hand side of (5) into a power series about the equilibrium radius r_0 retaining only the linear term in r. We then arrive at the equation of motion of small radial oscillations of a cylindrical bubble domain:

$$\ddot{r} + \frac{\beta}{m^*}\,\dot{r} + \frac{1}{2\pi N h m^*}\,\frac{\partial}{\partial r}\left(\frac{1}{r}\,\frac{\partial\epsilon_T}{\partial r}\right)_{r=r_0} (r - r_0) = 0 .\ ^{2)} \tag{6}$$

Experiments shows [2] that if the bubble density in a platelet placed in a homogeneous magnetic field is large enough for the bubbles magnetostatically to interact they will arrange themselves in a more or less (depending on the

---

[2]) It was shown that the total mass m_T and the total damping β_T are linear functions of r. Dividing equation (1) by m_T r appears in the denominator of the last term on the left-hand side of equation (5). The general form of the total spring constant of the linearized equation for small oscillations is then $k_T = \frac{\partial}{\partial r}\left(\frac{1}{r}\,\frac{\partial\epsilon_T}{\partial r}\right)_{r=r_0}$ as can be seen from equation (6). This is in contrast to expression (14) of the paper by Henry [8] which is incorrect.

coercivity of the platelet) ideal hexagonal lattice. We can deduce, that under ideal conditions this will be the final arrangement since for a given bubble density this arrangement gives the largest centre to centre distance between bubbles as required by the magnetostatic repulsive forces between the individual bubbles. We have, therefore, chosen just this arrangement (Fig. 1) for further investigations.

Let us, therefore, substitute in equation (6) for $N = 2/\sqrt{3}\, D^2 h$ the number of bubbles per unit volume and for ϵ_T the density of the total potential energy of a close packed bubble array (i.e. the energy per unit volume) $E_T = E_W + E_D + E_H$. The expressions for the individual terms of the energy density E_T and for their derivatives for a hexagonal lattice are given by Druyvesteyn and Dorleijn [6]. Before writing down these expressions we introduce the following dimensionless quantities:

$$x = \frac{r}{h}, \qquad s = \frac{D}{h}, \qquad \varkappa = \frac{H}{4\pi M_s}, \qquad \lambda = \frac{\sigma_W}{4\pi M_s^2 h}. \tag{7}$$

Equation (6) in the new variables can then be written

$$\ddot{x} + \frac{\beta}{m^*}\dot{x} + \frac{s^2\sqrt{3}}{4\pi h m^*}\frac{\partial}{\partial x}\left(\frac{1}{x}\frac{\partial E_T}{\partial x}\right)_{x=x_0}(x - x_0) = 0 . \tag{8}$$

The derivative of the total potential energy density is according to [6]

$$\frac{\partial E_T}{\partial x} = \frac{\partial E_W}{\partial x} + \frac{\partial E_H}{\partial x} + \frac{\partial E_D}{\partial x} = \frac{32\pi^2 M_s^2}{s^2\sqrt{3}}\left(\frac{\lambda}{2} + \varkappa x + \frac{s^2\sqrt{3}}{4}\frac{\partial E_D/\partial x}{8\pi^2 M_s^2}\right), \tag{9}$$

where

$$\frac{\partial E_D/\partial x}{8\pi^2 M_s^2} = \frac{4x}{s^2\sqrt{3}}\left(\frac{4\pi x^2}{s^2\sqrt{3}} - 1\right) - \frac{8x^2}{3\pi s^2}\sum_{m=-\infty}^{+\infty}\sum_{n=-\infty}^{+\infty}\frac{\left[1 - \exp\left\{-\frac{2\pi}{s}\left(m^2 + \frac{n^2}{3}\right)^{1/2}\right\}\right]}{m^2 + \frac{n^2}{3}} \times$$
$$\times J_0\left(\frac{2\pi}{s}x\left[m^2 + \frac{n^2}{3}\right]^{1/2}\right) J_1\left(\frac{2\pi}{s}x\left[m^2 + \frac{n^2}{3}\right]^{1/2}\right). \tag{10}$$

The summation in (10) has to be carried out for $m + n$ even except for $m = n = 0$. J_0 and J_1 are Bessel functions of first order.

Equation (8) is a linear differential equation of second order for the dimensionless bubble radius x and in principle the solution is quite simple. However, because of the cumbersome term $\partial E_T/\partial x$ we lose the possibility of solving equation (8) without a computer and the analytical expressions used in the discussion are complex. That is one of the reasons why we decided first to analyse the limiting case when the centre to centre distance D tends to infinity (it can be shown that the condition $D \gtrsim 15\,r$ is sufficient), i.e. to solve and discuss the case of an isolated bubble.

3.2 Isolated bubble

For an isolated bubble we introduce into (6) $N = 1$ and ϵ_T will be the total free energy of a single bubble. The expressions for the derivative of the individual components of the energy is given by Thiele [5]. After introducing the dimensionless quantities (7) and the dimensionless energy

$$\mathscr{E}_T = \frac{\epsilon_T}{16\,\pi^2\,M_s^2\,h^3} \tag{11}$$

we can write

$$\frac{\partial \mathscr{E}_T}{\partial x} = \frac{\partial \mathscr{E}_W}{\partial x} + \frac{\partial \mathscr{E}_H}{\partial x} + \frac{\partial \mathscr{E}_D}{\partial x} =$$

$$= \frac{\lambda}{2} + \varkappa\,x - \frac{4\,x^2}{\pi}\left[\left(\frac{4\,x^2}{1 + 4\,x^2}\right)^{-1/2} \mathrm{E}\left(\frac{4\,x^2}{1 + 4\,x^2}\right) - 1\right], \tag{12}$$

where E is the complete elliptic integral of second kind.

However, not even expression (12) is straightforward enough. Callen and Josephs [7] showed for $x \in \langle 0, 10 \rangle$, however, that the last term in equation (12) can with surprising accuracy be replaced by

$$\frac{\partial \mathscr{E}_D}{\partial x} \approx -\frac{x}{1 + \frac{3}{2}\,x}, \tag{13}$$

which makes the solution much more simple.

Using the simplified expression (13) we can write for (12)

$$\frac{\partial \mathscr{E}_T}{\partial x} = \frac{\lambda}{2} + \varkappa\,x - \frac{x}{1 + \frac{3}{2}\,x}, \tag{12a}$$

and introducing this and the dimensionless quantities (7) into equation (6) we get the equation for the radial oscillations of an isolated cylindrical domain:

$$\ddot{x} + 2\,b\,\dot{x} + \omega_0^2\,(x - x_0) = 0\,. \tag{14}$$

Here we used the notations

$$b = \frac{\beta}{2\,m^*} = \frac{2\,\pi\,\gamma^2\,M_s}{\mu}\sqrt{\frac{A}{K_u}} \tag{15}$$

and

$$\omega_0 = \omega_{0\,\max}\,\Omega(x_0, \lambda)\,, \tag{16}$$

where

$$\omega_{0\,\max} = \gamma\,M_s^2\sqrt{\frac{\pi^3}{2\,K_u}} \tag{17}$$

and

$$\Omega(x_0, \lambda) = 4\sqrt{\lambda\left(\frac{3}{(1 + \frac{3}{2}\,x_0)^2} - \frac{\lambda}{x_0^2}\right)}\,. \tag{18}$$

The solution of the linear differential equation (14) depends on the relation between the values of the coefficients b and ω_0. We have to distinguish the following cases.

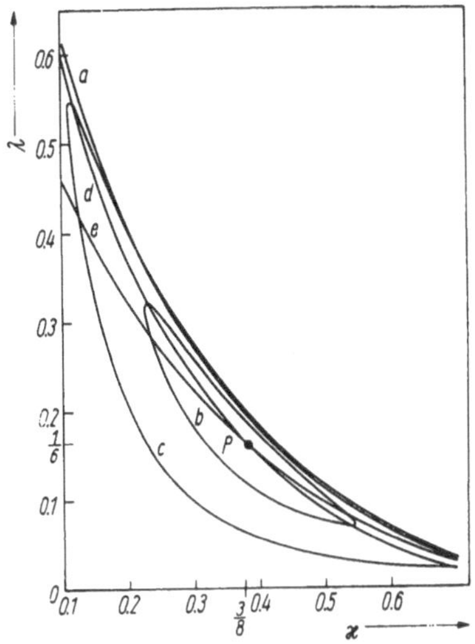

Fig. 2. Regions of different solutions of equation (14) in the $\varkappa$–λ plane. Curves a to e as in Fig. 4

1. absolutely unstable ($\omega_0^2 < 0$) corresponding to the conditions of radial instability of the bubble (collapse), (19)
2. aperiodic motion ($0 < \omega_0^2 < b^2$), and
3. periodic motion ($0 < b^2 < \omega_0^2$).

Since the coefficients b and ω_0 are functions of the material parameters and in addition ω_0 is a function of the bubble radius r_0 and as a consequence also of the external field H (see equation (20) below) we have to discuss the individual cases of the solution of (14) depending on the values of the various parameters.

The most interesting case seems to be the discussion of the solution of (14) as a function of the relation between $\varkappa$ and λ (Fig. 2) and between x_0 and λ (Fig. 4). The remaining material parameters which affect the values of ω_0 and b and which thus influence the conditions for the individual solutions are included in Fig. 2 and 4 in the form of parameters.

Even though for given material parameters and thickness h the only directly variable parameter is the external field $\varkappa$ we chose λ for the discussion of the results in terms of x_0 and λ rather than $\varkappa$. There are several reasons for this.

The equation of motion is solved in the variable x and similarly the calculations involving the statics of bubbles [5], [6], [7] are presented in an analogous way. The interpretation using the radius of the bubble is technically more straightforward. Last but not least we also had to take into account the fact

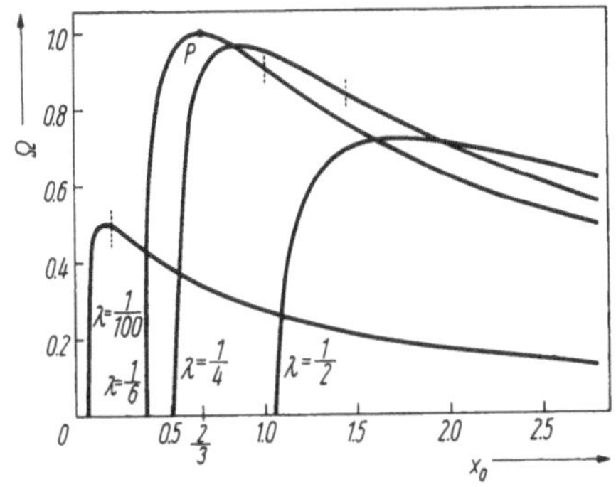

Fig. 3. Family of curves showing $\Omega(x_0)$ with λ as a parameter. The vertical dashed lines show the boundaries of elliptic instability

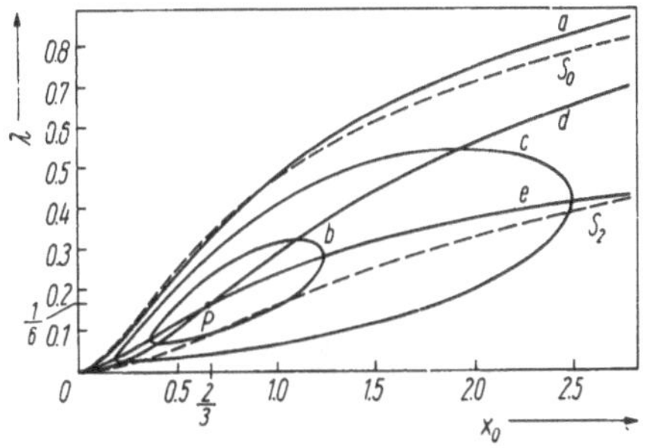

Fig. 4. Regions of different solutions of equation (14) in the x_0–λ plane. (a) collapse boundary; (b) boundary between aperiodic and periodic solution for $C = 1.1$ (The quantity C is explained in equation (26)); (c) same as (b) for $C = 1.5$; (d) maximum of function $\Omega(x_0, \lambda)$ with respect to x_0; (e) maximum of function $\Omega(x_0, \lambda)$ with respect to λ. S_0 collapse boundary according to Thiele [5], S_2 boundary of elliptic instability according to [5]

that the relation combining λ, $\varkappa$, and x_0 is given by the application of the condition (1a) to the case of the isolated bubble (12a), i.e.

$$\frac{\lambda}{2} + \varkappa\, x_0 - \frac{x_0}{1 + \frac{3}{2}\, x_0} = 0\,, \tag{20}$$

and is somewhat cumbersome. Therefore, we used it only for the transformation of the family of curves from the x_0–λ plane to the $\varkappa$–λ plane. The curves denoted by the same letters in Fig. 2 and 4 mutually correspond and we do not present the analytic expressions used for the curves in Fig. 2.

Let us now investigate the value of the resonance frequency ω_0. Equation (16) shows that the frequency of the oscillations of an isolated bubble is not only a function of the material parameters, but also of the equilibrium radius x_0 about which the bubble oscillates. The dependence of this frequency ω_0 on the values of x_0 and λ is given by the function $\Omega(x_0, \lambda)$ in the x_0–λ plane. Plots of this function are presented in Fig. 3 and 4.

Fig. 3 shows plots of $\Omega(x_0, \lambda)$ as a function of x_0 with λ as a parameter. The curves which are vertical sections represented by horizontal lines ($\lambda = \text{const}$) in the three-dimensional surface $\Omega(x_0, \lambda)$ of Fig. 4 quite clearly show the shape of the function $\Omega(x_0, \lambda)$ and thus also of $\omega_0(x_0, \lambda)$. If in Fig. 4 we move along a line $\lambda = \text{const}$ from larger values of x_0 towards smaller ones $\Omega(x_0, \lambda)$ reaches a maximum for a certain x_0 after which there is a sudden decrease to zero at the value of x_0 corresponding to the collapse of the bubble.

From the extrema of $\Omega(x_0, \lambda)$ we get

$$\frac{\partial \Omega}{\partial x_0} = 0 \Rightarrow \lambda = \frac{9}{2}\, \frac{x_0^3}{(1 + \frac{3}{2}\, x_0)^3} \tag{21}$$

and

$$\frac{\partial \Omega}{\partial \lambda} = 0 \Rightarrow \lambda = \frac{3\, x_0^2}{2\,(1 + \frac{3}{2}\, x_0)^2}\,. \tag{22}$$

Expression (21) is represented in Fig. 4 by curve d, expression (22) by curve e. The intersection of both curves at point $\text{P} \equiv (x_0 = \frac{2}{3}, \lambda = \frac{1}{6})$ is the location of the absolute maximum of the surface $\Omega(x_0, \lambda)$ with $\Omega(\text{P}) = 1$. According to (16) the frequency of the radial oscillations will also reach a maximum at point P corresponding to $\omega_0 = \omega_{0\,\max}$. For any other equilibrium radius x_0 or value of λ below curve *a* (for which $\Omega = 0$) we have a resonance frequency smaller than $\omega_{0\,\max}$.

The next thing of interest are the lines of constant frequency in the x_0–λ plane, i.e. values of x_0 and λ for which

$$\Omega(x_0, \lambda) = \Omega = \text{const}\,. \tag{23}$$

Substituting (18) into (23) we arrive at an analytical form of the isofrequency curves in the x_0–λ plane:

$$\lambda_{1,2} = \frac{3\, x_0^2}{2\,(1 + \frac{3}{2}\, x_0)^2} \pm \frac{1}{2} \sqrt{\frac{9\, x_0^4}{(1 + \frac{3}{2}\, x_0)^4} - \frac{x_0^2\, \Omega^2}{4}}\,. \tag{24}$$

The curves b and c in Fig. 4 are such curves for $\Omega^{-1} = 1.1$ and $\Omega^{-1} = 1.5$, respectively.

The values of the maximum frequency of the oscillation of an isolated bubble for a number of representative materials are shown in Fig. 5. This figure follows

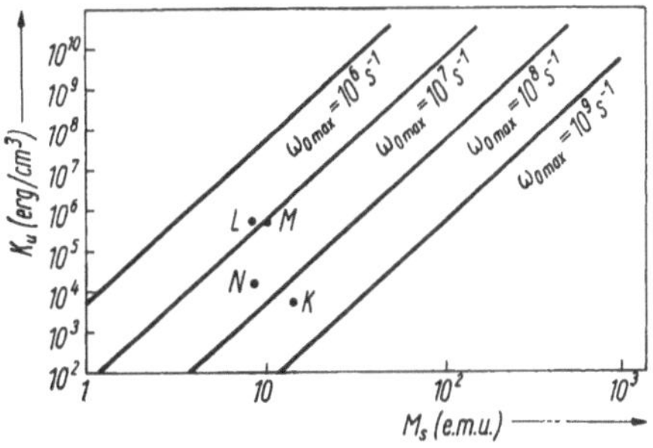

Fig. 5. Values of maximum resonance frequency $\omega_{0\,max}$ for various materials. K: $Y_3Ga_xFe_{5-x}O_{12}$ ($x = 1.2$) [12]; L: $YFeO_3$ [13], [14]; M: $DyFeO_3$ [13], [7]; N: $Sm_{0.55}Tb_{0.45}FeO_3$ [13], [15]

from equation (17) and shows lines of constant frequency $\omega_{0\,max}$ as a function of the material constants K_u and M_s and thus permits to determine $\omega_{0\,max}$ for an arbitrary material. The points in Fig. 5 mark the position of a number of actual materials (see the caption of Fig. 5).

Let us now return to the discussion of the conditions (19) for different solutions of (14).

The unstable solution ($\omega_0^2 < 0$) of equation (14) is represented by the region above the curve where Ω vanishes (denoted by a in Fig. 4) and expressed analytically by

$$\lambda = \frac{3\,x_0^2}{(1 + \frac{3}{2}\,x_0)^2}\,. \tag{25}$$

In this region the bubbles are radially unstable and for no values x_0, λ a stable isolated bubble exists having corresponding parameters.

If we reduce the diameter of a stable bubble along the line $\lambda = \text{const}$ (by increasing the external field) we arrive at a minimum radius x_0 which lies on curve a. On a further increase of the field the bubble collapses.

In the region below curve a $\omega_0^2 > 0$ and the solution of (14) can be either periodic or aperiodic. The damping constant b is determined only by material parameters and for a given material is uniquely determined. Let us introduce a new material constant

$$C = \frac{\omega_{0\,max}}{b} = \frac{M_s\,\mu}{2\,\gamma}\sqrt{\frac{\pi}{2\,A}}\,; \tag{26}$$

the boundary between the periodic and aperiodic solution of (14) given by $\omega_0 = b$ can then be expressed after substituting (16) into (26) by

$$\Omega(x_0, \lambda) = C^{-1} \tag{27}$$

which for $\Omega^{-1} = C$ is equivalent to (23).

For $\Omega^{-1} = C$ (24) defines the boundary between regions in the x_0–λ plane corresponding to periodic (inside region) and aperiodic (outside region) solutions of (14).

The range of values x_0 for which a value λ exists so that (14) has a periodic solution can be found by a simple graphical method. Let us draw in Fig. 4 a line through the origin having a slope $1/2\,C$. The points of intersection of this line with curve a determine the range sought.

It is found that for $C < 1$ no intersection will be found. In this case the whole region below curve a corresponds to the aperiodic solution. For $C = 1$

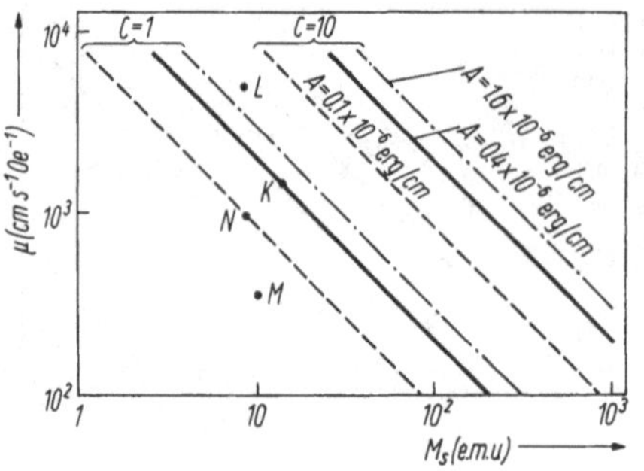

Fig. 6. Values of the constant C for various materials according to equation (26). Points K, L, M, N as in Fig. 5

the line just touches curve a in one point corresponding to point P and for $C > 1$ there exists a region of periodic oscillations about the point P. This region is the larger the constant C is. Curves b and c show the boundary of these regions for $C = 1.1$ and $C = 1.5$, respectively. A similar region for $C = 10$ fills nearly the whole area below curve a of the Fig. 4.

The value of the constant C can be determined from Fig. 6 as a function of the mobility μ, the saturation magnetization M_s and the exchange constant A which is a parameter ($A = 0.1 \times 10^{-6}$, 0.4×10^{-6}, 1.6×10^{-6} erg cm^{-1}). The points shown in the plot mark the position of a few representative materials (see the caption to Fig. 5).

4. The Range of Validity of the Model Used

Discussing the stability of a bubble domain Thiele [5] considers small radial deformations, the circular mode ($n = 0$), the elliptical mode ($n = 2$), and higher modes with a higher number of spatial periodicities. The subject of interest in the present paper is only the zero order mode, i.e. only circular radial oscillations. If we consider the stability with respect to deformations of higher order then we have to limit the region of stable solution by Thiele's curve S_2 (see Fig. 4) which determines the boundary of elliptic instability.

This limitation will necessarily also be seen in Fig. 3. For the individual parameters λ it is shown as a vertical dashed line. For x_0 right of this line we have the region of elliptic instability and our considerations which are only valid for circular radial oscillations are senseless.

Our calculations are also burdened by the approximation introduced by Callen and Josephs for the derivative of the demagnetizing energy (13) instead of Thiele's exact expression. However, for reasons of simplicity and lucidity we preferred the above solution to the expression of Thiele which is more exact but cumbersome especially since the numerical results are for the case of small bubbles practically the same. (Compare curve a in Fig. 4 with Thiele's curve S_0 which is also shown.)

Further we have to point out that Becker's model for the inertial mass of the wall is valid only for small wall velocities. The criterion here is, as Becker himself points out, that the normal component of magnetization produced by the motion of the wall should be small compared to the saturation magnetization M_s [9] (see also [16]).

As far as the equation of motion of the hexagonal close packed array is concerned we calculated the minimum of the energy determining the stable state of the system only from the derivative of the demagnetizing energy with respect to the radius. The distance between the bubbles is only a parameter in all considerations. In many experimental cases and especially in bubble devices where the permalloy structure determines the stable positions of the bubbles this procedure is fully justified.

5. Conclusion

At present an intensive research in the field of technology of bubble devices is carried out with the view to attain the highest possible speeds for information transfer and storage in bubble memories. The increasing bubble velocities require the investigation of their dynamics with respect to their inertial properties which would enable us to clarify new effects and, on the other hand, on the basis of the knowledge of wall dynamics it is possible to expect new means for increasing their mobility.

The present paper contains the complete formulation of the problem of radial circular oscillations of bubble domains both for the case of a hexagonal close packed array of bubbles as well as for an isolated bubble. The equations containing the inertial and the damping terms are non-linear and the solution is carried out on the basis of non-linear oscillation theory in the approximation of small oscillations by linearizing the fundamental equations of motion. The solution of this linearized equation is discussed in the light of the material parameters and the results are displayed in a number of illustrative plots.

Acknowledgement

We would like to thank J. Fousek and V. Kamberský for some valuable remarks and suggestions which greatly improved the clarity of presentation.

References

[1] C. Kooy and U. Enz, Philips Res. Rep. **15**, 7 (1960).
[2] J. Kaczér and R. Gemperle, Czech. J. Phys. **B11**, 510 (1961).
[3] A. H. Bobeck, Bell Syst. tech. J. **46**, 1901 (1967).
[4] P. C. Michaelis, J. appl. Phys. **39**, 1224 (1968).
[5] A. A. Thiele, Bell Syst. tech. J. **48**, 3287 (1969).
[6] W. F. Druyvesteyn and J. W. F. Dorleijn, Philips Res. Rep. **26**, 11 (1971).
[7] H. Callen and R. M. Josephs, J. appl. Phys. **42**, 1977 (1971).
[8] G. R. Henry, J. appl. Phys. **42**, 3150 (1971).
[9] R. Becker, J. Phys. Radium **12**, 332 (1951).
[10] W. Döring, Z. Naturf. **3A**, 373 (1948).
[11] C. Kittel and J. K. Galt, Solid State Phys. **3**, 534 (1956).
[12] D. M. Heinz, P. J. Besser, J. M. Owens, J. E. Mee, and G. R. Pulliam, J. appl. Phys. **42**, 1243 (1971).
[13] U. F. Gianola, D. H. Smith, A. A. Thiele, and L. G. van Uitert, IEEE Trans. Magn. MAG **5**, 558 (1969).
[14] A. H. Bobeck, E. G. Spencer, L. G. van Uitert, S. C. Abrahams, R. L. Barns, W. H. Grodkiewicz, R. C. Sherwood, P. H. Schmidt, D. H. Smith, and E. H. Walters, Appl. Phys. Letters **17**, 131 (1970).
[15] F. C. Rossol and A. A. Thiele, J. appl. Phys. **41**, 1163 (1970).
[16] E. Schlömann, Appl. Phys. Letters **19**, 274 (1971).

(Received January 10, 1972)

phys. stat. sol. (a) **10**, 631 (1972)

Subject classification: 10.2 and 11; 1.1; 1.4; 22.8.1

Institute for Low Temperature and Structure Research, Polish Academy of Sciences, Wrocław (a), and Institute of Physics, Polish Academy of Sciences, Warsaw (b)

X-Ray Examination of the Real Structure of γ-Irradiated $NaClO_3$ Single Crystals

By

J. A. Stępień (a), J. Auleytner (b), and K. Łukaszewicz (a)

Using two-circle diffractometer and oscillating film spectrometer the real structure of $NaClO_3$ single crystals irradiated by γ-rays at various doses up to 2.8×10^8 rad is examined by following the changes of the lattice constant and the broadening of diffraction profiles. The analysis of the results shows that in the initial stage of radiolysis the changes observed are due to the strains introduced in the lattice by the decomposition products. At higher doses mosaic blocks are formed which gradually undergo disorientation. At the dose $\approx 7 \times 10^7$ rad the strains are partially relaxed and at $\approx 2 \times 10^8$ rad NaCl crystallites oriented parallel to the $\langle 100 \rangle$ directions of the host lattice appear. A tentative explanation of the above phenomena is given.

Mit einem Zweikreis-Diffraktometer und einem oszillatorischen Filmspektrometer wurde die Realstruktur von $NaClO_3$-Einkristallen, die mit γ-Strahlen bei verschiedenen Dosen bis zu $2{,}8 \times 10^8$ rad bestrahlt wurden, durch die Beobachtung der Gitterkonstante und Verbreiterung der Beugungsprofile untersucht. Die Analyse der Ergebnisse zeigt, daß im Anfangszustand der Radiolyse die beobachteten Änderungen durch die Spannungen hervorgerufen werden, die in das Gitter durch die Zerfallsprodukte eingebracht wurden. Bei höheren Dosen werden Mosaikblöcke gebildet, die graduell einer Fehlorientierung unterliegen. Bei Dosen $\approx 7 \times 10^7$ rad relaxiieren die Spannungen teilweise und bei $\approx 2 \times 10^8$ rad treten NaCl-Kristallite auf, die parallel zur $\langle 100 \rangle$-Richtung des Wirtsgitters orientiert sind. Es wird eine mögliche Erklärung der Phänomene angegeben.

1. Introduction

Ionizing radiation introduces into the ionic crystals various defects [1, 2] responsible, among others, for the essential changes of their real structure and lattice parameters. These latter effects were examined by the X-ray methods in alkali halide crystals by e.g. [3, 4].

In spite of the rather ample research work on the radiolysis of more composed inorganic salts no detailed X-ray investigations of the radiation damage in these materials were undertaken, so far. As it is known, in irradiated salts, e.g., of the $Me^{I}ClO_3$ type various new chemical species are formed cp. [2, 5 to 7], among which the gaseous products are mostly responsible for the appearence of heavy strains, leading in the extreme to the total destruction of the crystal.

The present work aims at an examination of the changes in the real structure of $NaClO_3$ single crystals irradiated by γ-rays in a wide dose range.

2. Experimental

$NaClO_3$ crystals were grown from an aqueous solution by evaporation at room temperature. The obtained single crystals usually possessed the shape of regular slabs with well developed {100} faces. The crystals were irradiated with γ-rays from a Co^{60} source (MRCh-γ-100) at the dose rate $\approx 10^6$ rad h^{-1}, in dry air at room temperature.

Two different measuring methods were employed:

a) In the first one a two-circle single crystal diffractometer was applied. The size of the crystal examined was $\approx 0.5 \times 0.2 \times 0.2$ mm³. Using CuKα radiation the rocking curves of the 080, 027, 045, and 025 diffraction profiles were measured. In this method the intensity of X-rays reflected from the rotating crystal depends on its position given by the angle φ and on the position of the detector given by the angle 2θ. Following measuring procedures were applied permitting to record two different functions:

$$I_1(2\theta) = \int_{\varphi_1}^{\varphi_2} I(2\theta, \varphi)\, d\varphi ,$$

and

$$I_2(\varphi) = \int_{\theta_1}^{\theta_2} I(2\theta, \varphi)\, d\theta ,$$

where $\varphi_2 - \varphi_1 = \Delta\varphi = 2°$ denotes the rotation range of the crystal during the measurement performed for a stationary detector with a narrow slit; $\theta_2 - \theta_1 = \Delta\theta$ denotes the slit width used in the measurement for stationary crystal adjusted at a definite angle φ. The measuring steps of the parameters φ and 2θ were 0.1 and 0.05°, respectively. The measurements were carried out at constant temperature 22.1 ± 0.1 °C for two $NaClO_3$ crystals.

Using CuKα radiation a series of Weissenberg diagrams of the $0kl$ layer were taken for the sample irradiated with successively increasing γ-radiation doses.

b) In the second method the angular distribution function of the mosaic blocks in irradiated crystals was examined by the oscillating film spectrometer [8] using an X-ray point source [9]. In these measurements 16 samples of linear dimensions of a few mm, irradiated by various γ-radiation doses were used. The profiles of photographed 020 diffraction lines ($CuK\alpha_1$) were measured in a Zeiss microphotometer. The measurements were made at room temperature without stabilization.

3. Results and Discussion

3.1 Single crystal diffractometer method

3.1.1 Changes of lattice constant

The lattice constant of irradiated $NaClO_3$ crystal was calculated by the centroid method [10] for 080, 027, and 045 reflections. In the initial stage of γ-irradiation (up to about 15 h) the centroids of well separated $CuK\alpha_1$, and $CuK\alpha_2$ lines were estimated. For longer irradiations the centroid of not separated doublet of these lines was determined.

The plot of the values of lattice constant a against the dose of γ-radiation is shown in Fig. 1. It is evident that a significant linear increase of the lattice constant takes place in the initial stage of radiolysis up to about 25 h. For longer irradiations the slope of the curve continuously decreases and beginning with

Fig. 1. Effect of γ-radiation on the lattice constant a

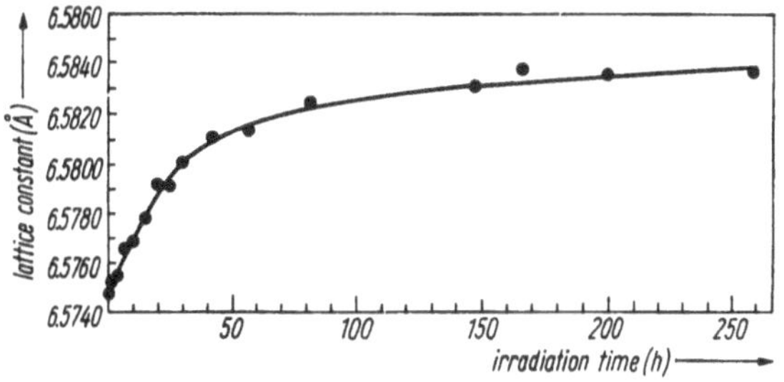

125 h the value of the lattice constant a is kept on the same level within the limits of experimental error.[1])

Separate experiments have shown that the weak X-radiation used in these measurements does not induce any observable changes of the lattice constant of the examined crystals.

3.1.2 X-ray reflection profiles

Based on experiments described in Section 2a), it has been ascertained that with increasing radiation dose the diffraction lines undergo broadening, their intensity decreases and the centroids shift towards smaller values of the angle θ. In Fig. 2 and 3 the plots are shown reflecting the radiation dose effect on the halfwidths of the functions $I_1(2\theta)$ and $I_2(\varphi)$ for 080, 027, 045, and 025 diffrac-

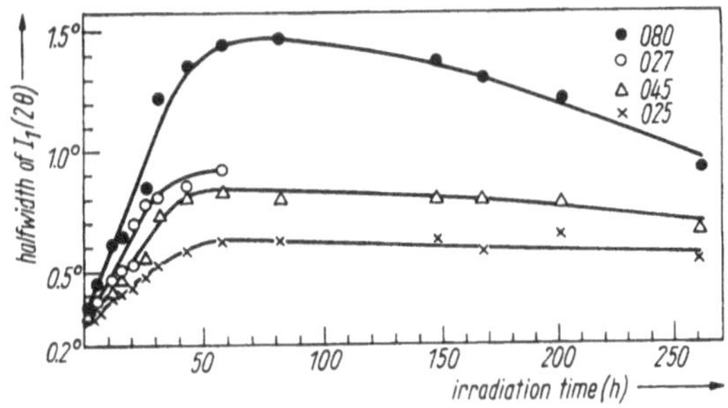

Fig. 2. Dependence of the halfwidth of the function $I_1(2\theta)$ for four chosen diffraction profiles on irradiation time

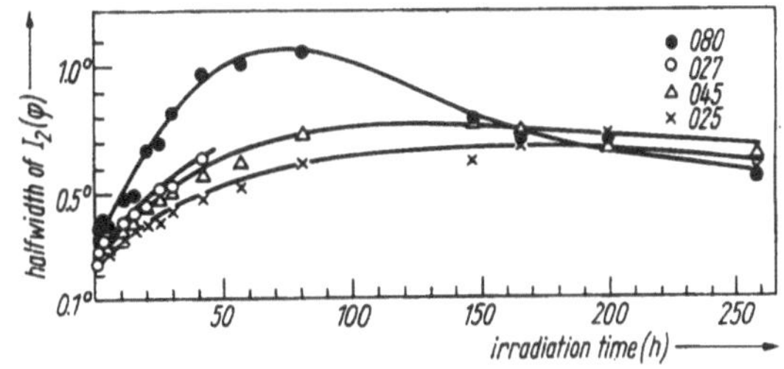

Fig. 3. Dependence of the halfwidth of the function $I_2(\varphi)$ for four chosen diffraction profiles on irradiation time

---

[1]) It should be emphasized that a striking similarity exists between the plot shown in Fig. 1 and those reflecting the changes of optical absorption in the 420 and 305 nm bands for the crystal irradiated within the same dose range. The optical density in the bands mentioned increases linearly up to about 20 h of γ-irradiation reaching the values of 25 and 37 cm^{-1}, respectively. With increasing irradiation time the slope of both curves becomes smaller and above 100 h increases linearly again with the slope about eigth times smaller than that observed in the initial stage of radiolysis [11].

tion lines. In the initial stage of radiolysis one observes a rapid increase of the halfwidth of the function $I_1(2\theta)$ reaching for the high-angle reflection a maximum at ≈ 70 h. For higher doses the value of the halfwidth decreases. The halfwidth of the function $I_2(\varphi)$ behaves in a similar way.

The function describing the shape of the reflection line presents the convolution of a few functions depending on the operating conditions, i.e., on the beam divergence, which in turn is a function of the focus size, slit width, and accuracy of adjusting the crystal in the required position, as well as on the properties of the crystal itself, i.e., on the shape of the sample, the strains, and the size and angular distribution of the mosaic blocks [12].

Under the experimental conditions described in Section 2a) the apparatus function was the same for all the measurements. It was therefore anticipated that the halfwidth of the function $I_1(2\theta)$ should depend only on the strains existing in the crystal and the size of the mosaic blocks. On the other hand, the halfwidth of the function $I_2(\varphi)$ should depend on the strains, and the size and angular distribution of the mosaic blocks.

The halfwidth of the diffraction line related with the presence of the strains in the crystal is directly proportional to tg θ, while the halfwidth related with the size of the mosaic blocks is inversely proportional to cos θ. The strains exert a larger influence on the diffraction profiles with higher indices, while the size of the mosaic blocks affects much stronger the lines with lower indices [13].

The measurements of the halfwidths of the mentioned four diffraction profiles of irradiated $NaClO_3$ permit to ascertain the following facts. In the initial stage of radiolysis (up to about 11 h) only strains are introduced into the crystal — the halfwidths of the profiles $I_1(2\theta)$ and $I_2(\varphi)$ are changing linearly with tg θ of the corresponding reflections (Fig. 4 and 5). At higher radiation doses —

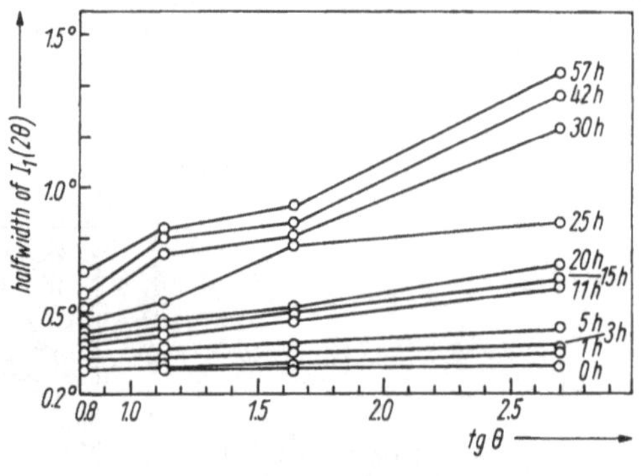

Fig. 4. Dependence of the halfwidth of $I_1(2\theta)$ for the same diffraction profiles as in Fig. 2 on tg θ for various radiation doses

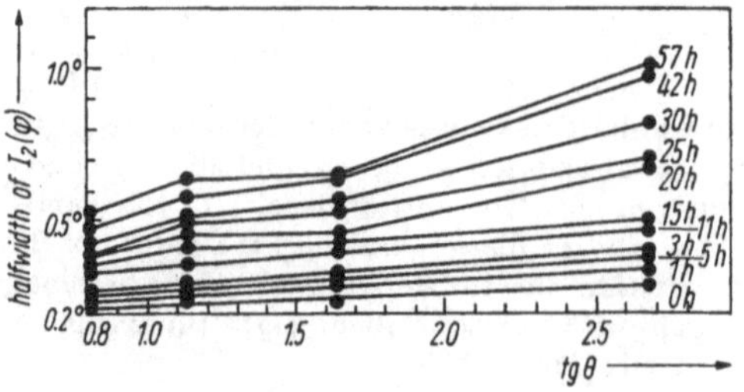

Fig. 5. Dependence of the halfwidth of $I_2(\varphi)$ for the same diffraction profiles as in Fig. 3 on tg θ for various radiation doses

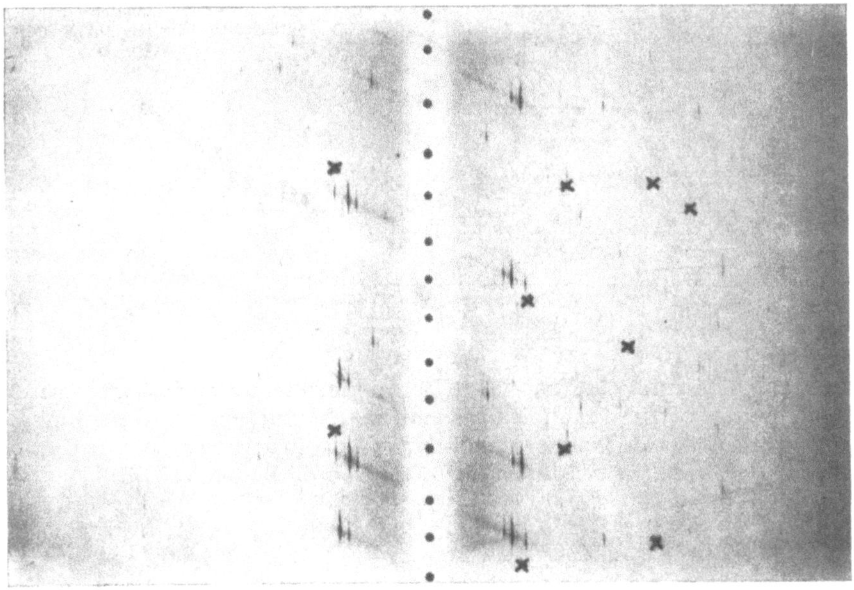

Fig. 6. Weissenberg diagram of the 0*kl* layer of a heavily irradiated $NaClO_3$ single crystal; 020, 022, 040, 042, 051 reflexions from NaCl crystallites are marked by crosses

beginning with 15 h — a distinct deviation from the above-mentioned linearity is observed (Fig. 5), pointing to the increasing disorientation of the mosaic blocks. At the dose corresponding to 20 h (Fig. 4) the influence of other factors affecting the broadening of diffraction profiles is evident. After reaching some critical radiation dose (about 6×10^8 rad) the single crystal undergoes complete destruction into small crystallites.

3.1.3 Final product of radiolysis

In the Weissenberg diagrams of the crystal irradiated with high doses (about 200 h) new low-angle reflections arising from the presence of a crystalline NaCl phase are observed (Fig. 6). It has been established that these crystallites are displaying definite orientation in respect to the host lattice (overlap of the main crystallographic directions of both phases). The growth of these NaCl crystallites proceeds in the bulk of the $NaClO_3$ crystal and not only in its superficial layer [14].

3.2 Oscillating film spectrometer method

3.2.1 Analysis of 020 diffraction profile

In Fig. 7 the changes are plotted of the halfwidths of the functions $G(x)$, $H(x)$, and $f(x)$ describing the shape of the diffraction line, in dependence on the γ-radiation dose.

The function $G(x)$ obtained for the oscillating crystal and the stationary film represents the convolution of the functions describing the shape of the incident, primary beam depending on the diffuseness of the spots induced by the strains and the size of the mosaic blocks [15, 16]. In the case of the film oscillating

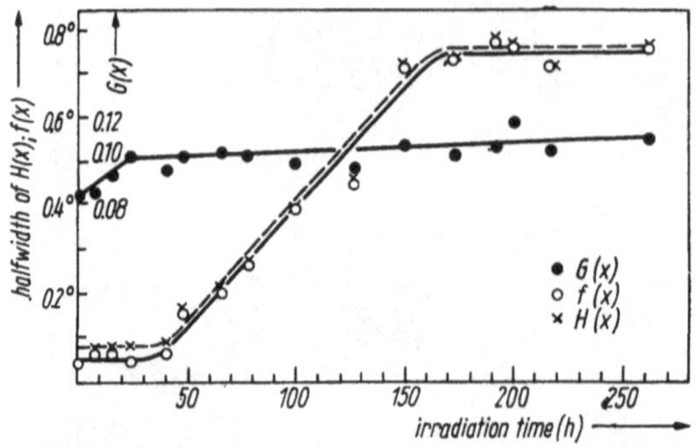

Fig. 7. Effect of γ-radiation on the halfwidths of the functions $G(x)$, $H(x)$, and $f(x)$

along with the crystal the function $H(x)$ obtained experimentally constitutes the convolution of the function $G(x)$ and of the function $f(x)$ describing the angular distribution of the mosaic blocks. The width of the $f(x)$ function can be appreciated from the simplified formula given by Bedyńska [15]

$$\vartheta = \Delta_2^2 + \Delta_3^2 - \Delta_1^2 - \Delta_4^2 ,$$

where Δ_1 halfwidth of the function $G(x)$, Δ_2 halfwidth of the function $H(x)$, Δ_3 halfwidth of a function reflecting the angular distribution of the beam incident from the distance $1/2\ R$, Δ_4 halfwidth of the function describing the intensity distribution of the incident beam at the site of the crystal.

As shown in Fig. 5 the linear rise of the disorientation of the mosaic blocks is observed for irradiations between 40 and 150 h. On comparing this observation with that presented in Section 3.1.2, according to which the strains induced in the initial stage of radiolysis are most probably brought about by the gaseous oxygen, it seems reasonable to assume that accumulation of oxygen taking place at higher doses is responsible for the appearance of the mosaic blocks and then for their disorientation.

3.2.2 *Changes of dislocation density*

Based on the knowledge of the halfwidth of the function $f(x)$ one can calculate the dislocation density in the crystal [16] for densities higher than $10^5\ \text{cm}^{-2}$. For lower dislocation densities this method yields rather rough estimates. Table 1 presents dislocation densities estimated in this way as a function of radiation dose.

The above calculations were based on Cottrell's model [17], anticipating a random distribution of dislocations, and a Burgers vector module $b = \frac{a\sqrt{2}}{2} \times \times (1/2\langle 110 \rangle)$. It has been found that in non-irradiated crystals the dislocation

Table 1

D (h)	0	8	16	24	41	48	66	79
ϱ (10^3 cm^{-2})	1.2	1.1	1.7	1.2	0.9	3.0	4.3	6.3
D (h)	100	127	150	173	192	200	217	261
ϱ (10^3 cm^{-2})	10	11	12	11	12	12	11	12

density is small, being constant up to an about 40 h lasting irradiation. For higher doses it increases from about 3.0×10^3 up to 1.2×10^4 cm^{-2}. For irradiations lasting more than 100 h the dislocation density remains practically constant

4. Final Remarks

The radiolysis products contained in irradiated $NaClO_3$ crystals induce several interesting changes in the real structure of the host lattice. In dependence on the γ-radiation dose various effects appear. So, within the range up to about 2.5×10^7 rad the lattice constant increases linearly with the dose, owing to the formation of a solid solution of radiolysis products in the $NaClO_3$ lattice. Here, the strains introduced into the lattice by the gaseous oxygen can play an important role, cf. [5].

At higher doses beginning with about 4×10^7 rad the formation of mosaic blocks is observed followed by an increasing degree of disorientation of the blocks related with the increasing dislocation density.

In the crystalline lattice of $NaClO_3$ one might expect the formation of some new specific dislocations. It is possible that owing to accumulation of interstitial oxygen small prismatic dislocation loops are formed. A similar process is anticipated in irradiated alkali halide crystals in which such loops are expected to be formed from the interstitial halogen [18]. The above defects might be in turn responsible for the formation of the walls of the mosaic blocks, undergoing disorientation with the increasing amount of oxygen accumulated at these sites during radiolysis. At the end of this stage one should expect the appearance of the first NaCl crystallites.

Within the dose range beginning with about 1.5×10^8 rad the angular distribution of the mosaic blocks does not undergo any observable changes. Moreover, at the same time some strain relaxation is observed, too. The above phenomenon can be explained on the basis of formation of pores in the host lattice arising from the appearance of NaCl crystallites. (It should be noticed that because of rather large difference in the density of both compounds the volume of the NaCl phase produced from an equivalent quantity of $NaClO_3$ is smaller by about 35%.)

The NaCl crystallites are strictly oriented, their $\langle 100 \rangle$ directions being parallel to the $\langle 100 \rangle$ directions in $NaClO_3$. This phenomenon is an example of endotaxy. On considering the crystalline structures of both compounds one can anticipate that owing to the removal of the oxygen atoms from the $NaClO_3$ molecule the closest neighbouring Na^+ and Cl^- ions need only a small shift to yield an NaCl nucleus strictly oriented in the $\langle 100 \rangle$ direction of the host lattice.

Acknowledgements

The authors are indebted to Dr. J. Z. Damm for suggesting the problem, and for valuable discussions. The authors wish to express their thanks to Mr. K. Godwood, Mr. Z. Janiszewski, and Mr. S. Sobociński for their help in performing the measurements on the oscillating film spectrometer and the discussion.

References

[1] J. H. Schulman and W. D. Compton, Color Centers in Solids, Pergamon Press, Oxford/London 1962.

[2] W. W. Sviridov, Photochemistry and Radiation Chemistry of Solid Inorganic Compounds, Vyshaga Shkola, Minsk 1964 (in Russian).

[3] R. Balzer, H. Peisl, and W. Waidelich, phys. stat. sol. **28**, 207 (1968).
[4] G. J. Crowe, W. Fuchs, and D. A. Wiegand, Internat. Symp. Colour Centres in Alkali Halides, Abstract No. 47, Consiglio Nazionale delle Ricesche, Rome 1968.
[5] A. Czerwonko and J. Z. Damm, Nukleonika (Warszawa) **10**, 531 (1965).
[6] C. Ramasastry and S. B. S. Sastry, Solid State Commun. **5**, 799 (1967).
[7] V. Räsänen and R. Esselström, Ann. Acad. Sci. fennicae, Ser. A, **6**, No 256 (1967).
[8] J. Auleytner, Acta phys. Polon. **17**, 11 (1958).
[9] J. Auleytner, Acta phys. Polon. **18**, 81 (1959).
[10] A. J. Wilson, Mathematical Theory of X-Ray Powder Diffractometry, Phillips Technical Library, Eindhoven 1963.
[11] J. Z. Damm, private communication.
[12] H. P. Klug and L. E. Alexander, X-Ray Diffraction Procedures for Polycrystalline and Amorphous Materials, John Wiley & Sons, New York 1954.
[13] Z. Bojarski and T. Bołd, Prace IH **22**, 115 (1970) (in Polish).
[14] Z. Brynikowski, private communication.
[15] T. Bedyńska, Acta phys. Polon. **23**, 4 (1963).
[16] T. Bedyńska and J. Chmielewska, Acta phys. Polon. **26**, 2 (1964).
[17] A. H. Cottrell, Dislocations and Plastic Flow in Crystals, Clarendon Press, Oxford 1953.
[18] L. W. Hobbs, M. J. Gorringe, and D. Pooley, Internat. Symp. Colour Centres in Ionic Crystals, Reading, England 1971 (p. 146).

(Received January 6, 1972)

phys. stat. sol. (a) **10**, 639 (1972)

Subject classification: 10.2; 11; 22.5.2

Department of Physics, University of Fort Hare, South Africa

Annealing of Defects Produced by Room Temperature Irradiation in Nominally Pure KCl Crystals (I)

By

J. R. SERETLO

The results of investigations are given of the annealing of defects produced by RT irradiation in nominally pure KCl crystals. A one-to-one correspondence between the annealing of the F- and V_3-bands establishes the latter as due to interstitial centres. An estimate of the oscillator strength of the V_3-band is given. The kinetics order and activation energy for the F–V_3 reaction are deduced from the results.

Es werden Ergebnisse über die Ausheilung von Gitterstörungen, die durch Röntgenbestrahlung bei Zimmertemperatur in nominell reinem KCl erzeugt wurden, beschrieben. Die Ausheilkurven der F- und V_3-Absorptionsbanden sind stark korreliert, was auf Zwischengitterstörungen als Ursprung der V_3-Banden hinweist. Die Oszillatorenstärke der V_3-Bande wird angegeben und die kinetische Ordnung und Aktivierungsenergie der F–V_3-Reaktion berechnet.

1. Introduction

It is now believed that there is one efficient primary process which is responsible for production of Frenkel defects in alkali halide crystals at all temperatures [1, 2]. The existence of the H-centre, which is known to be an interstitial atom, has been firmly established by optical absorption and ESR experiments in KCl at temperatures below 55 °K [3]. Faraday and Compton [4] studied the growth curves of these centres, as well as F-centres, and concluded that they are complementary defects. α-centres, which are produced in large concentrations at liquid helium temperature (LHeT), have as their complements interstitial halogen ions which are known to give rise to an absorption band on the long wavelength side of the α-band [5, 6].

Although in recent years results of investigations of room temperature (RT) irradiation products have provided incontrovertible evidence for the existence of interstitial centres [7 to 9], our knowledge of the optical absorption bands to which they give rise is not on as firm a foundation as we would like it to be. In particular the occurrence of the V_3-band peaking at 212 nm in KCl has been a subject of controversy: Lüty [10] and subsequently Fritz et al. [11] have identified this band as a U-band arising from hydride ions resulting from the decomposition of OH^- ions during irradiation; however, the results of Sonder et al. [12] and Christy and Phelps [13] indicate that this band could be due to interstitials.

We thought that a study of the annealing of the absorption bands produced by RT irradiation in KCl would throw light on the nature of the defects responsible for the uv spectral region bands. Moreover, by extending these annealing studies to deformed samples in which coloration is known to be much more enhanced than in undeformed ones [14], valuable information about the nature of the defects which absorb in the uv region can be obtained.

2. Experimental Details

Samples of thickness varing between 0.8 and 1.2 mm were cleaved from nominally pure "random ingot" KCl crystals purchased from Harshaw Chemical Company in 1967. A few samples were cleaved from a crystal which had been deformed 6.4% between the jaws of a vice.

The samples were either irradiated with γ-rays from a cobalt source of strength 2400 Ci with sample-to-source distance of 6 cm or with X-rays obtained from a Philips generator operated at 41 kV and 20 mA, with samples placed about 5 cm from the tungsten cathode of the X-ray tube. All irradiations were carried out at RT.

A Gallenkamp muffle furnace which, with some adjustment, could control temperature to within ±1.5 °C at the highest temperature used (250 °C) was used for annealing the samples. We also performed experiments in which the samples were simultaneously illuminated with light and warmed up in the spectrophotometer. For this purpose, a stainless steel block with a heating coil encased in it was used as a heater and the temperature could be controlled to better than ±1 °C.

We carried out both isochronal and isothermal anneals on the samples. The former was performed in steps of 10 to 20 °C at successively higher temperatures. At the end of each 15 min anneal pulse the sample was quenched to RT by placing it on a broad aluminium plate. All optical measurements were made in a Unicam SP 500 spectrophotometer at RT. The isothermal anneals were made at preselected temperatures and the samples were given anneal pulses of total times $2\frac{1}{2}$, 5, 10, 20 etc. min, with optical absorption measurements taken in between anneals.

In all cases care was taken not to expose irradiated samples to light other than the one used for optical measurements.

3. Methods of Analysis

The details of methods employed in analysing the data have been given elsewhere [15]; only relevant expressions will be quoted.

3.1 Arrhenius method

Use is made of the isothermal annealing data to obtain kinetics order of reaction, p. The optical absorption coefficient at the peak of a band of interest (which is proportional to the concentration of defects responsible for the band) is related to the duration of annealing t, by the relations

$$\ln (\Delta\alpha_0/\Delta\alpha) = K\,t \qquad \text{for} \qquad p = 1 \tag{1a}$$

and

$$(\Delta\alpha)^{1-p} - (\Delta\alpha_0)^{1-p} = K\,c\,(p-1)\,t \qquad \text{for} \qquad p > 1\,, \tag{1b}$$

where $\Delta\alpha_0 = \alpha_0 - \alpha_\infty$ and $\Delta\alpha = \alpha - \alpha_\infty$; α_0 and α are the values of the absorption coefficient at the peak of the band at the start of an isothermal anneal programme and after annealing for a total time t, respectively; α_∞ is the asymptotic value, K is the rate constant, and c is a constant relating the concentration of defects to the height of the absorption band. For $p = 1$, a plot of

$$\ln (\Delta\alpha_0/\Delta\alpha) \quad \text{vs.} \quad t \tag{2a}$$

should yield a straight line; and similarly if $p > 1$, a straight line will result if the following plot is made:

$$(\Delta\alpha)^{1-p} \text{ vs. } t\,. \tag{2b}$$

The activation energy E is obtained from the isochronal anneal data as follows: Once p is known, the following plots should give straight lines:

$$\ln\{\ln(\Delta\alpha_{i-1}/\Delta\alpha_i)\} \text{ vs. } \frac{1}{T_i} \quad \text{for} \quad p = 1 \tag{3a}$$

and

$$\ln\{(\Delta\alpha_i)^{1-p} - (\Delta\alpha_{i-1})^{1-p}\} \text{ vs. } \frac{1}{T_i} \quad \text{for} \quad p > 1\,. \tag{3b}$$

Here $\Delta\alpha_i = \alpha_i - \alpha_\infty$; α_i and α_∞ are the values of the absorption coefficient at the peak of a band after the i-th anneal at a temperature T_i and at the end of an annealing stage, respectively. The slope of either (3a) or (3b) is E/k, where k is Boltzmann's constant, and hence E can be obtained.

3.2 Meechan and Brinkman (M and B) method

Data from two samples with closely similar histories, one annealed isothermally and the other isochronally, are compared. The activation energy E is calculated from the slope of the straight line given by

$$\ln \Delta\tau_i = B - E/kT_i\,, \tag{4}$$

where B = const, $\Delta\tau_i$ time interval which, during the isochronal anneal at a temperature T_i, results in a reduction of defect concentration from the same initial value to the same final value as that which occurs during the i-th anneal at a fixed temperature T_a.

We obtained the best slopes of straight lines 3 and 4 and their "standard" errors by using the method of least squares [16].

4. Results

4.1 Isochronal annealing behaviour of defects

4.1.1 F-centres

Fig. 1 shows typical isochronal anneal curves for γ-irradiated samples. Sample H I was γ-irradiated for about twice the time of irradiation for samples H III and H VII. Also shown is the plot for a sample, H X, which was deformed 6.4% and given the same dose of radiation as samples H III and H VII. Clearly deformation has resulted in a marked enhancement of coloration.

In Fig. 2 similar curves are shown for X-irradiated samples. Like in the previous case, we varied the dose given to various samples.

From these curves it can be seen that F-centres anneal out in two distinct stages, I and II, occurring approximately in the temperature ranges 300 to 410 °K and 410 to 500 °K, respectively. The lower temperature annealing stage, stage I, is much smaller than the high temperature stage. Further in lightly coloured samples, H III and H VII (Fig. 1), stage I can be resolved into two smaller stages.

4.1.2 F-aggregate centres

We tried to monitor the F-aggregate centres by observing the associated bands between anneals. The R- and N-bands were too small to give any meaning-

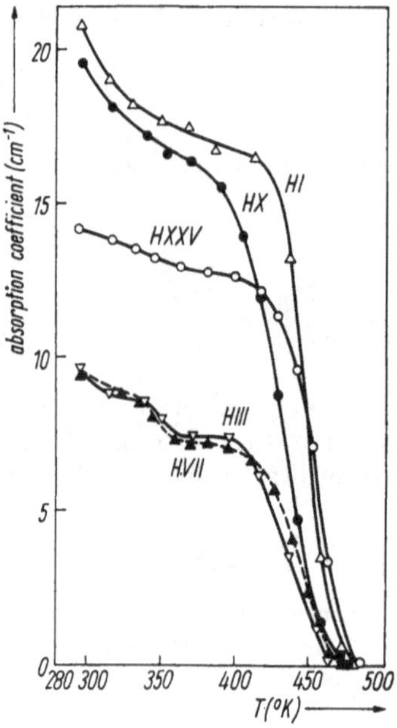

Fig. 1. The isochronal anneal curves of the F-band peak height in γ-irradiated KCl crystals. Samples H III, H VII, and a 6.4% deformed sample H X were given the same dose of γ-radiation

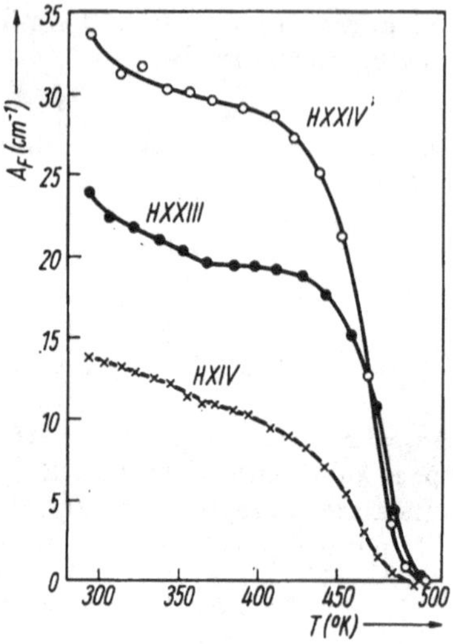

Fig. 2. The isochronal annealing behaviour of the F-band peak height in X-irradiated samples with varying doses of X-radiation

ful results in some heavily irradiated samples and completely absent in lightly irradiated ones. In Fig. 3 we show the annealing behaviour of M-centres in a number of samples. It can be seen that in the temperature range corresponding to stage I of F-centres the concentration of M-centres increases — we shall henceforth refer to this as the "M-centre growth stage". Corresponding to the F-centre stage II, we have the "M-centre decay stage".

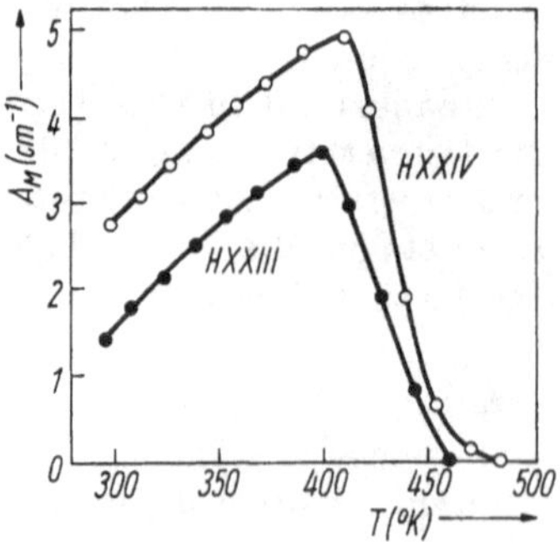

Fig. 3. The isochronal anneal characteristics of the M-band peak height in X-irradiated samples

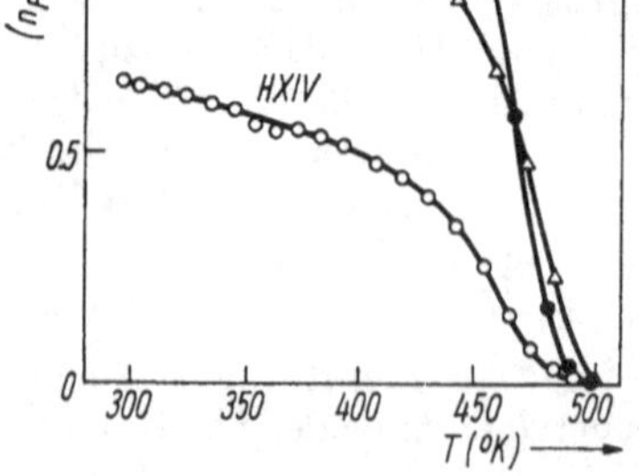

Fig. 4. The isochronal anneal data plot for total vacancy centres ($n_F + 2\,n_M$) in X-irradiated samples

4.1.3 Total vacancy centres (F + 2 M)

We thought it might be helpful to study the annealing behaviour of all the measureable vacancy centres (F + 2 M). For this purpose we used Smakula's relation in the form [17]

$$n_F = 4.5 \times 10^{15} \alpha_F$$

and

$$n_M = 4.1 \times 10^{15} \alpha_M$$

for the concentration of F- and M-centres, respectively. The numerical factors include, amongst others, the refractive indices of KCl at wavelengths corresponding to the F- and M-band peaks, half-widths, and oscillator strengths.

Fig. 4 shows such curves for X-irradiated samples H XIV, H XXIII, and H XXIV. Note that in the temperature range 300 to 410 °K very few of these centres have been "lost". The 410 to 500 °K range is characterized by a prominent annealing stage.

4.1.4 V_3-Centres

In Fig. 5 we show absorption bands in the uv spectral region, taken after isochronal anneals at temperatures indicated in the figure. Two bands are just resolved: a very prominent one peaking at 212 nm (V_3-band) and a less prominent one at 230 nm (V_2-band). The figure also shows that these two bands anneal out at almost the same rate. Another noteworthy feature about this figure is that even after anneals at temperatures as high as 520 °C where the vacancy centres (F- and M-centres) have completely disappeared, there remains some residual absorption in this region. We treated it as a "baseline" and subtracted it from readings obtained after anneals at lower temperatures. We must, however, stress that this residual absorption varied from sample to sample, and also depended on the type of radiation, for example, in X-irradiated samples,

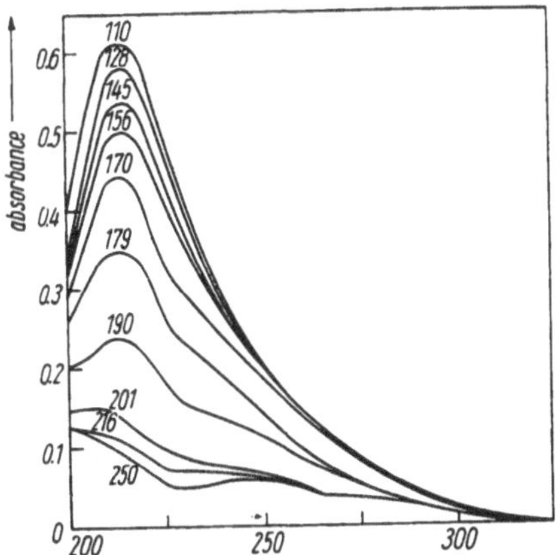

Fig. 5. The ultraviolet absorption bands measured at RT after isochronal anneal pulses at successively higher temperatures in a γ-irradiated sample

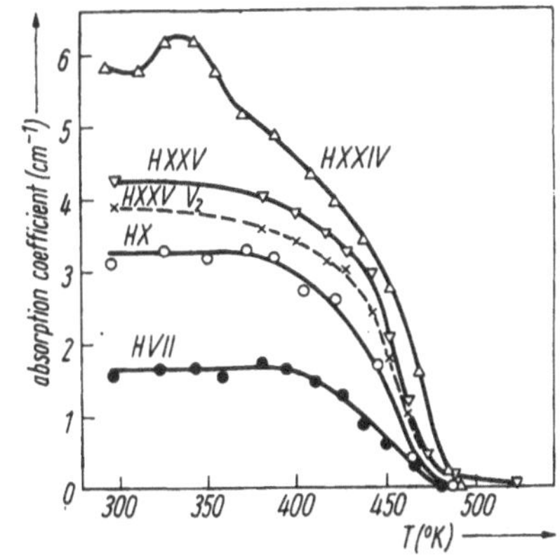

Fig. 6. The isochronal annealing behaviour of the ultraviolet absorption bands as measured from the peak height at 212 nm. The dashed curve was obtained by measuring the height at 230 nm for sample H XXV. The curves H VII and H X are plots of data from two samples one deformed 6.4% (H X) and the other undeformed (H VII) and given the same dose of γ-radiation

where it appeared to be a long wavelength tail of some absorption band peaking below 200 nm, it was found to be much more pronounced than in γ-irradiated samples.

Fig. 6 shows some isochronal annealing curves for the V_3-band in a number of specimens. Also included is the curve (dashed line) for the overlapping V_2-band which behaves similarly. No attempt was made to correct for this overlapping. Curves H X and H VII represent data from two samples, one deformed 6.4 % prior to irradiation and the other undeformed, given the same γ-ray dose. Evidently deformation not only enhances the F-band but also the V_3-band. It was not possible to determine with any confidence whether the V_2-band displays a similar deformation-induced enhancement as it was completely obliterated by the much more prominent V_3-band. The absorption bands taken in between any two successive anneals of the deformed samples (not shown) show the V_2-band hidden below the V_3-band even after anneals at temperatures as high as 445 °K where the F-centre concentration has been reduced to about 50% of its initial value in most samples.

4.1.5 Simultaneous annealing and illumination in the M-band

The data from samples which had been simultaneously annealed and illuminated in the M-band are plotted in Fig. 7. In Fig. 7a we show the annealing

a
b

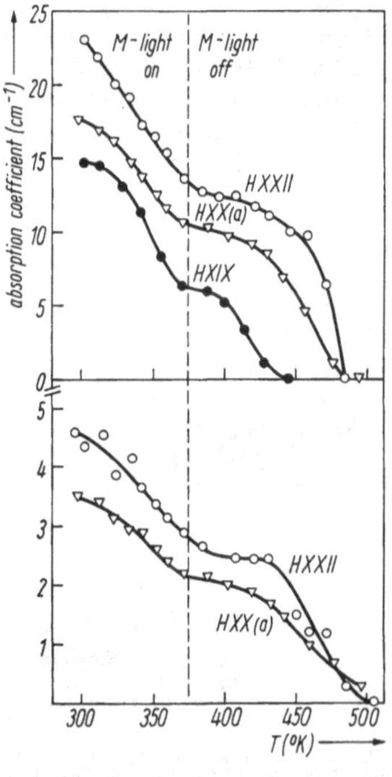

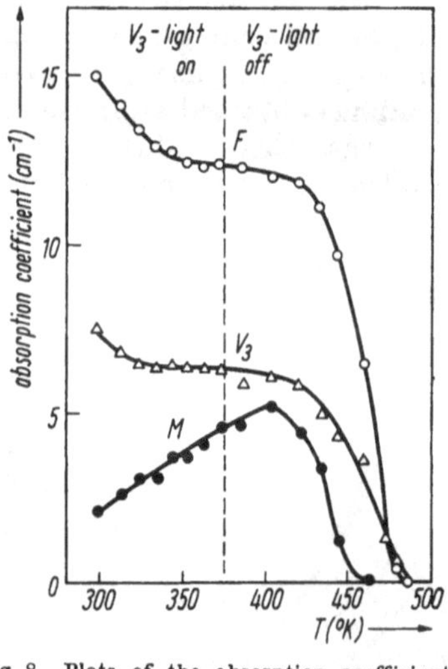

Fig. 8. Plots of the absorption coefficients at the peaks of the F-, V_3-, and M-bands vs. temperature for the γ-irradiated sample H XX (b), which was simultaneously annealed and illuminated in the V_3-band up to the temperature indicated by a vertical dashed line, and subsequently annealed in the dark

Fig. 7. a) Plots of the absorption coefficient at the peak of the F-band vs. temperature for samples that were simultaneously annealed and illuminated in the M-band up to the temperatures indicated by a dashed vertical line, and subsequently annealed in the dark. Samples H XXII and H XIX were X-irradiated and H XX (a) was γ-irradiated. The absorption coefficients for H XX (a) and H XIX were scaled up by factors 1.25 and 2, respectively.

b) Similar plots for the V_3-band (212 nm) in samples H XXII and H XX (a)

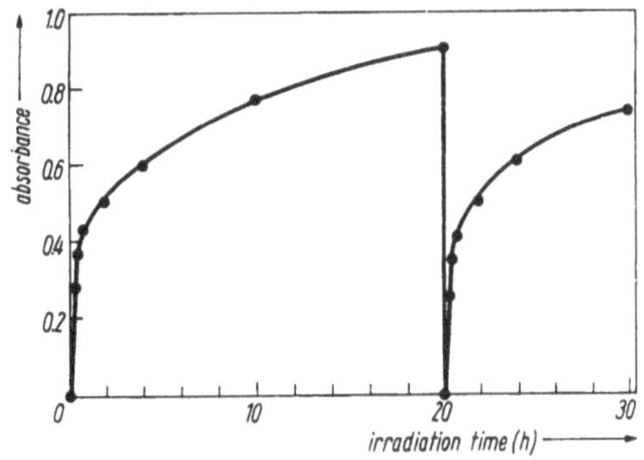

Fig. 9. The growth curve of the F-band interrupted by 1/2 h thermal anneal at 530 °C after X-irradiation of total time 20 h

behaviour of F-centres and in Fig. 7b for the V_3-centres. These experiments were conducted in the spectrophotometer with a slit width of 1 mm and the wavelength set at 820 nm. A Kodak filter with maximum transmission at 820 nm was used. Because of the fact that M-centres begin to decay around 400 °K (see Fig. 3), illumination of samples with this light was discontinued at about 380 °K. Note that both the F- and V_3-centres decay in two stages, corresponding to F-centre stages I and II of samples annealed in the dark.

The M-band was virtually suppressed by this treatment.

We repeated these experiments with the V_3-light in place of the M-light. Curves "F" and "V_3" in Fig. 8 show the effect of this treatment on the F-, M-, and V_3-centres. Comparison of Fig. 8 with Fig. 1 to 3 shows that the F- and M-centres are not affected by the V_3-light during anneals. However, the curve for the V_3-centres shows a slight initial drop (Fig. 8) which was not observed in samples annealed in the dark.

4.1.6 Growth curves

To establish whether F-centres are being annihilated or merely depopulated of their electrons during stage II annealing, we studied their growth curves interrupted by thermal annealing at about 500 °K, a temperature at which all vacancy centres disappear. Fig. 9 shows such a curve. It can be seen that the post-anneal growth curve is very similar to the pre-anneal curve, showing that the sample had been returned to its pre-irradiation condition by the anneal treatment.

4.1.7 F vs. V_3 curves

Because of the one-to-one correspondence between the annealing of F- and V_3-centres in F-centre stage II of samples that had been annealed in the dark (see Fig. 1, 2, and 6) and in both stages for those which were simultaneously annealed and illuminated in the M-band, it was decided to plot F vs. V_3 curves. We show these curves for a number of samples in Fig. 10. It can be seen that we obtain linear plots which have almost the same slopes. We used data from samples which had been γ-irradiated for this purpose because of uniform coloration produced by this type of radiation.

We calculated the F-to-V_3 ratio from the slopes of these plots and the values so obtained are indicated in the diagram.

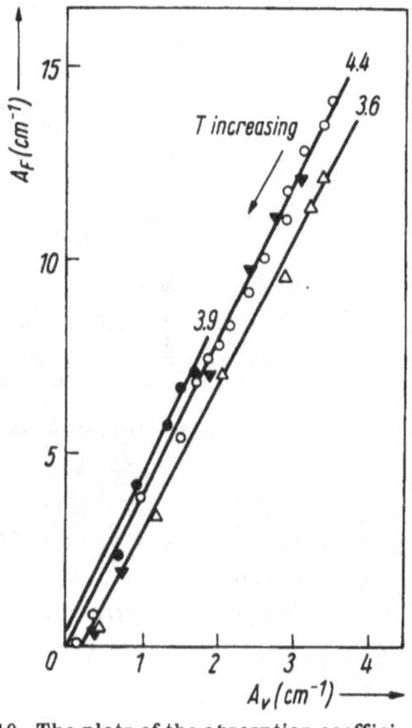

Fig. 10. The plots of the absorption coefficients at the peak of the F-band versus the absorption coefficient at the peak of the V_3-band at successively higher temperatures (increasing temperature is indicated by an arrow). The average slope obtained is 4. △, ● samples H XXV and H VII, respectively, annealed in the dark. ▼, ○ samples H XX (b) and H XX (a), which were simultaneously annealed and illuminated in the V_3- and M-band, respectively, up to ≈ 380 °K and then in the dark

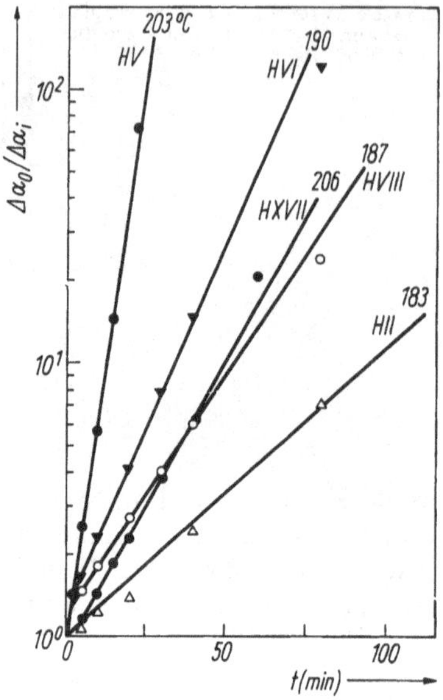

Fig. 11. Isothermal anneal plots, i.e. ln $\{\Delta\alpha_0/\Delta\alpha_i\}$ vs. time for samples H II, H VI, H VIII (γ-irradiated), and H XVII (X-irradiated). The anneals were carried out at temperatures indicated

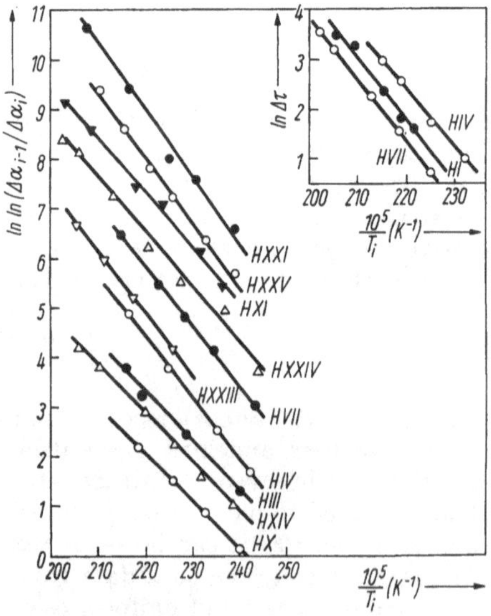

Fig. 12. Arrhenius-type plots for F-centre stage I. The Meechan and Brinkman plots are inserted

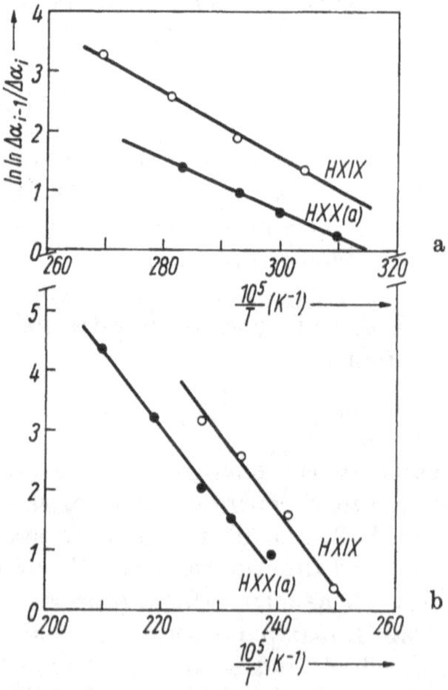

Fig. 13. a) Isochronal anneal (Arrhenius-type) for F-centre stage I in samples that had been simultaneously annealed and illuminated in the M-band; b) similar plots for stage II in the same samples

4.2 Kinetics and energetics

4.2.1 Samples annealed in the dark

The kinetics order for annealing stages was obtained by plotting the isothermal annealing data using the equations (2) (see Section 3). Fig. 11 shows these curves for stage II in a number of samples annealed in the dark, indicating that this stage follows first order kinetics. This fact is further confirmed by plots of isochronal anneal data (equation (3a), Section 3), which are expected to be linear if an appropriate kinetics order p has been chosen. These straight lines are shown in Fig. 12. We have deliberately shifted them (and others to follow) by various amounts along the ordinate to avoid overlapping.

The activation energy values were calculated from the slopes of these straight lines (Fig. 12). These and the other pertinent information have been collected in Table 1. The error limits indicated were calculated from the errors in slopes as described previously (see Section 3).

The activation energy values were checked by using M and B method. The insert in Fig. 12 shows linear plots obtained by combining isochronal and isothermal anneal data from samples indicated. The agreement in activation energy values obtained for any one sample by the two methods is very good (compare columns 4 and 5, Table 1).

Table 1

specimen No.	radiation	initial α_F (cm^{-1})	F-centre stage II activation energy (eV)		M-centre decay stage
			Arrhenius method	M and B method	
H I	γ-rays	20.7	—	1.10 ± 0.12	1.22 ± 0.05 (M and B)
H III	γ-rays	9.6	0.86 ± 0.09	—	—
H IV	γ-rays	12.7	1.06 ± 0.03	1.00 ± 0.06	—
H VII	γ-rays	9.6	1.04 ± 0.03	1.09 ± 0.03	—
H XI	X-rays	15.1	1.12 ± 0.06	—	—
H XIV	X-rays	13.8	0.87 ± 0.04	—	—
H XXI	X-rays	14.5	1.19 ± 0.09	—	1.38 ± 0.02 (Arrh.)
H XXIII	X-rays	24.1	1.10 ± 0.02	—	—
H XXIV	X-rays	33.7	0.98 ± 0.06	—	0.86 ± 0.04 (Arrh.)
H XXV	γ-rays	14.1	1.14 ± 0.05	—	—
H X*)	γ-rays	19.6	0.83 ± 0.02	—	—
Mean value			1.04	1.06	

*) Deformed sample. Activation energy not included for purposes of obtaining the mean value.

Also shown in Table 1 (column 6) are activation energies for "M-centre decay stage", which like F-centre stage II followed first order kinetics.

4.2.2 Samples irradiated with M-light during anneals

In Fig. 13a Arrhenius plots are shown (equation (3a), Section 3) for stage I of samples that had been simultaneous annealed and illuminated in the M-band. Clearly the data yield straight lines indicating that this stage follows first order kinetics. In Fig. 13b we show similar plots for stage II of these samples, which also follows first order kinetics.

The activation energy values for these two stages are shown in Table 2.

Table 2

sample No.	irradiation	initial α_F (cm^{-1})	activation energy (eV)	
			stage I	stage II
H XIX	X-rays	7.3	0.38 ± 0.03	1.09 ± 0.1
H XX	γ-rays	14.1	0.44 ± 0.04	1.15 ± 0.06
		mean	0.41	1.12

Clearly the activation energy values for stage II of these samples agree very well with those obtained for the same stage in samples which were annealed in the dark.

5. Summary of Results

(i) F-centres in "pure" KCl crystals irradiated at RT decay in two stages, I and II, occurring in the temperature range 298 to 410 °K and 410 to 500 °K, respectively.

(ii) The M-centres show a "growth" stage between 298 and 410 °K, and then disappear along with F-centres between 410 and 500 °K.

(iii) The plots of total vacancy centres (F + 2M) vs. temperature show that a negligible quantity of these centres disappears in the temperature range 298 to 410 °K indicating that it is characterized by F-aggregation process. However in the temperature range 410 to 500 °K, the vacancy centres are annihilated.

(iv) The decay processes for both F- and M-centres in the temperature range 410 to 500 °K are characterized by the same kinetics order and comparable activation energies, indicating that there is perhaps one process operative for both types of centres.

(v) The annealing behaviour of V_3-centres is very similar to that of "total vacancy centres", i.e. they show no decay in the temperature range 298 to 410 °K, but disappear in the 410 to 500 °K range, indicating that they are complements of the vacancy centres.

(vi) The close correlation between vacancy and V_3-centres is demonstrated by the plots of F-centre concentration against V_3-centre concentration. The plots are linear through the origin and all have nearly identical slopes, irrespective of irradiation time, annealing or deformation of the sample prior to irradiation. (The slight variations in slopes from one specimen to another could be due to different M-centre concentrations which are thought to complete for interstitials with F-centres — see paper II.)

(vii) Simultaneous irradiation of samples with M-light and pulse annealing result in the suppression of F-aggregation process in F-centre stage I; instead it induces simultaneous decay of the F- and V_3-centres. This reaction follows first order kinetics with a reduced activation energy of 0.4 eV. This decay stage is followed by a second one (between 410 and 500 °K) which also follows first order kinetics with an activation energy around 1 eV.

6. Discussion

The one-to-one correspondence between the annealing behaviour of the F-, M-, and V_3-bands in the F-centre stage II (410 to 500 °K) on the one hand and the close similarity between the annealing characteristics of the V_3-centres on the other hand, indicate that the V_3-band is due to the interstitial centres, which are complements of the vacancy centres. This point is further confirmed by a nearly constant F-to-V_3 ratio which we obtained from the isochronal annealing data for a number of samples given different doses of γ-irradiation (see Fig. 10). Moreover, the fact that deformation-induced enhancement of the F-band is accompanied by a similar enhancement of the V_3-band cannot be explained by invoking the idea that hydride ions are responsible for the V_3-band as suggested by some investigators [10, 11]. It is difficult to see how deformation can result in a high concentration of hydride ions.

We have succeeded in inducing the simultaneous decay of the F- and V_3-centres at lower temperatures (298 to 410 °K) by shining the M-light on the samples during isochronal pulse anneals. We interpret this decays as being due to the reaction between the interstitial centres (V_3-centres) and the α-centres produced by M-light.

In recent years a large body of information concerning RT irradiation produced interstitials has been accumulated. A number of experimental results are consistent with the idea that interstitials produced occur in cluster form [12]. Sonder and Walton [18] have estimated that these clusters consist of 100 to 1000 Cl atoms in samples with an F-centre concentration of $\approx 10^{18}$ cm^{-3}.

Comins [19] has suggested that the absorption by a cluster is proportional to the total number of interstitials in it with perhaps some perturbation from the cluster, and not to the number of clusters. This idea is consistent with our experimental results, viz. a constant F-to-V_3 ratio obtained from samples with differing concentrations of these defects.

We made use of this fact to calculate the oscillator strength for the V_3-centre absorption. Using Smakula's equation in the form

$$N f = 0.87 \times 10^{17} \frac{\mu}{(\mu^2 + 2)^2} W_{1/2} \alpha ,$$

where symbols have their usual meaning, for both the F- and V_3-bands, and the fact that

$$\alpha_F = 4 \alpha_V ,$$

as well as other pertinent data shown in Table 3 below, we obtain

$$f_V = 0.5 f_F .$$

Table 3

	μ	$W_{1/2}$
F	1.49	0.35 eV
V_3	1.66 *) [20]	0.81 eV

*) This is the refractive index at 214 nm.

$W_{1/2}$ for the V_3-band was measured from the absorption bands shown in Fig. 5.

A model to account for the kinetics order and activation energy values for the F-centre stage II has been developed and for its discussion will be referred to part II of the paper.

7. Conclusion

The results presented here clearly show that the V_3-band is due to interstitial centres. Although it is believed that these centres occur in clusters, there are indications that they absorb as single entities. Assuming this to be the case, the oscillator strength for their absorption was calculated.

Acknowledgements

The author wishes to thank Professor P. T. Wedepohl and Dr. J. D. Comins for useful discussion and suggestions during the preparation of this paper and also Dr. R. J. Keddy for performing γ-irradiations.

I am also indebted to the Council of Scientific and Industrial Research (Pretoria) for partly financing the project.

References

[1] D. POOLEY, Proc. Phys. Soc. **87**, 245 (1966).
[2] H. N. HERSH, Phys. Rev. **148**, 928 (1966).
[3] W. KÄNZIG and T. O. WOODRUFF, J. Phys. Chem. Solids **9**, 70 (1918).
[4] B. J. FARADAY and W. D. COMPTON, Phys. Rev. **138**, A893 (1965).
[5] W. GEBHARDT, J. Phys. Chem. Solids **23**, 1123 (1962).
[6] A. BEHR, H. PEISL, and W. WAIDELICH, Phys. Letters (Netherlands) **24A**, 379 (1967).
[7] J. S. NADEAU, J. appl. Phys. **33**, 3480 (1962); **34**, 2248 (1963); **35**, 1248 (1964).
[8] W. A. SIBLEY and E. SONDER, Phys. Rev. **128**, 540 (1962); J. appl. Phys. **34**, 2364 (1963).
[9] R. BALZER, H. PEISL, and W. WAIDELICH, phys. stat. sol. **15**, 495 (1966).
[10] F. LÜTY, Adv. Phys. **17**, 93 (1968).
[11] B. FRITZ, F. LÜTY, and J. ANGER, Adv. Phys. **17**, 93 (1968).
[12] E. SONDER, W. A. SIBLEY, J. E. ROWE, and C. M. NELSON, Phys. Rev. **153**, 1000 (1967).
[13] R. W. CHRISTY and D. H. PHELPS, Phys. Rev. **124**, 1053 (1961).
[14] J. H. SCHULMAN and W. D. COMPTON, Color Centres in Solids, Chap. VI, Pergamon Press, 1963.
[15] J. R. SERETLO and P. T. WEDEPOHL, J. Phys. C (Solid State Phys.) **2**, 18 (1969).
[16] H. MARGENAU and G. N. MURPHY, The Mathematics of Physics and Chemistry, 2nd edition, Van Nostrand Co., 1959 (p. 504).
[17] W. A. SIBLEY and J. R. RUSSELL, Phys. Rev. **154**, 831 (1967).
[18] E. SONDER and D. WALTON, Phys. Letters (Netherlands) **25A**, 222 (1967).
[19] J. D. COMINS, private communication.
[20] Internat. Critical Tables VII, McGraw-Hill Publ. Co., 1930 (p. 14).

(Received July 30, 1971)

phys. stat. sol. (a) 10, 651 (1972)

Subject classification: 20.3; 10.2; 22.2.1

Institute of Semiconductors, Academy of Sciences of the Ukrainian SSR, Kiev

Effect of Heat Treatment on the 0.93 eV Luminescence Band in p-GaAs

By

F. M. VOROBKALO, K. D. GLINCHUK, and A. V. PROKHOROVICH

It is shown that the intensity of the 0.93 eV emission band (the concentration of the 0.93 eV radiative centres) in p-GaAs can be increased and subsequently decreased by annealing. Evidence is presented that the centre giving rise to the 0.93 eV emission, may be associated with interstitial copper atoms and arsenic vacancies.

Показано, что интенсивность полосы излучения 0,93 eV (концентрация 0,93 eV излучающих центров) в p-GaAs может быть увеличена, а затем уменьшена путем отжига. Приведены данные, указывающие на то, что центр, ответственный за полосу люминесценции 0,93 eV, может быть связан с междуузельными атомами меди и вакансиями мышьяка.

1. Introduction

A study of the effect of annealing on the intensity of different luminescence bands is a powerful instrument for the identification of radiative centres in GaAs. Previously, a study of the effect of annealing on the intensity of the emission bands at $h\nu = 1.37$ [1, 2] and 1.49 eV [3] in p-type GaAs and at $h\nu = 1.20$ [4], 0.96 to 1.02, 1.2, 1.3 [5], 1.22, and 1.49 eV [6] in n-type GaAs lead to the identification of the origin of radiative centres in these crystals. In the present work we investigate the nature of the defect responsible for the 0.93 eV band in p-type GaAs, by studying the effect of heat treatment on the intensity of low-energy emission at $h\nu = 0.93$ eV. A broad low-energy band at $h\nu = 0.93$ eV is observed in weakly doped p-type GaAs [7]. The characteristics of this low energy emission band were investigated in [7]. However, the specific defects, giving rise to this band, were unknown.

2. Experimental

The investigations were made on zinc doped p-type GaAs crystals with hole concentrations of about 3×10^{16} to 2×10^{17} cm^{-3}, for melt grown crystals of this doping usually exhibit relatively strong emission of intensity I_0 at $h\nu = 0.93$ eV. The heat treatment was performed in quartz ampoules, placed in a horizontal two-zone furnace, over a wide range of temperatures (135 to 850 °C) under several annealing conditions (vacuum, constant pumping, As atmosphere, Ga treatment, copper diffusion, different temperature gradients in sealed and unsealed quartz ampoules). Most of these environments are usually used to control the arsenic vacancies or the surface impurity concentration. Copper diffusion can reveal information about the defect because it reacts relatively easily with the defects. After heat treatment the ampoules were quickly quenched to room temperature.

Strongly absorbed light ($1/k = 1\ \mu m$) was used to excite the photoluminescence. Prior to the photoluminescence measurements the samples were etched in a solution of 3 H_2SO_4, 1 H_2O_2, and 1 H_2O. All spectra were taken at 77 °K in the region of linear dependence of the steady-state photoluminescence intensity on the intensity of excitation. Special experiments showed that heat treatment of GaAs samples does not lead to a significant increase of the recombination rate of charge carriers [8]. So in this case the changes in the photoluminescence intensity I upon heat treatment are proportional to the changes in the concentration of radiative centres [8, 9]. As the diffusion length of electron–hole pairs was small enough [1 to 2 μm], the photoluminescence gave information about the concentration of radiative centres near the GaAs surface. The emission intensity profiles (the ratio of the 0.93 eV emission intensity after annealing (I) to the 0.93 eV emission intensity before annealing $I_0 - I/I_0$ as a function of the distance from the GaAs surface) were obtained by a successive lapping of GaAs.

3. Results

Fig. 1 shows results of heat treatments on the intensity of the 0.93 eV luminescence band in p-type GaAs. One can see the following main facts: 1. The intensity of the emission band at $h\nu = 0.93$ eV can be increased by heat treatment at $T \leqq 600$ °C. 2. At $T \leqq 400$ °C this increase depends markedly on the annealing conditions: the increase is rather strong during annealing in a two-zone furnace with a temperature gradient along the quartz ampoule; no increase or small increase in the 0.93 eV luminescence intensity is observed without temperature gradients in the sealed quartz ampoule. 3. The intensity of the emission band at $h\nu = 0.93$ eV can be greatly decreased (this band nearly disappears) by heat treatment at $T \geqq 600$ °C.

Fig. 2 shows the results at different annealing temperatures of the dependence of the 0.93 eV emission intensity (the concentration of the 0.93 eV radiative

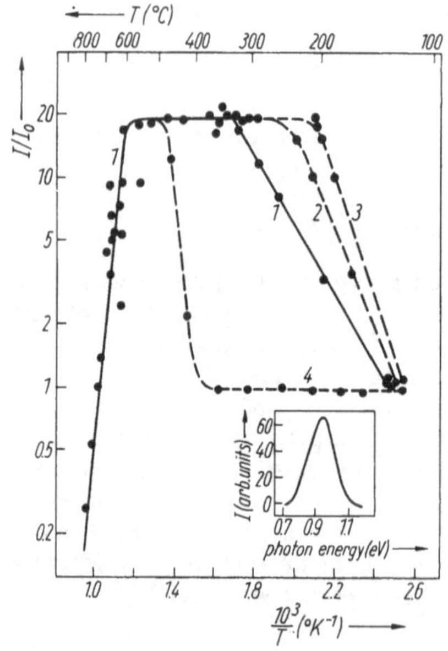

Fig. 1. Dependence of 0.93 eV emission intensity on annealing temperature for p-GaAs. (1 to 3) temperature gradient in the furnace (the low-temperature end of sealed quartz ampoule is at 20 °C); (4) no temperature gradient in the sealed quartz ampoule. Annealing at $T \geqq 500$ °C results in a weak dependence on the temperature gradients in the ampoule or GaAs environments. Duration of annealing (1) 10 min, (2, 4) 1 h, (3) 10 h. The inset shows the 77 °K photoluminescence spectra for lightly doped p-GaAs

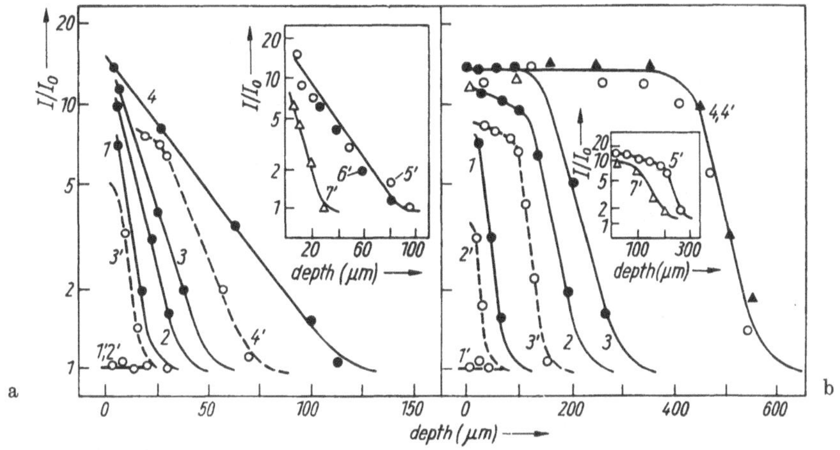

Fig. 2. 0.93 eV emission intensity profiles for p-GaAs at different annealing temperatures. a) heat treatment in an ordinary quartz ampoule with no intentional copper diffusion, b) heat treatment with copper diffusion. Annealing temperatures: (1, 1') 250 °C, (2, 2') 350 °C, (3, 3') 400 °C, (4, 4') 500 °C, (5' to 7') 450 °C. Duration of annealing 10 min. (1 to 4) the low-temperature end of quartz ampoule is at 20 °C; (1' to 7') no temperature gradient in the sealed quartz ampoule; (6') Ga treated; (7') in As atmosphere

centres) on the distance from the GaAs surface. The following facts are clearly seen: 1. The intensity of the 0.93 eV emission band is maximum near the sample surface and diminishes in the interior below the surface. 2. Copper diffusion increases markedly the intensity of the 0.93 eV emission band in the interior of the GaAs samples, i.e. increases the concentration of the 0.93 eV radiative centres and their "diffusion length". 3. At low annealing temperatures ($T \leqq \leqq 400$ °C) a temperature gradient along the quartz ampoule also increases the intensity of the 0.93 eV emission band and the "diffusion length" of the 0.93 eV radiative centres. 4. Ga treatment has no effect on the intensity of the 0.93 eV emission band. 5. As atmosphere diminishes the intensity of the 0.93 eV emission band and "the diffusion length" of the 0.93 eV radiative centres.

Fig. 3 shows the results at different annealing times of the 0.93 eV emission intensity profiles (of the "diffusion" of the 0.93 eV radiative centres). One can

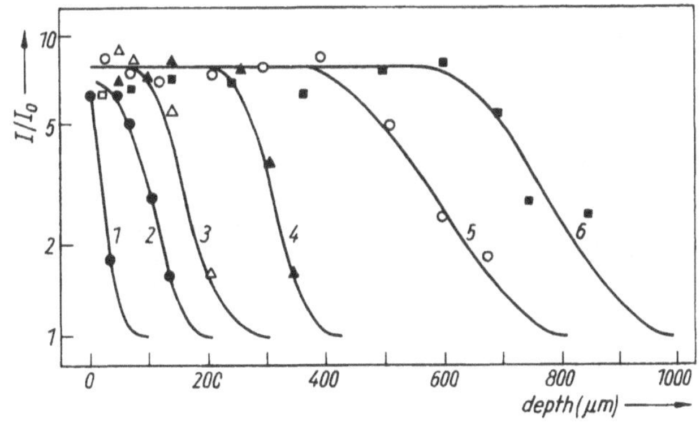

Fig. 3. 0.93 eV emission intensity profiles for p-GaAs as a function of annealing time at 350 °C. (1, 2) heat treatment at 350 °C in an ordinary quartz ampoule with no intentional copper diffusion, (3 to 6) heat treatment at 350 °C with copper diffusion. Annealing time (3) 10 min; (1, 4) 45 min; (5) 3 h; (2, 6) 8 h. The low-temperature end of the quartz ampoule is at 20 °C

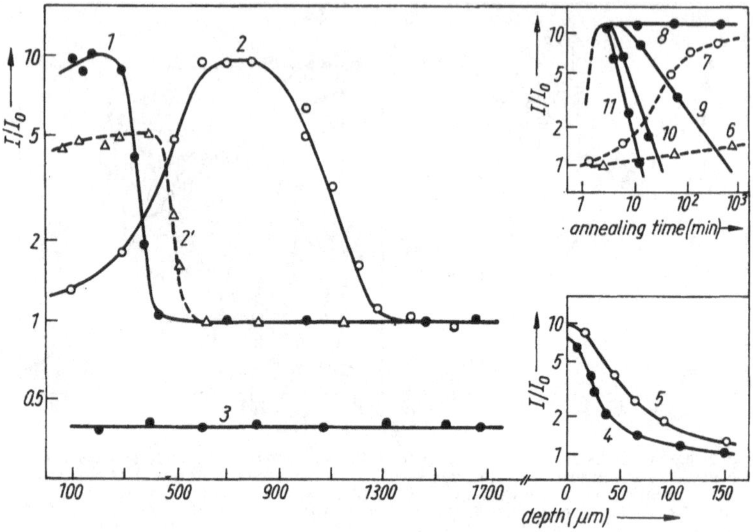

Fig. 4. 0.93 eV emission intensity profiles for p-GaAs as function of annealing time at 550 to 600 °C. (1 to 3) heat treatment in vacuum at 550 °C with copper diffusion; (2') heat treatment in As atmosphere at 550 °C with copper diffusion; (4, 5) heat treatment in vacuum at 600 °C in an ordinary quartz ampoule with no intentional copper diffusion; annealing time (1) 3 min; (2, 2', 4) 10 min; (3) 30 min; (5) 1 h. In the upper inset the 0.93 eV emission intensities at the depth $x = 0$ as a function of annealing time at different annealing temperatures are plotted for p-GaAs; temperature of heat treatment: (6) 130 °C, (7) 190 °C, (8) 350 °C, (9) 450 °C, (10) 500 °C, (11) 550 °C; heat treatment was performed in vacuum with copper diffusion; the low-temperature end of the quartz ampoule is at 20 °C

clearly see that the increase in annealing time leads to an increase in the "diffusion length" of the 0.93 eV radiative centres. One can also see that at any annealing time the copper diffusion serves to increase markedly the "diffusion length" of the 0.93 eV radiative centres. In both cases long annealing times could lead to a nearly uniform distribution of the 0.93 eV radiative centres in the GaAs volume diffusing from the surface.

Fig. 4 shows the results of the 0.93 eV emission intensity profiles as a function of annealing time at 550 to 600 °C. One can see the complex processes which are taking place in GaAs samples after high-temperature annealing with copper diffusion. At first ($t = 3$ min) the concentration of the 0.93 eV radiative centres increases due to the copper diffusion from the GaAs surface. Then ($t = 10$ min) their concentration is greatly decreased near the sample surface. At last ($t = 30$ min) the concentration of the 0.93 eV radiative centres is greatly reduced in the whole GaAs volume.

The results described above were typical for many samples of different doping, so they did not depend on any particular samples. But the induced concentration of the 0.93 eV radiative centres, their "diffusion length" and the temperature range, in which these effects are observed, could slightly differ for samples of different doping.

4. Discussion

These experiments on the effect of heat treatment on the 0.93 eV emission band in p-type GaAs establish that copper diffusion, As atmosphere, and temperature gradients in quartz ampoules change significantly the intensity of the 0.93 eV luminescence band, i.e. the concentration of the 0.93 eV radiative centres.

The fact that the copper diffusion increases the concentration of the 0.93 eV radiative centres testifies that the 0.93 eV radiative centre involves a copper atom. As in p-type GaAs at $T \leqq 500$ °C the dominant form of copper atoms is the interstitial (the interstitial copper solubility is much higher than the substitutional copper solubility [10]), the 0.93 eV radiative centre must involve an interstitial copper atom. But the isolated interstitial copper atom cannot give rise to the 0.93 eV emission band, it is known that radiative recombination via interstitial copper atoms leads to the emission band at $h\nu = 1.34$ eV [11]. At the same time it is known that the copper atoms react relatively easily with defects, especially with the As vacancies [2, 12]. So it is suggested that the centre giving rise to the 0.93 eV emission band is associated with copper atoms and arsenic vacancies. The simplest form of the centre would be a pair consisting of an As monovacancy and an interstitial copper impurity. Therefore it is proposed that the centres associated with the 0.93 eV band, are $V_{as}Cu_i$ pairs, where V_{as} denotes a negatively charged As vacancy [13] and Cu_i denotes a positively charged interstitial copper impurity [10, 11]. Then the increase in the intensity of the 0.93 eV emission band after heat treatment at $T \leqq 600$ °C is connected with the increase in the concentration of copper atoms which could react with the dissolved arsenic vacancies [14]. The magnitude and the rate of this increase should be mostly controlled by the solubility and the diffusion process of interstitial copper atoms and arsenic vacancies. Estimates show that the concentration ($N = 10^{14}$ to 10^{15} cm^{-3}) and the diffusion coefficient ($D = 10^{-8}$ to 10^{-5} cm^2/s at $T = 200$ to 500 °C) of the 0.93 eV radiative centres are close to the values of solubility and diffusivity of interstitial copper atoms [10]. This strongly supports the point of view that the 0.93 eV radiative centres include interstitial copper atoms.

The problem is much more complex when we try to prove directly the role of As vacancies in the formation of the 0.93 eV radiative centres. As it was stated above, As atmosphere, Ga treatment, and temperature gradients in the furnace are usually used to control the As vacancy concentration (As atmosphere [1 to 4, 12, 13] diminishes V_{as}, Ga treatment [1 to 4], and temperature gradient [15, 16] increase V_{as}). In our experiments these environments really influenced the concentration of the 0.93 eV radiative centres. One could judge from these experiments that the As vacancies also form the 0.93 eV radiative centres. But the main difficulty is that because of the very low dissociation (equilibrium) pressure of arsenic (P_{as}) over GaAs at 500 °C (as one could judge from the extrapolated values of P_{as} at $T \leqq 500$ °C [16, 17]) the evaporation of GaAs is very small and could not lead to the creation of a large concentration of arsenic vacancies which is needed to explain our results. Besides the "diffusion coefficient" of the 0.93 eV radiative centres is much higher than one would expect for the vacancy diffusion [3, 4, 13, 18], so As evaporation, Ga treatment, and arsenic atmosphere should not influence significantly the vacancy concentration in GaAs interior.

So the intensity of the 0.93 eV emission band is determined by the As vacancy concentration and the interstitial copper concentration (the dissolved arsenic vacancy concentration may be great enough in melt-grown GaAs crystals [14]). At $T \geqq 600$ °C the concentration of vacancy-interstitial pairs $V_{as}Cu_i$ decreases. This leads to the decrease in the concentration of the 0.93 eV radiative centres, i.e. the intensity of the 0.93 eV emission band after heat treatment at $T \geqq$ $\geqq 600$ °C. The exact mechanism leading to a decrease in the pair concentration is not known now. But we think that a simple pair dissociation is not the dominant one. We propose that in the course of high-temperature annealing $nV_{as}mCu_i$ complexes containing two or more copper atoms and As vacancies ($n, m = 2, 3 \ldots$) are predominantly built up. These complexes no longer act effectively as the 0.93 eV radiative centres, so the intensity of the 0.93 eV emission band decreases. At low annealing temperatures the rate constant for the complex formation is much less than the rate constant for the pair formation, so the $V_{as}Cu_i$, pairs are predominantly formed. At high annealing temperatures the rate constant for the complex formation is only slightly less than the rate constant for the pair formation, so at the initial stages of annealing $V_{as}Cu_i$ pairs are formed; at the final stages of annealing $nV_{as}mCu_i$ complexes are formed; this leads to a decrease of the 0.93 eV emission band intensity. The study of the annealing kinetics of GaAs crystals at different annealing temperatures supports this point of view (Fig. 4). The effect of complex formation is more pronounced in crystals heat treated with copper diffusion (Fig. 4). Copper diffusion serves to create near the GaAs surface complexes containing two or more copper atoms, so the intensity of the 0.93 eV emission band near the GaAs surface decreases (Fig. 4). The effect of complex formation is less pronounced in crystals heat treated with copper diffusion in As atmosphere (Fig. 4), because the presence of a condensed arsenic-rich phase which could dissolve most of the copper, reduces the copper concentration within the gallium arsenide [10]. Note here, that the complex formation in GaAs crystals and their electrical activity may be in many respects similar to the formation and electrical activity of oxygen–silicon complexes in heat treated silicon (only those silicon–oxygen complexes, which possess less than five bound oxygen atoms appear to act intensively as donor centres) [19].

References

[1] C. J. Hwang, J. appl. Phys. **39**, 4307 (1968).
[2] C. J. Hwang, J. appl. Phys. **39**, 4313 (1968).
[3] C. J. Hwang, J. appl. Phys. **39**, 1654 (1968).
[4] B. Tuck, phys. stat. sol. **29**, 793 (1968).
[5] J. I. Alferov et al., Fiz. Tekh. Poluprov. **1**, 1375 (1967).
[6] C. J. Hwang, J. appl. Phys. **40**, 4584 (1969); **40**, 4591 (1969).
[7] F. M. Vorobkalo, K. D. Glinchuk, and A. V. Prokhorovich, phys. stat. sol. (a) **1**, K109 (1970).
[8] F. M. Vorobkalo, K. D. Glinchuk, and N. M. Litovchenko, Fiz. Tekh. Poluprov. **4**, 847 (1970).
[9] F. M. Vorobkalo, K. D. Glinchuk, and A. V. Prokhorovich, phys. stat. sol. (a) **7**, 135 (1971).
[10] R. H. Hall and J. H. Racette, J. appl. Phys. **35**, 379 (1964).
[11] T. N. Morgan, M. H. Pilkuhn, and H. Rupprecht, Phys. Rev. **138**, A1551 (1965).
[12] J. Furukawa and C. D. Thurmond, J. Phys. Chem. Solids **26**, 1535 (1965).
[13] E. Munoz, W. Snyder, and J. Moll, Appl. Phys. Letters **16**, 262 (1970).
[14] R. M. Logan and D. T. Hurle, J. Phys. Chem. Solids **32**, 1739 (1971).
[15] M. F. Millea and D. F. Kyser, J. appl. Phys. **36**, 308 (1965).
[16] F. E. Roberts, phys. stat. sol. **22**, 373 (1967).
[17] J. R. Arthur, J. Phys. Chem. Solids **28**, 2257 (1967).
[18] B. Tuck, phys. stat. sol. (b) **45**, K157 (1971).
[19] W. Kaiser, H. L. Frisch, and H. Reiss, Phys. Rev. **112**, 1546 (1958).

(Received December 3, 1971)

phys. stat. sol. (a) **10**, 657 (1972)

Subject classification: 13.3; 20; 21; 21.1.1

Institute of Metal Physics, Academy of Sciences of the Ukrainian SSR, Kiev

A New Method of Investigation of Electron Structure in Metal Surfaces with the Aid of Positron Annihilation

By

I. YA. DEKHTYAR and V. I. SILANTEV

A method is given for the investigation of momentum distribution of γ-quantum annihilation from the metal surfaces. The method uses the interaction between positronium atoms and surface atoms. Possible mechanisms of this interaction are discussed and the conditions under which the method may be applied are defined. The method was used for the investigation of the momentum distribution of photon annihilation from the surface of polycrystalline iron and aluminium, and single-crystalline molybdenum. It is shown that this distribution is parabolic for Fe and Al and differs significantly from the conduction electron distribution in the bulk metal. For Mo crystals it was concluded that there are two states for electrons on the surface of the (100) plane. It is assumed that the experimental definition of the electrons density gradient on the metal–vacuum interface may serve as a criterion which allows to estimate some physical properties of metal and alloy surfaces. The possibilities of the new method are disccussed.

Излагается метод изучения импульсного распределения аннигиляционных гамма-квантов с поверхности металлов. В методе используется взаимодействие атомов позитрония с атомами поверхности. Рассматриваются возможные механизмы этого взаимодействия и обосновываются условия применения метода. Метод использован для изучения импульсного распределения аннигиляционных фотонов от поверхности поликристаллического железа и алюминия и монокристалла молибдена. Показано, что это распределение подчиняется параболической зависимости для Fe и Al и существенно отличается от импульсного распределения электронов проводимости в объеме металла. Для монокристалла Mo наблюдается два состояния для электронов на поверхности грани (100). Делается заключение, что экспериментальное определение градиента электронной плотности на границе металл-вакуум может быть критерием, позволяющим оценивать некоторые физические свойства поверхности металлов и сплавов. Обсуждаются перспективы использования нового метода.

1. Introduction

It is well known that many metal properties are determined by the distribution of the electron density on the surface, i.e. thermo-electric emission, adsorption properties which are defined by the character of interatomic interaction between substrate atoms and adsorbent atoms, catalytic properties, scattering of waves of different types, etc.

Explanations of many bulk properties of metal electrons are based on the fundamental ideas about their electron structure found with the aid of different resonance methods, γ-resonance spectroscopy, X-ray spectroscopy, and others.

But the physics of surfaces till now has in principle no corresponding method enabling us to define the distribution of the electron density at the surface of samples under investigation.

The truth is that in some cases attempts were made to explain the interatomic bonds and the charge transition connected with the interaction between the atom and the metal surface [1]. But these explanations are based on theoretical ideas which have no experimental evidence.

Some attempts are known to use physical methods for the investigation of the electron structure of the surface.

Thus, for example, in [2] the annihilation lifetime of positrons was determined in the fine-grain silicon powders. The observed lifetime of 2.50×10^{-10} s was assumed to be due to positions which interact with electrons in a bulk sample, but the authors have attributed another value of 5.2×10^{-10} s to an oxide film on the powder surface.

It is easy to show that the latter suggests the formation of an oxide layer of considerable thickness on the surface, but not the formation of a monolayer. If the atomic concentrations on the surface and in the bulk of a crystal are equal to 2×10^{14} and 10^{22} cm^{-3}, respectively, and the radius of a powder particle is denoted by r, then we get for the relative number of positron annihilation processes on the surface (N_s) and in the volume (N_v) (assuming equal probability of annihilation)

$$\frac{N_s}{N_v} = \frac{6 \times 10^{-8}}{r};$$

for $r = 2 \times 10^{-4}$ cm

$$\frac{N_s}{N_v} \approx 3 \times 10^{-4}.$$

Such a negligibly small contribution of positron annihilation on the surface makes, in principle, this method inapplicable for the investigation of surfaces.

In [3] the Mössbauer effect was used to study the adsorption of Fe^{57} on different faces of tungsten single crystals. The information obtained there refers indirectly to the electron states of adsorbent atoms and does not elucidate directly the electron states on the surfaces of W.

The method outlined below is free from these disadvantages. The aim is to obtain direct evidence about the electron structure of the crystal surface. This method together with the method of slow electron diffraction, which gives information only about the distribution of atoms on the surface, will in future improve our knowledge of the physical properties of surfaces.

2. Essence of the Method

The method developed by us was proposed for the first time in [4]. It is connected with the use of positron annihilation on the metal surface and in its neighbourhood.

As has been mentioned above free positrons which usually possess high energy before their penetration into the crystal are not suitable for the investigation of the annihilaion with electrons of the metal surface. However, it is possible to increase the annihilation rate if one uses the interaction with the surface atoms of positronium. Fortunately, positronium (Ps) cannot exist in the bulk metal.

According to [5] the probability that a quasi-bound state analogous to positronium will appear in the metal is determined by the condition that

$$\tau_0 = \frac{2}{\pi\, a_H\, k_F} > 2.85\,, \tag{1}$$

where $\boldsymbol{k}_F$ is the wave vector on the Fermi surface, a_H is the Bohr radius. It was found further that even for metals with low electron density (high values of the parameter $r_s = r_0/a_H$, where r_0 is the radius of a unit electron sphere) the mentioned criterion is not fulfilled.

In fact, the values of r_s corresponding to the region of real electron densities of metals lie in the limits $2 \leqq r_s \leqq 5.5$. Approximately we have

$$k_F = \frac{1.917}{a_H\, r_s} \qquad \text{and} \qquad \tau_0 \approx 0.33\, r_s\,. \tag{2}$$

Hence the values of τ_0 corresponding to real electron densities of metals lie in the limits $0.67 < \tau_0 < 1.83$.

Thus the criterion $\tau_0 > 2.85$ may be considered as prohibition of positronium formation and existence in the bulk metal.

But we may assume that on the surface of a metal conditions are realized favourable for the existence of Ps, because the surface is characterized by an electron density essentially lower and differs from the volume by the magnitude of screening effects.

From equation (1) we can get at least an approximate estimation of the expected electron density on the metal surface which is necessary for condition (1).

The estimate shows that if electrons on the metal surface possess quasi-impulses $\hbar\, k_F < 0.44 \times 10^{-19}$ g cm s^{-1}, then the positronium in this region will have a finite lifetime. From (1) and (2) for the existence of Ps on the surface it is necessary that $r_s > 8.55$. For example, for a metal with $r_s = 3$ the ratio of bulk electron density (n_v) to surface electron density (n_s) is $n_v/n_s = (8.55/3)^3 \approx 23$. This agrees with the distribution of the electron density at the transition from the bulk to the surface [6].

The probability of two-quantum annihilation for a positron carried to the surface by the Ps atom was estimated under the assumption that the equation for the factor increasing the annihilation rate [7] is valid.

From this estimate it follows that this probability grows by some times at the surface.

The factor which increases the annihilation rate depends on the degree of bonding of the annihilating electrons. Therefore, despite the approximate character of the estimates, the physical sense of the result may be expressed as follows: This factor is some times higher for valence electrons than for electrons bound more strongly [8]. In the same way, the annihilation probability for electrons at the crystal surface should be higher than for valence electrons which are characterized by tighter binding inside the metal.

This allows to hope that Ps may turn to be an effective "probe" which would enable to obtain characteristics of electron states at the metal surface.

As to the interaction of positronium with the metal surface such an interaction can be supposed to fulfil at least the following mechanisms [9]:

1. At that instant when Ps hits the metal surface annihilation with electrons can take place (pick-off annihilation).

2. At the interaction of Ps with paramagnetic metal particles the ortho–para conversion can take place (transformation of Ps from triplet to singlet state). — Interactions of the type of chemical reactions liberating a positron (e^+) from Ps are also possible.

The mechanism of pick-off annihilation decreases sharply the lifetime of ortho-Ps and the process of the ortho–para conversion is realized by a direct spin exchange with paramagnetic atoms. The spin of the positron in para-Ps can interact with an "alien" electron and give information about its state by registration of the two-quantum annihilation. Both processes, the ortho–para conversion and the pick-off annihilation, lead to shortening the lifetime of the triplet state of Ps. The positron formed in the dissociation of Ps at the moment of hitting the metal surface can penetrate the subsurface layer (depth of the order of some monolayers) at least in $\approx 10^{-14}$ s and annihilate there.

Moreover the possibility cannot be excluded that the free Ps atom tunnels near the metal surface and interacts with the electrons of Friedel oscillations.

Hence it may be suggested that the interaction of Ps with the metal surface should manifest itself as more or less complex spectrum of two-quantum annihilation due to e^+-annihilation with both the surface electrons and the electrons of the subsurface layer of the metal.

Investigations [9] have shown that there is a high probability of positronium formation in gases. For example, if a positron ray passes through argon at atmospheric pressure with a superposed static electric field (≈ 500 V/cm), the probability of Ps formation is $\approx 90\%$. This fact can be used in experiments. Here we have the favourable fact that the metal surface preliminarily purified will be pure in the atmosphere of pure argon, which condition is extermely important for the investigation of metal surfaces by the method of positron annihilation.

3. Experimental Procedure

Experiments were carried out in the automatic equipment UA-64 which made it possible to study the angular distribution of two-quantum positron annihilation with angular resolution 0.45 or 0.9 mrad. The angular setting, the angular interval, and the counting time were given by the programming machine. The sample and the Na^{22} source with an activity of 5 mCi were put into a vacuum chamber. The method of screened emitter was used.

The sample and the source were separated by a metallic grid electrically isolated against the chamber. The arrangement of blinds, sample, grid, source, and Pb screens is shown schematically in Fig. 1.

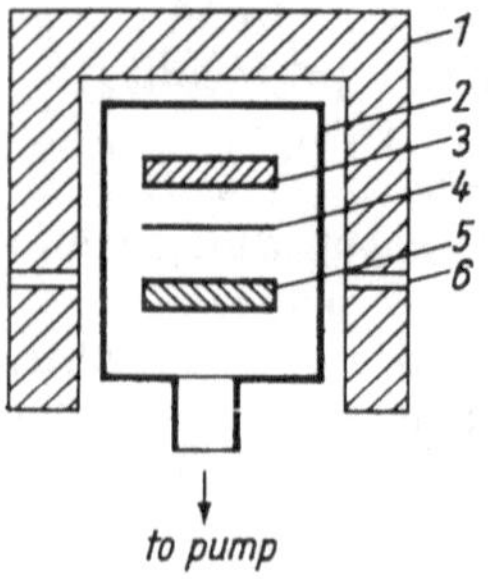

Fig. 1. The scheme of a vacuum chamber in the equipment UA-64. 1 Pb screen; 2 vacuum chamber; 3 source $Na^{22}Cl$; 4 metal grid; 5 sample; 6 slit for γ-ray exit

The chamber was evacuated by a zeolite pump (up to 10^{-5} Torr) connected with the chamber through the valve. The other valve served for the inert gas supply into the chamber. The chamber was successively evacuated and filled with purified argon three times. After such "washing" and filling with argon an electric field ($\approx$ 500 V/cm) was applied between the grid and the sample.

The angular correlation curves for annihilation photons were recorded both for the electric field being present and absent. Thus the character of changes in the shape of the angular correlation curve due to interaction between positronium and sample surface could be considered.

Polycrystalline samples of iron (99.98%) and aluminium (99.99%) were preliminarily annealed in a vacuum of 10^{-6} Torr at 1100 °C (Fe) and 500 °C (Al) for two hours to relieve stresses. The samples were discs with 15 mm diameter and $\approx$ 2 mm thickness. After thermal working they were carried into the chamber of the annihilation equipment. The surface was not subjected to an additional purification in the chamber.

An Mo single crystal (99.99%) with 12 mm diameter, $\approx$ 2 mm thickness, and (100) surface was also investigated. The sample surface in the chamber was cleaned by bombardment by argon ions. But the degree of purification was not controlled.

It must be noted that under the conditions of our experiment we could not guarantee the absence of an adsorbed gas film on the sample surface. But this fact is of no importance for the definition of the possibilities of the method and the particularities of the electron states at real metal surfaces.

However, the physically pure surface is of utmost necessity to solve different problems of physical electronics which are connected with the investigation of electron emission and adsorption. Hence the method is planned to be improved in subsequent experiments.

4. Results and Discussion

Experimental results for polycrystalline iron are given in Fig. 2. An enlarged curve of the angular distribution is shown in Fig. 3. These data clearly indicate the presence of two types of momentum distribution. The wide momentum distribution corresponds to positron annihilation with conduction electrons in the bulk of the sample. The narrow distribution is to be attributed to atoms of the singlet positronium which annihilate with electrons at the surface of the sample. This suggestion is supported by the fact that the full width at the half-maximum of momentum distribution for gaseous argon is equal to $\theta_{1/2} \lesssim 2$ mrad while in our case we have for the narrow distribution $\theta_{1/2} \approx 2.9$ mrad.

It should be noted that the angular-distribution curve of annihilation photons obtained for the same iron sample in argon atmoshpere without electric field shows only one wide momentum distribution which corresponds to positron annihilation with conduction and core electrons in the volume.

The observed (Fig. 2 and 3) narrow momentum distribution must not be identified with the annihilation of singlet positronium in the "bubble" state. For the first time this narrow distribution was postulated by Ferrell [10]. The idea is that in a liquified inert gas the positronium atoms form bubbles of such a size that hydrostatic forces are balanced with the pressure of positronium atoms.

The bubble state of positronium atoms in liquid He, H_2, Ne, Ar, Kr, and Xe

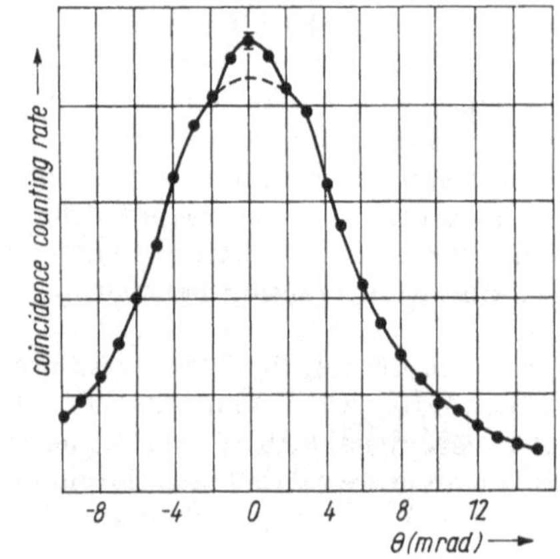

Fig. 2. Angular-correlation distributions $N(\theta)$ for polycrystalline iron. Angular resolution is 0.9 mrad

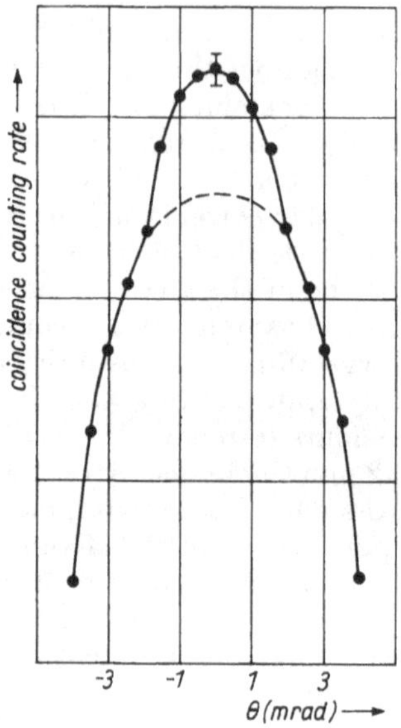

Fig. 3. The same as in Fig. 2 with double statistics for the angular interval 0 to $\pm$ 4 mrad

was observed in [11] to [13]. It was shown in [13] that the positronium exists in liquid, but not in solid Ar, Kr, and Xe.

The narrow component for liquid Ar is characterized by the value $\theta_{1/2}$ equal to 2.1 mrad in [12] or 2.3 mrad in [13]. These values are much lower than that which is observed for the data shown in Fig. 3. This indicates that in our experiments the bubble state of positronium was not formed.

The presence of a momentum distribution significantly broadened with lower intensity may perhaps be ascribed to the interaction of positronium atoms with electrons at the metal surface.

As soon as the metal under investigation is ferromagnetic, it is reasonable to assume that the annihilation of thermalized positronium with surface electrons is fulfilled by one of the possible mechanisms, for example, by means of pick-off annihilation. In this case metal surface electrons for which the momentum distribution can in principle be described by the model of a Fermi gas participate in the annihilation.

As it is well known the Fermi distribution of quasi-momenta corresponds to the parabolic curve of angular correlation.

The analysis of the angular correlation curve of annihilation photons basing on the difference between curves for narrow and broad distributions (Fig. 3) shows that it is described by a parabola $N(\theta) = N_0\,(\theta_{m,s}^2 - \theta^2)$ where N_0 is the counting rate of coincidences for $\theta = 0$, $\theta_{m,s} = 2.0$ mrad is the extreme angle characterizing the momentum distribution of electrons at the surface of a polycrystalline metal.

The extreme angle characterizing the momentum distribution for positron annihilation with conduction electrons in bulk polycrystalline iron is $\theta_{m,v} = 5.0$ mrad. This quantity was found from the parabolic part of the broad component of the angular distribution (Fig. 2).

Hence $\theta_{m,s} < \theta_{m,v}$ and this, according to equation (1), must favour the existence of Ps on the surface. Really, in Section 2 the estimate showed that electrons at the metal surface must possess quasi-momenta with limiting value $\hbar k_F \lesssim \lesssim 0.44 \times 10^{-19}$ g cm s^{-1}. Then from $\hbar k_F = m c \theta$, where m is the electron mass and c the light velocity, we find $\theta_{m,s} \approx 1.6$ m rad. Taking into account the approximate character of this estimate, it should be deduced that this quantity is in fair agreement with $\theta_{m,s}$ found from experiment.

On the other hand, since the extreme value of a quasi-momentum for the Fermi distribution is proportional to $n^{1/3}$, the measured values of $\theta_{m,s}$ and $\theta_{m,v}$ allow to estimate the electron density gradient in the interface metal–vacuum:

$$\Gamma = \frac{n_v}{n_s} = \left(\frac{\theta_{m,v}}{\theta_{m,s}}\right)^3. \tag{3}$$

In our case $\Gamma = 16$, which is close to the expected value (see Section 2).

The analysis of the angular distribution curve for polycrystalline Al (Fig. 4) shows that for the narrow momentum distribution of electrons $\theta_{1/2} \approx 2.6$ mrad and $\theta_{m,s} = 2$ mrad, while for the wide one $\theta_{m,v} = 5.4$ mrad. In this case the narrow distribution also differs from Ps annihilation in argon gas. That is why we believe that it is due to the interaction of Ps with the electrons at the Al surface which leads to the annihilation from the singlet state: $^1S_0 \rightarrow 2\,\gamma$-quanta. From the results obtained follows for Al $\Gamma \approx 20$ which is also a reasonable value (Section 2).

Contrary to polycrystalline specimens of Fe and Al the results obtained for Mo single crystals (Fig. 5) indicate a complex nature of the narrow momentum

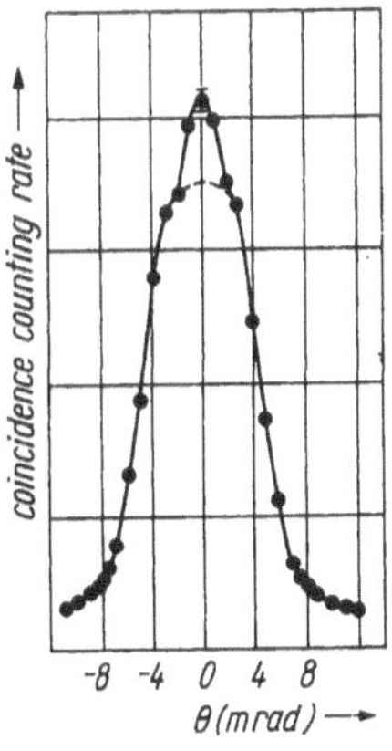

Fig. 4. Angular-correlation distributions $N(\theta)$ for polycrystalline aluminium. Angular resolution is 0.9 mrad

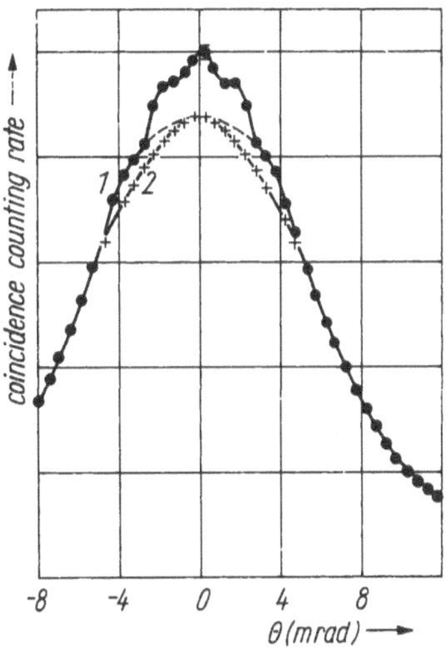

Fig. 5. Angular-correlation distributions $N(\theta)$ for the (100) face of a molybdenum single crystal. Angular resolution is 0.45 mrad. The dashed line (1) is the extrapolation of the parabolic part; curve (2) was obtained without electric field in argon and is normalized in the maximum to the dashed line

distribution for which two values of θ_s are typical: $\theta_{m1,s} = 1.7$ mrad and $\theta_{m2,s} = 2.7$ mrad. This suggests that there are at least two electron states on the surface of the (100) face in Mo. But to find out their nature the investigations of single crystals should be conducted in more detail. From the parabolic part of the broad momentum distribution we obtain

$$\theta_{m,v} \approx 4.7 \text{ mrad} .$$

This allows to evaluate $\Gamma = 21$.

From the relative values of the areas below the narrow and broad distributions (Fig. 2) one can conclude that in spite of the fact that under the conditions of our experiment the formation of almost 80% positronium atoms should be expected, only a small part of them annihilates from the singlet state as a result of interaction with electrons at the metal surface.

A large part of the positrons formed by the dissociation of Ps atoms penetrates during its lifetime into the crystal bulk giving a perceptible contribution to the annihilation with conduction and core electrons. Hence it is suggested that if one applies the proposed method to the investigation of heavier metals (tungsten, rhenium, tantalum, etc.), which are of interest for electronics, the effect will become higher (all other factors being the same) with the relative fraction of positrons annihilating at the crystal surface.

This is due to the fact that the scattering of positrons at metal surfaces is increased with increasing atomic number of the metal [14]. This is one of the reasons that the area below the angular correlation curve for metals with a high atomic number is decreased. For the same reason it may be assumed that the annihilation rate of Ps at the crystal surface will be higher for metals with higher atomic numbers. This is confirmed by the results for Fe and Mo (Fig. 2 and 5).

We believe that an experimental definition of the value may serve as a criterion which allows to estimate some physical properties of metal and alloy surfaces. This criterion must be able to characterize the state of the metal surface and, particularly, its electron emission and adsorption properties. But this requires more experimental data for different metals and alloys as well as the development of a model for the electron structure of the metal surface.

Referring to the latter the theory should overcome some difficulties which arise from the fast decrease of the electron density near the surface and from the loss of translational symmetry.

In [6] the electron structure of the metal surface is described by a model where positive ions are replaced by a uniform semi-infinite positive charge density. The calculation is carried out by a self-consistent method based on the general theory of a non-homogeneous electron gas which takes into account the exchange and correlation effects.

Fig. 6 shows the theoretical results [6] for a self-consistent electron density near the metal surface for $r_s = 2$ and $r_s = 5$. The distance x is measured in units of the Fermi wavelength $2\pi/k_F$ from the surface of a positive backgound. It can be seen that the Friedel oscillations of the electron density near the surface inside the metal are quite distinct for metals with lower electron density in the bulk ($r_s = 5$), but at $x = -2\pi/k_F$ the electron density attains a value typical for the bulk metal. Hence near the surface the electron density is distributed rather inhomogeneously.

The problem of the electron density distribution at the metal surface was discussed in some papers [1]. The model of a charged double layer is widely

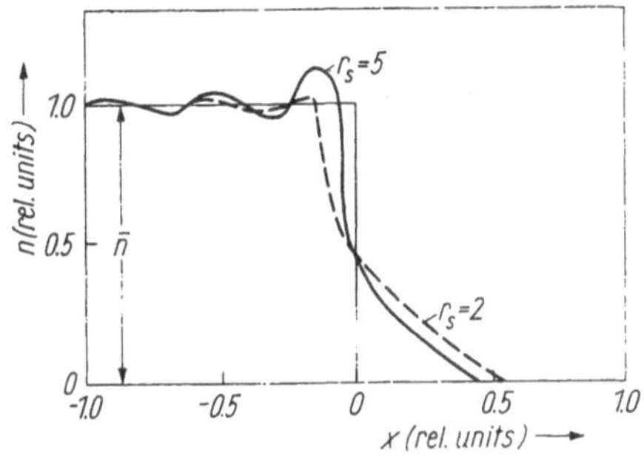

Fig. 6. The charge density distribution near the metal surface [6]

used. It is applied to the explanation, e.g., of the well-known fact that the electron work function depends on the crystallographic structure.

But up to now it is not clear how the double layer depends on the atomic structure of different crystallographic planes on the metal surface. Therefore at present it is difficult to give quantitative characteristics of the electron gas state at the surface of a solid. Even more difficult is the explanation of different kinds of reactions taking place at the surface. In particular the questions of the interatomic bond at the surface defining such phenomena as adsorption may be correctly elucidated only if the dynamics of the electron gas at the pure surface is known. The importance of these problems for the physical electronics can hardly be overestimated.

Meanwhile the existing ideas concerning the electron work function from the point of view of the electronic physics at the solid surface are far from being perfect [1]. Still obscure are those models of adsorption interaction which are based on the ideas of a double layer charge at the surface.

By the method proposed in the present work characteristics of the electron density distribution at the metal surface can be obtained. These characteristics may serve as a real base needed for the verification and further development of the above-mentioned theoretical considerations.

The method may be applied to the investigation of surfaces of metal single crystals before and after adsorption of metal atoms. It will be useful to determine more precisely the mechanism of their interaction and to formulate quantitative criteria. But for this more experimental and theoretical investigations are needed.

Acknowledgement

The authors would like to thank V. T. Adonkin for his assistance in the experimental work.

References

[1] T. N. Rhodin, R. L. Gerlach, and E. W. Plummer, in: Structure et propriétés des surfaces des solides, Editions du Centre National de la Recherche Scientifique, Paris 1970.
[2] A. Gainoti and C. Ghezzi, Phys. Rev. Letters **24**, 349 (1970).
[3] I. W. Burton, H. Frauenfelder, and R. P. Godwin, Applications of the Mössbauer Effect in Chemistry and Solid State Physics, IAEA, Vienna 1966.
[4] I. Ya. Dekhtyar, Preprint No. 9, Akad. Nauk Ukr. SSR, Kiev 1971.

[5] A. Held and S. Kahana, Canad. J. Phys. **42**, 1908 (1964).
[6] N. D. Lang and W. Kohn, Phys. Rev. **B 1**, 4155 (1970).
[7] I. P. Carbotte, Phys. Rev. **155**, 197 (1967);
A. Hatano, H. Kanazawa, and Y. Mizuno, Progr. theor. Phys. (Kyoto) **34**, 875 (1965).
[8] E. I. Wool and I. P. Carbotte, Phys. Rev. **164**, 985 (1967).
[9] V. I. Goldanskii, Fizicheskaya khimiya pozitrona i pozitroniya, Izd. Nauka, Moskva 1968.
[10] R. A. Ferrell, Phys. Rev. **108**, 167 (1967).
[11] L. O. Roellig and T. M. Kelly, Phys. Rev. Letters **18**, 387 (1967).
[12] C. V. Briscoe, S. I. Choi, and A. T. Stewart, Phys. Rev. Letters **20**, 493 (1968).
[13] P. G. Varlashkin, Phys. Rev. **A 3**, 1231 (1971).
[14] H. H. Seliger, Phys. Rev. **88**, 408 (1952).

(Received November 9, 1971)

phys. stat. sol. (a) 10, 667 (1972)

Subject classification: 13.4 and 14.3.1; 1.5; 14.3.4; 22.9

Institute of Physics, Academy of Sciences of the Ukrainian SSR, Kiev (a) and Chemistry Faculty of Chernovtsy University (b)

On Trap Distributions in Diphenylpolyene Films

By

N. I. GANUSHAK (b), N. I. GRITSENKO (a), and M. V. KURIK (a)

Injection currents in diphenylbutadiene (DPB), diphenylhexatriene (DPHT), and diphenyloctatetraene (DPOT) films obtained by thermal vacuum deposition have been studied. From temperature measurements of the space-charge-limited currents (SCLC) the trap parameters are determined and the current carrier mobility are calculated in films of the studied substances. The experimental results show that the change of the film structure causes trap energy distribution changes in the forbidden bands of these substances. These traps are supposed to be of a structural nature.

Es wurden Injektionsströme in Diphenyl-Butadien (DPB)-, Diphenyl-Hexatrien (PPHT)- und Diphenyl-Oktatetraen (DPOT)-Schichten, die im Vakuum thermisch aufgedampft wurden, untersucht. Aus den Temperaturmessungen der raumladungsbegrenzten Ströme (SCLC) werden die Haftstellenparameter bestimmt und die Beweglichkeiten der Ladungsträger der untersuchten Substanzen berechnet. Die experimentellen Ergebnisse zeigen, daß die Änderung der Schichtstruktur eine Änderung der Haftstellen-Energieverteilung in der verbotenen Zone dieser Substanzen verursacht. Es wird angenommen, daß diese Haftstellen von struktureller Natur sind.

Injection currents in dielectrics and in high-resistivity semiconductors are very sensitive to the capture levels in a given medium by which the injected carriers are trapped. Hence, investigation of SCLC features allows to study the important parameters of these traps, that is, their concentration and distribution in the forbidden band. This information is contained in the current–voltage characteristics (CVC) of SCLC, their form being determined by the trap parameters.

The physical nature of traps in organic semiconductors is not yet clear. Usually they are connected with structural defects or with chemical impurities, although there is no reliable experimental confirmation of these models [1]. It is, therefore, interesting to find out the influence of different controlling factors, especially the structure of the specimen, on the injection currents in organic substances. For this purpose we studied the CVC of SCLC of organic semiconductor films, obtained by vacuum thermal deposition of the substances in question under different conditions. Three representatives of diphenylpolyenes were studied: diphenylbutadiene (DPB), diphenylhexatriene (DPHT), and diphenyloctatetraene (DPOT). They are of similar crystalline structure [2, 3] and their molecular structure changes regularly.

The measurements were carried out on "sandwich"-like samples, the organic film thickness being 0.5 to 5 μm. The structure of organic films depends on the thermal deposition conditions of the substances, that is, on substrate temperature, deposition rate, vacuum in the system, and other factors. The substrate temperature influenced the film structure most. As a rule, at film deposition

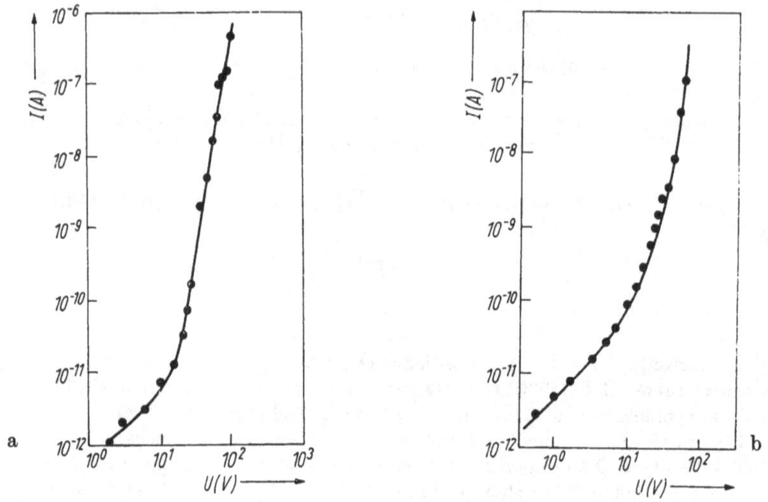

Fig. 1. Typical static current–voltage characteristics at room temperature: a) DPOT films of type I, b) DPB films of type II

on a substrate being at room temperature, a polycrystalline structure is formed; the microcrystals of this structure are clearly seen in polarized light through a microscope (films of the first type). In DPB films the biggest microcrystals up to some tens of μm are produced. At deposition on a substrate cooled to about 100 °K the film structure is more disordered (type II films), which is indicated by an absence of microcrystals at a 1000-fold magnification. Type I and II DPB films and DPHT and the DPOT films of type I have been studied. Injecting contacts of Ag and Au were used in all cases. Before measurements the samples were subjected to a long-time thermal annealing at 333 to 353 °K and a conditioning in a static electric field of 10^4 V/cm. This procedure favoured the formation of ohmic contacts and improved the reproducibility of CVC. The measurements were made in steady state conditions after the relaxation drop of current.

It was found that the electric field strength at which a nonlinear CVC begins, is approximately identical for films of different structures. The type I film characteristics, however, differed appreciably from those of type II.

The CVC's of type I films of all three studied substances are similar. In Fig. 1a a typical CVC of a DPOT film with silver electrodes is shown. The characteristics are linear up to field strengths of $\mathcal{E} = (3 \text{ to } 5) \times 10^4$ V/cm, then the transient

Table 1

The average parameters of the studied type I films, obtained by SCLC measurements

substance	m	$T_c \pm 100$ (°K)	kT_c (eV)	H_0 (cm^{-3})	E_f (eV)	μ (cm^2/Vs)
DPB	7.2 to 6.5	1950	0.17	5×10^{16}	0.6 to 0.17	4×10^{-5}
DPHT	5 to 4.2	1200	0.1	8×10^{16}	0.53 to 0.42	10^{-5}
DPOT	5.5 to 3.9	1400	0.12	3×10^{17}	0.48 to 0.33	8×10^{-7}

region follows and at $\mathscr{E} > 8 \times 10^4$ V/cm a strong voltage dependence of current is observed. In this region the CVC follows the dependence $I \sim U^m$ where $m > 2$.

For type I films the index m in this voltage region usually does not change with voltage rise and is approximately constant for each substance (Table 1).

The CVC's of type II DPB films, plotted in a double-logarithmic scale, have on index m which smoothly increases with voltage. A typical CVC is shown in Fig. 1 b.

The temperature dependence of CVC's has also been studied. A family of CVC's of type I DPB films with silver electrodes, measured at different temperatures, is shown in Fig. 2. The slope decreases from 7.2 to 6.5 with increasing temperature, that is, these characteristics are getting close to each other when voltage increases. The CVC's of type II films, measured at different temperatures, show a similar behaviour.

In the nonlinear CVC region the current is a strong function of the film thickness. Fig. 3 shows an example of the current vs. thickness dependence for DPHT films. This dependence follows a $I \sim d^{-9.5}$ dependence. The strong dependence of current on voltage and sample thickness allows us to treat the CVC's of type I films in terms of SCLC with an exponential trap distribution [4 to 7]. If the exponential trap distribution in the forbidden band of the semiconductor has the form $h(E) = \frac{H_0}{kT_c} \exp\left(-\frac{E}{kT_c}\right)$, the CVC of the SCLC

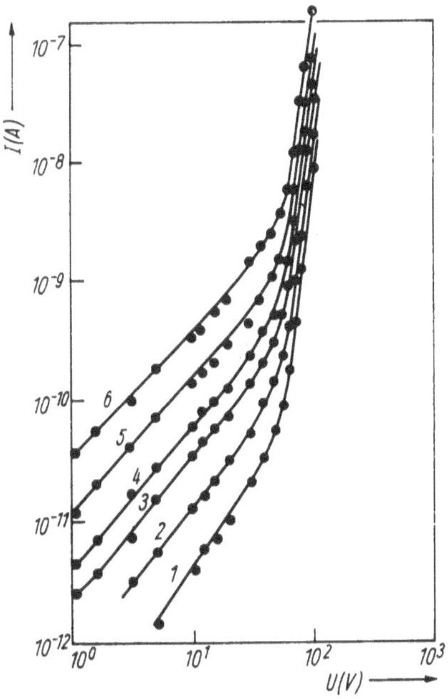

Fig. 2. Family of static CVC's of type I DPB films for different temperatures. The film thickness is $d = 5$ μm. The curves 1 to 6 refer to temperatures of 295, 306, 318, 324, 335, and 348 °K, respectively

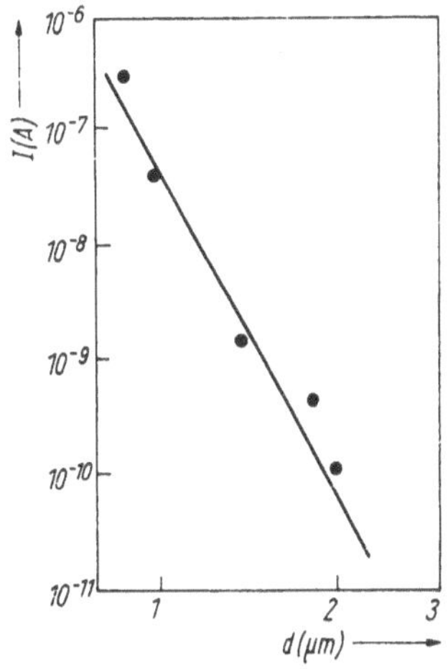

Fig. 3. The current vs. film thickness dependence

is as follows (5):

$$j = \mu N_0 e_0^{1-l} \left(\frac{\varepsilon \varepsilon_0 l}{H_0 (l + 1)}\right) \left(\frac{2l + 1}{l + 1}\right) \frac{U^{l+1}}{d^{2l+1}}, \tag{1}$$

where j is the current density, N_0 the density of states in the corresponding band, μ the mobility of current carriers, e_0 the elementary charge, T_c the characteristic temperature determining the steepness of the exponential distribution ($T_c > T$), $l = T_c/T$ (T is the absolute temperature), H_0 is the total trap density, ε the dielectric constant of the substance, $\varepsilon_0 = 8.85 \times 10^{-12}$ F/m, d is the sample thickness, and U the applied voltage.

The form of the CVC of the SCLC is defined by the type of trap distribution near the quasi Fermi level E_f since the injected carriers are captured by traps located near E_f. The current magnitude is defined by the concentration of carriers that are in thermodynamical equilibrium with traps. This carrier concentrations is $n \sim \exp(-E_f/kT)$. Hence, the temperature dependence of SCLC allows to determine the position of the quasi-Fermi level and its voltage dependence. The limits of change with voltage for the studied substances are given in Table 1.

The theoretical voltage dependence of the position of the quasi-Fermi level in the case of SCLC in a dielectric with an exponential trap distribution has the form [5]

$$E_f = kT_c \ln \left[\frac{(l + 1)^2 e_0 H_0 d^2}{l (2l + 1) \varepsilon \varepsilon_0 U}\right]. \tag{2}$$

Fig. 4 gives a comparison of the theoretical dependence (2) and the experimental points for DPB films of 5 μm thickness. The theoretical solid curve is calculated for $U = 100$ V and an average value of $H_0 = 5 \times 10^{16}$ cm^{-3}. The close agreement between the theoretical dependence and the experimental points confirms the injection mechanism of the current in the samples studied and allows one to determine from (2) the total trap concentration H_0 in the forbidden band. By making use of (1), we estimated the carrier mobility in the films. All the calculated parameters are given in Table 1. For calculations $\varepsilon = 3$, $N_0 = 4 \times 10^{21}$ cm^{-3} were taken.

The experimental CVC's of type II DPB films in lg (I/U) vs. U coordinates in the nonlinear region are straight lines (Fig. 5). Such dependence is charac-

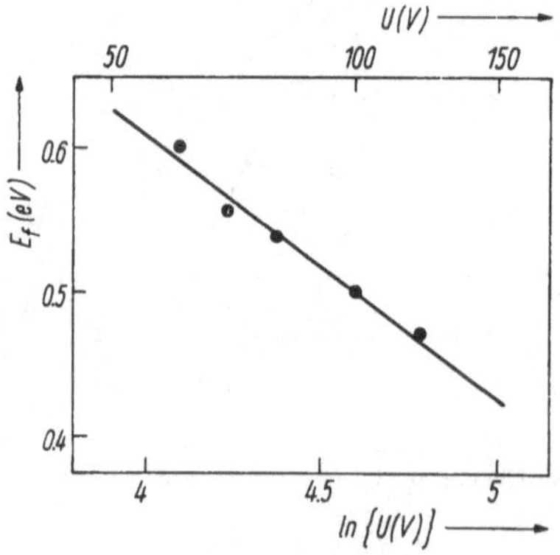

Fig. 4. The dependence of quasi-Fermi level position on the applied voltage (d = 5 μm)

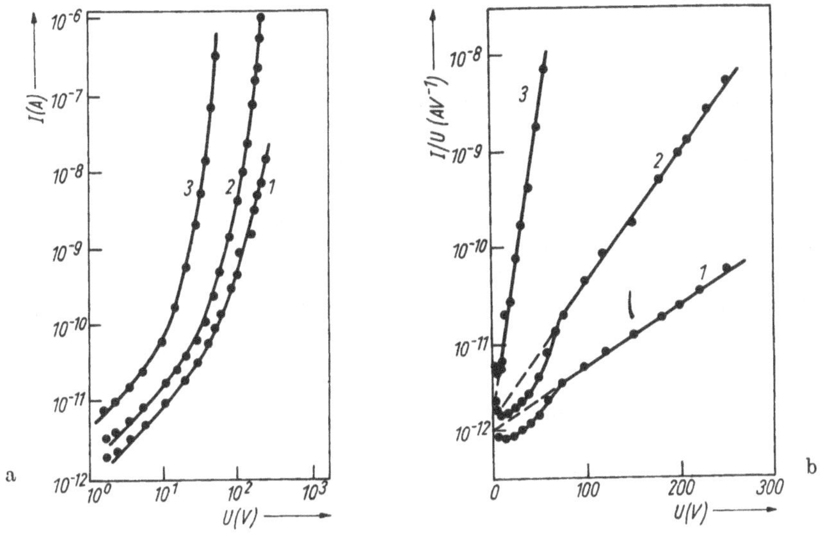

Fig. 5. Dark static CVC's of type II DPB films in a) a double-logarithmic scale and b) in lg (I/U), U coordinates. The curves 1 to 3 refer to thicknesses d = 5, 3, 1 μm, respectively

teristic of SCLC in substances with a uniform trap distribution in the forbidden band. The expression for the CVC of the SCLC in this case is [4, 8]

$$I \sim U \exp(\alpha U)\,, \tag{3}$$

where

$$\alpha = \frac{\varepsilon\,\varepsilon_0}{e_0\, d^2\, N_t^*\, kT}\,. \tag{4}$$

Here N_t^* is the trap density per unit energy.

The approximation of the experimental CVC by the expression (3) shows that within the shift limits of the quasi-Fermi level E_f ($\Delta E_f = 0.1$ to 0.14 eV) the quasi-continuous trap distribution in the forbidden band is nearly uniform. The trap concentration near the quasi-Fermi level calculated from the experimental curves is $N_t^* = (3 \times 10^{16}$ to $2 \times 10^{17})$ cm^{-3} eV^{-1}, which is of the same order of magnitude as the trap concentration in the DPB films of type I.

Thus, when the degree of the film disorder increases, the energy distribution of trap changes, their concentrations are approximately similar. It is plausible to assume the capture levels to be connected with the inner disorder of the substance. The capture centres are regions of disorder with more dense structure [9], which include some tens of molecules. The carrier capture in the centre of such a dislocation will prevent the subsequent capture of a carrier by a molecule from the same region because of the Coulomb repulsive forces [10].

Let us note that formulas (1) and (3) correspond only to the limiting cases of a quasi-continuous trap distribution, and their real distribution can only be more or less close to one of these cases. The interpretation of SCLC is made in terms of these two approximations because of the comparatively simple mathematical description of these models, for which analytical expressions of CVC are obtained.

For traps of structural origin resulting from the change of polarization energy, a Gaussian distribution [9] is more real, which in reasonable limits is close to the exponential one.

With increase of the organic film disorder the polarization energy straggling of the current carriers changes, which results in a distinct variation of the trap distribution, the latter in a small energy interval can be taken as a uniform. Sometimes in the case of type I films it appears possible to approximate the CVC by expression (3), the trap concentrations obtained from both approximations being of the same order.

The above results allow one to assume that the real quasi-continuous trap distribution in the polycrystalline films of the studied substances is close to an exponential one. The change of film structure gives rise to a change of the trap spectrum. No regularities have been found, however, in the change of trap distribution at passing from one substance to the other in the studied sequence of the related compounds. For this reason the traps in the studied substances are supposed to be of structural nature. At passing from DPB to DPOT the tendency of lowering the current carrier mobility in these substances can be observed.

References

[1] W. HELFRICH, in: Physics and Chemistry of the Organic Solid State, Vol. 3, Wiley, New York 1967.
[2] W. DRETH and E. H. WIEBENGA, Recueil Trav. chim. Pays-Bas **72**, 39 (1953).
[3] W. DRETH and E. H. WIEBENGA, Acta cryst. **8**, 755 (1955).
[4] A. ROSE, Phys. Rev. **97**, 1538 (1955).
[5] P. MARK and W. HELFRICH, J. appl. Phys. **33**, 205 (1962).
[6] A. SUSSMAN, J. appl. Phys. **38**, 2738 (1967).
[7] E. A. SILINSH and L. F. TAURE, phys. stat. sol. **32**, 847 (1969).
[8] J. L. HARTKE, Phys. Rev. **125**, 1177 (1962).
[9] E. A. SILINSH, phys. stat. sol. (a) **3**, 817 (1970).
[10] J. SWORAKOWSKI, Mol. Cryst. and liquid Cryst. **11**, 1 (1970).

(Received February 7, 1972)

See Short Note by U.A. DURASOVA et al.

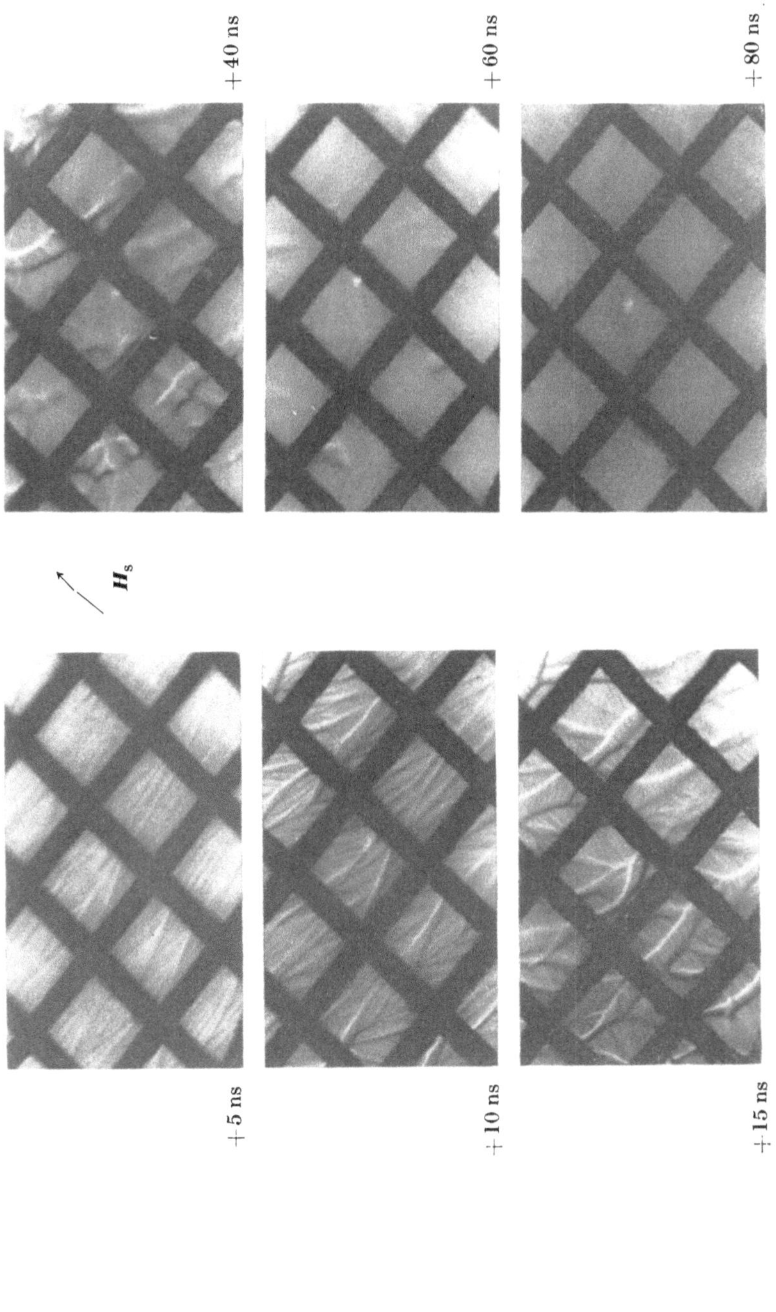

Fig. 1. Microphotographs of the dynamic domain structure. The numbers indicate the time in nanoseconds at which the microphotograph was taken, measured from the initial application of the pulse field. The period of the grid is 30 μm

Short Notes

phys. stat. sol. (a) 10, K101 (1972)

Subject classification: 18.2.1; 21.1.1

Faculty of Physics, Moscow State University

Dynamic Domains Arising during Incoherent Rotation in Thin Magnetic Films

By

U.A. DURASOVA, I.S. KOLOTOV, O.S. KOLOTOV, and R.V. TELESNIN

The aim of the present note is to discuss the dynamic domains which arise during incoherent rotation taking place when a pulse magnetic field is applied at an angle to the easy axis of the thin magnetic film. The dynamic domain structures arising during incoherent rotation were first investigated by Kryder and Humphrey (1 to 3). They used a 10 ns exposure time Kerr magneto-optic camera with a resolving power of about 10 μm. Here we used the method of stroboscopic Lorentz microscopy (4). In apparatus with effective exposure time of about 2 ns has been developed (5, 6). This apparatus allowed the observation of some new peculiarities of the dynamic domains.

A typical example of incoherent rotation is seen in the sequence of microphotographs (see Fig. 1) which were obtained for a film having the following parameters: dimensions 2x3 mm^2, thickness 290 Å, coercive force 1.8 Oe, anisotropy field 3.5 Oe, angular dispersion $\alpha_{90} = 1.8^{\circ}$. The film was obtained by thermal evaporation of a 83Ni-17Fe permalloy.

The switching field H_s = 7 Oe was applied at an angle of 7° to the easy axis of the film. This angle was greater than the angular dispersion. Therefore in the initial state all local magnetization vectors were oriented to one side relative to the direction of the applied field.

It can be seen from the 5 and 10 ns microphotographs that the switching process starts with the rotation of the local magnetization. This rotation is not coherent and results in a distortion of the magnetization ripple and its transformation into dynamic domains. The magnetization configuration taking place at times $t \lessapprox 5$ ns can still be presented as a longitudinal magnetization ripple. But the next microphotograph taken at t = 10 ns shows that the appearance of a transverse component of the magnetization ripples takes place. The appearance of the transverse components results in large stray fields, which retard the rotation. Then domain walls appear separating the regions of the film with the sharpest difference in the magnetization direction. The configuration of the domain walls and their positions are not strongly fixed. However, it can be also seen that further switching of the film is not completed by any oriented moving of the domain walls as it was previously supposed by Stein (7).

It can also be seen that in contrast to Stein's suggestion (7) and Kryder and Humphrey's results (1 to 3), the dynamic domains have a very complicated form and are not similar to the stripe domains whereas the stripe-like domains indeed form during the pulse switching of the same film along its easy axis (8).

Therefore the difference in the dynamic domains arising during the switching of the film along its easy axis or at an angle to the easy axis is due to the difference in the behaviour of the magnetization ripples. When the field is applied along the easy axis - the case of bidirectional incoherent rotation - the formation of the dynamic domain structure can be regarded as a simple development of the longitudinal magnetization ripple. When the field is applied at an angle to the easy axis the transverse components of the magnetization ripples, whose relaxation rate is probably less than the rate of magnetization rotation, have an important influence on the formation of the dynamic domains. To a first approximation, the dynamic domains arising in this case can be represented by a superposition of the longitudinal and transverse magnetization ripples.

It can also be seen that after the domains have formed a rotation of the local magnetization and gradual contraction of the domains with delayed magnetization occurs. The magnetization rotation on this stage was earlier observed by Kryder and Humphrey (2, 3) as change of the picture contrast of different regions of the investigated film. The 15 and 40 ns microphotographs show that similar to the case of bidirectional incoherent rotation (8), a destruction of domain walls takes place in this stage. Delayed domains eventuelly assume the shape of isolated islets which gradually disappear with time.

References

(1) M.H. KRYDER and F.B. HUMPHREY, in: Fizika magnitnykh Plenok, Irkutsk 1968 (p. 316).

(2) M.H. KRYDER and F.B. HUMPHREY, J. appl. Phys. 40, 2469 (1969).

(3) M.H. KRYDER and F.B. HUMPHREY, J. appl. Phys. 41, 1130 (1970).

(4) V.I. PETROV, G.V. SPIVAK, and O.P. PAVLUCHENKO, Izv. Akad. Nauk SSSR, Ser. fiz. 31, 362 (1967).

(5) U.A. DURASOVA, I.S. KOLOTOV, and O.S. KOLOTOV, Pribory i Tekhnika Eksperimenta, Nr. 3, 233 (1970).

(6) I.S. KOLOTOV and O.S. KOLOTOV, Pribory i Tekhnika Eksperimenta, Nr. 1, 230 (1971).

(7) K.U. STEIN, Z. angew. Phys. 20, 36 (1965).

(8) U.A. DURASOVA, I.S. KOLOTOV, O.S. KOLOTOV, V.I. PETROV, G.V. SPIVAK, and R.V. TELESNIN, Czech. J. Phys. B21, 451 (1971).

(Received January 24, 1972)

phys. stat. sol. (a) 10, K105 (1972)

Subject classification: 18.2.1

Physics Department, The University, Salford, Lancashire

Electron Diffraction by Single Domain Walls

By

G.F. COX and M.I. DARBY

The defocussed mode of Lorentz electron microscopy has been employed by several workers, e.g. (1, 2), to observe single domain walls in ferromagnetic thin films. The interpretation of the diffraction patterns obtained in this way is made using the wave optical theory. Assuming that the domain wall is straight and infinite in length, the electron diffraction problem reduces to one dimension, and the diffracted wave amplitude $\psi(u)$ in the viewing plane (u, v, Z), distance Z from the film, takes the following form within the Fresnel approximation (small u):

$$\psi(u) = \psi_0 \int_{-\infty}^{\infty} \exp\left\{-i\left[\alpha x^2 + \beta u x + \phi(x)\right]\right\} dx,$$

where $$\alpha = \frac{k}{2}\left\{\frac{1}{L} + \frac{1}{Z}\right\}, \quad \beta = -\frac{k}{Z}, \quad \phi(x) = -\frac{et}{\hbar}\int_0^x B_y(x) dx.$$

Here L distance from source to film, t film thickness, k electron wavenumber, and $B_y(x)$ is the y-component of the magnetic induction.

By choosing a simple analytical form for the variation of $B_y(x)$ through a wall to fit the observed diffraction intensities, information has been obtained about the wall structure, particularly about wall widths (2). It would be supposed that the amplitude is more sensitive than the intensity to the phase function ϕ, and here this is shown explicitly to be the case. Further, if the complex amplitude is known, it is shown that the phase function may be obtained directly from it by Fourier inversion.

The practical problem of obtaining the complex amplitude from experimental diffraction measurements is difficult, and it has not yet been solved. However it seems worthwhile asking what results are to be expected, and in particular how sensitive they are to the form of $\phi(x)$.

In order to compute the amplitude, three analytical forms have been considered for the variation of $B_y(x)$ through a convergent wall. These are:

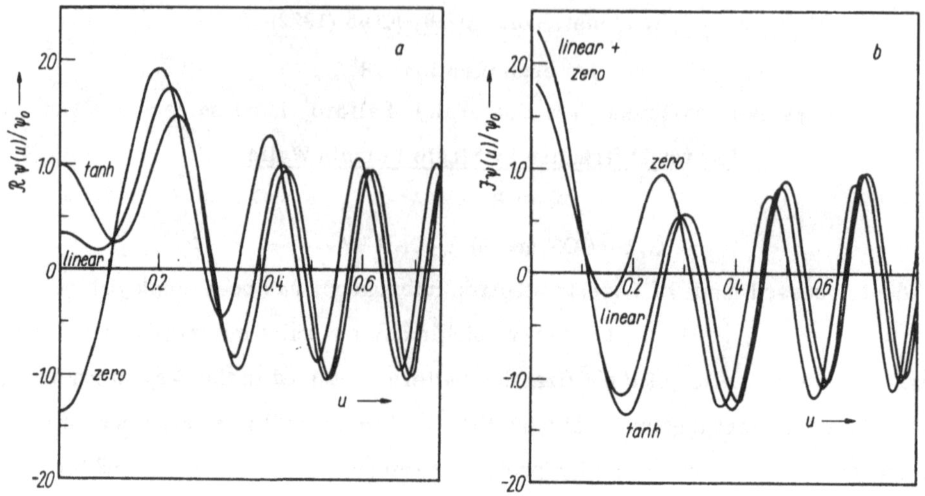

Fig. 1. a) The real (ℜ) and b) the imaginary (ℑ) parts of the amplitude function $\psi(u)$ for a single convergent domain wall, obtained using the models discussed in the text. The units of u are 10^{-7} m. The parameters employed were $L = 0.102$ m, $Z = 4\times10^{-3}$ m, $\delta = 10^{-8}$ m, $2\pi/k = 3.8\times10^{-12}$ m

(i) zero wall; $B_y(x) = B_o (x > 0)$; $B_y(x) = - B_o (x < 0)$,

(ii) linear wall; $B_y(x) = - B_o (x < \delta/2)$; $B_y(x) = 2 B_o x/\delta$

$(-\delta/2 \leqq x \leqq \delta/2)$; $B_y(x) = B_o (x > \delta/2)$,

(iii) tanh wall; $B_y(x) = B_o \tanh(2x/\delta)$,

where 2δ is the wall width. It is assumed for simplicity that $|x|$ extends to infinity, although in practice the maximum value of $|x|$ will be limited. However the basic conclusions should not be affected. For the zero and linear walls, the amplitude functions are obtained in terms of Fresnel integrals, and these have been evaluated using series expressions(3). In the third case it was found necessary in calculating $\psi(u)$ to use numerical integration over part of the region of x. Typical results for the amplitudes are shown in Fig. 1, and it can be observed that the most pronounced differences between the models occur near the centre of the intensity distribution. In particular it is found that in this region the sign of the real part of ψ is different for the zero wall and the tanh wall.

The phase function Φ can be obtained directly from the Fourier transform $\overline{\psi}(q)$

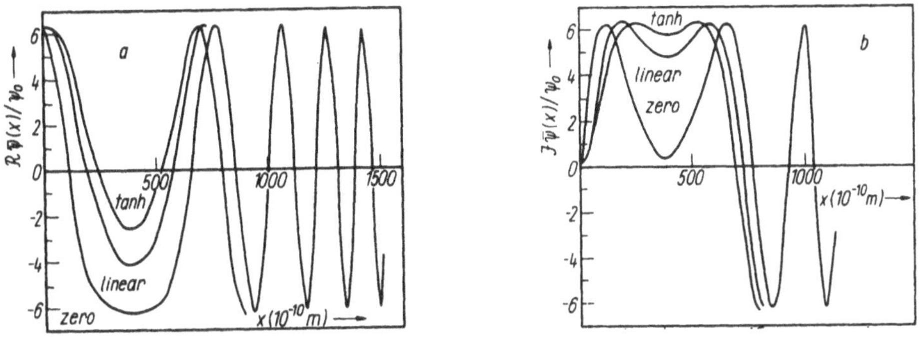

Fig. 2. a) The real and b) the imaginary parts of the Fourier transforms $\overline{\psi}(x)$ of the amplitudes shown in Fig. 1; $x = q/\beta$

of $\psi(u)$. It is readily shown that

$$\overline{\psi}(q) = \int_{-\infty}^{\infty} \psi(u) e^{iqu} du = 2\pi \psi_0 \exp \left\{ -i \left[\alpha q^2 / \beta^2 + \phi(q/\beta) \right] \right\} ,$$

where $q = \beta x$. Typical curves for the real ($\mathfrak{R}$) and imaginary ($\mathfrak{I}$) parts of $\overline{\psi}(q)$ are shown in Fig. 2, and it can be seen that the most marked differences between the models occur near the origin. The phase variation through a domain wall is given by

$$\phi(x) = \tan^{-1} \left\{ - \frac{\mathfrak{I}\overline{\psi}(x)}{\mathfrak{R}\overline{\psi}(x)} \right\} - \alpha x^2 .$$

Hence if $\psi(u)$ can be obtained from observed diffraction patterns, its Fourier inversion may be achieved, see for example (4), and $\phi(x)$ determined. The present work indicates the form of the amplitude and transformed functions to be expected.

References

(1) D. WOHLLEBEN, J. appl. Phys. 38, 3341 (1967).

(2) D.C. HOTHERSALL, Phil. Mag. 20, 89 (1969).

(3) J. BOERSMA, Math. Comput. 14, 380 (1960).

(4) J.W. COOLEY and J.W. TUKEY, Math. Comput. 19, 297 (1965).

(Received January 13, 1972)

phys. stat. sol. 10, K109 (1072)

Subject classification: 9; 14.3.3

Odessa State University

Two-Step Diffusion Process for EPIC Technology

By

V.K. BASHENOV, S.A. GARYAINOV, A.K. MEDNIKOV, and V.I. SOLOSHENKO

In semiconductor device technology there is a well-known EPIC technology decreasing considerably the active areas of the semiconductor devices (1). Modern device fabrication techniques have been directed toward a two-step diffusion process involving a deposition step and a redistribution step. Slices of semiconductor materials are first exposed to a constant impurity atom density at high temperature. Next, the impurity atom source is removed, and diffusion is continued under the presumed condition that additional impurity atoms can neither enter nor leave the semiconductor material. The two-step impurity diffusion into a semi-infinite crystal has already been treated by Kennedy and Murley (2). In this note we consider the two-step impurity diffusion into a finite crystal; we deal here with such a crystal in EPIC technology.

The impurity atom distribution in a finite crystal is calculated by solving the one-dimensional diffusion equation subject to a particular set of initial and boundary conditions. Our finite crystal is presumed to have the completely impermeable surface at the "pocket" depth ($x = b$). The first part of the diffusion problem ($0 < t < t_1$) is approximated by a semiconductor material of finite geometry ($0 \leqq x \leqq b$), that is initially free of impurity atoms of the particular type under consideration. This type of impurity is presumed to have a diffusion constant D_1 during the deposition step, and also a constant concentration N_o upon the external surface ($x = 0$). The flux normal to the completely impermeable surface must be zero; this boundary condition implies a zero impurity atom density gradient at the internal surface ($x = b$).

The Green 's function solution for this deposition step is given by (3)

$$G_1(x,t\,;x',t') = \frac{2}{b}\sum_{n=0}\sin(s_n\pi\frac{x}{b})\,\sin(s_n\pi\frac{x'}{b})\;\exp\left[-s_n^2\,\pi^2 D_1(t-t')b^{-2}\right] \tag{1}$$

and the distribution resulting from this diffusion process is

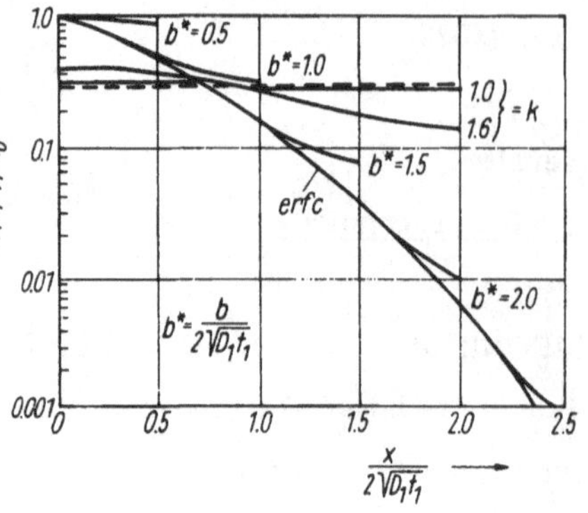

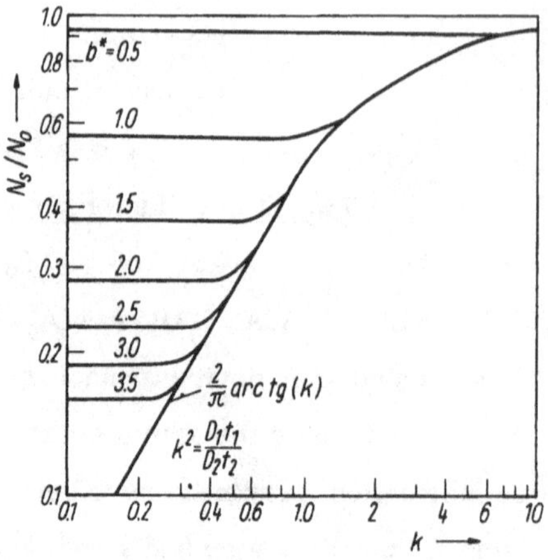

Fig. 1 Fig. 2

Fig. 1. Impurity atom distribution after a deposition step as the "pocket" depth b^* is varied. The impurity profile after the two-step diffusion with k = 1.6 and 1.0 for $b^* = 2$ is also indicated

Fig. 2. Impurity atom concentration upon the semiconductor surface after a two-step diffusion process into the finite crystals with a completely impermeable internal surface

$$N_1(x,t_1) = D_1 N_o \int_0^{t_1} \left(\frac{dG_1}{dx'}\right)_{x'=0} dt'$$

$$= N_o \left\{1 - \frac{2}{\pi} \sum_{n=0} \frac{1}{s_n} \sin\left(s_n \pi \frac{x}{b}\right) \exp\left[-s_n^2 \pi^2 D_1 t_1 b^{-2}\right]\right\}, \qquad (2)$$

where

$$s_n = n + \frac{1}{2}. \qquad (3)$$

Numerical methods have been used to calculate equation (2) throughout a range of values most frequently encounterned in the fabrication of semiconductor devices. Fig. 1 presents the results of this calculation in a normalized form. As it is seen from Fig. 1 for

$$b > 2\sqrt{D_1 t_1}$$

the impurity profile is nearly the well-known profile for a semi-infinite crystal (or a completely penetrating surface of finite crystal), the erfc-distribution.

The second part of the two-step diffusion process is initiated at $t = t_1$ from an impurity distribution given by equation (2). This second diffusion is not necessarily conducted at the same temperature as the first diffusion; we must therefore introduce a new diffusion constant D_2. Because impurity atoms can neither enter nor leave the crystal during this second diffusion ($t_1 < t < t_2$), the flux normal to the external surface must also be zero. Thus, the Green's function solution for this redistribution step is given by (reference (3))

$$G_2(x,t\,;\,x',t') = \frac{1}{b}\left\{1 + 2\sum_{m=1} \cos(m\pi \frac{x}{b}) \cos(m\pi \frac{x'}{b}) \exp\left[-m^2 \pi^2 D_2 (t-t')b^{-2}\right]\right\} \quad (4)$$

and the final impurity profile is

$$N_2(x,t_2\,;\,t_1) = \int_0^b G_2 N_1(x', t_1)dx'$$

$$= N_o\left\{1 - \frac{2}{\pi^2}\sum_{n=0}\left(\frac{1}{s_n}\right)^2 \exp(-s_n^2 \pi^2 D_1 t_1 b^{-2}) \times\right.$$

$$\left.\times\left[1 + 2\sum_{m=1} \frac{1}{1-(m/s_n)^2} \cos(m\pi \frac{x}{b}) \exp(-m^2 \pi^2 D_2 t_2 b^{-2})\right]\right\} \quad (5)$$

We introduce the following abbreviations:

$$b^* = b/2\sqrt{D_1 t_1} \text{ and } k = \sqrt{D_1 t_1/D_2 t_2}\,. \quad (6)$$

The impurity profile after the two-step diffusion into the finite crystal with $b^* = 2$ is illustrated in Fig. 1. The limit of equation (5) as k approaches zero is the constant impurity profile.

The surface concentration N_s versus k for various "pocket" depths is compared in Fig. 2 with the corresponding dependence for the two-step diffusion into the semi-infinite crystal which is given by (reference(2))

$$N_s = \frac{2N_o}{\pi} \operatorname{arctg}(k). \quad (7)$$

As it is seen from Fig. 2 for

$$k \leqq \operatorname{tg}(\sqrt{\pi}/2b^*) \quad (8)$$

the surface concentration N_s in a finite crystal with internal impermeable surface does not depend on k and is given by

$$N_s = \frac{N_o}{\sqrt{\pi}\, b^*} \tag{9}$$

for all values of $b^* \geqq 1$. Besides, the impurity distribution is homogeneous, and the impurity concentration in any point is equal to N_s; this result is extremely important for the choice of the diffusion regime when fabricating semiconductor devices with the help of EPIC technology.

In summary it is important to note that the case with poorly penetrating surface may be analyzed with the data of Fig. 2 if instead the actual "pocket" depth we should use the effective depth $b_{eff} > b$.

References

(1) R.M. WARNER and J.N. FORDEMWALT, Eds., Integrated Circuits: Design Principles and Fabrication, McGraw Hill Publ. Co., New York 1965.

(2) D.P. KENNEDY and P.C. MURLEY, Proc. IEE 52, 620 (1964).

(3) H.S. CARLSLAW and J.C. JAEGER, Conduction of Heat in Solids, Oxford University Press, London 1959.

(Received January 24, 1972)

phys. stat. sol. (a) 10, K113 (1972)
Subject classification: 10.1; 21.1

Institute of Solid State Physics,
Academy of Sciences of the USSR, Chernogolovka

Mechanism of the Low Temperature Plastic Deformation in Zinc Single Crystals

By
Ya.M. SOIFER and V.G. SHTEINBERG

As previously reported (1), the microscopic plastic behaviour of zinc single crystals for basal and pyramidal slip is controlled by the Peierls mechanism. The corresponding activation parameters obtained are given in Table 1.

The aim of this paper is to investigate the significance of the Peierls mechanism in the macroflow behaviour of zinc. The temperature dependence of the critical resolved shear stress (CRSS) for zinc single crystals deformed by basal and pyramidal glide was studied. The specimens of required orientation were tested in compression with the strain rates $5x10^{-5}$ and $2x10^{-4}$ s^{-1} in the temperature range 4.2 to 300 ^oK.

The results of these measurements are shown in Fig. 1. As it is seen the temperature dependence of CRSS for pyramidal slip is much stronger than for basal slip.

Using the well known relation

$$\dot{\varepsilon} = bnv , \qquad (1)$$

where $\dot{\varepsilon}$ is the strain rate, b is the length of Burgers vector, n is the mobile dislocation density, v is the dislocation velocity, and taking the data of Table 1 to determine v, one can estimate the mobile dislocation density which corresponds to the CRSS at a given temperature.

Table 1

slip system	kink energy U_k(eV)	activation area A^* (b^2)
basal (0001) $\langle 11\bar{2}0 \rangle$	0.035	40
pyramidal $\{11\bar{2}2\}\langle 11\bar{2}3 \rangle$	0.16	30

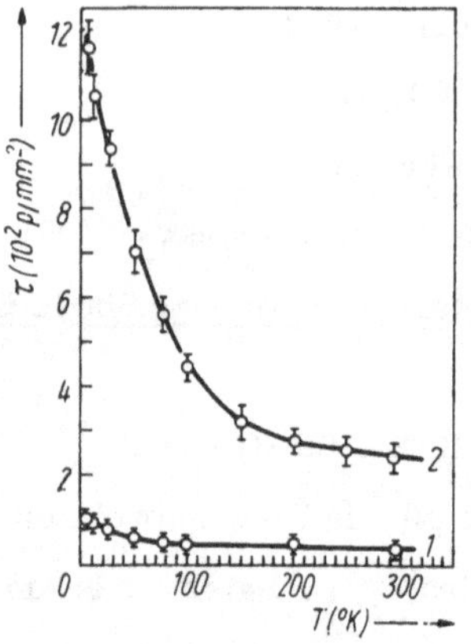

Fig. 1

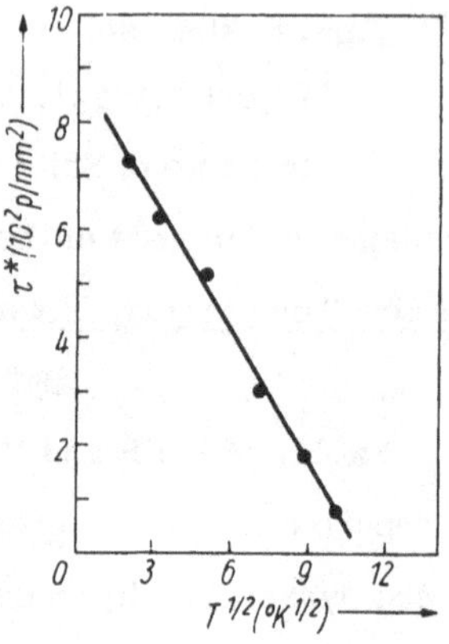

Fig. 2

Fig. 1. The critical resolved shear stress (flow stress) vs. temperature for zinc single crystals:(1) basal slip (0001) $\langle 11\bar{2}0\rangle$, $\dot{\varepsilon} = 5\times10^{-5}\ s^{-1}$; (2) pyramidal slip $\{11\bar{2}2\}\ \langle 11\bar{2}3\rangle$, $\dot{\varepsilon} = 2\times10^{-4}\ s^{-1}$

Fig. 2. Thermal flow stress vs. $T^{1/2}$ for pyramidal slip

Such estimates for pyramidal glide result in $n \approx 10^5\ cm^{-2}$, which coincides with the value obtained by etch pit measurements of the pyramidal dislocation density at the yield point.

According to Guyot and Dorn (2) the variation of the flow stress with temperature for the Peierls mechanism can be written as

$$\tau^* = \tau_p \left[1 - \left(\frac{T}{T_c}\right)^{1/2}\right], \tag{2}$$

where τ^* is the thermal flow stress, τ_p is the Peierls stress, and the critical temperature T_c is given by

$$U_k = \frac{kT_c}{2} \ln \frac{nZab^2\nu}{2w^2\dot{\varepsilon}}, \tag{3}$$

where U_k is the kink energy, $Z = n^{-1/2}$, ν is the Debye frequency, w is the width of a critical pair of kinks.

To compare the present results with equation (2) the values of τ^* for pyramidal slip have been plotted against $T^{1/2}$ and as seen from Fig. 2 the points are well fitted by a straight line. Using $T_c \approx 120\ ^oK$ obtained from Fig. 2 the kink energy can be estimated. Taking reasonable values for n, Z, w, we get U_k = 0.12 to 0.15 eV, that coincides with the value of U_k determined from microstrain measurements (Table 1).

The above comparison permits the conclusion that for the pyramidal glide the Peierls mechanism is the most probable mechanism of low temperature deformation.

For basal slip the value of n estimated from equation (1) is equal to $\approx 10^2\ cm^{-2}$, i.e., at least by a factor 10^4 smaller than the basal dislocation density usually observed. This discrepancy and the weak temperature dependence of the CRSS leads to the conclusion that the macroscopic plastic behaviour for basal slip is not determined by the Peierls mechanism.

References

(1) Ya.M. SOIFER and V.G. SHTEINBERG, phys. stat. sol. (a) 6, 409 (1971).

(2) P. GUYOT and I.E. DORN, Canad. J. Phys. 45, 983 (1967).

(Received January 26, 1972)

phys. stat. sol. (a) 10, K117 (1972)
Subject classification: 5; 21.1.1

Department of Physics, Banaras Hindu University, Varanasi

Quadrupole Splitting in Deformed Stainless Steel

By

B.P. SRIVASTAVA, H.N.K. SARMA, and D.L. BHATTACHARYA

A typical Mössbauer spectrum, using stainless steel as an absorber, shows a single line structure, which means that the charge due to electrons around the Fe nucleus in stainless steel is spherically symmetric.

A broadening of the Mössbauer line has been observed by Lagunov et al. (1) in a deformed copper which contained Co^{57} as impurity. They attributed this to the quadrupole splitting of the nuclear levels of the Mössbauer atoms near the lattice defects. Splitting of the line was not, however, observed which may be due to the small concentration and random distribution of the Mössbauer atoms in the host lattice (copper).

In the present work, we have performed the Mössbauer experiment with a deformed (bent along a line) stainless steel (type 310) foil as an absorber and Co^{57} in Pd matrix as the source. The observation of Mössbauer spectrum was first made with an ordinary stainless steel 310 as shown in Fig. 1. The same foil was then folded along a line and straightened to its original shape thereby producing a line defect in the foil. The γ-rays were collimated to straddle this line of deformation. The Mössbauer spectrum of the deformed foil was found typically of the form given in Fig. 2. The isomer shift is found to remain unchanged at (0.30 ± 0.02)mm/s for both the observations of annealed and deformed specimens. As may be seen in Fig. 2, the line is split with peak positions at (0.20 ± 0.02)mm/s and (0.40 ± 0.02)mm/s. The line is also broadened. This observation shows evidence of a characteristic quadrupole splitting. The splitting can only mean that along the deformation line iron atoms no longer remain in their original charge environment. Preston and Gerlach (2) have predicted that if the iron atoms produce an asymmetric distortion of the lattice, then the Mössbauer line will be broadened. This broadening may be observed in a case, where these atoms produce an asymmetric distortion of the lattice, locally. In the present work the atoms along the bent line are dislocated, thereby producing an asymmetric distortion of the lattice along that line. Thus the original cubic crystal

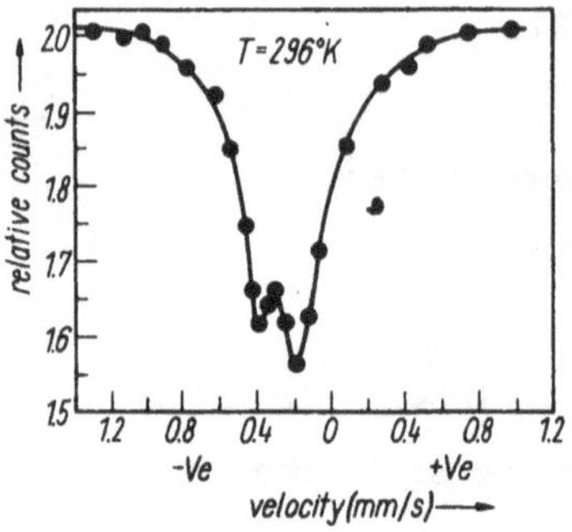

Fig. 1

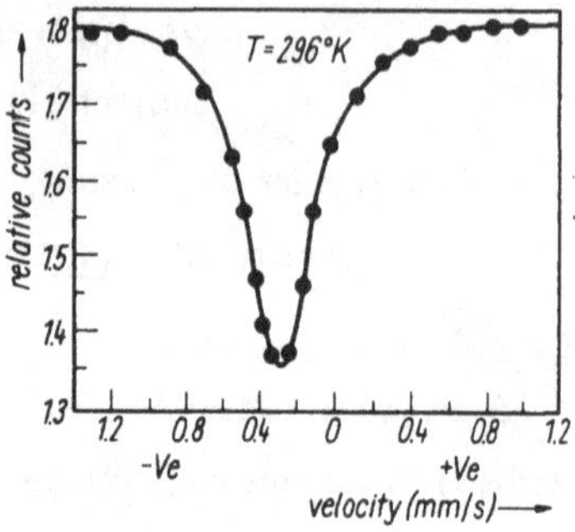

Fig. 2

Fig. 1. Mössbauer spectrum of stainless steel (310)

Fig. 2. Mössbauer spectrum of deformed (bent along a line) stainless steel (310)

symmetry is now destroyed along the line which destroys the spherical nature of the charge distribution at the nuclei of the atoms located just near the line. Thus instead of observing a single Mössbauer line, which shows typically a spherical charge distribution at the nucleus of the Mössbauer atom, we are observing a double line, which is an evidence of an asymmetric charge distribution along the line. The two peaks are asymmetric (Fig. 2) which confirms the displacement of the iron atoms along the line which gives rise to an anisotropic crystalline field on iron nuclei (Goldanskii-Karyagin effect (3)).

References

(1) V.A. LAGUNOV, V.P. POLOZENKO, and V.A. STEPANOV, Zh. eksper. teor. Fiz. Pisma 12, 136 (1970).

(2) R.S. PRESTON and R. GERLACH, Phys. Rev. B3, 1519 (1971).

(3) V.I. GOLDANSKII, E.E. MAKAEV, and V. KHRAPOV, Phys. Letters (Netherlands) 3, 344 (1963).

(Received February 3, 1972)

phys. stat. sol. (a) 10, K119 (1972)
Subject classification: 9; 22.1.1

Institute of Nuclear Sciences "Boris Kidrič", Beograd

Solubility of Sodium in Germanium

By
V. SPIRIĆ and M. STOJIĆ

From the elements of the group of alkali metals lithium has already been shown to be a fast diffusing interstitial impurity in germanium, behaving as a donor element (1). The properties of the other elements of this group acting as impurities in Ge have not been examined. In this note the solubility of sodium in Ge is given together with the method used for introducing sodium into Ge.

The diffusion of sodium into Ge was performed in an evacuated closed quartz tube, where a germanium sample and a smaller quartz tube with metallic sodium inside were placed. Thus, the introduction of sodium into Ge by means of the diffusion was carried out from a constant vapour source. The crystal faces were protected from the etching by liquid sodium. The quartz tube was heated in a vertical furnace and after the diffusion process the tube was dropped into water. The cooling of the Ge samples to room temperature lasted several seconds. The diffusion was carried out sufficiently long to achieve the saturation of Ge samples by sodium. The etching of quartz tubes by sodium took place during the diffusion.

The solubility of sodium in Ge was determined by means of the radioactive isotope Na^{24} measuring the γ-specific activity of the samples after the diffusion process and estimating the ratio of Na^{23} to Na^{24}. The dependence of solubility vs. temperature was followed up to 800 ^{o}C and the results are shown in Fig. 1.

At the beginning the concentration of sodium increases reaching the maximum of about 10^{18} cm^{-3} at 750 ^{o}C and at higher temperatures decreases. The explanation of this retrograde shape of temperature dependence was given by Reiss et al. on the basis of the theory of electrolytes applied to the impurities in semiconductors (2). The obtained solubility of sodium in Ge is about one order of magnitude smaller than the reported solubility of lithium in Ge (3). That could be explained, if sodium acts as an interstitial impurity in Ge, in view of greater ionic radius of sodium (4).

In order to determine the electrical activity of sodium in Ge, the electrical con-

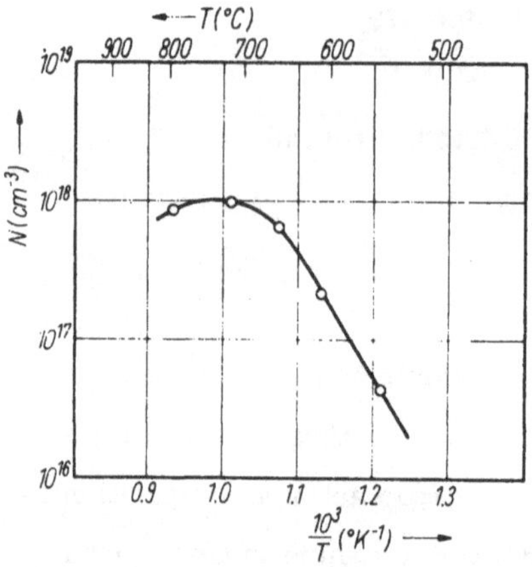

Fig. 1. Solubility of sodium in germanium

ductivity of samples was measured at room temperature before and after diffusion, using the four-point probes method (5). The concentration of induced thermal acceptors from the changes of sample specific conductivity before and after heating without sodium was determinded.

In some samples doped with sodium a pronounced change of the conductivity type from the original p- to n-type was observed. The donor concentration is much lower than the total sodium concentration. This fact could be explained by the precipitation of sodium during the cooling of the samples as in the case of sodium in Si (6).

However, in several samples there was no change of the conductivity type p- to n-type, although the same experimental conditions and materials were applied, except the quartz in which the diffusion was performed out. Thus, at present a reliable conclusion about the electrical activity of sodium atoms in Ge cannot be done.

References

(1) C.S. FULLER and J.C. SEVERIENS, Phys. Rev. 95, 21 (1954).

(2) H. REISS, C.S. FULLER, and F.J. MORIN, Bell Syst. tech. J. 35, 535 (1956).

(3) E.M. PELL, J. Phys. Chem. Solids 3, 74 (1957).

(4) K. WEISER, Phys. Rev. 126, 1427 (1962).

(5) F.M. SMITS, Bell Syst. tech. J. 37, 711 (1958).

(6) L. ŠVOB, Solid State Electronics 10, 991 (1967).

(Received February 7, 1972)

phys. stat. sol. (a) 10, K121 (1972)

Subject classification: 5; 21.1.1

Laboratorio MASPEC del CNR, Parma (a),
and Istituto Fisica, Gruppo Nazionale Struttura della Materia, Università Parma(b)

A Study on Some Composition of the Fe-Te System by Mössbauer Spectroscopy

By

V. FANO (a) and I. ORTALLI (b)

Iron and tellurium form compounds with different phases depending on the composition and the temperature. In (1 to 3) are reported some state diagrams which do not always coincide in some composition ranges. Recently, the study of the nature of some of these phases has also been carried out with Mössbauer spectroscopy (4 to 6). In (5) and (6), for compositions of $FeTe_{0.90}$ (tetragonal structure PbO type) and $FeTe_2$ (orthorhombic, marcassite type of structure), respectively, the appearance of two lines has been attributed to a quadrupole splitting, differently to what has been reported in (4) for Fe_2Te_3 compositions, (assuming a structure like NiAs) where only a peak has been found.

In this note we have attempted a Mössbauer analysis of stoichiometric compositions (Fe_2Te_3) equal to, or compositions, ($FeTe_{0.95}$ and $FeTe_{1.97}$) nearly equal to those mentioned above: these should all fit into the same region of the phase diagram.

The Mössbauer apparatus used was of electromagnetic type working at constant acceleration; Co^{57} in Pd was used as source and a proportional counter was used as detector. The absorbers were cooled in a simple cryostat system (7) for variable temperatures.

The substances with which the absorbers were prepared by mixing uniformly the specified amounts with a different thermoplastic binder, were synthesized, as in (8), and analyzed at room temperature and at 110 ^{o}K (Fig. 1). The spectra at room temperature (Fig. 1a, b, c) show the existence of an asymmetric doublet, which can be attributed with difficulty to the existence of a one-quadrupole splitting. In fact, the spectra can be resolved by a fit of four Lorentzians of FWHM 0.42 mm/s, attributed to a double quadrupole splitting with variable intensity. In effect, the existence of two possible types of iron can also be seen from the spectra reported in (5). The

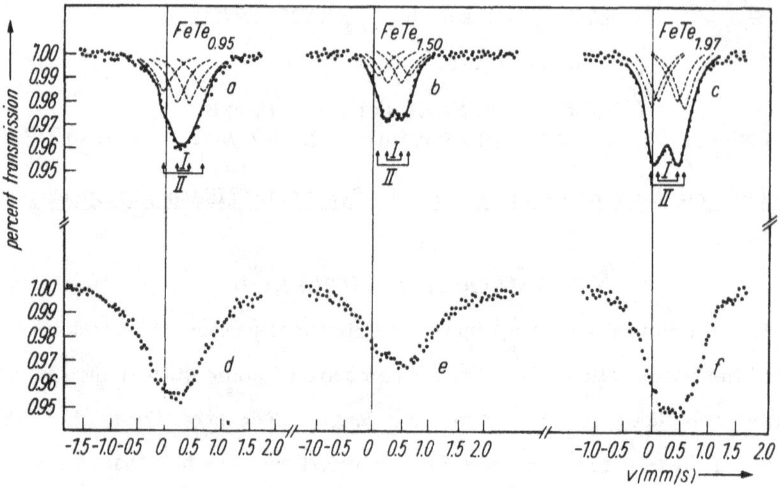

Fig. 1. Transmission Mössbauer spectra: a), b), c) room temperature; d), e), f) 110 °K. The solid line indicates the fit by a sum of four Lorentzian curves

isomeric shift for both quadrupole splittings may be considered equal (Table 1). This value is compatible with a state of covalent bonding with hybridization state of $3d^2 4s4p^3$ type. In fact, in the NiAs structure and in its derivatives, both orthorhombic and marcassite, iron is always in a position with similar boundings, in an octahedral symmetry bounding field with a different degree of distortion. This can ex-

Table 1

Mössbauer isomer shift (δ) relative to Co^{57} in Pd source and quadrupole splittings (ΔE) at room temperature

		δ (mm/s)	ΔE(mm/s)
$FeTe_{0.95}$	I	+0.29 ± 0.04	0.33 ± 0.06
	II	+0.29 ± 0.04	0.58 ± 0.06
$FeTe_{1.50}$	I	+0.34 ± 0.04	0.25 ± 0.06
	II	+0.33 ± 0.04	0.54 ± 0.06
$FeTe_{1.97}$	I	+0.31 ± 0.04	0.21 ± 0.06
	II	+0.31 ± 0.04	0.70 ± 0.06

plain the different values of the quadrupole splittings At 110 ^{o}K (Fig. 1d, e, f) the spectra are already different. Therefore, measurements should be made in a wider range of temperatures and compositions, so that we can identify, with certainty, both quadrupole splittings we have seen.

References

(1) G. GRØNVOLD, H. HARALDSEN, and J. VIHOVDE, Acta Chem. Scand. 8, 1927 (1954).

(2) P. LEWELLYN and T. SMITH, Proc. Phys. Soc. 74, 65 (1959).

(3) S. CHIBA, J. Phys. Soc. Japan 10, 837 (1955).

(4) P. MANCA, J.P. SUCHET, and G.A. FATSEAS, Ann. Phys. (France) 1, 621 (1966).

(5) J. SUWALSKI, J. PIEKOSZEWSKI, and J. LECIEJEWICZ, Proc. Conf. Appl. Mössbauer Effect, Tihany 1969.

(6) A.A. TEMPERLY and H.W. LEFÈVRE, J. Phys. Chem. Solids 27, 85 (1966).

(7) I. ORTALLI and R. DEL PRATO, Rapporto Interno Istituto Fisica, Parma 1969.

(8) I. ORTALLI, V. FANO, and P. GIBART, Proc. XVI. Coll. AMPÈRE, Bucarest 1970 (p. 1075).

(Received February 7, 1972)

phys. stat. sol. (a) 10, K125 (1972)

Subject classification: 19; 22.2.2

Sektion Physik der Karl-Marx-Universität Leipzig

Dynamic Nuclear Polarization in GaP

By

K. GAST, R. BÖTTCHER, S. WARTEWIG, and K. WERNER

A report is given on experiments of dynamic nuclear polarization of both phosphor and gallium nuclei of Te-doped GaP crystals at 27 K observed for the first time. There are only a few papers known dealing with EPR investigations of impurities in GaP (1 to 3) that are somehow rather contradictory.

The single crystals studied have been grown from a nonstoichiometric melt doped with Te. At 27 K we found an isotropic EPR line with a g-value of 1.997 ± 0.001 attributed to the donor electrons of the Te impurities.

In order to gain further information about the behaviour of the impurities, three GaP samples of different impurity concentrations were chosen for nuclear polarization measurements.

All three samples showed a dynamically enhanced ^{31}P nuclear polarization by the solid state effect. A sensitive enhancement of the ^{71}Ga nuclear polarization could be found only in sample No. 2. The experimental results are presented in Table 1 and in Fig. 1 and 2.

The distance of the polarization extrema of all samples is almost the same and amounts to about 50 G. The greatest solid state effect occurs at an electron spin concentration of 10^{18} cm^{-3}. The saturation behaviour corresponds to that of an in-

Table 1

sample No.	concentration of unpaired electrons (cm^{-3}) $\pm$ 35 %	ΔH (G)	$E_{^{31}P}$	$E_{^{71}Ga}$
1	$6x10^{17}$	43 ± 4	6 ± 1	-
2	$1x10^{18}$	45 ± 4	18 ± 1	6 ± 1
3	$3x10^{18}$	45 ± 4	4.5 ± 1	-

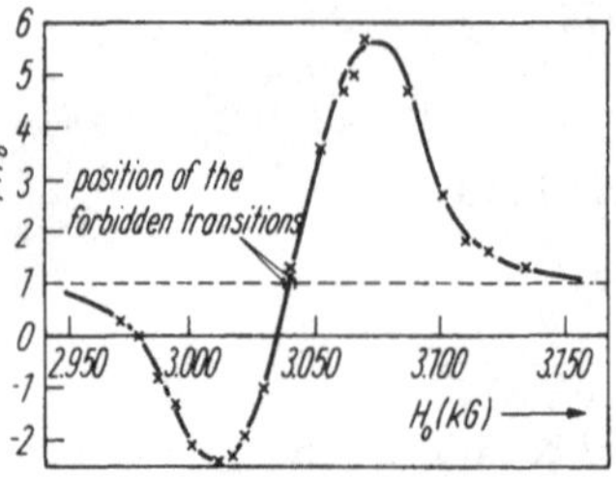

Fig. 1

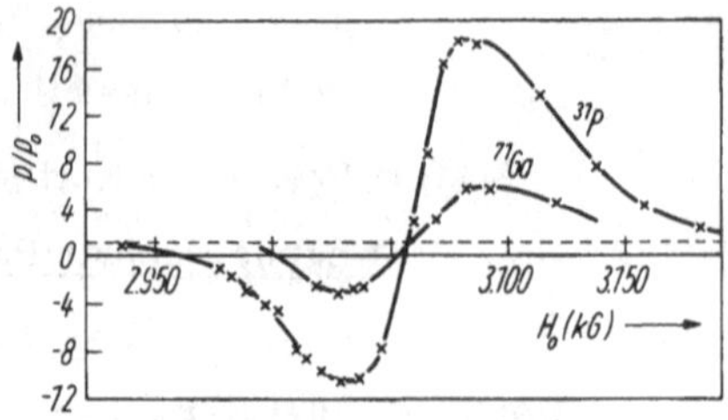

Fig. 2

Fig. 1. ^{31}P nuclear polarization vs. magnetic field (solid state effect); sample No. 1 - position of forbidden transitions with respect to EPR line centre at g-value = 1.997

Fig. 2. ^{31}P and ^{71}Ga nuclear polarization vs. magnetic field (solid state effect); sample No. 2

homogeneously broadened line. The maximum microwave power available of 4 W yielded for all three samples a total saturation of the EPR line; however, the power necessary varied (sample No. 2: 0.1 W, No. 3: 1.5 W). This means that there exists a strong dependence of the spin-lattice relaxation time T_{1e} of the electrons of the different samples.

By observation of the enhanced ^{31}P nuclear polarization decay to the value of thermal equilibrium while switching off the microwave power, it was also possible to determine the nuclear spin-lattice relaxation time for sample No. 2 to T_{1n} = = (45 $\pm$ 5)s. This time is strongly dependent on the impurity concentration.

Due to the possibility of additional quadrupole relaxation (nuclear spin 3/2) the dynamic polarization of ^{71}Ga nuclei is reduced by one third. From the ratio of the enhancement factors for ^{31}P and ^{71}Ga conclusions thus may be drawn as to the contribution of the quadrupole relaxation.

No one of the samples investigated revealed any additional Overhauser effect. Therefore, our measurements illustrate in a significant manner that the sole occurrence of the solid state effect gives evidence of a static dipole-dipole interaction between nuclei and electrons and of a sufficiently strong localization at the temperature and impurity concentrations mentioned. On the other hand the fact that the lineshape of the EPR spectrum is Lorentzian around the centre of the line and Gaussian at the wings ($|H_0 - H| > 40$ G) allows the suggestion that both localized and delocalized unpaired electrons should be existent.

Further detailed investigations are in progress.

The experiments were performed by means of a homemade double resonance spectrometer (3 kG; 8.6 GHz; 4 MHz and 5.3 MHz Ga and P resonance frequency, respectively) and a JEOL-EPR spectrometer JES-3BQ, the samples being in a neon bath at 27 K.

We are grateful to Dr. Jacobi, VEB SMF, Freiberg, for supplying the crystals. This studies have been done within the "Arbeitsgemeinschaft A^{III}-B^{V}-Halbleiter" of the Karl-Marx-Universität Leipzig.

References

(1) R. TITLE, Phys. Rev. 154, 668 (1967).

(2) G. LANCASTER, BRSG Conf., Coventry 1970.

(3) S. HARALDSON and C.G. RIBBING, J. Phys. Chem. Solids 30, 2419 (1969).

(Received February 7, 1972)

phys. stat. sol. (a) 10, K129 (1972)
Subject classification: 8; 18.4; 21.1.1

The Research Institute for Iron, Steel, and Other Metals, Tôhoku University, Sendai

On New Nonferromagnetic B.C.C. Cr-Fe-Mn Invar Alloys

By
K. FUKAMICHI and H. SAITÔ

Invar and Elinvar alloys are widely used as materials for precision measurement instruments, but the ferromagnetism of the alloys often constitutes a barrier to their practical applications. The elastic properties of metals and alloys are remarkably affected by microscopic structural changes (i.e. fiber structure) and fractional changes in magnetic properties. Accordingly, since around 1966, it has been known that the Elinvar property can be obtained in antiferromagnetic alloys (1, 2), and even in paramagnetic pure metals (Nb, Pd) (1) and alloys (Nb-Zr, Au-Pd) (3, 4). On the contrary, there is much difficulty in obtaining nonferromagnetic Invar alloys because the Invar property originates from the spontaneous volume magnetostriction due to ferromagnetism. Antiferromagnetic alloys also exhibit the Invar property by their spontaneous volume magnetostriction. However, the smallest thermal expansion coefficient obtained with antiferromagnetic alloys is about $8x10^{-6}$ for a Fe-Mn alloy (5), which is too large to be used as practical materials. This phenomenon is regarded as an important subject of study called the "Invar effect" in the area of materials science.

In recent years it has been made clear that Cr is antiferromagnetic (6), and a considerable effort has been devoted to studies on the antiferromagnetic properties of the alloys in Cr-base solid solution (7 to 14). In these studies the additional elements are transition metals in neary all cases, and experiments have been carried out in regard to electrical resistivity ϱ, susceptibility χ, thermoelectric power S, the pressure effect on the Néel point T_N, neutron diffraction, etc. The physical properties thus obtained have been discussed with the one band model or the two band model.

Newmann and Stevens(7), Suzuki (8), Ishikawa et al. (9), Bastow (10), and other workers (11, 12) have investigated Cr-Fe and Cr-Mn alloys in the primary solid solution phase and reported the presence of anomalies in α, ϱ, S, and Young's mo-

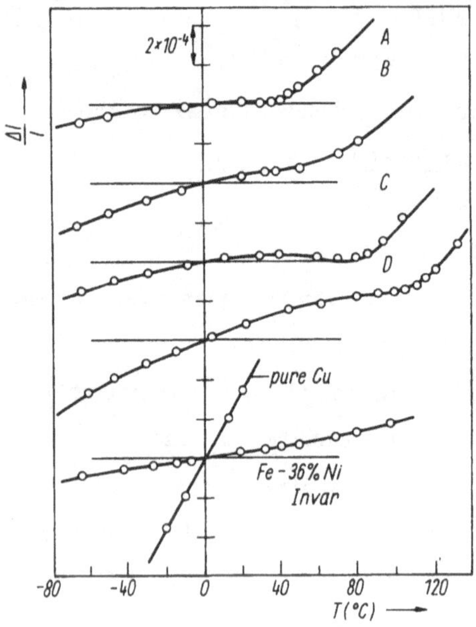

Fig. 1. Thermal expansion curves of Cr-Fe-Mn alloys, Cu, and Fe-36%Ni Invar alloy: (A) 5.53%Fe-0.54%Mn, (B) 5.51%Fe-1.04%Mn, (C) 4.33%Fe-0.54%Mn, (D) 4.23%Fe-1.01%Mn (at%)

dulus at temperatures in the neighbourhood of T_N. The magnetic phase diagrams have been derived by magnetic measurements and neutron diffraction analyses, a remarkable anomaly in thermal expansion has been detected at temperatures below T_N despite the difference in the relationship between the spin arrangement and temperature for these two alloy systems. However, to our knowlege, no studies have ever been reported on the primary solid solution alloys of the Cr-Fe-Mn ternary system.

The present authors have for the first time discovered that the ternary alloys of optimum composition reveal the Invar property over a relatively wide temperature range centering around room temperature.

The specimens were melted in argon atmosphere with an arc furnace. It has been confirmed from the results of magnetic measurements and X-ray analysis that these alloys are the antiferromagnetic b.c.c. phase. Fig. 1 shows the thermal expansion curves for several kinds of Cr-Fe-Mn alloys. For comparison, the thermal expansion curves of Cu and Fe-36%Ni Invar alloy are also shown in the figure. As can be seen from the figure, the value of α for specimen (A) is about 1.0×10^{-6} in the range of -40 to 40 $^{\circ}$C, which is smaller than 1.2×10^{-6} for the Fe-36%Ni Invar alloy. It is also shown that with decreasing Fe concentration, the Invar region shifts toward the high temperature range, and the value of α becomes larger with increasing Mn concentration. The value of χ decreases gradually with increasing temperature, while it becomes larger with Fe concentration. For example, the value of χ for specimen (A) is approximately 1.1×10^{-5} e.m.u./g in a magnetic field of 1500 Oe at room temperature.

Therefore, the effect of magnetism of these alloys for the practical applications can completely be disregarded.

Similar results have also been obtained on several kinds of other Cr-base ternary alloys in the primary solid solution range. The details of the studies will be reported elsewhere.

References

(1) S. STEINEMANN and M. PETER, Société Suisse de Chronométrie No. 1/2 (1968).

(2) H. MASUMOTO, S. SAWAYA, and M. KIKUCHI, J. Japan Inst. Met. 35, 723 (1971).

(3) H. ALBERT, F. ASSMUS, and I. PFEIFFER, Solid State Commun. 4, 519 (1966).

(4) H. MASUMOTO and S. SAWAYA, J. Japan Inst. Met. 33, 121 (1969).

(5) Y. TANJI and Y. SHIRAKAWA, J. Japan Inst. Met. 34, 897 (1970).

(6) G.E. BACON, Acta cryst. 14, 823 (1961).

(7) M.M. NEWMANN and K.W.H. STEVENS, Proc. Phys. Soc. 74, 290 (1959).

(8) T. SUZUKI, J. Phys. Soc. Japan 21, 442 (1966).

(9) Y. ISHIKAWA, S. HOSHINO, and Y. ENDOH, J. Phys. Soc. Japan 22, 1221 (1967).

(10) T.J. BASTOW, Proc. Phys. Soc. 88, 935 (1966).

(11) Y. HAMAGUCHI and N. KUNITOMI, J. Phys. Soc. Japan 19, 1849 (1964).

(12) K. SCHRÖDER, M.J. YESSIK, and N.P. BAUM, J. appl. Phys. 37, 1019 (1966).

(13) S. KOMURA and N. KUNITOMI, J. Phys. Soc. Japan 20, 103 (1965).

(14) W.C. KOEHLER, R.M. MOON, A.L. TREGO, and A.R. MACKINTOSH, Phys. Rev. 151, 405 (1966).

(Received February 15, 1972)

phys. stat. sol. (a) 10, K133 (1972)
Subject classification: 14.3.3; 13.4; 22.1.2

VEB Werk für Fernsehelektronik, Berlin

Recombination at Iron Atoms and at Thermally Generated Lattice Defects in n-Silicon

By

K. THIESSEN and G. ZECH

In the investigation of the recombination properties of iron atoms in n-silicon (1) a strong dependence of the capture cross section of the iron centres on the cooling rate after thermal treatments was found.

The iron was diffused into silicon in such a way that the maximum solubility corresponding to the respective temperature was attained.

From this material diodes were prepared on which the switching time t_{rr} was measured that resulted when the diodes were switched from the forward into the reverse directions. According to Lax and Neustadter (2) the minority carrier-lifetimes result from the switching times. They were controlled by lifetime measurements according to the procedure of Goodman (3).

The dependence of the switching time on the diffusion temperature and on the cooling rate is shown in Fig. 1.

To determine the portion of the lattice disorder freezed-in during the cooling process in the recombination of the minority carriers, material was investigated which had no iron centres, but was thermally treated in the same way. The switching times measured in dependence on the cooling rate are shown in Fig. 2.

The comparison with the iron-doped material suggests that for cooling rates of $>$ 300 deg/s the freezed-in lattice disorder will have influence on the recombination behaviour. Further, the graph shows that there is an interaction between the iron centres and the lattice disorder freezed-in, because the minority carrier lifetime is longer at a large cooling rate and in the presence of iron atoms than in material thermally treated at the same time, without iron.

To confirm this observation the iron concentration in silicon was varied by the diffusion time at a constant diffusion temperature and cooling rate (see Fig. 3).

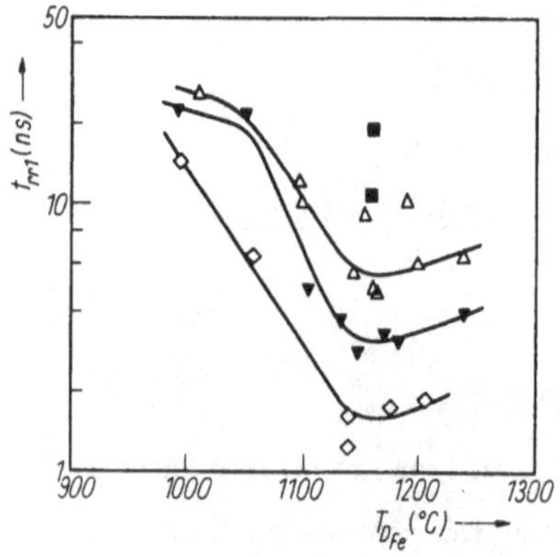

Fig. 1

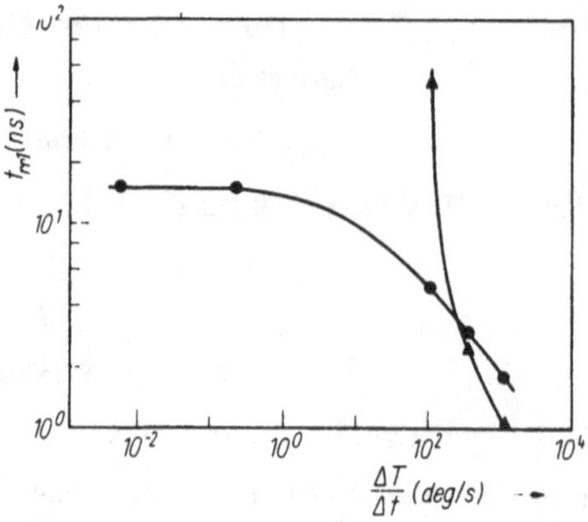

Fig. 2

Fig. 1. Dependence of the switching time of the diodes on the diffusion temperature of iron; ■ cooling rate 13 deg/min, △ cooling rate 100 deg/s, ▼ cooling rate 320 deg/s, ◇ cooling rate 1100 deg/s

Fig. 2. Effect of the cooling rate on the switching time; ▲ diodes without iron, annealed at 1165 to 1200 °C, ● diodes with iron, diffused at 1165 °C

With increasing iron concentration in silicon the lifetime also increase in this case, until an approximate saturation range - equivalent to attaining the maximum solubility of the iron - is obtained.

Silicon with an oxygen concentration of $\geq 10^{18}$ cm^{-3} was used. It is suggested that the recombination centre freezed-in at a large cooling rate consists in a lattice defect-oxygen complex. By the incorporation of iron the lattice defect is occupied by the iron atom and the recombination-efficient complex is neutralized. Thus, also at a high cooling rate the recombination properties of the iron in silicon dominate.

The dependence of the minority carrier-capture cross section of the iron centres in n-silicon deduced from the experimental results is depicted in Fig. 4.

For cooling rates < 1 deg/s the capture cross section σ is constant. For rates > 1 deg/s it can be described by the dimensional equation

$$\ln \sigma = \left[1 + \frac{a}{\lg(\sigma_o/10^{-16})} (\lg \Delta T/\Delta t)^b\right] \ln \sigma_o$$

with a = 0.13 and b = 1.78, where σ_o is measured in cm^2, T in deg, and t in s.

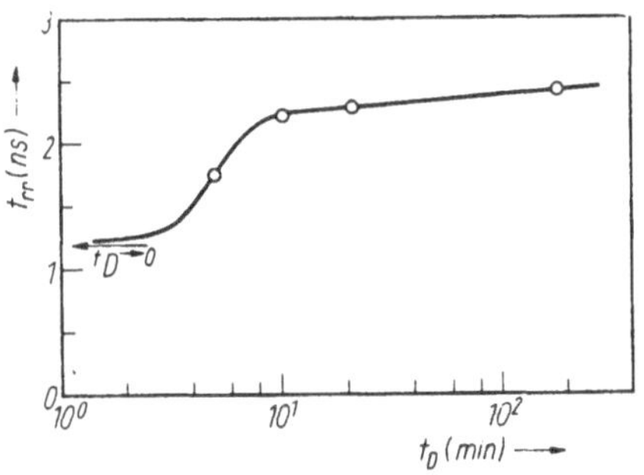

Fig. 3

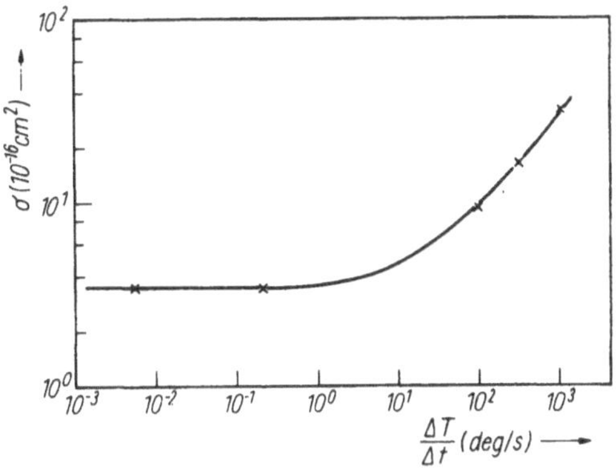

Fig. 4

Fig. 3. Effect of the diffusion time on the switching time; cooling rate 320 deg/s, Fe-diffusion temperature 1190 ^{o}C

Fig. 4. Dependence of the capture cross section on the cooling rate; x experimental values

When comparing the values of Collins and Carlson (4) ($\sigma_p = 3\mathrm{x}10^{-16}$ cm^2) and Klose (5) (σ_p = 5.8 to 7.7x10^{-16} cm^2) with the experimental results one finds that these values would correspond to cooling rates < 47 deg/s. This is just the range which is experimentally realized frequently, because in cooling a sample in the diffusion furnace rates < 20 deg/s are obtained often, and when taking the sample out of the furnace a cooling rate in the range of < 100 deg/s is attained.

In conclusion it should be noted that all cooling rates are valid in the range between the diffusion temperature and about 800 ^{o}C, because this temperature range is decisive for the attained final state. With a cooling rate of 20 deg/s up to 800 ^{o}C and subsequently with 1100 deg/s in the range between 800 and 20 ^{o}C a capture cross section was obtained which corresponds to a sample cooled with a rate of 20 deg/s according to Fig. 4.

References

(1) K. THIESSEN and G. ZECH, phys. stat. sol. (a) 10, K53 (1972).

(2) B. LAX and S.F. NEUSTADTER, J. appl. Phys. 25, 1148 (1954).

(3) A.M. GOODMAN, J. appl. Phys. 32, 2550 (1961).

(4) C.B. COLLINS and R.O. CARLSON, Phys. Rev. 108, 1409 (1957).

(5) H. KLOSE, Dissertation, Humboldt-Universität zu Berlin, 1968.

(Received March 9, 1972)

phys. stat. sol. (a) 10, K137 (1972)
Subject classification: 10.1 and 14.1; 21

Chemistry Division, Bhabha Atomic Research Centre, Trombay, Bombay

Recovery of Deformed Zirconium above Room Temperature

By

B.M. PANDE, M.S. ANAND, and R.P. AGARWALA

Studies on the recovery (1 to 3) of irradiated and deformed zirconium are very few and most of them have been carried out at low temperatures. Desalvo and Ziguani (4) studied the recovery of the deformed zirconium (99.95%) above room temperature, but the results did not give any definite indication of a point defect recovery peak, although the expected temperature ranges for the recovery as per the melting point, T_m (i.e. $\approx 0.16\ T_m$, $\approx 0.22\ T_m$, and $\approx 0.30\ T_m$) would be above room temperature for this metal. Recently, Neely (5) showed the existence of the third stage in zirconium at $0.13\ T_m$ by dose dependence experiments, although in his earlier work (3) he had assigned this recovery as due to the second stage and expected the third stage at 420 $^{\circ}$K. In view of the reported third stage at comparatively low temperature, it was considered interesting to investigate other recovery stages above room temperature.

The present note reports on the resistometric recovery of deformed zirconium above room temperature. The samples were prepared from thin foils of zirconium (200 and 250 μm thick) where deformation was carried out by rolling from 5 mm to the desired thicknesses in various steps and precooling the samples in liquid nitrogen. Zirconium used in these experiments was 99.99% pure and was supplied by Semielements, U.S.A. The resistivity measurements of the samples were performed in liquid nitrogen, using the standard potentiometric techniques. The isochronal annealing experiments (time interval 30 min and temperature interval 20 deg) were carried out from 52 to 450 $^{\circ}$C. The temperature was controlled to ± 1 deg. Fig. 1 shows the isochronal annealing spectra from 52 to 450 $^{\circ}$C for two samples cold rolled to different thicknesses. There is a continuous recovery of the resistivity throughout the temperature range and the resistivity of the two samples becomes nearly the same beyond 320 $^{\circ}$C, indicating thereby that the effect of cold work

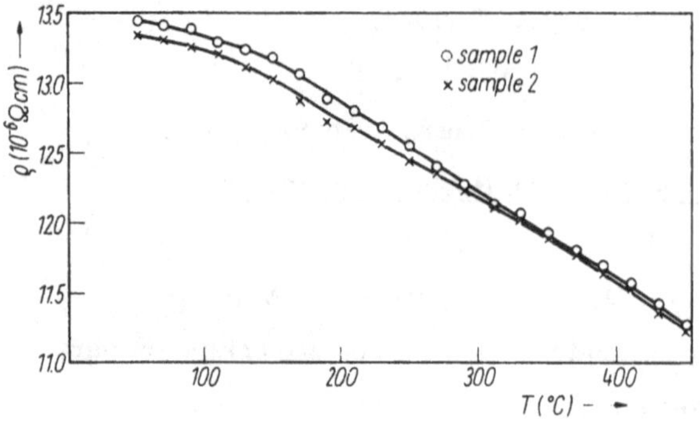

Fig. 1. Recovery curves for specimens rolled down to 250 μm (sample 1) and 200 μm (sample 2)

is nearly over by that temperature. The total change in the resistivity ($\Delta\varrho$) on annealing is 2.75×10^{-6} Ω cm and 2.64×10^{-6} Ωcm respectively in the two cases.

The absence of the defect recovery peaks above room temperature in the present case seems to be similar to the recovery of deformed iron as observed by Cuddy (6). It is seen that in iron on small deformation (15%), there is only a continuous change in the resistivity in the range 110 to 520 °K. Only when more deformation than this amount is carried out, the third stage at 210 °K became well defined, but even then the resistivity recovery remained continuous after this stage showing the absence of higher recovery peaks. The third stage in iron at 0.12 T_m seems to occur at much lower temperature than in other metals. In zirconium too, the third stage of recovery is at similarly lower temperature (0.13 T_m) as demonstrated by Neely (5) and the present observation of the absence of higher peaks is similar to the recovery curve of iron (6) beyond third stage. The existence of the third stage in zirconium at such a low temperature, similar to that seen in iron, in spite of its higher melting point ((1852 ± 5) °C) could perhaps be due to the fact that both have nearly similar crystallographic phase transition temperature, i.e. 863 °C for zirconium and 910 °C for iron.

The continuous recovery observed above room temperature in this investigation seems to be due to the recovery of dislocations formed during the deformation. Although the specific resistivity of dislocations in zirconium is not known, the total change in resistivity ($\Delta\varrho = 2.75\times10^{-6}$ and 2.64×10^{-6} Ωcm respectively for the two samples), would certainly lead to a high dislocation density, perhaps higher than the dislocation density of 2.5×10^{10} lines/cm^2 estimated by Cuddy for 200% deformation in iron. It is further substantiated by the observation that even in 10% deformed zirconium the dislocation density (7) is as high as 4.8×10^{9} lines/cm^{2}. Furthermore, the anomalous results of diffusion (8) in zirconium have been explained on the basis of high dislocation densities of the order of 10^{10} lines/cm^{2}. The high dislocation density in deformed zirconium is probably hindering the higher recovery stages leading to a continuous recovery, unlike in the case of rhenium (9) where high dislocation density seems to have no predominent influence on the annealing kinetics of cold-worked specimens.

Thanks are due to Dr. J. Nihoul for his comments.

References

(1) M.L. SWANSON, A.F. QUENNEVILLE, and H. SCHULTZ, Appl. Phys. Letters 6, 49 (1965).

(2) M.L. SWANSON, Canad. J. Phys. 44, 3241 (1966).

(3) H.H. NEELY, Canad. J. Phys. 46, 321 (1968).

(4) A. DESALVO and F. ZIGUANI, J. nuclear Mater. 20, 108 (1966).

(5) H.H. NEELY, Radiation Effects 3, 189 (1970).

(6) L.J. CUDDY, Acta. metall. 16, 23 (1968).

(7) P. MUKHOPADHYA, S. BANERJEE, and R. KRISHNAN, Proc. Symp. Materials Science Research, Department of Atomic Energy, India 1, 352 (1970).

(8) M.C. NAIK and R.P. AGARWALA, J. Phys. Chem. Solids 30, 2330 (1969).

(9) H. VANDENBORRE, L. STALS, and J. NIHOUL, phys. stat. sol. 35, 1009 (1969).

(Received December 1, 1971)

phys. stat. sol. (a) 10, K141 (1972)

Subject classification: 14.3; 22.2.1; 22.2.3

Institute of Radiophysics and Electronics,
University College of Science and Technology, Calcutta

Generalized Einstein Relation for Semiconductors in the Presence of a Magnetic Field

By

A.N. CHAKRAVARTI and D.P. PARUI

In a recent work, Lindholm and Ayers (1) established that the conventional form of the Einstein relation for the diffusivity-mobility ratio holds true only for non-degenerate semiconductors and not for degenerate semiconductors as erroneously suggested (2) and later pointed out (3) elsewhere. They demonstrated (1) the error arising from the application of the conventional relation to degenerate semiconductors and also examined a generalized form of the Einstein relation that holds for degenerate semiconductors. However, it is still to be examined whether the Einstein relations for both non-degenerate and degenerate semiconductors remain unaffected in the presence of a magnetic field. In the present note we shall derive the generalized Einstein relation that will be applicable in general to semiconductors in the presence of a magnetic field. We shall also use this relation for obtaining the diffusivity-mobility ratio as functions of the field strength and the extent of zero-field degeneracy for highly degenerate specimens of semiconductors like InSb and InAs in which the energy bands are strongly quantized at very low temperatures in the presence of a large magnetic field.

According to the principle of detailed balance, the net electron current density in a semiconductor at thermal equilibrium must be zero. Thus we can express the electron current density as

$$J_n = \frac{\sigma_n}{e} \frac{\partial \zeta_n'}{\partial x} = - e n_o \mu_n \varepsilon_n + \frac{\sigma_n}{e} \frac{\partial \zeta_n}{\partial x}$$

$$= - e n_o \mu_n \varepsilon_n - e D_n \frac{\partial n_o}{\partial x} = 0 , \qquad (1)$$

where e is the electron charge, σ_n the conductivity, ζ'_n the electrochemical potential, n_o the electron concentration, D_n the diffusion coefficient, μ_n the mobility, x a spatial coordinate, ε_n the electric field, and $\zeta_n = \eta kT$, $\eta\,(=(\zeta'_n - E_c)/kT))$ being the normalized difference between the Fermi energy and the lowest energy in the conduction band.

Equation (1) gives us

$$D_n/\mu_n = -\left(\frac{1}{e}\right) n_o \frac{\partial \zeta_n}{\partial x} \Big/ \frac{\partial n_o}{\partial x}$$

$$= \left(\frac{1}{e}\right) n_o \frac{d\zeta_n}{dn_o} . \qquad (2)$$

Now, the electron concentration n_o in the presence of a magnetic field can be written (4) as

$$n_o = N_c \sum_{n=o}^{\infty} \theta F_{-1/2} \left[\eta - (n + \tfrac{1}{2})\theta\right] , \qquad (3)$$

where N_c is the effective density of states, $\theta = \hbar\omega_o/kT$ (h and ω_o being Planck's constant and the cyclotron frequency respectively), F is the symbol for Fermi-Dirac integrals, and n is a positive integer or zero. Thus using equations (2) and (3), we can express the generalized Einstein relation for the diffusivity-mobility ratio in semiconductors in the presence of a magnetic field as

$$D_n/\mu_n = \left(\frac{kT}{e}\right) \sum_{n=o}^{\infty} \frac{F_{-1/2}\left[\eta - (n + \frac{1}{2})\theta\right]}{F_{-3/2}\left[\eta - (n + \frac{1}{2})\theta\right]} \qquad (4)$$

(the properties of the Fermi-Dirac integrals being such that $\frac{d}{d\Phi} F_j \; \Phi = F_{j-1} \; \Phi$). Equation (4) holds also for holes if one interprets η as the normalized difference between the Fermi energy and the highest energy in the valence band.

In the case of highly degenerate semiconductors, the reduced Fermi energies η and η_o in the presence and absence of a magnetic field, respectively, are both much greater than zero and we can write (reference (4))

$$F_{-1/2} \left[\eta - (n + \tfrac{1}{2})\theta\right] \approx 2\,\pi^{-1/2} \left[\eta - (n + \tfrac{1}{2})\theta\right]^{1/2} , \quad \left[\eta - (n + \tfrac{1}{2})\theta\right] > 0 . \qquad (5)$$

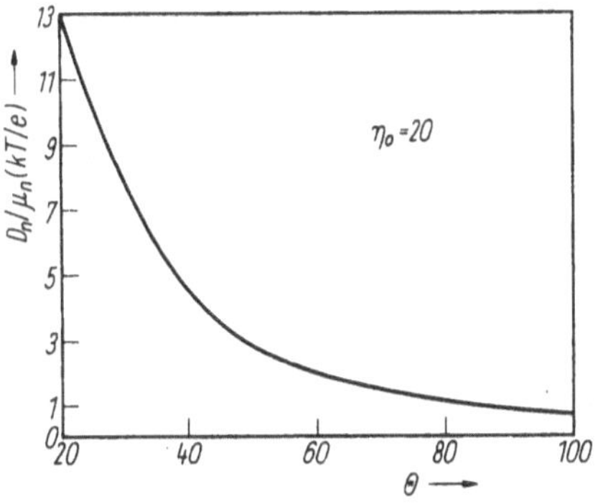

Fig. 1

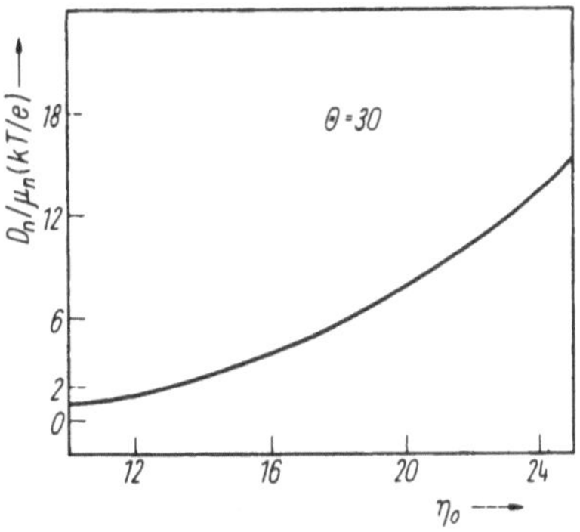

Fig. 2

Fig. 1. Magnetic field dependence of the diffusivity-mobility ratio for degenerate specimens of semiconductors like InSb and InAs having a given zero-field value of the reduced Fermi energy at any given low temperature

Fig. 2. Dependence of the diffusivity-mobility ratio on the extent of zero-field degeneracy in semiconductors like InSb and InAs at any given low temperature in the presence of a large magnetic field

Equations (4) and (5) may, therefore, be combined to get

$$D_n/\mu_n \approx \left(\frac{2\,kT}{e}\right) \sum_{n=o}^{\infty} \left[\eta - (n+\frac{1}{2})\theta\right] \quad . \tag{6}$$

In this connection, it would also be interesting to examine the case of strictly non-degenerate semiconductors placed in a magnetic field. For such semiconductors, η and η_o are both much smaller than zero and the Fermi-Dirac integral $F_{-1/2}\left[\eta - (n+\frac{1}{2})\theta\right]$ can simply be expressed (4) as $\exp\left[\eta - (n+\frac{1}{2})\theta\right]$. The diffusivity-mobility ratio is, therefore, obtained from equation (4) as

$$D_n/\mu_n = kT/e \tag{7}$$

which is the conventional Einstein relation. It appears then that the presence of a magnetic field results in a modification of the diffusivity-mobility ratio only for semiconductors which are degenerate or nearly degenerate. Incidentally, certain simplifications could also be done for degenerate specimens of semiconductors like

InSb and InAs. In these semiconductors at very low temperatures in the presence of a large magnetic field ($\hbar \omega_0/kT \gg 1$), most of the conduction band electrons occupy the lowest Landau sub-band and, therefore, we may take n = 0 to a fairly good approximation. Moreover, for degenerate semiconductors in general, the reduced Fermi energy in the absence of a magnetic field is related (4) to that in the presence of a large magnetic field by

$$\left[\left(\frac{\eta}{\theta}\right) - \frac{1}{2} \right]^{1/2} = \frac{2}{3} \left(\frac{\eta_0}{\theta}\right)^{3/2} , \tag{8}$$

where n is assumed to be zero and η and η_0 are both taken to be much greater than zero. Hence, using equations (6) and (8), we can determine the magnetic field dependence of the diffusivity-mobility ratio for degenerate specimens of semiconductors like InSb and InAs having a given zero-field value of the reduced Fermi energy as shown in Fig. 1. Similarly, the dependence of the diffusivity-mobility ratio on the extent of zero-field degeneracy in the above semiconductors in the presence of a given large magnetic field can be determined as shown in Fig. 2.

Acknowledgements

The authors are indebted to Prof. J.N. Bhar for his keen interest in the work. One of the authors (D.P.P.) is also grateful to the University Grants Commission of India for the award of a research fellowship.

References

(1) F.A. LINDHOLM and R.W. AYERS, Proc. IEE 56, 371 (1968).

(2) R.W. LADE, Proc. IEE 52, 743 (1964).

(3) W.B. BERRY, Proc. IEE 53, 188 (1965).

(4) J.S. BLAKEMORE, Semiconductor Statistics, Pergamon Press, London 1962 (p. 87).

(Received March 21, 1972)

phys. stat. sol. (a) 10, K145 (1972)

Subject classification: 10.1; 21.1

Institute of Physical Chemistry, Polish Academy of Sciences, Warsaw

On the Possibility of Revealing Low Angle Boundaries and Dislocations in Nickel Single Crystals by Electrolytic Decoration with Copper

By

J. PIELASZEK

The electrodeposition of small quantities of copper on nickel single crystals from the diluted solution of Cu^{2+} ions in 1 N H_2SO_4 (aqueous) has been investigated. The electrolysis was carried out during 19 h in solution of 10^{-5} to 10^{-6} mol/l between a Pt wire as anode and a nickel single crystal as cathode. The cathodic current density was 20 mA/cm^2. The copper deposited on nickel was practically undetectable by X-ray diffraction methods. In the present paper the metallographic observations of the nickel surface after the deposition of copper are described.

The nickel single crystals were oriented by the Laue method and cut parallel to (111) crystallographic planes by the spark-erosion technique. After cutting the samples were chemically and electrolytically polished till the surface to be examined was optically smooth. Then, the exit points of dislocations were observed when revealed by chemical etching in a solution of 80 ml concentrated H_3PO_4 + 20 ml H_2O + 5 ml concentrated HNO_3 + 5 g $FeCl_3$ (1). This solution revealed dislocations only on the (111) and (100) planes. The subsequent observations with the metallographic microscope under the magnification of 200 to 800 x were as follows:

1. After previous etching of the crystal sample the subsequent electrolysis changed the optical contrast on the etch-pits. This effect could be attributed to local deposition of minute quantities of copper. If the Cu deposit had formed on nickel a uniform layer, its thickness under the experimental conditions would not exceed some Å units and therefore the relative changes between the optical contrast on the etched points and the surface could not be observed.
2. When the crystal was polished electrolytically and then etched to reveal only the low angle boundaries (without etch-pits on individual dislocations), dark "points" distributed on the grain boundaries and throughout the whole surface were observed

after the electrolysis. The distribution of these points on the surface was similar to that of the previously observed etch-pits, i.e. the areas with higher density of etch-pits corresponded to those with higher density of dark "points".

We conclude that during the electrolysis under the described conditions small quantities of copper were electrodeposited preferentially on the exit points of dislocations and on the grain boundaries.

Nickel samples with orientation (110) were prepared in the same way. In the literature, as far as known to the author, there are no information on the chemical etchant for revealing dislocations on this crystal orientation of nickel. After the electrolysis under the conditions described the formation of similar dark "points" was observed and explained as a nonuniform, i.e. local, deposition of copper. The average surface concentration of the observed points was of about 2×10^{6} cm^{-2}. The density of dislocations as calculated from the X-ray measurements using an oscillating film method (2) was 10^{7} dislocations/cm^{2}. Taking into account that the selective electrodeposition of copper "decors" presumably only the dislocations which form a large angle with the sample surface and that the estimation of the dislocation density by the X-ray method needs additional arbitrary assumption concerning the model of dislocation distribution (here the Cottrell's one) conformity of the above two results could be regarded as satisfactory.

Copper when "decorating" the exit points of dislocations (or diffusing into) is "blocking" them at the same time. In consequence the diffusional properties of the metal will be changed. That phenomenon was successfully used (3) for the verification of the influence of dislocations on the stability of β-nickel hydride (4, 5).

The possibility of the dislocation "decoration" by other metals was also considered. For example preliminary investigations already proved that similar results may be obtained in the case of the "decoration" with silver.

Therefore, it could be expected that under proper electrolysis conditions the presented method may be adopted for revealing the exit points of dislocations on the surface of different metals.

References

(1) E. JUN-ICHI, Japan J. appl. Phys. 8, 964 (1969).

(2) J. AULEYTNER, Acta Phys. Polon. 17, 111 (1958); X-Ray Methods in the Study of Defects in Single Crystals, Pergamon Press, 1967.

(3) J. PIELASZEK, in preparation.

(4) A. JANKO, Bull. Acad. Polon., Ser. sci. chim. 8, 131 (1960); Naturwissenschaften 47, 225 (1960).

(5) B. BARANOWSKI and M. ŚMIAŁOWSKI, Internat. J. Phys. Solids 12, 206 (1959); Bull. Acad. Polon., Ser. sci. chim. 7, 663 (1959).

(Received October 27, 1971)

phys. stat. sol. (a) 10, K149 (1972)
Subject classification: 1.1; 10.2

Polytechnic Institute of Brooklyn, Brooklyn, New York

Comment on "Multiple-Beam Borrmann Effect in Perfect Crystals of Germanium and Silicon" by T.I. Borodina et al.

By
T.C. HUANG

Borodina et al. (1) recently reported a procedure for the determination of double Borrmann (3-beam) X-ray absorption coefficients. Unfortunately they have overlooked a number of important factors and their results are therefore unacceptable.

First, in their description (and presumably in their calculations) of the so-called ($1\bar{1}1$, $20\bar{2}$) "3-beam" Borrmann effect, they apparently did not realize they were actually dealing with a 4-beam case. Simple geometrical considerations show that when 000, $1\bar{1}1$, and $20\bar{2}$ are simultaneously on the Ewald sphere, they must be accompanied by the $3\bar{1}\bar{1}$ reciprocal lattice point. The four reciprocal lattice points 000, $1\bar{1}1$, $20\bar{2}$, and $3\bar{1}\bar{1}$ lie in a plane whose intersection with the Ewald sphere is a circle with the $3\bar{1}\bar{1}$ lattice vector as diameter. It is therefore essential to include the $3\bar{1}\bar{1}$ in all calculations and it is pointless to discuss the "3-beam" absorption coefficient for the $1\bar{1}1$, $20\bar{2}$ case; no such case exists in cubic crystals.

Second, in multiple Borrmann diffraction the portion of the dispersion surface relevant to a discussion of 3-beam diffraction is only a few arc seconds wide. The divergence of the incident beam used in (1) is apparently several degrees in extent. Even when great pains are taken to restrict the divergence of the incident beam, it is generally more than enough to excite the entire 3-beam portion of the dispersion surface. In a recent paper (2) it was shown that "3-beam" absorption coefficients vary rapidly with angle when the incident beam deviates from the exact 3-beam Bragg angle. In one instance they increase from 20 to about 200 cm^{-1} when $\Delta\theta_{Bragg}$ reaches 10" of arc. Calculated 3-beam absorption coefficients, such as thosed used in (1), refer only to values applicable at the exact 3-beam point. Clearly, there seems to be little point in comparing such calculated values with experimental measurements which, of necessity, cover wide ranges of absorption coefficients.

In a recent thesis, Feldman (3) reported that the measured value of the 3-beam absorption coefficient for the 111, $1\bar{1}1$ (CuK α_1 radiation transmitted through perfect Ge crystals) was about 45 cm^{-1} compared with 18.9 cm^{-1} calculated. Feldman also found that the average calculated 3-beam absorption coefficient, determined by integrating the calculated transmitted intensity over an area of the dispersion surface near the exact 3-beam point (out to regions where $\mu = 10\ \mu_{min}$), yielded a value of 43 cm^{-1} in satisfactory agreement with the theory.

Also, in 3-beam diffraction six tie points are excited yielding six absorption coefficients. It is well known (4) that the average of the six coefficients should equal μ_o, the "normal" absorption coefficient. It is therefore difficult to understand why the average of the six absorption coefficients listed by (1) is somewhat less than half the accepted value of μ_o.

Incidentally, the film technique described by the authors is not suited for the proposed measurements inasmuch as, unless resolution of the order of a fraction of a second of arc can be achieved, the 3-beam film reflection spot also represents an average over a wide range of absorption coefficients. A number determined even by the ingenious method of measurement proposed by the authors would have no physical meaning.

References

(1) T.I. BORODINA, A.A. KATSNELSON, and V.I. KISSIN, phys. stat. sol. (a) 3, 105 (1970).

(2) S. BALTER, R. FELDMAN, and B. POST, Phys. Rev. Letters 27, 307 (1971).

(3) R. FELDMAN, Ph. D. Dissertation, Polytechnic Institute of Brooklyn, 1969.

(4) T. JOKO and A. FUKUHARA, J. Phys. Soc. Japan 22, 597 (1967).

(Received December 20, 1971)

phys. stat. sol. (a) 10, K151 (1972)

Subject classification: 7 and 14.3; 22.2.3

Department of Applied Physics (a) and Department of Nuclear Engineering (b), Faculty of Engineering, Hokkaido University, Sapporo

Ultrasonic Amplification in Electron-Hole Plasma

By

K. TANAKA (a), Y. ABE (b), and A. ODAJIMA (a)

Although amplifications of ultrasound in semiconductors have been studied extensively (1), no work in the avalanche-breakdown region (2) has been reported. The purpose of the present note is to investigate this characteristics in n-InSb at 77 $^{\circ}$K.

The sample used in our experiments was undoped n-InSb whose carrier concentration and mobility were 2.2×10^{14} cm^{-3} and 5.2×10^{5} cm^{2}/Vs, respectively. A bar specimen cut from an ingot with dimensions of 1x1x10 mm^{3} has its long axis along the ⟨110⟩ direction so as to propagate the piezoelectrically active shear wave. An Y-cut quartz transducer was bonded to one end of the bar to generate and detect 285 MHz ultrasound. In order to produce the dc electric field, short pulses of width 2 μs and repetition frequency 20 Hz were applied to the specimen through In contacts. From the current-voltage characteristics of the sample, it was found that the threshold electric field for the avalanche breakdown was nearly 240 V/cm and above this

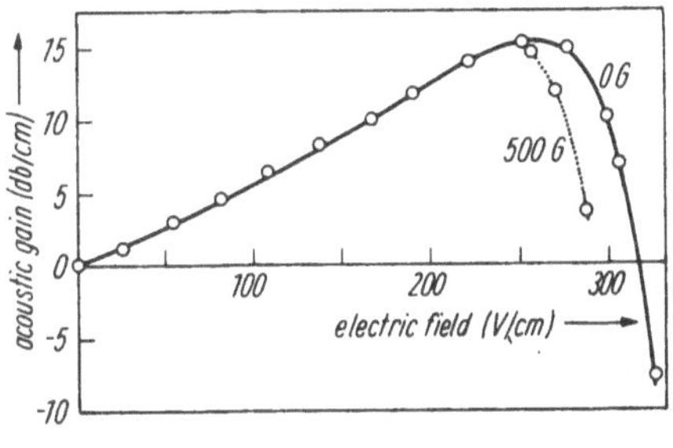

Fig. 1

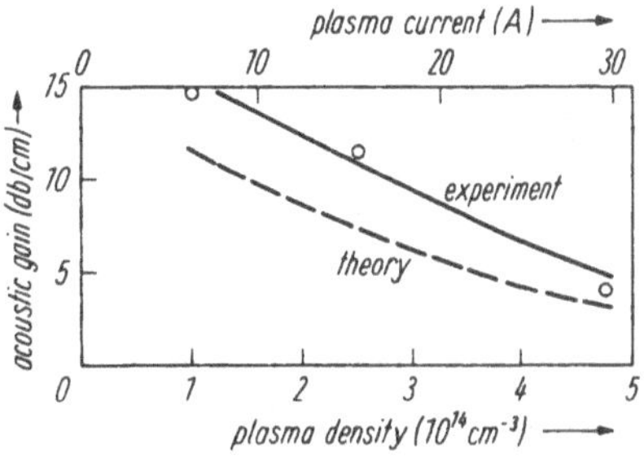

Fig. 2

Fig. 1. Electric-field dependence of acoustic gain for 0 G and the longitudinal magnetic field of 500 G. The output signals are expressed in db/cm by taking the output without electric field as reference

Fig. 2. Comparison between the theory and the experiment for magnetic field of 500 G voltage the longitudinal magnetic field of 500 G inhibited the occurrence of pinching (3).

As shown in Fig. 1, the magnetic field of 500 G has no effect on the ultrasound amplification characteristics up to 240 V/cm. In the avalanche-breakdown region the decreasing of the acoustic gain, which depends on the magnetic field, is obtained.

In order to compare the above result in the avalanche breakdown for 500 G with theory, we use the theoretical result which is applicable to electron-hole plasma. Equation (15) of (4) is used under the following estimations: I. The drift velocities of electrons and holes are equal to 4×10^7 cm/s (5) and 8×10^5 cm/s, respectively. II. Both the electron and hole temperatures are equal to 250 ^{o}K (6). Fig. 2 shows that agreement between the theory and the experiment is reasonable. For B = 0 G, the increase of Akhieser loss (7) due to magnetothermal pinching (8) should be taken into account. Theoretical calculation in this case is rather complicated, and is the subject for a future study.

References

(1) M.C. STEELE and B. VURAL, Wave Interactions in Solid State Plasmas, McGraw-Hill Publ. Co., 1969 (p. 133).

(2) M. GLICKSMAN and M.C. STEELE, Phys. Rev. Letters 2, 461 (1959).

(3) A.G. CHYNOWETH and A.A. MURRAY, Phys. Rev. 123, 515 (1961).

(4) Y. ABE and N. MIKOSHIBA, Japan J. appl. Phys. 7, 881 (1968).

(5) M. TACANO and S. KATAOKA, Japan J. appl. Phys. 29, 1241 (1970).

(6) B.D. OSIPOV and A.N. KHVOSHCHEV, Soviet Phys.-J. exper. theor. Phys. 16, 833 (1963).

(7) K. TANAKA, Y. ABE, and A. ODAJIMA, to be published.

(8) B. ANCKER-JOHNSON, Proc. 9th Internat. Conf. Phys. Semicond., Moscow 1968 (p. 813).

(Received February 23, 1972)

phys. stat. sol. (a) 10, K153 (1972)
Subject classification: 8; 21.4

Department of Experimental Physics, Faculty of Science,
Šafarik University, Košice

Thermal Conductivity of Thulium in the Temperature Range from 0.5 to 4.5 K

By
Š. JÁNOŠ and A. FEHER

Data on the thermal conductivity of thulium in the range from 4 to 300 K have been presented in (1). The determination of the thermal conductivity of thulium below 4 K has not been reported yet.

We have carried out an investigation of the thermal conductivity of polycrystalline thulium in the range from 0.5 to 4.5 K. The degree of sample purity was 99.9%. A steady-state method was used. The temperature was measured with two Allen-Bradley resistors (20 Ω, 1/8 W), by means of two ac bridges with a power dissipation of less than 10^{-10} W in a thermometer. The distance between the thermometers on the sample was 2.11 cm and the form factor was 158 cm^{-1}. Waiting periods of the order 20 to 30 min were necessary before each thermal conductivity measurement. The data obtained are plotted in Fig. 1. The thermal conductivity of thulium (in $mWcm^{-1}\ deg^{-1}$) can be separated into a linear (αT) and a quadratic (βT^2) term, with $\alpha = 1.06$ mW $cm^{-1} deg^{-2}$; and $\beta = 0.245$ mW $cm^{-1} deg^{-3}$. The linear term represents an electronic contribution from impurity scattering. The quadratic term in the thermal conductivity of thulium is of the same order of magnitude as that observed in other heavy rare-earth metals (2). The analysis presented in (2) suggests

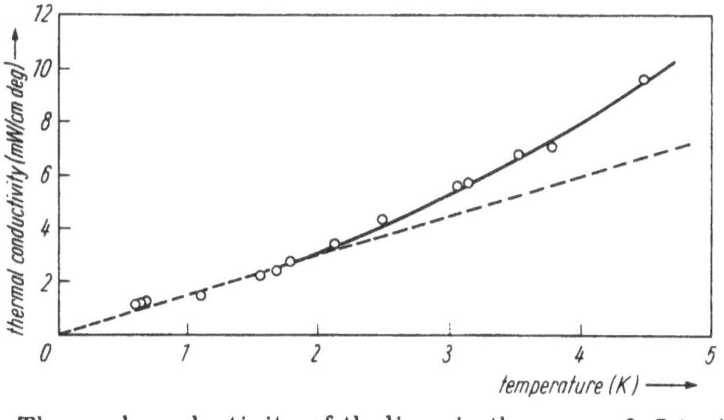

Fig. 1. Thermal conductivity of thulium in the range 0.5 to 4.5 K

that the dominant part of the quadratic term is due to phonon conduction limited by electron scattering with possibly some resistance due to dislocation scattering. The interpretation of the quadratic term, however, requires more experimental evidence and more detailed knowledge about electron-dislocation and phonon-dislocation scattering.

References

(1) L.R. EDWARDS and S. LEGVOLD, Phys. Rev. 176, 753 (1968).

(2) R. RATNALINGAM and J.B. SOUSA, J. Low Temp. Phys. 4, 401 (1971).

(Received March 17, 1972)

Pre-printed Titles

Papers to be published
in "physica status solidi (a) - applied research"
Vol. 11, No. 1

Review Article

L.N. ALEKSANDROV, Novosibirsk
Structure and Properties of Transition Layers Formed in the Epitaxy Process

Original Papers

V.M. POLOVOV, CH. V. KOPETSKII, G.I. KULESKO, and V.N. MATVEEV, Chernogolovka
Dislocations Energy in Deformed Copper Crystals

A. SOLER and F. RUEDA, Noordwijk
F-Centre Photoelectric Materials as Image Storage Targets

D. WEINMANN and H. VOGT, Köln
Modulation of the Millimeter Wave Absorption in $NaNO_2$ by Polarization Reversal

N. REZLESCU, E. CUCIUREANU, C. IOAN, and E. LUCA, Iassy
Time Variation of the Electrical Conductivity in Spinel Ferrites

A.V. FERRIS-PRABHU, Essex Junction, Vermont
Maximum Tunnelling Distance in MNOS Devices. Theory

E. PINK, Oxford
The Role of Impurities in Low Temperature Softening Phenomena

M. ZOUAGHI, Strasbourg
Near Infrared Optical and Photoelectric Properties of Cu_2O (II)

B.SH. BARZHALOV and E.L. LUTSENKO, Leningrad
Thermostimulated Conductivity in Organic Dyes (Rhodamine 6G)

J.M.D. COEY and D. KHALAFALLA, Winnipeg, Manitoba

Superparamagnetic γFe_2O_3

O. BOSTANJOGLO and K. RÖHKEL, Berlin

Superferromagnetism in Gadolinium Films (II)

D.J. MORGAN and M.J. ADAMS, Cardiff

Quantum Noise in Semiconductor Lasers

L.A. ALMASOV and I.M. DYKMAN, Kiev

Hot Electron Kerr Effect in Semiconductors with Non-Parabolic Dispersion Law of Energy

J. PEBLER, F.-W. RICHTER und E. VOGT, Marburg

Untersuchung der magnetischen Phasenübergänge der geordneten ternären $Fe(Pd_xPt_{1-x})_3$-Mischkristallreihe

G. HAURET, J.P. CHAPELLE, and L. TAUREL, Orsay

Brillouin Scattering in Sodium Nitrite at Room Temperature

B. NØST, B.C. LARSON, and F.W. YOUNG, Jr., Oak Ridge, Tennessee

Anomalous X-Ray Transmission Study of Low-Dislocation-Density Aluminum Crystals

M.R. WINSTONE and R.D. RAWLINGS, London

Dynamic Strain Ageing in an Iron-0.3 at% Niobium Alloy and an Aluminium/Niobium Steel

I. SAXL and F. KROUPA, Řež and Prague

Relations between the Experimental Parameters Describing the Steady-State and Transient Creep

T. MALIS, D.J. LLOYD, and K. TANGRI, Winnipeg, Manitoba

Dislocation Generation from Grain Boundaries in Nickel

N.S. GOLOSOV, L.YA. PUDAN, and L.E. POPOV, Tomsk

Resistance to the Motion of Dislocations in the $L1_2$ Superstructure Caused by Short-Range Order

P. CHAUDHARI, M. SHATZKES, and A.F. MAYADAS, Yorktown Heights, New York and East Fishkill, New York
On the Structure Factor of Large Angle Boundaries

J.L. PAUTRAT and J.C. PFISTER, Grenoble
Electronic Properties of the Si-SiO_2 Interface as a Function of Oxide Growth Conditions (I)

W. BOLLMANN and H. HENNIGER, Jena
Concentration and Mobility of Fluorine Ion Vacancies in CaF_2

C.T. SAH and H.S. FU, Illinois, Urbana
Current and Capacitance Transient Responses of MOS Capacitor (I)

J. KUČERA, B. MILLION, and J. PLŠKOVÁ, Brno
Diffusion of Zinc in Cu and in Cu-Zn Solid Solutions

U. KÜMMEL, E. KRAUTZ, and R. KNÜTTER, München and Graz
Thermoluminescence of Zn_2SiO_4:Mn

M.J. WITCOMB and A.V. NARLIKAR, Lancaster
Magnetic Irreversibility and Microstructure of a Nb-40 at% Ti Alloy

J.R. SERETLO, Fort Hare, South Africa
Annealing of Defects Produced by Room Temperature Irradiation in KCl:Ca and KCl:Sr Crystals (II)

K. NAUKKARINEN, T.O. TUOMI, and U. KORHONEN, Otaniemi
Dislocation Networks in Vapor Grown Selenium Single Crystals as Seen by X-Ray Topography

D. SHINODA, Kawasaki
Magnetic Properties of $Co_{1-x}Fe_xSi$, $Co_{1-x}Mn_xSi$, and $Fe_{1-x}Mn_xSi$ Solid Solutions

B.T. KOLOMIETS, T.N. MAMONTOVA, E.A. SMORGONSKAYA, and A.A. BABAEV, Leningrad
Study of the Mechanism of Radiative Recombination in Vitreous and Monocrystalline Arsenic Selenide

M. DOUMENG, B.T. KOLOMIETS, E.A. LEBEDEV, V.C. SHPUNT et I.A. TAKSAMI, Villeurbanne

Interdépendance des effets de commutation et de mémorisation et du comportement thermique d'une serie de verres chalcogenures

P.N. MURGATROYD, Bristol

On the Theory of the Transition from Space-Charge-Limited Current to Electrode-Limited-Current

F. STÖCKMANN, Karlsruhe

Electrostatic Potentials in Photoconductors

U. HOFMANN, Dresden

Properties of High-Permeability Nickel-Iron-Tungsten Alloys

R.F. LAITANO and M. MARTINELLI, Roma

The Role of Sm and Mn as Activators in Calcium Sulphate and Lithium Tetraborate

Z. MÁLEK, M. POLCAROVÁ, J. ŠTREJBLOVÁ, and J. JANTA, Prague

The Effect of Growth Rate on the Defect Structure and Dielectric Properties of TGS Single Crystals

S. BORNMANN, A. SCHÖNECKER, and W. HAUBENREISSER, Jena and Dresden

Magnetostatic Mode Pattern and Mode Spectrum Analysis in Finite Ferromagnetic Circular Disks

H. PFLEIDERER and H. REHME, München

Spike Cathode for Field Emission

Papers to be published

in "physica status solidi (b) - basic research"

Vol. 51, No. 1

Review Article

S. WANG, C.K. MALUTTE, and M. MATSUURA, Waterloo

Electronic Polarization Effect on Electron States in Insulators

Original Papers

M.S. BAPNA and M. MESHII, Evanston, Illinois

An Elastic-Modulus Effect Calculation of the Interaction of Point Defects with Extended Dislocations in F.C.C. Metals

A. OPPELT, E. DORMANN, and K.H.J. BUSCHOW, Darmstadt and Eindhoven

NMR Study of Ferromagnetic Gd Intermetalic Compounds with CsCl Structure

E.E. KOCH and A. OTTO, München

Optical Anisotropy of Anthracene Single Crystals for Photon Energies from 4.5 to 11.5 eV

F.R. KESSLER and K. DETTMER, Braunschweig

Separation of the e_1- and E_1-Interband Transitions near 2 eV in Germanium by Thermoabsorption

M. IKEYA, Nagoya

Neutral Manganese Centers at the Anion Site in NaCl

L. GENZEL and T.P. MARTIN, Stuttgart

Infrared Absorption in Small Ionic Crystals

H. VAN CONG and G. MESNARD, Villeurbanne

Some Thermomagnetic and Adiabatic Effects in Heavily Doped Semiconductors at Low Temperatures for Weak Magnetic Field

L. GENZEL and T.P. MARTIN, Stuttgart

Lattice Dynamics of MgO Microcrystals

K.N. SHRIVASTAVA, Simla, India

Raman Relaxation Times in Electronic Systems

R.K. SINGH and K.S. UPADHYAYA, Varanasi

The Dynamical Behaviour of Silver Chloride with Modified Shell Model

J. HAFNER and H. NOWOTNY, Wien

Pseudopotential Calculation of Binding Energies, Structures, and Lattice Constants of Simple Metals

G. PAASCH, H. ESCHRIG, and W. JOHN, Dresden

Work Function and Surface Structure of Simple Metals

H. TRINKAUS, Darmstadt

On Determination of the Double-Force Tensor of Point Defects in Cubic Crystals by Diffuse X-Ray Scattering

H. WERHEIT, A. HAUSEN, and H. BINNENBRUCK, Köln

Optical Anisotropy of β-Rhombohedral Boron from 0.4 to 16 μm

Ø. RA, Trondheim

On the Use of the Green's Function Method in Shell-Model Dynamics

A.S. DAVYDOV and A.A. SERIKOV, Kiev

Energy Transfer between Impurity Molecules of a Crystal in the Presence of Relaxation

TH. NATTERMANN, Leipzig

On the Influence of Screening on the Ferroelectric Curie Point

H.D. KOSWIG, U. RETTER, and W. ULRICI, Berlin

Jahn-Teller Effect on Transition Metal Ions in Silver Halides (II)

H. BÖTTGER, Berlin

On the Localization of Vibrational States in a Disordered Solid

W. GÜTH, Stuttgart

Electronic States of Dislocations in Germanium

A.B. WOLBARST, Hanover, New Hampshire

High Pressure ENDOR Search for the K'-Center

J. BRAEKHUS and J. LOTHE, Oslo

Dislocation Jog and Kink Interactions in Ionic Crystals

J. STANKOWSKI and M. MAĆKOWIAK, Poznań

EPR of the Four Copper-Ion Complex with Spin S=2 in Triglycine Fluoroberylate Monocrystal

Y.F. LOS, Kiev

Coherent Magnetoelastic Effects in Ferromagnets with Impurities

V.N. ABAKUMOV, M. SCHLEIFSTEIN, and I.N. YASSIEVICH, Leningrad and Berlin

The Effect of Phonon Heating on Photoelectric Phenomena (II)

G. VAUBEL and L. PETER, Berlin

Intermolecular Triplet Energy Transfer in Naphthalene-Pentacene Mixed Crystals via Higher Excited Triplet States

A.A. KATSNELSON, Moscow

Relation of Long-Range Interatomic Interactions to the Sign and Type of Short-Range Ordering in Alloys

S.P. SRIVASTAVA and H.D. BIST, Kanpur

Infrared Active Two-Phonon Processes in Some Alkali Iodides

B. SCHNEIDER, Hamburg

Effect of Crystalline Anisotropy on the Magnetostatic Spin Wave Modes in Ferromagnetic Plates (I)

M. TAUT and G. PAASCH, Dresden

The Change of Model Potentials in Alloying

E.H. BRANDT, Stuttgart

Ginsburg-Landau Theory of the Vortex Lattice in Type-II Superconductors for all Values of $\varkappa$ and B

E. BURZO, I. URZU et J. PIERRE , Bucarest et Grenoble

Résonance électronique de spin et susceptibilité magnétique des composés GdCu et GdAg

J.E. HEBBORN and C.M. PLACE, London

Some Comments on Kubo and Obata's Treatment of the Orbital Susceptibility of Electrons in Metals

B. ESSER, Moscow

On the Theory of Electroabsorption in Disordered Semiconductors

J. MONECKE, Dresden

Conduction in Narrow Energy Bands

L. CSER, V. GOROBCHENKO, and I. LUKASHEVICH, Budapest and Moscow

Investigation of Hyperfine Field Distributions in the Fe-Sn Alloys by Mössbauer Effect

19. Magnetic Resonance
20. Optical Properties
    - 20.1 Spectra. Optical Constants (X-Ray Spectra See 20)
    - 20.2 Lasers
    - 20.3 Luminescence (X-Ray Spectra See 20)
21. Metals. Alloys
    - 21.1 Transition Metals (Group Sc to Zn) and Their Alloys
        - 21.1.1 Fe and Fe Alloys
    - 21.2 Alkali Metals
    - 21.3 Alkali-Earth Metals
    - 21.4 Rare-Earth Metals
    - 21.5 Actinides
    - 21.6 Noble Metals
    - 21.7 Semi-Metals (Semiconducting Alloys See 22.7)
22. Semiconductors and Ionic Crystals
    - 22.1 Elements
        - 22.1.1 Germanium
        - 22.1.2 Silicon
        - 22.1.3 Group VI Elements
    - 22.2 III–V Compounds
        - 22.2.1 Arsenides
        - 22.2.2 Phosphides
        - 22.2.3 Antimonides
        - 22.2.4 Mixed Crystals
    - 22.3 IV–IV, II–V, and III–VI Compounds
        - 22.3.1 Mixed Crystals
    - 22.4 II–VI Compounds
        - 22.4.1 Sulphides
        - 22.4.2 Selenides
        - 22.4.3 Tellurides
        - 22.4.4 Mixed Crystals
    - 22.5 Halides
        - 22.5.1 Silver Halides
        - 22.5.2 Alkali Halides
        - 22.5.3 Alkali-Earth Halides
        - 22.5.4 Mixed Crystals
    - 22.6 Simple Oxides (CdO, ZnO See 22.4)
        - 22.6.1 Mixed Crystals
    - 22.7 Semiconducting Intermetallic Compounds
    - 22.8 Tertiary and Higher Compounds
        - 22.8.1 Salts of Oxy-Acids (Carbonates, Phosphates, Silicates Including Mica etc.)
        - 22.8.2 Spinels, Garnets, Ferrites
    - 22.9 Organic Semiconductors
23. Solidified Gases

Attention

The preceding classification scheme is closely akin to the scheme used by Physics Abstracts to classify the various fields of solid state physics. A paper will either be classified under one or more of the subsections or, if none of these are relevant, it will just appear under the general section heading (i. e. sections and subsections are equivalent in this sense).

The *subject classification* given at the head of each paper begins with the main subject matter of the paper, and this is followed, if necessary, by figures referring to subsidiary subject matter and the type of solids investigated.

The papers published in this issue are listed in the order of their subject classification at the end of the table of contents.

Manuscripts and letters for physica status solidi (b) — basic research and physica status solidi (a) — applied research

should be sent to

DDR, 102 Berlin, Neue Schönhauser Str. 20

or

Internationaler Buchversand GmbH, 1 Berlin 19, Eschenallee 21.

The object of "physica status solidi (b)" and "physica status solidi (a)" is to publish rapidly important and topical results concerning fundamental research in the field of solid state physics and aspects of application and experimental techniques of fundamental solid state research, respectively. The issues of "physica status solidi (b)" and "physica status solidi (a)" will be published on the 1st and the 16th day of each month, respectively. Either series will contain the following categories:

1. Review articles written by arrangement with the editors. They will be remunerated.
2. Original papers selected from those which have not been submitted for publication elsewhere, after being reviewed. (The latest date for acceptance of contributions is the 8th and the 23rd day of each month for "physica status solidi (b)" and "physica status solidi (a)", respectively, 50 days before the delivery of the journal.)
3. Short notes published in the off-set part of the journal. They are, if necessary after being reviewed, selected by the editorial staff. (The latest date for acceptance of manuscripts for this part is the 12th and the 27th day of each month for "physica status solidi (b)" and "physica status solidi (a)", respectively, 16 days before the delivery of the journal.)
4. Pre-printed titles of papers to be published in the next issue of both series.

Editorial Information

Manuscripts should be submitted preferentially in English. Besides, papers in German, French, and Russian are admitted. Authors are asked to observe the following rules:

1. Manuscripts have to be written as concisely as possible. As most, 20 and 3 pages (28 lines per page; abstracts, tables, references, captions of figures included) are admitted for original papers and short notes, respectively. All manuscripts should be type-written on one side only. Two copies are required. Badly arranged manuscripts are returned.
2. The author's name, initials, and institution should be given.
3. An abstract of about ten lines in English and a second one in German, French, or Russian are required for original papers.
4. Formulae should not be written too small and with pencil. Separate lines for formulae are desirable. Indistinct writing will lead to a return of the manuscript.
5. Drawings should be enclosed on separate sheets of paper; photographs should be submitted, on white paper, only for review articles and original papers. Duplicates are required.
6. Captions for the figures should be submitted on separate sheets of paper, and special instructions with regard to the size of illustrations should be given on these.
7. References should be numbered in order of appearance and should be given, on a separate sheet, in the form used in physica status solidi.
8. Units, symbols, etc. should be used in accordance with the recommendations given by the IUPAP.
9. Each paper should contain a proposal for the subject classification according to the classification scheme appearing on the second and third cover of each issue.

The authors will be sent galley proofs of review articles and original papers. Proof reading of short notes and pre-printed titles will be done by the editorial staff.

For each original paper and short note a total of 75 reprints is available free of charge. The number of reprints free of charge per review article is 25. Requests for additional reprints (against payment) should be given together with the submitted manuscript. **The deadline for orders of additional reprints is the 15th and the 30th day of each month** for "physica status solidi (b)" and "physica status solidi (a)", respectively, two weeks before the delivery of the journal.

Kolibri 360 GmbH
Ferdinand-Jühlke-Straße 7
99095 Erfurt, Deutschland
produktsicherheit@kolibri360.de